KU-093-865

£21·70

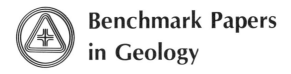

Benchmark Papers
in Geology

Series Editor: Rhodes W. Fairbridge
Columbia University

PUBLISHED VOLUMES

RIVER MORPHOLOGY / *Stanley A. Schumm*
SLOPE MORPHOLOGY / *Stanley A. Schumm* and *M. Paul Mosley*
SPITS AND BARS / *Maurice L. Schwartz*
BARRIER ISLANDS / *Maurice L. Schwartz*
ENVIRONMENTAL GEOMORPHOLOGY AND LANDSCAPE
     CONSERVATION, VOLUME I: Prior to 1900; VOLUME II: Urban
     Areas; VOLUME III: Non-Urban Regions / *Donald R. Coates*
TEKTITES / *Virgil E. Barnes* and *Mildred A. Barnes*
GEOCHRONOLOGY: Radiometric Dating of Rocks and Minerals / *C. T.
     Harper*
MARINE EVAPORITES: Origin, Diagenesis, and Geochemistry / *Douglas
     W. Kirkland* and *Robert Evans*
GLACIAL ISOSTASY / *John T. Andrews*
GLACIAL DEPOSITS / *Richard P. Goldthwait*
PHILOSOPHY OF GEOHISTORY: 1785–1970 / *Claude C. Albritton, Jr.*
GEOCHEMISTRY OF GERMANIUM / Jon N. Weber
GEOCHEMISTRY AND THE ORIGIN OF LIFE / *Keith A. Kvenvolden*
GEOCHEMISTRY OF WATER / *Yasushi Kitano*
GEOCHEMISTRY OF IRON / *Henry Lepp*
GEOCHEMISTRY OF BORON / *C. T. Walker*
SEDIMENTARY ROCKS: Concepts and History / *Albert V. Carozzi*
METAMORPHISM AND PLATE TECTONIC REGIMES / *W. G. Ernst*
SUBDUCTION-ZONE METAMORPHISM / *W. G. Ernst*
PLAYAS AND DRIED LAKES: Occurrence and Development / *James T.
     Neal*
PLANATION SURFACES: Peneplains, Pediplains, and Etchplains / *George
     Adams*
SUBMARINE CANYONS AND DEEP-SEA FANS: Modern and Ancient /
     J. H. McD. Whitaker

Additional volumes in preparation

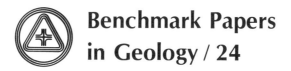

Benchmark Papers
in Geology / 24

A BENCHMARK® Books Series

SUBMARINE CANYONS
AND DEEP-SEA FANS
Modern and Ancient

Edited by

J. H. McD. WHITAKER
The University, Leicester, England

Dowden, Hutchinson & Ross, Inc.

STROUDSBURG, PENNSYLVANIA

Distributed by

HALSTED
PRESS

A division of
John Wiley & Sons, Inc.

Copyright © 1976 by **Dowden, Hutchinson & Ross, Inc.**
Benchmark Papers in Geology, Volume 24
Library of Congress Catalog Card Number: 75-5540
ISBN: 0-470-93912-5

All rights reserved. No part of this book covered by the copyrights
hereon may be reproduced or transmitted in any form or by any
means—graphic, electronic, or mechanical, including photocopying,
recording, taping or information storage and retrieval systems—
without written permission of the publisher.

77 76   1 2 3 4 5
Manufactured in the United States of America.

LIBRARY OF CONGRESS CATALOGING IN PUBLICATION DATA

Main entry under title:

Submarine canyons and deep-sea fans.

     (Benchmark papers in geology ; 24)
     Bibliography: p.
     Includes indexes.
     1. Submarine valleys--Addresses, essays, lectures.
2. Alluvial fans--Addresses, essays, lectures.
I. Whitaker, John Harry McDonald, 1921-
GC84.S92   551.4'6084   75-5540
ISBN 0-470-93912-5

Exclusive Distributor: **Halsted Press**
A Division of John Wiley & Sons, Inc.

UNIVERSITY LIBRARY
2 0 JUL 1976
LANCASTER

75 014955

ACKNOWLEDGMENTS
AND PERMISSIONS

ACKNOWLEDGMENTS

THE COLSTON RESEARCH SOCIETY, UNIVERSITY OF BRISTOL—*Submarine Geology and Geophysics*
   The Rhône Deep-Sea Fan

ROYAL SCOTTISH GEOGRAPHICAL SOCIETY—*Scottish Geographical Magazine*
   On the Land Slopes Separating Continents and Ocean Basins, Especially Those on the West Coast of Africa

PERMISSIONS

The following papers have been reprinted or translated with the permission of the authors and copyright holders.

ACCADEMIA NAZIONALE DEI LINCEI
   Submarine Slides in the Jurassic of Belluno and Friuli

AMERICAN ASSOCIATION OF PETROLEUM GEOLOGISTS—*American Association of Petroleum Geologists Bulletin*
   Congo Submarine Canyon
   Congo Submarine Canyon and Fan Valley
   Deep-Sea Channels, Topography and Sedimentation
   Growth Patterns of Deep-Sea Fans
   Physiography and Sedimentary Processes of La Jolla Submarine Fan and Fan-Valley, California
   Tarzana Fan, Deep Submarine Fan of Late Miocene Age, Los Angeles County, California
   Upper Paleocene Buried Channel in Sacramento Valley, California

AMERICAN JOURNAL OF SCIENCE (YALE UNIVERSITY)—*American Journal of Science*
   Canyons off the New England Coast
   Origin of Submarine Canyons

ELSEVIER PUBLISHING COMPANY—*Marine Geology*
   The Structure and Origin of the Large Submarine Canyons of the Bering Sea
   Submarine Canyons of the Continental Margin, East Bass Strait (Australia)

THE GEOLOGICAL ASSOCIATION OF CANADA—*Flysch Sedimentology in North America*
   Deep Sea Sediments in the Lower Paleozoic Québec Supergroup

GEOLOGICAL SOCIETY OF AMERICA—*Geological Society of America Bulletin*
   Growth of the Bengal Deep-Sea Fan and Denudation in the Himalayas

Acknowledgments and Permissions

THE GEOLOGICAL SOCIETY OF JAPAN—*The Journal of the Geological Society of Japan*
A Fossil Submarine Canyon near the Southern Foot of Mt. Kano, Tiba Prefecture

THE GEOLOGICAL SOCIETY OF LONDON—*Quarterly Journal of the Geological Society of London*
The Geology of the Area Around Leintwardine, Herefordshire

THE GEOLOGICAL SURVEY OF JAPAN—*Geological Survey of Japan, Bulletin*
Stratigraphy and Sedimentation of the Upper Cretaceous Himenoura Group in Koshiki-jima, Southwest Kyushu, Japan

GULF COAST ASSOCIATION OF GEOLOGICAL SOCIETIES—*Transactions—Gulf Coast Association of Geological Societies*
Erosional Channel in the Middle Wilcox near Yoakum, Lavaca County, Texas

MACMILLAN JOURNALS LTD.—*Nature Physical Sciences*
A Deep-Water Sand Fan in the Eocene Bay of Biscay

MACMILLAN PUBLISHING CO., INC.—*Papers in Marine Geology, Shepard Commemorative Volume*
Sedimentation and Erosion in Scripps Submarine Canyon Head

SOCIETY OF ECONOMIC PALEONTOLOGISTS AND MINEROLOGISTS
Journal of Sedimentary Petrology
Shale Grit and Grindslow Shales: Transition from Turbidite to Shallow Water Sediments in the Upper Carboniferous of Northern England
Modern and Ancient Geosynclinal Sedimentation
Ancient Submarine Canyons and Fan Valleys
Depositional Trends of Modern and Ancient Deep-Sea Fans

JOHN WILEY & SONS, INC.—*Submarine Canyons and Other Sea Valleys*
Excerpts

SERIES EDITOR'S PREFACE

The philosophy behind the "Benchmark Papers in Geology" is one of collection, sifting, and rediffusion. Scientific literature today is so vast, so dispersed, and, in the case of old papers, so inaccessible for readers not in the immediate neighborhood of major libraries that much valuable information has been ignored by default. It has become just so difficult, or so time consuming, to search out the key papers in any basic area of research that one can hardly blame a busy man for skimping on some of his "homework."

This series of volumes has been devised, therefore, to make a practical contribution to this critical problem. The geologist, perhaps even more than any other scientist, often suffers from twin difficulties—isolation from central library resources and immensely diffused sources of material. New colleges and industrial libraries simply cannot afford to purchase complete runs of all the world's earth science literature. Specialists simply cannot locate reprints or copies of all their principal reference materials. So it is that we are now making a concerted effort to gather into single volumes the critical material needed to reconstruct the background of any and every major topic of our discipline.

We are interpreting "geology" in its broadest sense: the fundamental science of the planet Earth, its materials, its history, and its dynamics. Because of training and experience in "earthy" materials, we also take in astrogeology, the corresponding aspect of the planetary sciences. Besides the classical core disciplines such as mineralogy, petrology, structure, geomorphology, paleontology, and stratigraphy, we embrace the new fields of geophysics and geochemistry, applied also to oceanography, geochronology, and paleoecology. We recognize the work of the mining geologists, the petroleum geologists, the hydrologists, the engineering and environmental geologists. Each specialist needs his working library. We are endeavoring to make his task a little easier.

Each volume in the series contains an Introduction prepared by a specialist (the volume editor)—a "state of the art" opening or a summary of the object and content of the volume. The articles, usually some thirty to fifty reproduced either in their entirety or in significant extracts, are selected in an attempt to cover the field, from the key papers of the last century to fairly recent work. Where the original works are in foreign languages, we have endeavored to locate or commission translations. Geologists, because of their global subject, are often acutely aware of the oneness of our world. The selections cannot, therefore, be restricted to any one country, and whenever possible an attempt is made to scan the world literature.

To each article, or group of kindred articles, some sort of "highlight commentary" is usually supplied by the volume editor. This commentary should serve to bring that article into historical perspective and to emphasize its particular role in the growth of the field. References, or citations, wherever possible, will be reproduced in their entirety—for by this means the observant reader can assess the background material available to that particular author, or, if he wishes, he, too, can double check the earlier sources.

A "benchmark," in surveyor's terminology, is an established point on the ground, recorded on our maps. It is usually anything that is a vantage point, from a modest hill to a mountain peak. From the historical viewpoint, these benchmarks are the bricks of our scientific edifice.

RHODES W. FAIRBRIDGE

PREFACE

Submarine canyons are "steep-walled, sinuous valleys, with V-shaped cross sections, axes sloping outward as continuously as river-cut land canyons, and relief comparable even to the largest of land canyons. Tributaries are found in most of the canyons and rock outcrops abound on their walls" (Shepard, 1973a, pp. 305–306). They are among the largest excavational features on the face of the earth, some being greater in size than the Grand Canyon of the Colorado. Although they have been known for over a century, their relative inaccessibility to direct observation made it difficult for early workers to study them in detail. Gradual improvements in exploration techniques, especially since World War II, have yielded more and more data on their morphology, sedimentary fills, and probable histories. In the 1930s there were widely differing theories of origin for canyons, and while there is now more general agreement, there are still strong differences in opinion among specialists in this field of research, in particular about the relative importance of different cutting agents, such as turbidity currents, sand creep and sand flows, and of slumping, up-building of the walls, reexcavation of former canyons, and the significance of structure in the siting of canyons. These differences will become apparent in the papers reprinted in this volume.

Progress up to 1966 was well summarized by Shepard and Dill in their book *Submarine Canyons and Other Sea Valleys,* which all students of canyons should consult. Some excerpts from their book are reprinted here, including their comprehensive bibliography. To give historical perspective, three early papers are reproduced in this volume, followed by studies of Congo, Scripps, and the Bering Sea Canyons.

Deep-sea fans are "cone- or fan-shaped accumulations of terrigenous sediment off shore from most of the world's great rivers, extending down to abyssal depths. They are built up at the lower ends of nearly all submarine canyons" (Fairbridge, 1966, p. 870). They are cut by valleys, "many of which have bordering levees that are built above the fans, most have distributaries, but very few have tributaries, and the walls are locally steep and are cut into unconsolidated fan sediments" (Shepard, 1973a, p. 306). Fans and fan valleys have been the subject of intense study only in the last few decades. In this volume, Menard's important paper of 1955 is followed by detailed studies of the Rhône, La Jolla, and Bengal Deep-Sea Fans and a synthesis of growth patterns of fans by Normark.

Today, sedimentologists are endeavoring to interpret sedimentary successions in terms of their environments of formation, and it seemed appropriate to devote the second half of this book to inferred ancient examples of both submarine canyons and deep-sea fans and fan valleys. This could not have been attempted even fifteen years ago, but such has been the growth of interest in this aspect of marine sedimentation that now it has proved difficult to select from the wealth of examples: the final choice has aimed to give as varied a selection as possible in time and space. Ancient canyons are described from California, Texas, Australia, Japan, Italy, and Great Britain, ranging in age from Silurian to Tertiary. Ancient fans are chosen from California, eastern Canada, northern England, and Spain: these also range from Lower Paleozoic to Tertiary in age.

I should like to thank all the authors and publishers of the papers used in this volume for their unfailing courtesy and helpfulness. Many authors went beyond what was asked of them and have made suggestions for improving the book. I thank particularly D. W. Scholl, A. Bosellini, J. Lajoie, and U. von Rad, who have contributed addenda and modifications to bring their work up to date; H. Okada for translating parts of the two Japanese papers and H. C. Jenkyns for translating the Italian contribution; and V. J. Ansfield, Stephanie Hrabar, R. Kepferle, H. W. Menard, A. A. Meyerhoff, T. H. Nilsen, L. E. Redwine, F. P. Shepard, and D. C. Van Siclen for dealing with specific questions. The Series Editor (Rhodes W. Fairbridge), George H. Keller, and Sam Thompson III read the whole manuscript and made constructive comments. Francis P. Shepard and Sam Thompson III drew my attention to a number of references on modern and ancient canyons, respectively. M. Adachi, A. Iijima, Y. Kanie, T. Matsumoto, S. Mizutani, H. Okada, and K. Tanaka enlightened me on Japanese ancient canyons. On the production side, Sally Ward cheerfully and promptly typed the manuscript. I have enjoyed working with such helpful and understanding publishers, especially the Benchmark Production Editor, Bernice Wisniewski.

J. H. McD. WHITAKER

CONTENTS

Contents

CONTENTS BY AUTHOR

INTRODUCTION

MODERN SUBMARINE CANYONS

To begin a volume in four parts, of which the first section only is to deal with modern submarine canyons, seems rather like composing Tchaikovsky's First Piano Concerto—starting with a great flourish but with the danger of ending weakly. To avoid this, we have deliberately soft-pedaled this opening section, especially because it has had excellent treatment from Shepard and Dill in their book *Submarine Canyons and Other Sea Valleys* (1966, Paper 8), and we have given approximately equal weight to the sections on modern deep-sea fans, ancient canyons, and ancient fans, which have received much less attention in review articles. We have had to resist the temptation to discuss the many theories of origin for submarine canyons except where appropriate in the introductions to individual papers, and we have had to omit many fine articles for lack of space and to achieve balance among the four sections. We would have liked to include more of Shepard and Dill's book and to have reproduced Martin and Emery's detailed study of Monterey and Carmel Canyons (1967); Nelson and others on Astoria Canyon and Fan (1970); some of the papers by Stanley, Kelling, Fenner, Pratt, and others on the U.S. East Coast canyons; the French work led by Bourcart in the Mediterranean and the Bay of Biscay; and the recent studies of the Great Bahama Canyon (Andrews and others, 1970), among others.

To give a general coverage of the subject, we have chosen an early publication by Buchanan (1887, Paper 1) on the Congo Canyon and two papers from the 1930s with contrasting views on the origin of canyons, one by Shepard on the U.S. East Coast canyons and part of the often-cited paper by Daly which led to the development of the turbidity-current hypothesis of submarine erosion of canyons. We return to the Congo Canyon (Heezen and others, 1964, Paper 4; Shepard and Emery, 1973, Paper 5) and then see a good example of the detailed studies made in Scripps Submarine Canyon head by Scuba diving (Dill, 1964b, Paper 6), investigations that led to new ideas about ways in which canyons could be eroded. The world's largest and longest canyons in the Bering Sea are awe inspiring, and a well-illustrated paper by Scholl and others (1970, Paper 7) deals with these impressive features. Finally, Shepard and Dill's data on all canyons known up to 1966, with their comprehensive bibliography, concludes this section.

Active research on canyons using improved techniques is discovering the details of their morphology and the nature of their walls and sedimentary fills, leading to a better appreciation of the processes forming them. To draw attention to work not included in Shepard and Dill's book (1966, Paper 8) or published since then, we have included a selection of references grouped under the general but overlapping headings of *Techniques; General Studies of Canyons and their Sediments; Currents and Sedimentary Processes Operating in Canyons;* and *Structural Control.*

Techniques of studying canyons are well summarized by Shepard (1973a, Chap. 2) and the results illustrated by Heezen and Hollister (1971). They include the use of deep-tow vehicles with transducers (e. g., Normark, Paper 12) which show up minor irregularities; side scanning by Asdic (see Belderson and Stride, 1969; Kenyon and Belderson, 1973); seismic reflection profiling (e. g., Kelling and Stanley, 1970; Rona, 1970; Belderson and others, 1972); improved methods of photographing the walls and floors of canyons (e. g., Gennesseaux, 1966; Stanley and Kelling, 1968a; Heezen and Hollister, 1971) and observing them by underwater television (Stanley and others, 1972; Stanley and Fenner, 1973). There has been a great increase in Scuba diving (e. g., Dill, 1969) and the use of deep-diving vehicles (e. g., Ross, 1968b; Got and others, 1969; Keller and others, 1973; Stanley, 1974), sampling with oriented box cores (e. g., Bouma and Shepard, 1964; Bouma, 1965; Rees and others, 1968) and the measurement of bottom currents with free instruments (e. g., Shepard and Emery, Paper 5). ^{14}C dating (Reimnitz and Gutiérrez-Estrada, 1970) is an important tool in the study of canyon development. Russian oceanographers have used luminescent sand to trace sediment movement from the Inguri River into the Inguri Canyon, Black Sea (Yegorov and Galanov, 1966; Trimonis and Shimkus, 1970).

Studies of specific canyons and their sediments include: Jordan,

1951 (De Soto); Northrop, 1953 (Hudson); Creager, 1958 (Campeche); Zenkovich, 1958 (Black Sea canyons); Jordan and Stewart, 1961 (Florida); Northrop and others, 1962 (Veatch and Hydrographer); La Fond, 1964 (east coast of India); Reyss, 1964 (western Mediterranean); Ryan and Heezen, 1965 (Ionian Sea canyons); Hopkins, 1966 (Australian canyons, including some filled canyons); Harbison, 1968 (De Soto Canyon and its relation to salt diapirs); Martin and Emery, 1967 (Monterey and Carmel); Rona and others, 1967 (Hatteras and Pamlico); Varadachari, 1967 (Coromandel Coast); Von der Borch, 1967, 1968 (Australian canyons); Bouma and others, 1968 (Alaminos Canyon winds among salt diapirs); Bouysse and others, 1968 (Cap Breton Canyon, Bay of Biscay); Dietz and others, 1968 (Cayar); Glangeaud and others, 1968 (Rech Bourcart); Nesteroff and others, 1968 (Cap Breton); Rees and others, 1968 (magnetic fabric of sediments in La Jolla Canyon and Fan Valley); Simpson and Forder, 1968 (Cape); Stanley and Kelling, 1968a, 1968b, 1970 (Wilmington); Uchupi, 1968a, 1968b (Wilmington and other canyons); Beer, 1969 (suspended sediment in Redondo); Carlson and Nelson, 1969 (Astoria); Duplaix and Olivet, 1969 (Rech Bourcart, Golfe du Lion); Il'yin and Lisitsyn, 1969 (map of Atlantic canyons and discussion of their origin in relation to climate); Newton and Pilkey, 1969 (Hatteras); Von der Borch, 1969 (New Guinea); Andrews and others, 1970 (Great Bahama); Dulemba, 1970 (Corsica); Duncan and Kulm, 1970 (Astoria and Willapa); Kelling and Stanley, 1970 (Wilmington and Baltimore); Mauffret and Sancho, 1970 (Gulf of Valencia, western Mediterranean); Nelson and others, 1970 (Astoria Canyon and Fan); Pryor, 1970 (deepwater submarine canyon head in Tongue of the Ocean); Scholl and others, 1970, Paper 7 (Bering Sea Canyons); Trimonis and Shimkus, 1970 (Inguri Canyon, Black Sea); Allen and others, 1971 (Cap Breton, Europe's largest canyon); Beer and Gorsline, 1971 (Redondo); Bouma and others, 1971 (Alaminos); Dietz and Knebel, 1971 (Trou Sans Fond, Ivory Coast); Field and Pilkey, 1971 (Hatteras); Felix and Gorsline, 1971 (Newport, now "dead" and filling with fine organic-rich sediments); Gennesseaux and others, 1971 (Var Canyon, Alpes Maritimes); Lonardi and Ewing, 1971 (canyons of the Argentine Basin); Pequegnat and others, 1971 (bottom photographs of De Soto Canyon); Burke, 1972 (Niger Delta canyons); Duplaix, 1972 (heavy minerals in sands of the French Mediterranean canyons); Emery and Uchupi, 1972 (their figure 37 is a 1 to 4 million colored map with canyons named, from Grand Banks to Caribbean); Goedicke, 1972 (Lebanon); Moyes and others, 1972 (Gascogne I); Houbolt, 1973 (Gulf of Guinea); Shepard and Emery, 1973, (Congo); Trumbull and Garrison, 1973 (canyons south of Puerto Rico); Vasiliev and Markov, 1973 (two canyons in the Sea of Japan); Coulbourn and others, 1974 (Hawaii);

Herman, 1974 (canyons of the Arctic Seas); Stanley and others, 1974 (Cap Creus and Lacaze-Duthiers Canyons, western Mediterranean); Sonnenfeld, 1975 (Mediterranean and Black Sea).

At present there is much interest in current movements and sedimentary processes operating in canyons, with attempts to quantify the active processes. A symposium on this topic was held at the Geological Society of America annual meeting in 1974. These aspects are dealt with by: Hand and Emery, 1964 (turbidites off California); Inman and Murray, 1964 (auto-suspended sand currents); Sprigg, 1964 (slumping and buried canyons off Australia); Marlowe, 1965 (the Gully Canyon); Shepard, 1965 (canyons as funnels); Agarate and others, 1967 (Lacaze-Duthiers and Cap Creus Canyons reexcavating earlier channels); Martin and Rex, 1967 (submarine weathering in Carmel and Monterey); Trumbull and McCamis, 1967 (mass transport in Oceanographer Canyon); Hulsemann, 1968 (internal waves and associated currents); Ross,1968a, 1968b (Corsair and Lydonia); Scholl and others, 1968 (Bering Sea Canyons); Starke and Howard, 1968 (fluming out of ancestral stream-eroded canyon); Von der Borch, 1968 (subaerial cutting of southern Australian canyons in early Tertiary); Beer, 1969 (Redondo); Emery, 1969 (canyons as funnels); Moore, 1969 ("drowned" canyon heads); Shepard, 1969 (canyons as funnels: filling and reopening); Shepard and Marshall, 1969 (internal waves in La Jolla and Scripps); Von der Borch, 1969 (rim upgrowth and axial downcutting, New Guinea); Andrews and others, 1970 (Great Bahama Canyon: upbuilding of sides and reexcavation of old filled troughs); Inman, 1970 (currents in Scripps Canyon); Mathewson, 1970 (Molokai, Hawaii, upbuilding of walls); Reimnitz and Gutiérrez-Estrada, 1970 (Rio Balsas Canyon, Mexico); Rona, 1970 (Cape Hatteras, upbuilding of walls; canyon heads now buried); Winterer, 1970 (erosional entrenchment and backfilling of Coral Sea Canyons); Beer and Gorsline, 1971 (Redondo); Fenner and others, 1971 (Wilmington bottom currents); Gennesseaux and others, 1971 (currents in submarine valley of Var, off Nice); Montadert and others,1971 (upbuilding of Cap Breton Canyon); Palmer, 1971 (erosion of submarine outcrops, La Jolla); Reimnitz, 1971 (currents in Rio Balsas Canyon); Rowe, 1971 (Hatteras); Bouma and others, 1972 (two canyons off Magdalena River); Cannon, 1972 (wind effects on currents in Juan de Fuca Canyon); Ballard and others, 1973 (Oceanographer Canyon in preexistent feature presently being reexcavated); Drake and Gorsline, 1973 (suspended sediment in Hueneme, Redondo, Newport, and La Jolla Canyons); Gonthier and Klingebiel, 1973 (Gascogne I, Bay of Biscay); Keller and others, 1973 (bottom currents in Hudson Canyon); Robb and others, 1973 (three submarine valleys off Liberia modified by large slumps: one of the valleys may mark the termination of an oceanic fracture zone); Shepard, 1973b (Aguja Canyon off Colombia); Shepard and Marshall, 1973a, 1973b (currents in canyons); Shepard and others, 1974 (internal waves in six Californian Canyons). Biological processes in canyons are dealt with by Dillon and Zimmerman, 1970 (Block and Corsair Canyons); Stanley, 1971 (bioturbation and sediment failure); Stanley and Fenner, 1973 (Wilmington); Hoover and Bebout, 1974 (two canyons off the Magdalena River delta, northern Colombia); Rowe and others, 1974 (Hudson); Stanley, 1974 (Wilmington).

Papers dealing with structural control of canyons include: Bourcart, 1965 (faults off the Pyrenees); Litvin, 1965 (dislocations off Norway); Martin and Emery, 1967 (Carmel and Monterey Canyons follow faults); Yerkes and others, 1967 (Redondo controlled by mid-Pleistocene faulting); Normark and Curray, 1968 (Baja California); Bush and Bush, 1969 (Trincomalee Canyon, Ceylon); Scholl and others, 1970, Paper 7 (Bering Sea Canyons); Winterer, 1970 (Coral Sea Canyons located in structural depressions); Montadert and others, 1971 (faults controlling directions of Cap Breton, Aviles, and other Bay of Biscay canyons); Morelock and others, 1972 (Manzanares Canyon, Venezuela); Payne and Conolly, 1972 (Antarctic canyons and fan valleys).

MODERN DEEP-SEA FANS

As submarine canyons became better known and their deeper parts more clearly mapped, it became apparent that their lower ends were distinguished by cone- or fan-shaped accumulations of sediment. At first called submarine deltas, these are now usually termed deep-sea fans. Many were recorded on the charts of Heezen and others (1959). In an often-cited paper of 1955 (Paper 9), Menard defined these terms and gave a good account of the fans off the western United States. During the 1960s, a number of deep-sea fans were studied in great detail. We have selected two of these investigations, one on the Rhône Fan in the Mediterranean (Menard and others, 1965, Paper 10) and the other on La Jolla Fan off southern California (Shepard and others, 1969, Paper 11). By 1970, the growing amount of information on fans enabled Normark to produce a model for their development (Paper 12). Finally, since we have included the world's largest canyons in Part I, it seemed appropriate to include in this section the world's largest fan, the Bengal Deep-Sea Fan in the Indian Ocean, the subject of continuing studies by Curray and Moore (1971, Paper 13).

With the increasing academic and economic interest in the deep oceans, and the equipping of modern research vessels with sophisticated apparatus for sounding and sampling the sea bed, we may expect

5

rapid advances in our knowledge of the form and structure of the deep-sea fans off all the continental shelves of the world. Besides their intrinsic interest and possible future economic importance, this knowledge will provide a firmer basis for comparison with evidence of suspected or inferred fans from the geologic past.

Listed below are some recent references on modern fan systems that were not cited in Shepard and Dill (1966) or have appeared since then, arranged in order of publication date: Hand and Emery, 1964 (fan channels with levees and terraces); Wilde, 1965 (Monterey Fan); Shepard, 1966 (tight meander in Monterey fan valley); Pratt, 1967 (left-hooking fan valleys extending from seven U.S. East Coast canyons); Yerkes and others, 1967 (Redondo Fan with leveed channels); Rees and others, 1968 (magnetic fabric of sediments from La Jolla Canyon and Fan); Shepard and Buffington, 1968 (complex levees, terraces of landslide origin, and left hook of La Jolla Fan Valley); Winterer and others, 1968 (Delgada Fan); Carlson and Nelson, 1969 (Astoria Canyon and Fan sediments); Komar, 1969 (hooking of Monterey Fan Channel); Moore, 1969 (Monterey and Delgada Fans mainly Pleistocene in age); Normark and Piper, 1969 (deep-sea fan valleys, past and present); Piper and Marshall, 1969 (bioturbation of Holocene sediments, La Jolla Fan); Shepard and others, 1969, Paper 11 (La Jolla Fan and Fan Valley); Duncan and Kulm, 1970 (Astoria Canyon and Fan); Huang and Goodell, 1970 (East Mississippi Cone not built by turbidity currents); Nelson and others, 1970 (Astoria Canyon and Fan); Normark, 1970a, Paper 12 (growth patterns of fans); Normark, 1970b (channel piracy on Monterey Fan); Piper, 1970 (box cores from La Jolla Fan); Walker and Massingill, 1970 (slumps on Mississippi Fan); Bergantino, 1971 (upper and lower Mississippi Fan); Conolly and Cleary, 1971 (network of channels and levees on braided fan at distal margin of Hatteras Canyon); Curray and Moore, 1971, Paper 13 (Bengal Fan); Haner, 1971 (morphology and sediments of Redondo Fan and Fan Valley; lateral migration of channels; overlap of older by newer levees; buried channels); Horn and others, 1971 (canyons and fans of northeastern Pacific); Pequegnat and others, 1971 (bottom photographs of Mississippi Fan); Stanley and others, 1971 (Hudson, Wilmington, and Washington Megafans); Wong and others, 1971 (Rosetta and Damietta Fans on Nile Cone); Burke, 1972 (fans old and new off the Niger Delta); Davies, 1972 (Gulf of Mexico cones); Normark and Piper, 1972 (Navy Fan, off California); Ruddiman and others, 1972 (Maury Channel and Fan off Iceland); Wilhelm and Ewing, 1972 (Mississippi Cone); Baker and others, 1973 (nepheloid layer over Nitinat Fan); Nelson and Kulm, 1973 (a useful review of fans and deep-sea channels); Naini and Leyden, 1973 (Bengal Fan); Weiler and Stanley, 1973 (tectonic origin for fan-shaped Balearic Rise, western Mediter-

6

ranean); Bouma, 1974 (levees and channels on the Mississippi Fan); Clifton, 1974 (similarities and contrasts between deep-sea fan and estuarine deposits); Curray and Moore, 1974 (Bengal Fan and Geosyncline); Normark, 1974 (growth processes in deep-sea fans).

ANCIENT SUBMARINE CANYONS

Compared with the knowledge of modern canyons that steadily accumulated over more than a century, the study of ancient (pre-Pleistocene) canyons is much more recent. A possible buried canyon in Texas–Louisiana was described as such by Bornhauser (1948) and in 1949, W. G. Payne of the U.S. Geological Survey suggested that a buried channel of Cretaceous age in Alaska might be a fossil submarine canyon (Robinson, 1964). The 1950s saw a number of publications in which authors postulated ancient canyons through which sediment was assumed to have been funneled to accumulate in former depositional basins, but still only a few actual canyons were described. Two of these early descriptions are reprinted in this volume (Sato and Koike, Paper 14; and Hoyt, Paper 16). During the 1960s, increasingly detailed attempts to interpret sedimentary successions in terms of the paleoenvironment resulted in the discovery, in widely separated areas of the world, of well-authenticated submarine canyons. Many are of Tertiary age, several of Mesozoic, and a few of Paleozoic age. We have selected four papers from this period of publication, Papers 17 to 20, as representative of the progress being made in this field of study. During the present decade, many articles on ancient canyons have already appeared and it is now apparent that they are widely distributed in space and time, occurring in most systems as far back as the Precambrian. Paper 15, by Tanaka and Teraoka, deals with a recently described canyon from Japan and in the final contribution to this section, there is a review by Whitaker of most of the ancient canyons and fan valleys known up to the beginning of 1974 (Paper 21). References additional to those cited in that review paper are: Davis, 1953 (Markley Gorge, California); Sato and Koike, 1957, Paper 14 (Tiba, Japan); Sacramento Petroleum Association, 1962 (Markley Gorge); Safonov, 1962 (Princeton and Markley Gorges); Silcox, 1962 (Meganos Gorge and two Markley Channels); Sprigg, 1964 (buried channels and fjords, southern Australia); Webb, 1965 (late Miocene Stevens Sands, San Joaquin Valley, California); Hopkins, 1966 [filled canyons off southern Australia: one is 3 miles (5 km) wide and 2800 ft (850 m) deep]; Houtz and others, 1967 (filled canyons off New Zealand); Arleth, 1968 (up-dip closure by impermeable Markley Gorge fill); Safonov, 1968 [Princeton, Markley, and Brentwood (= Meganos) Gorges]; Silcox, 1968 (Meganos Channel); Gvirtzman, 1969 (Gaza Canyon, Israel, filled in late Miocene); Halbouty, 1969 (Yoakum Channel, Texas); Stride and others, 1969 (filled submarine canyon southwest of Ireland); Conolly and others, 1970 (buried valley–canyon–fjord system, southern Australia); Remane, 1970 (upper Tithonian fossil canyon, Vocontian Trough, southeastern France); Von der Borch and others, 1970 (southern Australian buried canyons); Benson, 1971 [Oligocene channel 15 miles (24 km) long and 3 to 6 miles (5 to 10 km) wide in Louisiana]; Ewing and Lonardi, 1971 (buried shelf canyons off Argentina); Fischer, 1971 (Meganos Channel); Morrison and others, 1971 (Princeton, Markley, and Meganos Gorges shown in cross sections); Friedmann, 1972 (Precambrian block in possible early Cambrian canyon); Redwine, 1972 (Tertiary Princeton submarine valley system beneath Sacramento Valley, California); Travers, 1972 (Markley submarine valley related to downthrow of Midland Fault and a possible trench margin); Enos, 1973 (submarine channel, Cretaceous, Mexico); Hsü and others, 1973 (upper Miocene and earlier beds cut by deep channels when Mediterranean dried up, channels filled with alluvial gravels and then marine Pliocene); Kieken, 1973 (Miocene canyon of Saubrigues, Aquitaine); Tanaka and Teraoka, 1973, Paper 15 (Kyushu, Japan); Busch, 1974 (Yoakum Channel); Picha, 1974 (two vast canyons with Eocene–Oligocene fills, 2 to 7 km wide, 25 km long, and over 1000 m deep, Czechoslovakian Carpathians); Seibold and Hinz, 1974, filled canyons up to 22 km wide off West Africa); Sonnenfeld, 1974 (Tertiary canyons, Mediterranean).

Future trends in the study of ancient canyons should be to determine more accurately their dimensions (length, width, depth, wall slopes, axial gradients, etc.) and the history of their cutting and filling, aided by detailed sedimentological studies and paleontological work which, with suitable fossil groups such as the foraminifera, may give useful data on depths at the time of canyon formation and filling. Since many old canyons are important as traps for hydrocarbons, which may accumulate within the canyon fill (e. g. Benson, 1971) or below it, we may expect increasing efforts to be applied to their discovery and detailed delimitation, both on land and on the continental shelves, where many buried canyons are being discovered.

ANCIENT DEEP-SEA FANS

The interpretation of sedimentary sequences in the geological column as ancient deep-sea fan deposits is even more recent than the recognition of ancient canyons. Menard (1960), studying the enormous volumes of sediment making up the modern fans off western North America, suggested that they may well have been forming since Tertiary times, and in the same year Sullwold described an ancient fan on land:

this work is reprinted here as Paper 22. Part of Walker's well-known paper of 1966 forms Paper 23: in it he made a convincing reconstruction of upper Carboniferous shelf, slope, and basin environments in the Pennine Region of northern England, and showed how slope channels funneled sediment from the shelf to accumulate on deep-sea fans as proximal and distal turbidites. Since then, the number of papers on ancient deep-sea fans has increased rapidly. From these we have selected two representative examples, one by Hubert and others (Paper 24) on a lower Paleozoic fan from Québec which is still being studied, and the other (Paper 25) by Kruit and others on a Tertiary fan from northwest Spain. A profusely illustrated and detailed analysis of Permian deep-sea fans of the Delaware Basin by Jacka and others (1968) is omitted for lack of space. LeBlanc (1972) outlined a submarine canyon-fan model of clastic sedimentation. The final paper (26), by Nelson and Nilsen, published in May 1974, usefully reviews the present position on the study of modern and ancient fans and the criteria by which they may be compared: their bibliography will be helpful to those readers wishing to pursue this topic.

References additional to those given in Nelson and Nilsen's paper include: Conrey, 1967 (early Pliocene fans, Los Angeles Basin); James, 1967, 1971, 1972, and James and James, 1969 (upper Ordovician fans and channels, Wales); Colburn, 1968 (upper Cretaceous fan, Diablo Range, California); Kepferle, 1968, 1969 (Mississippian fan channels, Kentucky); Wezel, 1968 (Oligocene to lower Miocene fans in Tunisia); Mutti, 1969 (inner to middle fan with valleys, mid-Tertiary, Island of Rhodes); Ricci Lucchi, 1969 (middle fan with valleys in Miocene of Italy); Van Hoorn, 1969 (upper Cretaceous fan, Spanish Pyrenees); Komar, 1970 (late Miocene Doheny fan channel, southern California); Baldwin, 1971 (lower part of submarine fan and channel of Cambrian and Ordovician age, southwest Vermont); Davis, 1971 (Pliocene of Ventura interpreted as fans, each at the foot of a submarine canyon); Fischer, 1971 (upper Paleocene canyon and fan, California); Piper, 1971 (middle Cambrian turbidite fan, Canadian Rockies); Piper and Normark, 1971 (Miocene fan and fan valley, southern California); Ansfield, 1972 (Eocene fans, northern Olympic Peninsula, Washington); Chipping, 1972 (late Cretaceous to Paleocene fan with fan channels, San Francisco Peninsula); Kruit and others, 1972, Paper 25 (Eocene fan, Bay of Biscay); Lowe, 1972 (fan-shaped rise, Cretaceous of Sacramento Valley, California); Mansfield, 1972 (late Mesozoic fan and fan channel west of Coalinga, California); Mutti and Ghibaudo, 1972 (Miocene outer fan environment, northern Apennines); Mutti and Ricci Lucchi, 1972 (submarine fan associations—inner, middle, and outer fan); Clarke, 1973 (Point of Rocks Sandstone, Eocene, California); Fischer, 1973 (late

Eocene proximal submarine fans, Western Transverse Ranges, California); Galloway and Brown, 1973 (upper Pennsylvanian fans of north-central Texas); Ghibaudo and Mutti, 1973 (inner fan with valley 4 km wide, 250 m deep, northern Apennines); Hall and Stanley, 1973 (levee-bounded submarine base-of-slope channels, lower Devonian, northern Maine); Lajoie and Chagnon, 1973 (Cambrian canyon and fan, Canadian Appalachians); Morris, 1973 (Carboniferous subsea cones, Ouachitas); Nilsen, 1973 (fans of Tejon Formation, Eocene, California); Nilsen and Clarke, 1973 (early Tertiary fans, central California); Nilsen and Simoni, 1973 (Eocene Butano Fan, California); Ricci Lucchi, 1973 (sedimentary sequence used to differentiate inner, mid, and outer fan, and fan-channel, environments, Miocene, Italy); Ricci Lucchi and Pialli, 1973 (major and minor fans, Miocene Marnoso-arenacea Formation, northern Apennines); Schlager and Schlager, 1973 (upper Jurassic fan, eastern Alps); Schüpbach, 1973 (filled slope valleys and fans, Pennsylvanian, New Mexico); Walker and Mutti, 1973 (inner, middle, and outer fan associations, middle fan channels and depositional lobe); Cruz and others, 1974 (Miocene prograding submarine fans fed through deep canyons); Hoover and Bebout, 1974 (large fan, active in Pliocene, now cut off from sediment source, northern Colombia); Mansfield, 1974 (late Mesozoic deep-marine fan with fan valleys extended westward from Sierra Nevada magmatic arc across an arc-trench gap); Mutti, 1974 (inner, middle, and outer fan deposits from circum-Mediterranean geosyncline); Nilsen and others, 1974 (late Paleocene to early Eocene fan, Vallecitos area, California); Pícha and Niem, 1974 (Ouachita Mountains); Schüpbach and Morel, 1974 (fans and channels in classical flysch, central Alps: inner and middle fan associations, suprafan depositional lobes, depositional lobes with few channels deduced as beds are traced from west to east); Yeats and others, 1974 (early Tertiary Poway Fan and Cone fragmented by Miocene rifting, southern California); Carter and Lindqvist, 1975 and in press (Oligocene fan complexes, southern New Zealand); Parker, 1975 (Paleocene fans, central North Sea); Hrabar, in press (late Precambrian Ravalli Group, Montana: the Revett quartzites are envisaged as subaqueous channel deposits and the argillites as subaqueous overbank, levee, and distal fan deposits).

Work continues actively on matching up ancient turbidite successions, with or without channels, with modern fans (Walker and Mutti, 1973), studies that attempt to differentiate inner, middle, and outer fan, suprafan, and fan-channel environments in increasing detail.

Part I

MODERN SUBMARINE
CANYONS

Editor's Comments
on Papers 1 Through 3

Ever since submarine cables were first laid, breakages have had to be located and repaired by vessels equipped for the purpose. One important outcome of the soundings made by these trouble-shooting ships was the discovery that cable breaks were often associated with deep, steep-sided cañons, now termed submarine canyons.

This short excerpt from Buchanan's 1887 paper (1) illustrates the discovery and delimitation of three such canyons, the "Bottomless Pit" off the Guinea Coast of Africa, Avon's Deep east of Lagos and the Congo Canyon, which was found to penetrate 20 miles (32 km) up the estuary of the Congo River. Buchanan also draws comparisons with the already known Cap Breton Canyon in the Bay of Biscay, said to be the world's first-discovered canyon, although the Hudson Canyon may prove to have this distinction (Dana, 1863).

In suggesting an explanation for the origin of the Congo Canyon, Buchanan was prophetic of one currently held view, that canyons develop by the up-building of sediment on either side while currents within the canyons prevent sedimentation (e. g., Shepard, 1952; Andrews and others, 1970; Rona, 1970). In recent years the "Bottomless Pit" has been restudied by Dietz and Knebel (1971) and the Congo Canyon by Veatch and Smith (1939), Heezen and others (Paper 4) and Shepard and Emery (Paper 5).

One of the readers of Buchanan's paper was F. -A. Forel, who studied the sublacustrine gorge cut through the Rhône sedimentary fan in Lake Geneva. Forel (1887) supported Buchanan's idea that the Congo Canyon was kept clear of sediment by a deep countercurrent flowing up the canyon while the areas on either side built up by sedimentation, and he contrasted Stassano's (1886) view that the Congo Canyon was a former subaerial river bed submerged during a subsidence of the African Continent. Forel (1885, 1887) was the first to postulate the existence of density currents (now termed turbidity currents), stating that these could erode a subaqueous ravine like the deep channel in Lake Geneva if the velocity, density, and gradient were great enough. Even if no erosion occurred, the channel would be kept open while the areas on both sides built up.

It is salutary to reflect that nearly a century ago, many of our current theories on the origin of submarine canyons were already launched. Among other early views were theories of submarine springs, fault control, and subaerial erosion during a sea-level lowering amounting to thousands of feet (see the review by Shepard and Dill, 1966, Chap. 16).

By the 1930s, however, two major theories had become established, one advocated by Shepard, who favored predominantly deep subaerial erosion while the continental margins were briefly but greatly uplifted, followed by rapid drowning, then filling, and reopening by landslides. The other theory was advanced by Daly, who argued for wholly submarine processes, with sea level lowered by 200 to 300 ft (60 to 90m), causing the stirring up of mud which initiated density currents; these in turn eroded the canyons. Daly's ideas (criticized by Shepard in 1937) were developed by Kuenen (1937, 1938), who began a series of experiments to test Daly's hypothesis and who eventually became convinced that turbidity currents are able to erode hard rock and to excavate major canyon systems.

During the 40 years that have elapsed since these papers were published, the opposing theories have come much closer together. Most students of canyons would now agree that some canyons are wholly or mostly subaerial in origin [such as those off west Corsica (Kuenen, 1953), off Japan (Nasu, 1964), and off Molokai, Hawaii (Mathewson, 1970)] while many others were initiated as subaerial river systems and drowned, being kept open or actively eroded by turbidity currents, grain flow, slumping, or other submarine processes. A recent revival of an old view is that some canyons may be kept open by current action while the shelf region on either side builds up by sedimentation, resulting in the increasing height of the canyon walls, as in the Great Bahama Canyon (Andrews and others, 1970), the canyons off Cape Hatteras (Rona, 1970), and those off southeastern New Guinea (Von der Borch, 1969). Reexcavation of older filled canyons is also suggested (Agarate and others, 1967; Shepard, 1969).

Shepard's 1934 paper (2) is one of the earliest of his many distinguished and continuing contributions to the knowledge of submarine canyons and expresses his views on the origin of canyons as held at that time. The article includes some of the first maps made by the U.S. Coast and Geodetic Survey of the U.S. East Coast continental slope.

Opposing views on canyon origin are given in Daly's cautious, well-thought-out article of 1936 (3), in which he argues for submarine erosion. In a prophetic footnote to page 36 he suggests that canyons may have been cut before the glacial period, and that slump masses should be found at the bottoms of canyons. He also suggests that submarine sliding may be important in developing canyons, a view widely held today.

Biographical Notes

John Young Buchanan was born at Dowanhill in Scotland in 1844 and died in 1925. He attended Glasgow University, then the universities of Marburg, Leipzig, and Bonn, and the Ecole de Médecine in Paris. He was appointed chemist and physicist to the *Challenger* expedition and made many marine investigations, especially on the chemistry and physics of seawater during the 3½-year circumnavigation of the globe. Other voyages took him to the Gulf of Guinea in connection with the planning of a route for a new telegraph cable, and he accompanied the Prince of Monaco on many cruises. He later worked on the glaciers of the High Alps.

His many scientific papers included notes on manganese nodules, the Mediterranean and its deep-sea deposits, and articles on lakes, lochs, the winding of rivers, volcanoes, and earthquakes. His interests ranged over chemistry, physics, oceanography, geography, zoology (he wrote on "The Daintiness of the Rat"), astronomy, and accidents to ships, airships and balloons (which he witnessed), and trains (in which he was involved as a passenger). His papers were collected into two volumes, *Comptes Rendus of Observations and Reasoning*, published in 1917, and *Accounts Rendered of Work Done and Things Seen*, issued in 1919. The latter includes his inaugural lecture as University Reader in Geography at Cambridge, where he lectured from 1899 to 1903.

Buchanan was Commandeur de l'Ordre de Saint Charles de Monaco and Vice-President du Comité de Perfectionnement de l'Institut Océanographique (Fondation Albert 1er Prince de Monaco) and he received the Gold Medal of the Royal Scottish Geographical Society in 1912.

Francis P. Shepard was born in 1897, in Brookline, Massachusetts. He received the A.B. degree from Harvard University and the Ph.D. from Chicago University. After teaching geology at the University of Illinois, he moved to Scripps Institution of Oceanography, where he became Professor of Submarine Geology. Since 1967 he has been Professor Emeritus of that institution. He has worked on many aspects of submarine geology, in particular on coastal changes, bay sedimentation, the continental shelves, submarine canyons, and deep-sea fans; three of his contributions to the literature on canyons and fans are reprinted in this volume.

His distinguished career in submarine geology in many parts of the world has been recognized by honorary membership of many professional organizations, and the Society of Economic Paleontologists and Mineralogists has inaugurated a Francis P. Shepard Medal for Excellence in Marine Geology. The "Shepard Volume" (Miller, 1964) lists his many publications up to 1964. His books include *Submarine Geology* (3rd edition), *The Earth Beneath the Sea* (2nd edition), and, with R. F. Dill, *Submarine Canyons and Other Sea Valleys,* extracts from which are reproduced in Paper 8.

Reginald A. Daly was born at Napanee, Ontario, in 1871. He received the A.B. degree at the Victoria University, Toronto, in 1891, the A.M. in 1893, Ph.D. in 1896, and an Honorary Sc.D. in 1942, all at Harvard University. Other honorary doctorates were awarded by the universities of Toronto and Cincinnati. After studying at Heidelberg and Paris, Daly became Geologist for the Canadian International Boundary Survey (1901–1907), Professor of Physical Geology at Massachusetts Institute of Technology (1907–1912), Sturgis-Hooper Professor of Geology (1912-1942) at Harvard, and then Emeritus Professor until his death in 1957.

Reginald Daly was a member of many geological societies, becoming President of the Geological Society of America in 1932. He wrote numerous papers on subjects which were then controversial, such as the glacial control of coral reefs, and was the author of several books, of which *The Floor of the Ocean* (1942) is especially relevant to our topic of submarine canyons.

1

Reprinted from *Scottish Geog. Mag.*, **3**, 217, 222–224 (1887)

ON THE LAND SLOPES SEPARATING CONTINENTS AND OCEAN BASINS, ESPECIALLY THOSE ON THE WEST COAST OF AFRICA. OBSERVED AND SURVEYED IN THE S.S. " BUCCANEER," BELONGING TO THE INDIA-RUBBER, GUTTA-PERCHA, AND TELEGRAPH WORKS COMPANY (LIMITED) OF SILVERTOWN.*

(With Diagrams.)

By J. Y. BUCHANAN.

[*Editor's Note:* Buchanan's paper commences with an account of the steep-sided shoals rising abruptly from deep water between Spain and the Canary Islands and the steep seaward slopes of these islands. He then describes soundings around the coast of the Gulf of Guinea with a view to discovering a suitable place for a new telegraph cable to be laid. This part of his paper is omitted.]

Along the Guinea Coast as far as Cape St. Paul the 100-fathom line is at an average distance of 15 miles from the shore; and on the south-west coast between the Niger and the Congo it is from 30 to 40 miles from the shore. At three localities there are remarkable exceptions. Off Grand Bassan the 100-fathom line approaches within a quarter of a mile of the shore, and the curving-in shorewards of the contour lines produces the very remarkable submarine gully known by the name of the " Bottomless Pit." From the *Buccaneer* five lines of soundings were run across it at distances of two miles between the lines. At one mile from the shore the width of the gully is under a mile, with a depth of 150 fathoms; at eight miles from the shore its width is 1½ miles, with a maximum observed depth of 327 fathoms; while two miles further seawards its width has increased to 4 miles, with a maximum observed depth of 452 fathoms. The accompanying sketch-plan (Fig. 1) of the contour lines gives a better idea of this extraordinary feature than can be conveyed by description. The bottom consisted everywhere of a soft dark-coloured mud, and the slopes of the sides averaged in many places 2000 feet per mile.

Another similar gully occurs to the eastward of Lagos, called Avon's Deep, but it does not reach so close to the shore as the one just described.

The most remarkable feature of this kind occurs at the mouth of the river Congo (Fig. 2). Here the gully penetrates inland up the river, a depth of 150 fathoms being found 20 miles within its mouth, while it was traced by the *Buccaneer* to a distance of nearly 100 miles seaward. It forms an immense submarine cañon, with steep mud sides, penetrating deeply into the land. The 100-fathom line runs at a distance of about 35 miles from the coast both north and south of the cañon; the 500-fathom line runs at about 60 miles, and the 1000-fathom line at about 80 miles from the coast. Of these contour lines, that of 100 fathoms penetrates 20

*Reproduced from a microfilm of the original provided by Brown University Library.

Fig. 1 THE BOTTOMLESS PIT

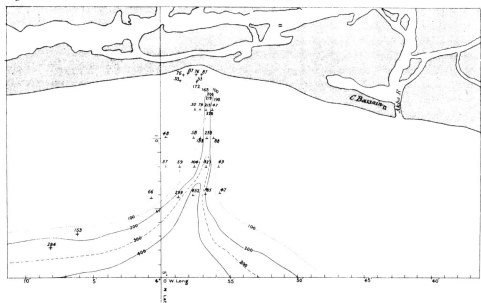

Fig. 2. SKETCH OF THE CONGO CAÑON

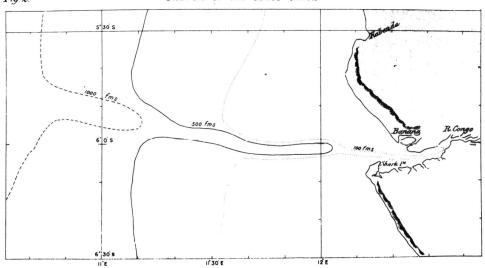

miles within the coast line, that of 500 reaches to within 10 miles, and that of 1000 fathoms to within 50 miles of it. At the mouth of the river, off Shark Point, the width of the gully is three miles, and the deepest sounding was 333 fathoms. The deepest sounding obtained in the river, off Banana Creek, was 242 fathoms. Thirty-five miles off the mouth of the river the width of the gully is six miles, and the maximum depth observed was 573 fathoms. The bottom of the cañon is here 3000 feet below the surface of the plateau in which it is cut.

It is difficult to furnish a satisfactory explanation of these gullies, but the Congo cañon is certainly connected with the river which flows over it, and the "Bottomless Pit" may be connected with the river Akba, which flows into the sea a few miles to the eastward of it. The Congo cañon is not due to erosion by the river, for the current, strong though it is, does not extend more than 20 fathoms from the surface, and is felt to a considerable distance out to sea, where it thins out considerably. For many miles off the mouth of the river the water has a dark reddish-yellow colour, but this forms only a thin layer, as the ship's propeller turns up the colourless salt water beneath. The existence and the persistence of the cañon are due to an agency which prevents the mud brought down by the river being deposited along its axis. This agency is probably the sea water running up the gully at the bottom and returning in the upper layers, mixed with the river water. A circulation in a vertical plane is thus produced, and in the axis of it, settlement of sediment is more difficult than on either side of it. The bottom of the cañon, or gully, would thus resemble more nearly the original form of the bottom of the sea before the Congo discharged its waves into it, than the flatter and shallower bottom on each side of it. In fact the cañon has been built up, not hollowed out. In how far a crack in the crust of the earth may have had to do with the depth or the position of this cañon, it is impossible to speak certainly, but the preservation and accentuation of it are certainly due to the prevention of the deposition of sediment.

The "Bottomless Pit" is situated only fourteen miles from the mouth of the large river Akba. As is usual on this coast, behind the beach of pebbles, heaped up and arranged by the combined action of the surf and the prevailing easterly current, there is an extensive lagoon formation connected with the river; and at the point where the "Bottomless Pit" approaches nearest to the beach, the width of the beach separating the lagoon from the sea is smallest, being little over a quarter of a mile. The mouths of these rivers frequently change their positions, and it is not improbable that the Akba may have entered the sea where the "Bottomless Pit" now is, and produced similar conditions to those existing at the mouth of the Congo. Since the date of the *Buccaneer's* visit, the lagoon at Porto Novo has opened a new communication with the sea.

In Europe, at the head of the Bay of Biscay, and on a coast resembling in some of its physical features the coast of the Gulf of Guinea, a gully or cañon quite like the Bottomless Pit penetrates close up to the shore

at Cape Breton, where, in former times, the river Adour entered the
sea.   This remarkable gully forms a deep indentation in the littoral
flat, and runs parallel with the range of the Pyrenees and the mountains of the north coast of Spain. It is not impossible that, in their
origin, these two features may have had some connection with each
other.   Much trouble was experienced with the first cable laid from
Bilbao over the deeper part of this gully, indicating some instability
in the features of the bottom in this region, even at depths of over
1000 fathoms.   Earthquakes are a frequent cause of rupture of cables
in the neighbourhood of the steep slopes connecting the continents with
the ocean beds.

[*Editor's Note:* Buchanan's paper concludes with a discussion of water
circulation, coast flats of the Guinea Coast, and erosional aspects which,
not being relevant to our subject, are omitted.]

2

Copyright © 1934 by the American Journal of Science (Yale University)

Reprinted from *Amer. Jour. Sci.*, Ser. 5, **27**(157), 24–36 (1934)

CANYONS OFF THE NEW ENGLAND COAST.*

FRANCIS P. SHEPARD.

INTRODUCTION.

During the summers of 1931 and 1932 I had the opportunity to accompany Coast Survey vessels during most of their charting of a 200 mile stretch of the continental slope off Georges Bank. This area extends east from a point 100 miles south by east from Cape Cod and 40 miles south of the Nantucket Lightship. Along this slope submarine canyons of awe-inspiring proportions were discovered. Prior to the Georges Bank survey there was almost no indication of these canyons, but at present some of them are more completely charted than any other deep submarine valleys in the world. There is every reason to believe that if these features were visible they would compare scenically with the most impressive canyons in the world.

Distribution and Character of the Canyons.

The general distribution of the submarine canyons along the 200 miles of continental slope is shown in Fig. 1. The scale of this contour map has prevented the drawing of details and except for the 300-foot contour line it was found advisable to use a 600-foot interval. It seems probable that there are various undiscovered valleys in the eastern portion of this map where soundings were less numerous than elsewhere. It will be observed that some of the valleys indent the continental shelf while others do not extend above the continental slope. In no case does a valley penetrate more than 12 miles into the continental shelf and the 50-fathom contour shows almost no relation to the valleys.

These submarine canyons are by no means limited to the upper parts of the slope. They all extend down to depths of at least 6000 feet below sea level and in no case does the survey attain great enough depths to show their outer termini. The greatest depth observed in one of these canyons is 8400 feet. Below the adjacent slopes they are cut to depths ranging up to about 4500 feet, 7 of them being incised more than 2500 and 24 others more than 1000 feet.

* The writer wishes to express his appreciation for privileges extended to him by the U. S. Coast and Geodetic Survey. Also thanks are due to Mr. George Cohee for the analysis of some samples of sediment, and to Dr. R. B. Stewart for an opinion of a fossil.

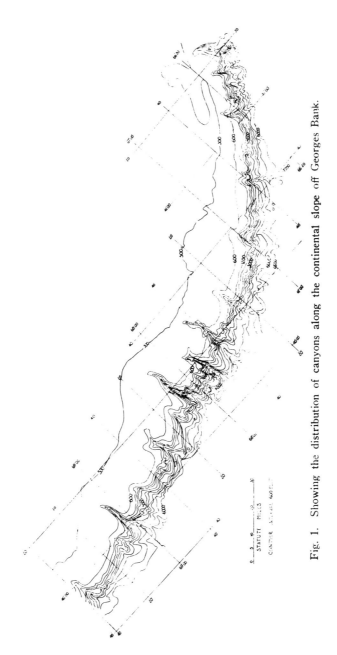

Fig. 1. Showing the distribution of canyons along the continental slope off Georges Bank.

STATUTE MILES

CONTOUR INTERVAL 600 FEET

21

The characteristics of the canyons are shown best in the maps of larger scale. An impressive group discovered during the summer of 1932 was surveyed in great detail (Fig. 2). Accurate determination of positions was made by dropping bombs at frequent intervals from the surveying ships. The sound from the explosions when received by two anchored station ships was relayed back by wireless and the time interval gave the distances. Where the sounding lines were well separated the contours are, of course, less accurate but their approximate nature was determined from the general character of the adjacent topography. Taken all in all the contours are probably as accurate as those of the older topographic maps of our western mountains. The established points are, of course, much more numerous in this marine survey, but this is offset by inability to see the features and by the fact that echo soundings on a slope somewhat minimize the depth and soften the relief. This last point makes it probable that these canyons are actually more impressive than is indicated by the map.

Resemblance to River Valleys.—The canyons have many features in common with those land valleys which have been cut by rivers. Their courses are undoubtedly sinuous probably more so than is shown on the maps. Curiously, most of the larger valleys have a gently curving Z-trend like that of the Hudson submarine gorge. The valley bottoms slope outward practically continuously and take directions such as might be expected of streams working down the slope towards the abyss. The tributaries come in with the dendritic pattern characteristic of land valleys. In cross section the submarine valleys are roughly V shaped, particularly in the upper portions and have very steep walls (Fig. 3).

Origin of the Canyons.

Faulting and Submarine Currents.—In recent years various authors have attributed submarine valleys either to faulting or to submarine currents.[1] If these canyons were due to faulting they should have the straight walls and trough shapes characteristic of fault troughs. Probable fault troughs are found at the base of the continental slopes in many places, but these trend parallel to the slopes and not at right angles as

[1] The writer has recently discussed the various hypotheses concerning submarine valleys giving rather complete references to the literature on the subject. Submarine Valleys, Geog. Review, **23**, 77-89, 1933.

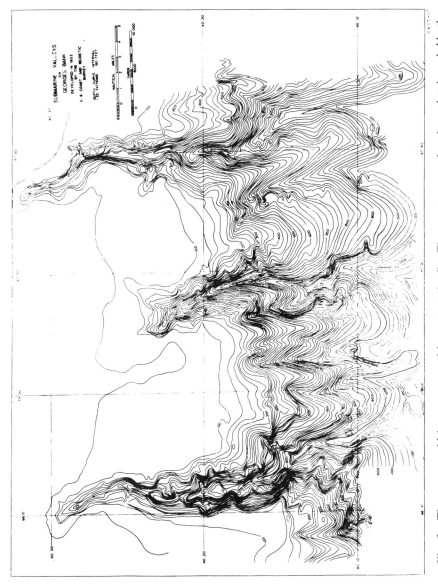

Fig. 2. The canyons which were surveyed in greatest detail. These contours are based on what is probably the best deep water survey in the world at the present date. The writer will be glad to send the original sheet with soundings to any one interested.

do the submarine canyons.   If the canyons were cut by submarine currents, such currents must have been unbelievably
powerful.   Judging from their high precipitous walls the

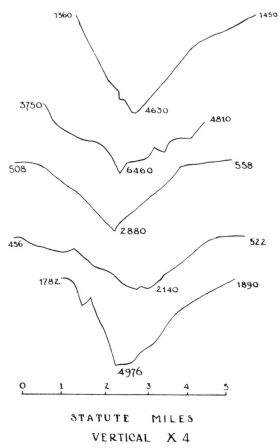

Fig. 3.   Showing the rough V-shaped transverse profiles characterizing
the canyons of figure 2.

canyons must have been cut in solid rock.   Such walls, if made
of unconsolidated material, would have caved in especially if
subjected to earthquake shocks like that of 1929.   Current
data show that at least the surface currents have the greatest
flow at right angles to the canyons on the outer portion of the
shelf in this vicinity.   Bottom forms on the shelf suggest the

TABLE I.

Showing the analyses of specimens in and around the submarine canyons. It will be observed that the $CaCO_3$ content is about the same in the base of the canyons as on the outer part of the adjacent continental shelf. Also it is noteworthy that there is more coarse material in the deepest sample within the canyons than in the others. The specimen from the side of a canyon was a consolidated sediment. Analyses were made by George V. Cohee.

	Lat. Long.	Depth (fathoms)	CaCO content	Gravel 2 mm. plus	Sand .25 mm.-2.0	Sand .074-.25 mm.	Clay-silt under .074 mm.	Heavy Minerals
Base of submarine canyon	40 16' 68 06'	725	7.55%	.67%	7.04%	27.75%	56.97%	.5%
Base of submarine canyon	40 19' 68 08'	520	1.25%		.49%	1.61%	89.40%	.02%
Base of submarine canyon	40 19.3' 68 07'	411	9.0%		10.3%	12.2%	68.0%	5.46%
Base of submarine canyon	40 18.4' 68 09.3'	256	2.4%		1.7%	4.4%	91.35%	.02%
Side of submarine canyon	40 23' 67 55'	150	40.0%	8.7%	48.7%	.6%		.01%
Shelf inside of canyons	40 25' 67 55'	77	7.91%		57.3%	33.6%	3.73%	7.0%
Shelf inside of canyons	40 33' 68 23'	50	6.02%	.3%	14.68%	70.01%	4.02%	1.8%

same trend.[2]   Furthermore, quiet conditions must exist on the bottoms of the canyons to allow the accumulation of the fine sediments which are found there.   These sediments consist predominantly of clay and silt (see Table I), which is in contrast to the coarse sediment characterizing the continental shelf inside.

Submerged River Valleys.—It seems to be a reasonable assumption that canyons like those described above were cut by rivers, but this explanation is fraught with many difficulties. If the continental shelf was much higher recently there should be evidence of this uplift on the lands.   Also the charts show that the continental shelf in this region has marginal depths which average about 70 fathoms.   These depths are characteristic of shelf margins over much of the world and suggest that the shelves possessing them developed under stable conditions.   If these canyons have been submerged for thousands of feet how can the shelf be in equilibrium?

Pre-shelf Submergence of the Canyons.—If the canyons were submerged in the remote past and the continental shelf was cut afterwards, the difficulties suggested above would cease to exist.   On the other hand a new difficulty would arise. As soon as the canyons had become submerged they would have been subject to filling by sediment from the surrounding lands.   During the cutting of the continental shelf the wave erosion and subaerial denudation would have provided vast quantities of sediment for the filling of the submerged canyons. Still later, when glaciers stood on the north side of Georges Bank,[3] there must have been an abundance of outwash carried into the submerged canyons by streams crossing the exposed continental shelves of glacial times.

Canyons Reopened by Landslides.—Filling of the canyons by sediment is not necessarily fatal to the idea that the canyons were cut prior to the shelf origin.   The valley fills might have been removed subsequently by landslides of a mud flow type. Conditions are particularly favorable for such action at the margin of the continental slope.   These slopes have here, as elsewhere, much the appearance of fault scarps and are subject to earthquakes.   The fine sediments which accumulated in the

[2] On the northern portion of the bank the currents trend northwest southeast producing elongate shoals in that direction, but this changes further south to northeast southwest.

[3] Proof that the glaciers extended onto Georges Bank was obtained from bottom samples of till obtained by the writer on trips with Coast Survey vessels.

valleys would slide easily under conditions of saturation. The gradients in the valley floors, averaging 300 feet per mile, are sufficiently steep to allow the sliding of muddy material.

Evidence which I have presented elsewhere[4] suggests that slides may have modified submarine valleys in recent years.

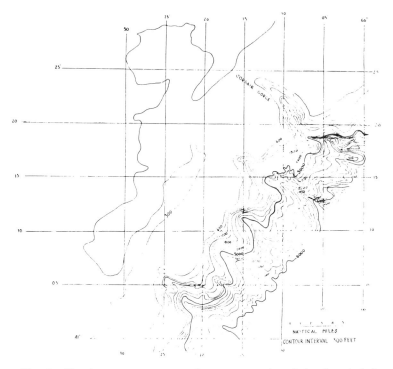

Fig. 4. Showing some canyons on the eastern portion of the slope including Corsair Gorge which is thought to have been opened by a landslide in 1929 at the time of the Grand Banks earthquake.

There is reason to believe that Corsair Gorge (Fig. 4) was opened at the time of the Grand Banks earthquake in 1929.

If large quantities of sediment have slid out of the valleys, there must be accumulations outside on the deep ocean floor and perhaps to some extent in the outer portions of the valleys. Where these accumulations had piled up into high mounds like those of landslides on land they might be found by closely

[4] Landslide Modifications of Submarine Valleys. Trans. Amer. Geoph. Union, pp. 226-230, 1932.

spaced soundings. Unfortunately, data are not available beyond the canyons to check this point, but there are some suggestions of hills in the outer canyons which may have originated in this way.

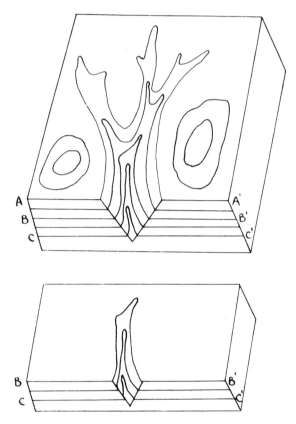

Fig. 5. Illustrating how the branching head of a canyon might be changed by levelling of the upper portions after submergence to level B-B'.

Mud flows moving into the outer canyons may have filled in the V-shaped bases and have left fairly smooth surfaces like that of the Slumgullion flow in Colorado. Some of the places. where there is evidence of a broadening of the base of the canyons may be explicable in this way.

Landslides as the Entire Cause.—I have been asked whether the submarine valleys could have been formed entirely by landslides, obviating the difficulty of the river cutting part of the hypothesis. I feel that there are various reasons for believing that this could not have been the case particularly in regard to the canyons under discussion. Their shapes are entirely different from the lunate scars which are left after the breaking away of a mass of rock or sediment from an escarpment. Since it appears highly probable that the canyon walls are made of rock, it becomes inconceivable that masses of solid rock

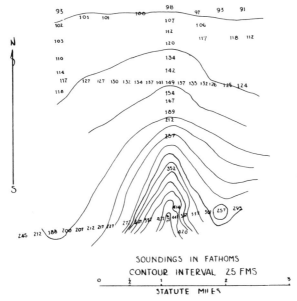

SOUNDINGS IN FATHOMS

CONTOUR INTERVAL 25 FMS

STATUTE MILES

Fig. 6. Showing the broadened and flattened head of a canyon which terminates below the outer edge of the continental shelf.

could have slid from the shelf and slope leaving these elongate, sinuous, steep-walled canyons. On the other hand it is perfectly reasonable to suppose that slides could carry away the loose sediment filling a rocky gorge.

Valley Heads.—The headward terminations of the submarine canyons differ from those of land canyons in that the former have fewer tributaries and more abrupt ends. This difference may be due to the erosion which is thought to have followed the canyon submergence and which would have cut away the valley heads and have left only their trunk portions (Fig. 5).

Another less usual type of canyon head is shown in Fig. 6. Here a V-shaped valley broadens out at its head into a gentle amphitheatre which does not truncate the continental shelf at all. A possible explanation of this case is that the inner portion of the valley is not yet cleared of sediment or that in the interval since a landslide evacuated this valley sediment has started to fill in its head.

Land Movements Preceding Landslides.—Under the circumstances it seems probable that the canyons were first cut by rivers, then filled with sediment, and finally reopened by landslides. It is not reasonable to suppose that the sea level could have been lowered 8000 feet or more allowing rivers to flow down these slopes so we must conclude that the lands were greatly elevated at some time during the past. At this time New England, or at least the continental shelf east of New England, must have been either a high plateau or a mountain range. It would appear from the steepness of the valley walls and from the high gradients that this land was not reduced to a low relief before submergence. The uplift must have been of relatively short duration.

Time of Canyon Cutting.

Possible Association with Appalachia.—The time when the canyons were excavated is not easy to determine. The area was probably greatly elevated during the late Paleozoic when the conglomerates of eastern Massachusetts and Rhode Island were being derived from a source to the east.[5] In later times there does not appear to be any indication that the area under discussion was greatly elevated. Accordingly, we are confronted with the possibility that these canyons were cut into the seaward flank of Appalachia. If such were the case the land mass must have been down-faulted before erosion could have greatly modified the canyons. They may have been maintained since then by repeated landslides which prevented the consolidation of the sediments accumulating in the valleys.

Old Continental Outlines.—If the above suggestion is correct, it would appear that at least in this area the old east slope of Appalachia is now represented by the continental slope. Here we have a suggestion that the old continental outline was not far different from that of the present provided we consider the true continental border which is the margin of the continental shelf. A narrower Appalachia would do away with the

[5] J. B. Woodworth. The Appalachians in Southern New England. Bull. Geol. Soc. of Amer., **34.** pp. 257, 1923.

difficulty of explaining the engulfment of a continental borderland into the oceanic abyss. The narrower Appalachia might
have provided the required amount of sediment for the
Appalachian geosyncline provided it were uplifted a sufficient
number of times. However, before carrying this speculation
too far we need more evidence that the canyons were formed
in these remote times. It seems not inconceivable that such
evidence might be found by oceanographic investigations.

Possible Tertiary Age.—It is also quite possible that the submarine canyons are much younger, perhaps of Tertiary age.
It has been suggested by Joly[6] that the continents are periodically rejuvenated. If whole continents are uplifted at certain times, valleys may be cut in their margins and later submerged. This wholesale uplift, unlike the elevation of marginal tracts, would not cause sediment to be washed inland.
Also, this explanation has some appeal because deep submarine valleys are found off all of the continents of the world.
Yet it is hard to believe that the continents could have been
uplifted en masse to elevations of one or even two miles above
the present levels.

Fossil Evidence.—Future hope for determination of the age
of submarine valleys may lie in the collection of bottom samples
particularly from the steep walls of canyons. Support of this
contention was found during the surveying of the canyons off
New England. Only a few samples were procured, but of
these one contained a poorly preserved fossil. Near the head
of one canyon four bottom samples were collected from the
same wall at depths ranging from 195 to 105 fathoms. In
each case fragments of the same rock formation were collected. These consisted of well rounded and sorted quartz
sand and small quartz pebbles cemented by calcium carbonate.
The latter made up about 50% of the formation. The fossil
collected at 195 fathoms was diagnosed by Dr. R. B. Stewart
as probably Tertiary in age. The formation is strikingly different from the aequeo-glacial sands and gravel which cover
the shelf adjacent to the canyon heads. The pureness of the
quartz and its well rounded character, together with the insignificant amounts of heavy minerals (Table I) are all proof
that the formation was not formed from material which
slumped down from the shelf above. It seems possible that
the formation was derived from the weathering of quartz-

[6] John Joly, Surface History of the Earth, p. 56, 1925.

bearing rocks during the submergence of the canyons. The deposition of calcium carbonate at the same time under conditions different from those now existing could have cemented this material together.

Conclusions.

We may conclude that the submarine canyons off New England are quite certainly of fluvial origin. The uplift which allowed the canyon cutting was clearly prior to the formation of the continental shelf in this vicinity and may have been as remote as the late Paleozoic. The canyon walls are almost certainly solid rock, indicating that the continental shelf here has been the product of erosion rather than of building up by sediments from the lands. The canyons are thought to have been filled with sediment in whole or in part, but are preserved in outline, by the evacuation of the sediment through the agency of landslides.

UNIVERSITY OF ILLINOIS
URBANA, ILL.

3

Copyright © 1936 by the American Journal of Science (Yale University)

Reprinted from *Amer. Jour. Sci.* Ser. 5, **31**(186), 401, 406–407, 416–419 (1936)

ORIGIN OF SUBMARINE "CANYONS."

REGINALD A. DALY.

ABSTRACT

Echo-sounding has increased to nearly one hundred the number of great valley-like trenches now known to fret the edges of the submerged continental slopes. No published explanation of these "canyons" by subaerial river-cutting seems adequate or permissible. A new working hypothesis is offered for discussion. During the Glacial stages of the Pleistocene Period the sealevel was lowered all over the world, in maximum from 200 to 300 feet. Then wind waves and tidal waves were everywhere breaking on the loose muds and sands of the continental shelves, and for long intervals of time not far from the upper limits of the continental slopes. The water so agitated was specially loaded with suspended sediment and therefore had effective density exceeding that of cleaner sea water elsewhere. The loaded water tended to slide down the continental slopes, along the bottom. The question arises as to whether the velocities of the density currents sufficed to have eroded the actual trenches. A definite answer cannot now be given, but the general hypothesis should be tested by appropriate observations. The possibility that in pre-Glacial time some "canyons" were cut by bottom currents is also considered.

[*Editor's Note:* In his opening pages, Daly states the problem of origin of submarine canyons and discusses the explanations offered. His own theory follows.]

A New Conception of Origin.

However, the more moderate eustatic oscillations of sea-level due to the fourfold Pleistocene glaciation and deglaciation of the lands represent a set of facts on which may be based a rational hypothesis concerning the development of the submarine trenches. Each of the four sets of ice-caps grew slowly and melted slowly, each process taking at least 25,000 years. Hence for more than 200,000 years, out of the million years or so that have elapsed since the Ice Age dawned, wind waves and tidal waves were beating on the mud and sand of the continental shelves—a condition utterly unlike that now ruling. These more or less mobile sediments had been built into embankments with widths measuring scores of kilometers and with depths averaging at least tens of meters. The volume of fine sediments was therefore enormous and sufficient to keep the tidal currents and storm waves of the lowered ocean well charged with solid particles for a large fraction of the 200,000 years. The waves were especially muddy because the depth of water on the outer, still submerged parts of the shelves was small. Then, too, the average storminess of the world was doubtless more pronounced during the Glacial epochs than at present.[5] Storms no more intense than those now affecting the shelves must have made the water overlying the continental slope (the fall-off of each shelf) much richer in suspended sediment than the water of similar location in pre-Glacial, Interglacial, or post-Glacial times. The tidal currents and gales of the twentieth century disturb the bottom of the North Sea so powerfully that sand is thrown up from depths of 40 to 50 meters to the decks of laboring ships. So long as sediment was "suspended" in the water on the Pleistocene shelves, that water was effectively denser than the clean water farther out to sea or the water below the zone of rapid stirring. There must have been a tendency for the weighted water to dive under the cleaner water, to slide along the gently inclined bottom of the shelf, and to flow still faster down the steeper continental slope. Since the solid particles kept settling out, the horizontal distance through which any such density current operated was limited. It is therefore important to remember throughout the discussion of the general hypothesis, that the belt of strong

[5] cf. Bryan, K.: This Journal, vol. 27, p. 251, 1934.

agitation by waves was, at the times of lowered sealevel, much nearer to the continental slope than now. In principle the imagined bottom current would be similar to the flow of ink or muddy water placed at the appropriate point in a tilted, partly filled glass of clear water. Each of those denser fluids slides down along the inclined "floor" of containing glass.

We thus picture a special kind of density current or convectional movement. Were these bottom currents strong enough to have excavated the submarine trenches now under discussion? The question is much too hard to answer off-hand, but some facts suggest that an affirmative answer may not prove to be entirely wild.

[*Editor's Note:* Daly goes on to discuss how much sediment can become suspended in strongly agitated water adjacent to the continental slope and the rate of flow and erosive power of a dense, muddy bottom layer. He then comments on the work of Forel on the deeply trenched sublacustrine part of the Rhône delta in Lake Geneva and the Rhine delta in Lake Constance.]

Conclusion.

If the revised conclusion of Forel is correct, deferred or differential deposition of silt in the growing deltas of the Rhone and Rhine rivers accounts for the strong radial furrowing of each delta to the depth of 200 meters below the lake surface. Because this mechanism demands relatively low velocities for the more localized bottom currents, it might appear easier to explain the submarine trenches in the same way. Yet there are serious objections. The Indus, Congo, and other trenches, including some along Georges Bank, are much too long to be credited to differential prograding of the corresponding banks during Pleistocene time.[20] Another difficulty is the steepness of the trench walls, too great to represent angles of rest for mud in actual deposition. Moreover, the process described does not well account for the pronounced

[20] The great Congo trough has special features which are not easily understood on any theory of origin for trenches in general. In this particular case the trench, with floor considerably more than 100 meters deep, extends well into an estuary, flanked with dry land. If the giant furrow was excavated by down-sliding bottom currents, it would seem necessary to assume extraordinary headward growth; yet there is no obvious reason why the Congo trench should have been so remarkably lengthened by recession of its upper limit. Again, the chart shows that the longitudinal profile is not continuously oceanward, the slope being reversed in at least one locus.

flaring at the mouths of some trenches. This feature, like the steepness of the walls, is explicable if the imagined bottom current erodes its bed, and in the very act takes up additional sediment and thus attains higher density and additional power to erode and transport.

An origin by erosion under the peculiar conditions of the Glacial epochs has been pictured in the light of certain facts and their logical consequences. Some of these need not be restated. Others, perhaps more liable to be lost to sight in an attempt to value the hypothesis, are: (1) The relatively small depth of water on the continental shelves during the four Glacial epochs;[21] (2) The absolute slowness with which muds sink in sea water; (3) The close proximity of the belt of shore breakers to each continental slope when the ice-caps were voluminous;[22] (4) The paroxysmal effects of major storms and of spring tides at the time, including extraordinary increase of suspended sediment and corresponding increase of potential energy in the mixture of water and sediment; (5) The probability that the "fore-set" sediments under the continental slope were close to the angle of rest, and hence in danger of sliding where dragged by a localized bottom current; (6) The increase of density of sea water by receipt of the silt that settles out of overlying river water as this, by its own momentum,

[21] Although the conditions of the Glacial epochs are emphasized in this paper, the possibility of some trenching of the continental slopes in pre-Glacial time is not thereby excluded from a full discussion of the general problem. According to both theory and observation, the continental shelves were truncated by wave erosion during each Glacial epoch. Hence, just before the first major glaciation began, the average depth of water on stable shelves may have been something like 25 meters smaller than at present. In other words, at that earlier time the shelves may have been more closely adjusted to wave-base, as defined for ordinary storms, than the shelves are now. If so, the sediments of the shelves may have been stirred up by exceptional, quasi-hurricane storms operating in pre-Glacial time. The question arises as to whether density (muddy-water) currents of prolonged, pre-Glacial time actually began the development of the trenches. The answer must be difficult and not to be satisfactorily made until the depths of wave-base for both ordinary and extraordinary storms and the relation of those depths to shelf profiles have been more accurately defined than has yet been done. The narrower the detrital shelf of pre-Glacial time, the greater would have been the likelihood of such earlier trenching.

[22] Shepard, F. P., explains the absence of trenches along the northern edge of Georges Bank (see Figure 1 of his paper, Bull. Geol. Soc. America. vol. 45, p. 281, 1934) by assuming glacial erosion there, sufficiently intense to have removed "all signs of stream erosion." However, is it not possible that the thick ice, responsible for the morainal material found on the northern part of the bank, prevented the trenching by bottom currents while the trenches along the ice-free, southern edge of the bank were developed?

forces its way out over the open ocean; (7) The self-accelera-
tion of the density currents by lateral addition of loaded
water on each shelf and within each deepening trench, and by
increase of the density where turbulent erosion of the bottom
added to the sediment in suspension; (8) The smallness of the
excess of density (probably no more than 0.004) required in a
bottom layer of water 60 to 100 meters thick, in order to cause
that layer to flow down a continental slope at the rate of two
to three kilometers per hour—an eroding rate; (9) The ex-
pectation that the erosional effects of the assumed bottom currents would rather closely resemble those due to rivers on the
land, even to the point of making systems of branching
trenches that recall the dentritic ground-plans of visible rivers.

On the other hand, the offered explanation of the trenches
is based on assumptions which are far from being certainties.
Perhaps the most vital of all is that the materials underlying
the continental slopes are weak enough to have permitted the
cutting of the trenches within a time interval no greater than
about 200,000 years. If Shepard's evidence that slumping
of, "landsliding" from, the trench walls is substantiated by
future charting, we should have direct proof of weakness.[23]
It is to be hoped that with ship and appropriate apparatus
several questions can be answered. What proportion of
unconsolidated sediment, like Stetson's late-Tertiary clay, is
characteristic of the trench walls? When was the lithification
at Georges Bank accomplished? Is the lithified material con-
cretionary and local, or does it constitute extensive layers
parallel to the bedding of the shelves? However, while awaiting the results of future investigations, geologists should value
the experience of Stetson, the pioneer among those who have
actually dredged samples from the trench walls. According
to his opinion, expressed verbally, much less than half of the
wall of any Georges Bank trench is endowed with the strength
of hard rock, the lithified material occurring in individual
layers separated by soft layers.

A second assumption calling for test by experiment or otherwise is that the imagined bottom currents retained sufficient
velocity in spite of the resistance offered by the overlying water
as well as by the sedimentary bottom. That the former kind
of resistance should not be over-emphasized in the problem
is suggested by the analogy at the Strait of Gibraltar.

[23] Shepard, F. P.: Geog. Review, vol. 23, p. 86, 1933.

A third principal question calling for future investigation relates to the apparent lack of submarine trenches along several long stretches of the continental slopes. According to Shepard this is the case off the Pacific coast of North America from Cape Mendocino almost to the mouth of the Columbia River, a distance of 550 kilometers. So far no trench has been reported off the Atlantic coast between Cape Hatteras and the Straits of Florida.[24] For the one instance it is natural to suspect the influence of the Gulf Stream. If in Glacial times this mighty current hugged the continental slope, it would have interfered with any transverse density currents that might otherwise have developed trenches. Or, if trenches were there actually excavated, the Gulf Stream of late-Glacial and post-Glacial time with its abnormal erosive power may have rubbed them out of the submarine topography. The absence of trenches off Washington State can hardly be attributed to a current sweeping the coast at the present time, and as yet there is no evidence of such a current during the Glacial epochs. In general, too, we need to know more about the conditions under which lithification of shelf sediments takes place. If local stretches of the shelves had already been lithified, the density currents of the Pleistocene could not have cut deep trenches there.

In view of the many uncertainties the idea of mud-control must now be rated as merely a working hypothesis. Yet its troubles seem incomparably less serious than those of the older explanations of submarine "canyons." On the other hand, it would be manifestly wrong to suppose all furrows crossing the continental shelves to have been excavated by marine density-currents. Here and there canyons and other types of valleys, cut by ordinary rivers in pre-Glacial time, have been drowned by strong, local subsidence of continental borders. Such old, rock-hewn valleys, if they had not been quite filled with fine-grained detritus, would naturally have attracted the water loaded with sediment during the Glacial epochs. The resulting, localized currents would have been likely to remove some of the filling of each of the old trenches, and thus have revived the open-valley form more or less completely. Illustrations of this speculative process might be looked for particularly along the coasts of California, Japan, and other uneasy parts of the continental borders.

[24] Shepard, F. P.: Zeit. für Geomorphologie, vol. 9, p. 99, 1935.

Editor's Comments
on Papers 4 and 5

4   HEEZEN, MENZIES, SCHNEIDER, EWING, and GRANELLI
    Excerpts from *Congo Submarine Canyon*

5   SHEPARD and EMERY
    Congo Submarine Canyon and Fan Valley

After three papers of historical interest, it seemed time to introduce a full-scale investigation of a major canyon using modern methods. For this we have chosen the Congo Canyon, already mentioned in Paper 1 as a place of repeated cable breaks. From 1887 to 1951, various survey ships made soundings in and around the canyon. A brief reconnaissance survey by Menzies in 1957 using precision depth recording, piston coring, and biological trawling was followed by Ewing's 1963 cruise in which additional surveys were made by magnetometer, seismic reflection profiler, and sea gravimeter, and photographs of the canyon floor were taken. The results of these two cruises are presented here and give a clear picture of the morphology of this great canyon (one of the world's largest), its sediments (including well-developed levees in its lower parts, where, as Shepard and Emery point out in the succeeding article, it is strictly a fan valley on the Congo Fan), and the effect of turbidity currents in eroding the canyon and breaking cables up to 120 miles (195 km) seaward of the river mouth. Breaks are related to times of greatest bed-load discharging from the Congo (or Zaire) River. The Congo Canyon is unique in that it heads well within the estuary of the river. [The physiographic diagram of the south Atlantic on pp. 1128–1129, the section on the sediment discharge of the Congo River and its relationship with cable breaks, and the tables detailing cable failures (pp. 1143–1148), are omitted.]

After this paper had been selected for reprinting, new work on the Congo Canyon was published by Shepard and Emery, usefully updating the studies on this canyon to the present. At the risk of being accused of favoritism toward the Congo, we include their paper here. In it, the authors differentiate between the Congo Canyon and leveed Fan Valley with distributaries, confirm the concept of origin largely through erosion and deposition by turbidity currents, and show, by seismic profiles, that the canyon has been cut through a belt of evaporite diapirs that are still actively rising (cf. De Soto Canyon, Harbison, 1968, and Alaminos Canyon, Bouma and others, 1968). A Savonius current meter was used to obtain a four-day record in the canyon head, which seems to have had an extensive recent fill.

Biographical Notes

Bruce C. Heezen was born in Vinton, Iowa, in 1924. He received the B.A. in 1948 at the State University of Iowa, the M.A. in 1952, and the Ph.D. in 1957 at Columbia University. His research and teaching has been centered on the Department of Geology and Lamont-Doherty Geological Observatory, Columbia University, where at present he is Associate Professor. He is an active member of many committees concerned with marine geology and geophysics, bathymetry of the oceans, the International Tectonic Map of the World, and the Law of the Sea. He is consultant to the U.S. Naval Oceanographic Office and the Naval Research Laboratory.

In 1964, Heezen was awarded the Henry Bryant Bigelow Gold Medal and Prize, and in 1973, the Cullum Geographical Medal from the American Geographical Society in recognition of his distinguished contribution to the knowledge of the earth beneath the oceans. He has published about 160 scientific papers, mainly on the world's oceans, and, with C. D. Hollister, a well-illustrated book entitled *The Face of the Deep* (1971), which was a nominee for the National Book Award in 1972.

Robert J. Menzies was born in Denver, Colorado, in 1923. He gained the B.A. in 1945 and M.A. in 1949 at the College of the Pacific and the Ph.D. at the University of Southern California in 1951. During a varied career he has worked as zoologist, biologist, and oceanographer at Scripps, Lamont, University of Southern California, Duke, and Florida State University, where he is now Professor of Oceanography: he is also Chief Scientist with the Offshore Ecology Program, Gulf Universities Research Consortium, and Chairman of its Biology Panel. A recipient of many awards, Menzies has published numerous papers on marine fauna (especially isopods, wood borers, and deep-sea *Neopilina*), on deep-sea biology, ecology and technology, hydrostatic pressure, and pollution. He is a joint author of a recent book on *Abyssal Environment and Ecology of the World Oceans* and editor of a volume on the *Effects of Hydrostatic Pressure on Living Aquatic Organisms.*

Eric D. Schneider was born in Wilmington, Delaware, in 1940. He was awarded the B.A. degree in 1962 at the University of Delaware and the Master's degree (1965) and Ph.D. (1969) at Columbia University, New York. He is Adjunct Professor, Graduate School of Oceanography, University of Rhode Island. From 1968 to 1971, Schneider was Director, Global Ocean Floor Analysis and Research Center, U.S. Naval Oceanographic Office, then Director, Office of Special Projects, Office of Research and Monitoring, Environmental Protection Agency, Washington, D.C., from 1971 to 1972. Currently he is Director, National Marine Water Quality Laboratory, Environmental Protection Agency, Narragansett, R.I. He has over thirty publications and books in the fields of marine geology, marine geophysics, and marine sedimentation. Specific topics of research involve the role of deep ocean currents in shaping the physiography of the deep sea floor and the evolution and early history of ocean basins. An active environmentalist, Schneider was formerly Director of the National Eutrophication Survey, a survey to determine the water quality of the United States' lakes and the impact of phosphorus on their quality.

William Maurice Ewing was born in Lockney, Texas, in 1906 and died in 1974. He studied at Rice Institute, Houston, Texas, and was awarded the B.A. degree in 1926, the M.A. in Physics in 1927, and the Ph.D. in Physics in 1931. He held positions at Woods Hole Oceanographic Institution and a number of universities, including Pittsburgh, Lehigh, Columbia (where for 13 years he was Higgins Professor of Geology and for 23 years, Director of the Lamont-Doherty Geological Observatory), University of Texas at Austin (where he was Professor of Geoscience since 1972), University of Texas at Galveston (where he was Chief, Earth and Planetary Sciences Division, The Marine Biomedical Institute, and Cecil H. and Ida Green Professor of Marine Sciences since 1972) and Rice University (Adjunct Professor of Geology since 1973).

Maurice Ewing was Vice President, then President, of both the American Geophysical Union (1953–1956; 1956–1959) and the Seismological Society of America (1952–1955; 1955–1957). He served on the Rice University Board of Governors and the U.S. Navy Oceanographic Advisory Committee. His high reputation brought him 10 honorary Sc.D. degrees and one LL.D., and from a list of at least 34 other honors we may note the world-wide esteem with which he was held by citing the following: Arthur L. Day Medalist, Geological Society of America, 1949; Agassiz Medal, National Academy of Sciences, 1955; William Bowie Medal, American Geophysical Union, 1957; Order of Naval Merit, Rank of Commander, Argentine Republic, 1957; Vetlesen Prize, Columbia University, 1960; Cullum Geographical Medal, American Geographical Society, 1961; Medal of Honor, Rice University, 1962; John J. Carty Medal, National Academy of Sciences, 1963; Gold Medal, Royal Astronomical Society, London, 1964; Vega Medal, Swedish Society for Anthropology and Geography, 1965; Sidney Powers Memorial Medal, American Association of Petroleum Geologists, 1968; Wollaston Medal, Geological Society of London, 1969; Alumni Gold Medal, Rice University, 1972; National Medal of Science, 1973; and the Walter H. Bucher Medal, American Geophysical Union, 1974. He was an honorary member of many scientific societies.

N. C. L. Granelli is with the Royal Argentine Navy.

Kenneth O. Emery was born in 1914. He received the B.S. degree in 1935 and the Ph.D. degree in 1941 at the University of Illinois. After working for the Illinois State Geological Survey and the University of California Division of War Research at San Diego, Emery became Assistant Professor, then Professor, at the University of Southern California (1945–1962). During this time, he also worked as geologist for the U.S. Geological Survey and as oceanographer with the Navy Ordnance Test Center at Pasadena. In 1962 he moved to Woods Hole Oceanographic Institution, where he is currently Senior Scientist. He is a member of, and has held office in, a number of scientific societies and has served on the Navy Research and Development Board and as delegate of the State Department to the Technical Advisory Group of the U. N. Economic Commission for Asia and the Far East. In 1969 Emery was awarded the Shepard Prize for Excellence in Marine Geology and two years later was elected to membership in the National Academy of Sciences and the American Academy of Arts and Sciences and became the recipient of the Prince Albert I[er] de Monaco Medal, France.

Kenneth Emery has researched in general marine geology (especially in physiography, sedimentation, and geophysics) in many parts of the world, and has authored or coauthored about 230 scientific publications, including five books, in marine geology and related fields.

4

Copyright © 1964 by the American Association of Petroleum Geologists

Reprinted from *Amer. Assoc. Petrol. Geol. Bull.*, **48**(7), 1126–1127, 1130–1143, 1148–1149 (1964)

CONGO SUBMARINE CANYON[1]

B. C. HEEZEN,[2] R. J. MENZIES,[3] E. D. SCHNEIDER,[2] W. M. EWING,[2] AND N. C. L. GRANELLI[4]

ABSTRACT

In May, 1957, a brief survey of the Congo Submarine Canyon was conducted from the Research Vessel VEMA. The canyon was found on the continental slope near the seaward limit of earlier surveys and traced to the west for 150 miles. Where the survey was discontinued, in depths of 2,200 fathoms, the leveed canyon was still a prominent feature. At the 550-fathom contour the canyon is about 5 miles wide at its rim and 500 fathoms deep. The canyon is V-shaped and echoes from the thalweg are usually recorded after the echoes from the steep walls. The outer parts of the canyon, in depths exceeding about 1,800 fathoms, are bounded by huge natural levees. In 1963 a 30-mile-long section of the canyon was surveyed 150 miles west of the 1957 study.

Four biological trawls and ten sediment cores were collected from the canyon region. One trawl in 2,140 fathoms contained abundant tree leaves and a rich fauna. The canyon cores contain silt, sand, and organic debris.

The canyon was discovered in 1886 by a cable route survey. Between 1887 and 1937 the Luanda-São Thomé cable broke 30 times in the canyon. Breaks occurred most frequently during months of maximum river discharge and consequently at the times of greatest bed-load discharge. Cable breaks were limited to periods of years when the river channel was undergoing major changes in position and depth. The cable breaks which occurred up to 120 miles seaward of the river mouth are attributed to turbidity currents generated at the river mouth at times of maximum bed-load transport. These turbidity currents flowed down the continental slope, eroding the deep slope canyon and building the natural levees of the continental rise, and eventually spread out on the Angola Abyssal Plain. In contrast to other great rivers, the Congo is not building a subaerial delta; virtually its entire bed load is being carried by turbidity currents via the Congo Submarine Canyon to the great Congo Cone on the floor of the Angola Basin.

INTRODUCTION

The Congo Submarine Canyon is situated at the mouth of the Congo River of the west coast of Africa (Fig. 1). It is one of the largest submarine canyons in the world. The maximum width is 5 miles and the maximum depth below the rim is 600 fm (Fig. 2b). The Congo Submarine Canyon is now known to be at least 500 miles long (exclusive of meanders). The heads of many submarine canyons are filled with sediment (Fisk and McFarlan, 1955). In contrast, the Congo Canyon

[1] Lamont Geological Observatory Contribution No. 717. Manuscript received, November 4, 1963.

[2] Department of Geology and Lamont Geological Observatory, Columbia University, Palisades, New York.

[3] Department of Zoology and Duke Marine Laboratory, Duke University, Beaufort, North Carolina.

[4] Servicio de Hidrografía Naval, Buenos Aires, Argentina.

This study was supported by the United States Navy, Office of Naval Research, the National Science Foundation, the Bell Telephone Laboratories, and the Argentine Hydrographic Office. Reproduction in whole or in part permitted for any purposes of the United States Government. The assistance of Marion Jacobs, G. L. Johnson, Marie Tharp and Captain V. R. Sinclair, USN (Ret.), is gratefully acknowledged. Discussions with Walter Bucher and Charles Drake were helpful. E. J. Devroey, J. Cl. De Bremaecker, and V. Van Straelen provided valuable information on the Congo River. Engineer-in-Chief of Cable & Wireless, Limited, C. J. V. Lawson, generously made available original cable repair reports.

extends 20 miles into the mouth of the Congo River with steep sides and a V-shaped profile. A delta is not being formed at the river mouth; instead the bed load of the river is ultimately transported through the canyon and deposited on the continental rise and the Angola Abyssal Plain on the west. Here an immense submarine distributary system was discovered. The transport of great quantities of silt and sand through the canyon at intervals must have eroded the deep canyon and formed an abyssal delta (Congo Cone).

PREVIOUS INVESTIGATIONS OF CONGO SUBMARINE CANYON

The canyon was first discovered in 1886 by J. Y. Buchanan (1888) aboard the British cable-route-exploring ship BUCCANEER (Fig. 2b). The objective of that survey was to select a route for the São Thomé-Luanda (Angola) submarine cable. Buchanan (1887, 1888) discovered three facts about the canyon: (a) its penetration into the Congo River mouth; (b) its great width and depth; (c) the fact that it extended at least to the 1,000-fathom contour. Buchanan correctly presumed that the canyon extends at least to the 2,000-fathom contour. Near shore the canyon has been extensively surveyed by the Congo and Angola governments principally because the axis of the canyon and river forms the international boundary, and changes in the position of the canyon axis have political significance (Devroey, 1946). Unfortunately, these surveys were never extended more than a few miles from the coast. Hull (1900, 1912, in Spencer, 1903) contoured the soundings available at that time. Wire soundings were made in the Congo Canyon by various vessels, including S.M.S. SPERBER [1910], S.M.S. PANTHER [1910], C.S. DUPLEX, C.S. PENDER, and other cable ships. In 1911 S.M.S. MÖWE made 12 sounding lines across the canyon between the river mouth and the shelf edge. Schott and Schulz (1914) present an excellent contour chart of the canyon based on the MÖWE soundings. Both Veatch and Smith (1939) and Shepard (1948) contoured the soundings available before World War II. In 1951 GALATHEA trawled and dredged in the canyon (Bruun, 1957) but the accompanying soundings have not been published.

1957 RECONNAISSANCE SURVEY

In May, 1957, Menzies, aboard the Research Vessel VEMA, had the opportunity to study the Congo Submarine Canyon, employing three techniques of exploration: (1) precision depth recording; (2) piston coring; and (3) biological trawling (Fig. 2). Additionally standard serial hydrographic work was done.

At the start of this exploration it was expected (1) that the canyon extends well past the 2,000-fathom contour; (2) that graded turbidite sands would be encountered in the floor and at the mouth of the canyon, evidencing submarine turbidity-current activity; (3) and that a relationship between turbidity-current activity and the standing crop of abyssal organisms would be seen. These expectations were all met in the course of the exploration. In addition, a study of the breakage of a submarine cable crossing the submarine canyon provides evidence of contemporary turbidity currents and thus a major clue to the origin of submarine canyons. In 1963, Ewing (VEMA Cruise 19) made a traverse across the western margin of the Angola Abyssal Plain (Figs. 2a, 7), employing the Precision Depth Recorder, magnetometer, seismic reflection profiler, sea-gravimeter, as well as obtaining photographs from the canyon floor, and several cores from the region.

TOPOGRAPHY

On VEMA Cruise 12 the Congo region was entered from the south (Fig. 2b) close to the 500-fathom contour. By chance the first crossing was approximately along one of the lines surveyed in 1886 by Buchanan. Once the canyon was found, attempts were made to keep as close to its axis as possible, following it always westward. The axis of the Congo Canyon trends 275° from the river mouth to 11° E; then gradually the canyon arches to the north assuming a trend of 295° T. At about $10\frac{1}{3}$° E the canyon abruptly bends toward the south, adopting a trend of 225° T. Near the 800-fathom contour, the deep V-shaped profile (Fig. 3) extends 600 fathoms below the adjacent sea floor. The bottom is very narrow, as indicated by the crossing of wall echoes above the floor echo; here the canyon is 5 miles in width. The surface water at this point was discolored, indicating the influence of the Congo River waters 90 miles distant from the mouth of the river. Cable ships have reported river water 120 miles seaward of the river mouth in the vicinity of the canyon. The surface salinity was recorded to be much less than that of the equatorial counter-current water. Here freshwater organisms from the Congo River became mixed with the marine organisms of the northward-flowing Bengela current. It is probable that at least part of the fresh-water diatoms recorded from the deposits on the Mid-Atlantic Ridge have their origin from this source (Kolbe, 1957).

In the vicinity of the 2,000-fathom contour the floor of the canyon lies near the general level of the adjacent continental rise. Extending outward 5 to 20 miles on either side of the various distributaries are well developed natural levees (Figs. 2, 3). The main distributary channel of the canyon shows a characteristic sharp V-shape; other channels are rounded and less distinct. In this region depositional features appear to exceed the erosional characteristics. The canyon was traced to 2,200 fathoms and was still quite evident; dropping 70 fathoms below its natural levees.

In 1963, Ewing (VEMA Cruise 19) made seven traverses of a flat-floored, 3-mile-wide, natural-leveed canyon between 05°34′E and 05°06′E, a distance of about 30 miles (Fig. 2a). The canyon floor dropped continuously westward at about 1:600 from 2,643 fathoms to 2,692 fathoms. The canyon walls are steep and are 10–20 fathoms high. This canyon appears to be a continuation of the southern or main branch of the Congo Canyon System. The survey thus extends the length of the

(*Text continued on page 1132*)

[*Editor's Note:* Figure 1 has been omitted.]

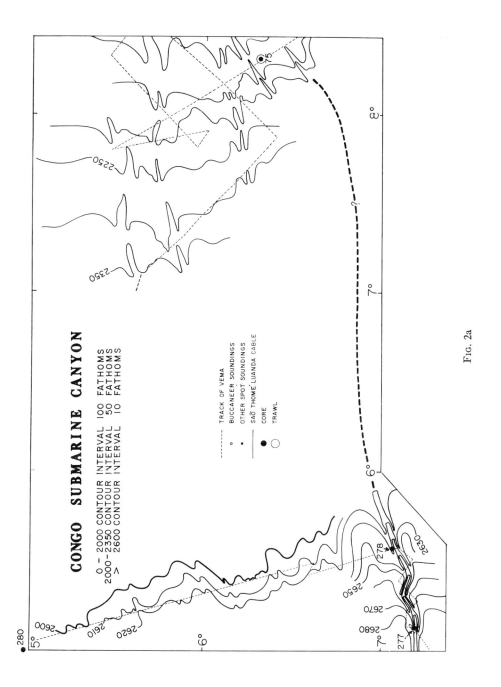

FIG. 2a

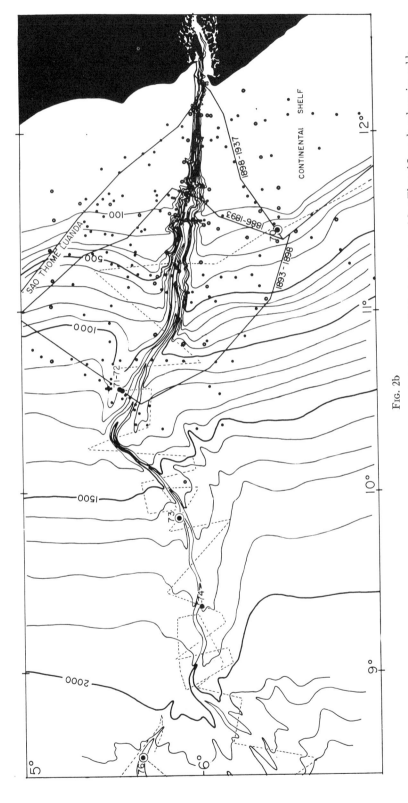

Fig. 2b

Fig. 2. (a, b)—Bathymetric chart of Congo Submarine Canyon. Soundings in fathoms at 800 fm/sec. Routes of Saõ Thomé-Luanda submarine cable shown by solid line. One degree latitude equals 60 nautical miles.

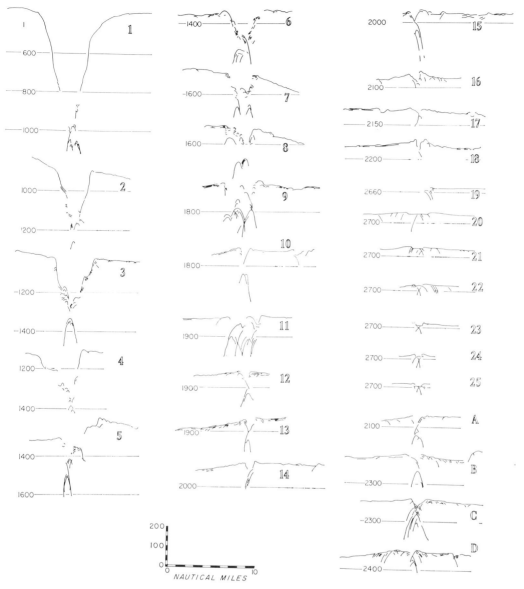

FIG. 3.—Tracings of Precision Depth Recorder echogram of Congo Submarine Canyon.
Depths in fathoms at 800 fms/sec.

canyon by 150 miles, making the entire proven length of the Congo Submarine Canyon 500 miles. Other branches of the canyon continue toward the west but they are much more poorly developed than the main channel.

SEISMIC REFLECTION PROFILE

Difficulty in recording sub-bottom echoes beneath deep-sea cones and on the landward side of abyssal plains is a common experience. This characteristic is generally attributed to numerous thin and highly reflective layers intercalated in the sedimentary column by turbidity-current deposition. In general, good penetration is obtained on gentle rises such as Bermuda Rise as well as at the seaward edges of the abyssal plains where in many places even the 12-kilocycle echo-sounding pulse returns from up to 20 fathoms below the bottom.

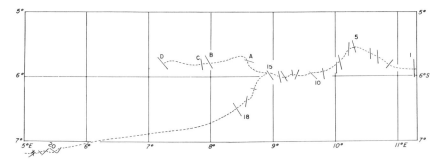

FIG. 4.—Location of profiles illustrated in Figure 3.

The VEMA-19 seismic reflection profile (Figs. 5, 6, 7) reveals relatively thick sediments up to 1.8 kilometers in thickness overlying the Walfisch Ridge and the Guinea Rise. The Guinea Rise is a broad, smooth arch which exhibits rather irregular mountainous sub-bottom topography beneath a 1-kilometer-thick mantle of sediment. On the Angola Abyssal Plain south of the Congo Canyon a remarkably transoral layer of sediment blankets the underlying relief. However, along a band about 150 miles on either side of the axis of the main channel of the Congo Canyon no sub-bottom penetration was obtained. This is not evidence that the sediment is thin in this area but simply an indication of the difficulty of obtaining reflections from beneath abyssal-cone type sediments. The whole central area of the Congo Cone is highly reflective. However, on the north side of the cone and in a slight depression that lies between the Congo Cone and the Guinea Rise prolonged confused echoes give a hint of closely spaced sub-bottom reflectors. There is a suggestion from the seismic reflection results that turbidity-current sedimentation in the northern part of the cone ceased some time ago, allowing the accumulation of a thin layer of transoral sediments above the lower and more reflective turbidites. This interpretation is supported by the sediment cores obtained in this area.

SEDIMENTS

Six piston cores were taken during VEMA Cruise 12 on the continental slope and continental rise in the vicinity of the Congo Canyon (Fig. 2). Cores were taken from the continental slope (V-12-70), the canyon wall (V-12-71, 72, 73), from the continental slope on the canyon rim (V-12-74), on the natural levees (V-12-75), from the floor of a

distributary channel on the continental rise (V-12-76).

Cores close to the axis of the canyon contain silts and sand beds (V-12-70, V-12-76); cores farther from the canyon axis (V-12-71, 72, 74) contain mainly hemipelagic silty clays with a few thin layers of silt. Megascopically, four sediment facies can be distinguished in the dried sediment. The first facies, comprising more than half of the sediment, is homogeneous silty lutite (Fig. 8). The silt, largely derived from the suspended sediment discharge of the Congo River, is mixed with the normal pelagic particle-by-particle deposition. A second lutite facies is described as crumbly, silty lutite. In the dried cores these lutites have a hackley broken surface which is distinct from the smooth surface of the dried homogeneous silty lutite. The third facies consists of graded silts and sands (V-12-73, 75, 76) which appear to be turbidites. These beds grade upward from fine sands to silts (in some places laminated). Three sand beds (one in V-12-75, and beds A and C in Core V-12-76; see Figs. 8, 9) contain leaves and twigs of dicotyledon land plants (identified by H. Becker, New York Botanical Gardens). A sample from the top of bed A in Core V-12-76 contains 68 per cent (by weight) leaves and twigs. Leaves and twigs have been found in turbidites originating from the Magdalena River (Heezen, 1956), from the Amazon (Locher, 1954) and in turbidites from the Hatteras Abyssal Plain. Bruun (1957) dredged coconuts and other vegetable matter from the Philippine and other trenches, material which may have floated to the site of deposition but which most probably was deposited by turbidity currents. The fourth facies is laminated silts (V-12-73, 74, 76). These beds consist of alternating lutite-rich and lutite-free laminae.

48

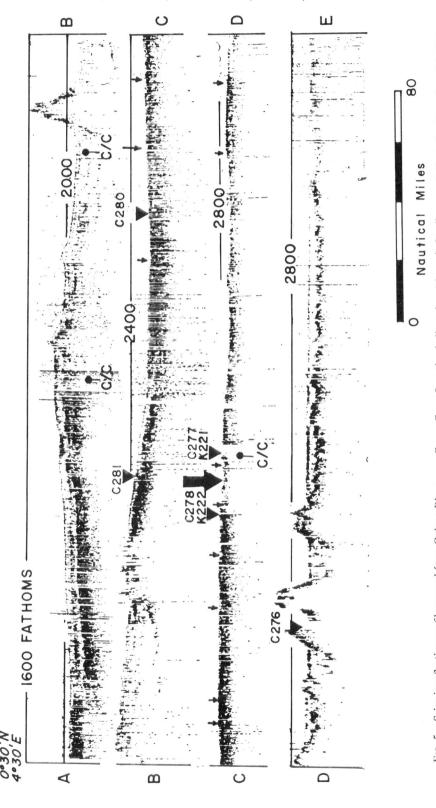

Fig. 5.—Seismic reflection profiler record from Guinea Rise across Congo Cone. Location indicated on Figure 7. Small arrows indicate location of small indistinct channels noted on PDR record. Large arrow indicates location of main branch of Congo Canyon system. Camera stations K221 and K222 and cores C276-C278, C280, C281 obtained on VEMA Cruise 19 indicated by dots. C/C indicates change of ship's course.

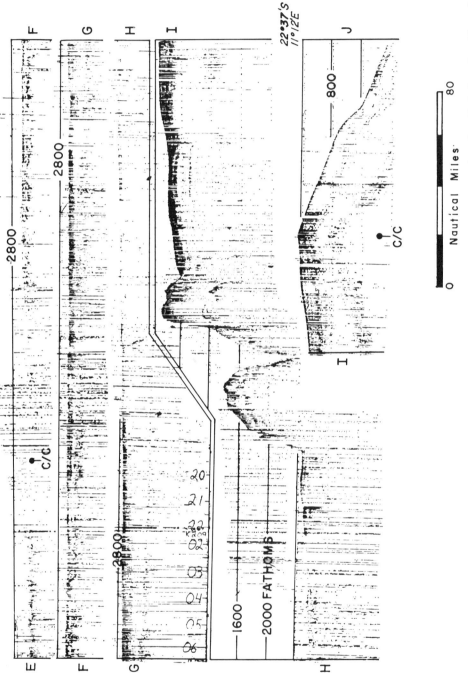

Fig. 6.—Seismic reflection profiler record from abyssal hills across Angola Abyssal Plain to Walfisch Ridge. Location of profile indicated on Figure 7.

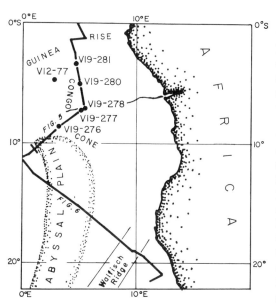

FIG. 7.—Location of VEMA Cruise 19 seismic
reflection profile and cores in Angola Basin.

Samples for mechanical and mineralogical
analyses were taken at the tops and the bottoms
of two of the turbidite sequences (beds A and C of
Core V-12-76; Fig. 9). Because of the short penetration of the coring apparatus it is suspected that
the deepest sand in Core V-12-76 was considerably
thicker than is indicated by the recovery. A single
sample was taken from the top turbidite sequence
(bed D) in the core because this bed was somewhat deformed. Beds A and C showed normal
grading in grain size and the size-distribution
curves are more peaked at the tops than at the
bottoms of the beds (Table II; Fig. 10).

Heavy-mineral separations were made from six
samples of Core V-12-76 (Table III). The percentage of heavy minerals seems to be inversely
proportional to the mean grain size (Tables II,
III).

Two mineral suites can be tentatively identified. Suite I is characterized by an abundance of
hematite and hematite aggregates in the heavy-
mineral separation and a preponderance of hematite-coated quartz grains in the light fraction.
This suite is also associated with concentrations
of leaf and twig fragments. Suite II is characterized by rounded black opaque heavy minerals and
a dominance of rounded clean and frosted quartz
grains. Residual traces of a former hematite coating are found only on a few grains.

In comparing the two mineral suites, it is noted
that the mineral species of Suite I contains many
loose aggregates of hematite and somewhat higher
concentrates of other easily altered minerals, such
as chlorite. This mineral suite is found in beds A
and C (V-12-76) and A' (V-12-73); beds that contain plant debris.

Since a large proportion of the sediment load of
the upper Congo River is deposited in the interior
Congo Basin above Leopoldville, much of the
sediment found in the braided streams of the
lower Congo must originate below Stanley Pool.
Red sands are common in the alluvial terraces,
erosion surfaces (King, 1962), marine terraces and
bedrock (Cahen, 1954; Furon, 1963) of the lower
Congo region. Red sands are in fact ubiquitous in
the Congo drainage basin. The 50 miles of the
Congo River cut across raised marine terraces of
Pleistocene age which for the most part consist of
semi-consolidated red ferruginous sandstones
(Mouta and O'Donnell, 1933; Cahen, 1954).

H.E.P. Cust (Anon., 1922), Commander of
H.M.S. RAMBLER during an 1899 investigation, found that:

"The nature of the bottom in the Congo River
is invariably sand, or in places hard clay, until the
deep gully is reached, when there is found everywhere a deep deposit of soft mud and decayed
vegetable matter, another proof of the tranquility
of the bottom water.

TABLE I. CORE LOCATIONS IN CONGO
CANYON AREA

	Lat.	Long.	Depth (Fathoms)	Length (Cm.)
V-12 70	06°29'S	11°26'E	240	610
V-12-71	05°38'S	10°41'E	1,210	1,072
V-12-72	05°38'S	10°40'E	1,131	490
V-12-73	05°54'S	09°53'E	1,635	979
V-12-74	06°00'S	09°20'E	1,845	910
V-12-75	06°19'S	08°19'E	2,145	683
V-12-76	05°43'S	08°30'E	2,137	330
V-12-77	04°48'S	02°45'E	2,747	1,078
V-19-276	08°58'S	02°52'E	2,798	348
V-19-277	07°10'S	05°08'E	2,700	1,344
V-19-278	07°04'S	05°34'E	2,637	996
V-19-280	04°56'S	05°00'E	2,610	1,594
V-19-281	03°19'S	04°39'E	2,430	1,787

CAMERA STATIONS IN CONGO CANYON AREA

Camera Station				No. of Frames
V-19-221	7°10'S	5°08'E	2,679	20
V-19-222	7°04'S	5°34'E	2,721	22

MEGASCOPIC DESCRIPTION OF DRY CORES

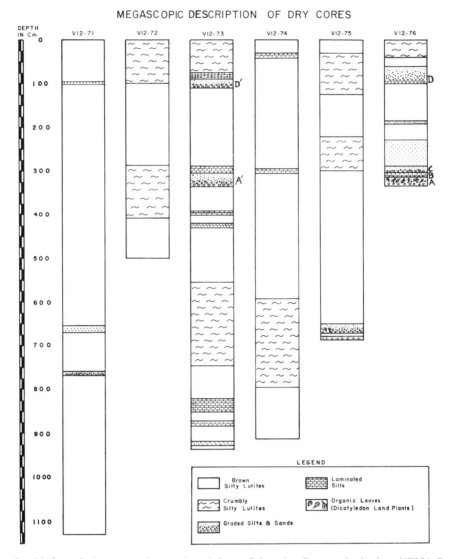

FIG. 8.—Graphic logs of piston cores from region of Congo Submarine Canyon obtained on VEMA Cruise 12.

"Whilst running a sectional line of soundings outside Banana creek, the bottom was invariably mud, but on one single occasion, at a depth of 56 fathoms, although mud was brought up, there were signs on the lead that it had struck a hard substance, the arming was dented and the lead slightly marked.

"The banks of the Congo, the numerous shoals in the river, and the bottom itself, were found during the survey to be invariably sand or hard clay.

"Mud is only met with in the small creeks in the upper portion of the river; and on the banks in the lower portion as far as the mangroves extend, which is from about Kissanga downwards.

"It appears that a very large proportion of the mud found in the lower part of the river in the deeper water is from the washings of the immediate neighbourhood, which is a vast expanse of mangrove swamp, with an innumerable network of creeks drained at every tide.

"The water of the river itself is found to be

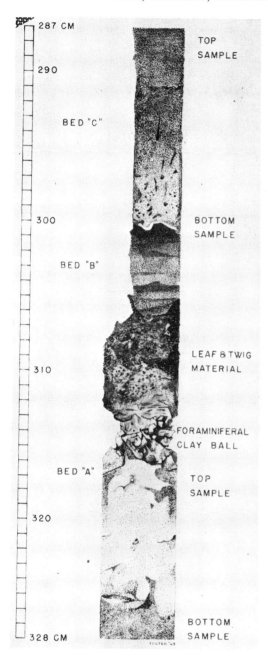

FIG. 9.—Turbidites in Core V-12-76.

deep water. It is stated that on grounding the anchor should never be let go; sooner or later the current will wash bank and ship together down the river until deep water is reached."

Due to the effects of vigorous wave action, clean sands characterize the Atlantic beaches of Angola, Congo, and Cabinda, despite the proximity of the red cliffs of marine terrace deposits which lie behind the beaches. The predominant northward long-shore drift in the region has produced Padrao Point, a sandspit which extends directly to the edge of the Congo Canyon. Two or more subsidiary canyons branch from the main canyon toward this point. Veatch and Smith (1939) note that this strong northerly long-shore current loosened and carried water pipes north from the Luanda. Trask (1955), Shepard (1951), and others have noted that long-shore drift often transports beach sands into the heads of submarine canyons.

The turbidites of mineral Suite I which contain leaf material and a preponderance of hematite and hematite-coated grains, probably are Congo River sediments. The minerals of Suite II which show much evidence of abrasion and transportation (beds B, D, D'; Figs. 8, 9) probably are long-shore drift sands which were swept into the canyon and later carried into deeper water by turbidity currents.

Both heavy-mineral Suites I and II contain appreciable percentages of pyrite. In many places the pyrite is a coating on another mineral and in some places it occurs inside diatom frustules. Gas bubbled profusely from Core V-12-73 when it was extruded on VEMA. Core V-12-70, taken on the continental slope south of the canyon, contains very dark silty clay with some mollusk layers. This core and the cores in the vicinity of the canyon show evidence of reducing conditions through their dark sediment color, pyrite mineralization, and the preservation of organic material which would be destroyed in an oxidizing environment.

Five cores were taken during VEMA Cruise 19 on or near the Congo Cone (Figs. 2a, 7) 150 miles west of the VEMA-12 suite of cores. These sediments represent two distinct facies of pelagic deposition.

The more northern facies consists of alternating layers of carbonate-rich gray lutites and black lutites (V-19-281, V-19-280) (Fig. 11) and the other facies consists of reddish brown abyssal lutite and is the dominant pelagic facies in the

heavily charged with sand. Vessels grounding on banks where the current is strong have had the sand piled up against one side of them nearly to the surface in a few hours, and then a sudden swirl of current has washed it away and left them in

TABLE II. MECHANICAL ANALYSES OF TURBIDITES OF CORE V-12-76: STATISTICAL PARAMETERS
(after Folk 1961)

		Mean Size	Mean Size mm	Sorting	Skewness	Kurtosis
Bed A	Top	3.480	.092	+.823	−.027	1.201
	Bottom	3.180	.291	+.305	−.047	1.262
Bed B		3.683	.085	+.651	−.001	1.667
Bed C	Top	3.470	.091	+.514	−.043	1.204
	Bottom	3.276	.101	+.474	−.003	.978
Bed D		2.780	.258	+.572	−.089	1.038

three southern cores (V-19-278, V-19-277, V-19-276).

The most northern of the cores studied (Core V-19-281) taken on the Guinea Rise, consists of approximately 18 meters of alternating bands of low-carbonate black lutite and gray lutite with carbonate percentages up to 80 per cent. Contacts between these distinct sediment types are in some places sharp and in some places gradational. In places the core is highly worm-burrowed. Radiolarians are found in parts of the core. Core V-12-77 taken somewhat west of V-19-281 contains a large percentage of siliceous material.

Core V-19-280 was taken about 100 miles south of Core 281 on the Congo Cone. The first 12-meter segment of this 16-meter core consists of alternating layers of light gray calcilutites and dark black lutites of the northern facies. The organic carbon content of the black lutite was determined to be 2.9 per cent. Professor William M. Sackett (personal communication) determined the C^{13}/C^{12} ratio of the organic ooze. The isotope ratio (−21.5 °/oo vs. PDB) suggests that the carbon is marine-derived, indicating high organic productivity in this area. Below 12 meters in this core is a distinctly different facies of reddish brown lutite containing little carbonate. The contact between these two facies is gradational over 20 centimeters. The brown lutite is intercalated by numerous sand and silt beds many of which are graded. Cross-bedding and parallel laminations due to lutite films are found in some of the coarser sand beds. In the vicinity of Core V-19-280 the Precision Depth Recorder echogram showed a sub-bottom reflection approximately 10–12 meters below the sediment water interface (.01–.015 sec.). This sub-bottom echo is probably from the contact between the two distinct sediment types found in Core V-19-280. The seismic reflection

profile record previously noted also showed shallow penetration (Fig. 5). Only one turbidite bed is found in the upper 12 meters of this core, whereas the lower 4 meters contain 68 turbidite beds. It can thus be concluded that turbidity current deposition at this site virtually ceased coincidentally with the beginning of deposition at this site of the northern gray-black lutite facies.

Cores V-19-288 and V-19-287 were taken from the northern bank of the canyon. These cores consist of thin turbidites intercalated in reddish brown lutite and resemble the sequence found below 12 meters in Core V-19-280. Both cores contain many turbidites consisting of fine sand to fine silts. Many of these beds are but laminae of fine silt, a centimeter to a few millimeters in thickness. In most places the beds are recognized only

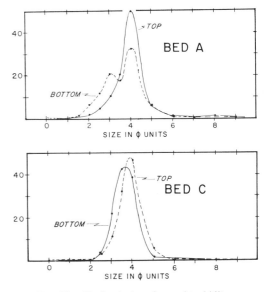

FIG. 10.—Mechanical analyses of turbidites in Core V-12-76.

TABLE III. MINERALOGY OF TURBIDITES IN CORE V-12-76

Mineral Suite	Bed		Leaves & Twigs	>90% Qtz. Hematite Coated	>50% Qtz. Frosted	Heavy Minerals*												% of Heavy Minerals in >62μ Fraction
						% Opaque ⊗				% Non-opaque ⊗								
						Hematite	Black Opaque	Leucoxene	Pyrite	Chlorite	Hornblende	Tourmaline	Staurolite	Epidote	Sillimanite	Others (<10% Each)	
I	A	Top	+	+	−	61	9	7	23								0.45
		Bott.	+	+	−	57	20	6	15	14	34	8	6	3	8	27*	0.38
	C	Top	+	+	−	65	22	4	8								0.38
		Bott.	+	+	−	62	4	6	27								0.41
II	B		−	−	+	39	46	2	12								0.73
	D		−	−	+	11	69	0	19	5	23	17	10	10	10	25†	.022

* Kyanite, garnet, zircon, rutile, orthopyroxene, biotite, zoisite, augite, and fiberolite.
† Same minerals as above but lacking zoisite and including topaz and sphene.
⊗ % of mineral grains counted >62 μ.

by slight changes of color, or from unequal shrinkage of the thin beds on drying. Core V-19-278 contained 440 such turbidite laminae in 10 meters. Some of the thicker silt and sand beds show grading and some contain microstructures such as cross-bedding and parallel laminations. At 798 cm. in Core V-19-279, a bed of light brown crystalline carbonate sand was found. This material consisted of small rhombohedrals of calcite and showed no evidence of organic deposition. X-ray diffraction indicated the material to be calcite with less than 5 per cent aragonite.

Core V-19-276 was taken in the abyssal hills southwest of the Congo Canyon. This core consisted of reddish brown lutite with a zone of hard indurated yellow lutite between 30 and 90 cm. The core shows no evidence of turbidity-current deposition.

All five cores contain a zone high in foraminifera in the top 14–25 cm. Below this zone, foraminifera are found mainly in the gray lutite facies and in some worm burrows. All the cores except V-19-276 showed evidence of a highly reducing environment; (1) each smelled of H_2S when opened on board ship; (2) each was stained by hydrotroilite; and (3) the lutites of each contained many small micro-crystals of pyrite.

The sediment pattern in the western part of the Angola Basin may be summarized as follows: (1) organic pelagic lutites in the northern equatorial region (either siliceous or carbonate), and (2) reddish brown abyssal lutite in the central and southern regions. Superimposed on this pelagic sedimentation pattern is the effect of the Congo Canyon which has introduced great amounts of silt and sand onto the Congo Cone. Turbidity current deposition apparently ceased in the northern part of the Congo Cone in mid-Pleistocene. A thin blanket of pelagic sediment has softened the topographic forms and obliterated the smaller channels in that area.

Bottom photographs were taken near the submarine canyon on VEMA Cruise 19 at the same positions as Core V-19-277 and Core V-19-278. The photographs (Fig. 12) indicate a generally muddy bottom with fairly abundant tracks and trails, and indicate a tranquil bottom unaffected by ocean currents.

55

BIOLOGICAL TRAWL RESULTS

Biological trawls were taken simultaneously with several cores (Fig. 2). The same techniques were used with each trawl; hence, the results are comparable among themselves. The standing crop of animals from the upper continental slope (trawl No. 21 at core location V-12-70) exceeded that of deeper trawls by a factor of 100 to 1000 as would be expected, but trawls taken at greater depths showed some interesting and unusual results.

Heezen, Ewing, and Menzies (1955) postulated that submarine turbidity currents might affect abyssal life in two ways: (1) to obliterate life by smothering it; (2) to feed life by providing avalanches of detrital nourishment from shallow water.

Trawl No. 23 was taken from a natural levee and although the water depth was more than 500 fathoms greater than that of trawl No. 22, there was no marked decrease in diversity and abundance of animal life as compared with trawl No. 22 which was taken from a similar situation farther up the continental rise and 90 miles closer to shore. Significantly lower was the population of trawl No. 24 from a sandy distributary floor, although the depth was the same as that of trawl No. 23.

A study was made of the ratio of detritus feeders to filter feeders. Filter feeders, it was reasoned, would be harmed while detritus feeders might be benefited by frequent turbidity currents. The proportion of detritus feeders appears to be sig-

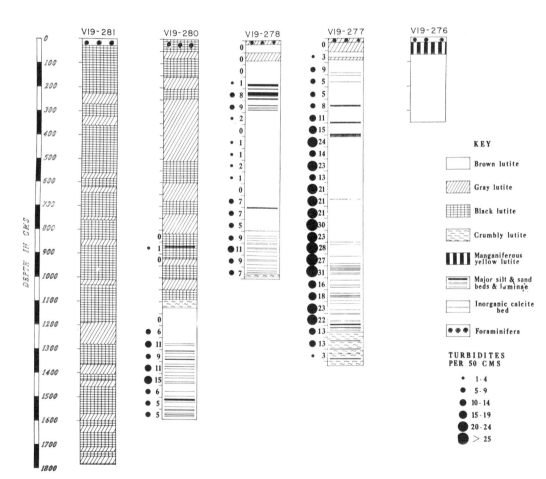

FIG. 11.—Graphic logs of piston cores obtained in vicinity of Congo Cone on VEMA Cruise 19. Locations shown on Figure 7. Positions and depths listed in Table I.

FIG. 12.—Sea-floor photographs of tranquil bottom obtained on natural levees of the Congo Canyon. (Upper photo) Station V-19-221 (18) 2679 fm. (Lower photo) Station V-19-222 (14) 2721 fm. Positions listed in Table I and Figure 5. Dimensions of photographed area approximately 6×9 feet. Note abundant evidence of bottom life and complete absence of current evidence.

nificantly higher in the Congo Canyon distributary system than in adjacent areas. Much more information is needed but the data are at least suggestive.

SUBMARINE CABLE FAILURES

The Saõ Thomé-Luanda (Angola) Cable (Fig. 2), originally laid in 1886 and abandoned in 1937, failed 30 times near the axis of the Congo Canyon.

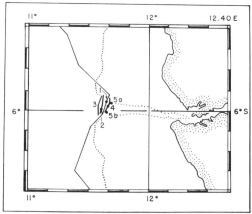

FIG. 13.—Cable failures in Congo Canyon; Cable Route 1, 1886–1894. Dots indicate location of cable breaks. Black line indicates cable replaced during repair.

The first route selected from Buchanan's survey was used from 1886 to 1893, during which time the cable failed five times (Fig. 13). The cable failed each time under tension and in most places was buried under sediment in the floor of the canyon. In 1893 it was decided to divert the cable into deeper water where failures are in general less common. The second route crossed the canyon in much deeper water 70 miles farther from the Congo River, but in this location the cable suffered eight breaks between 1893 and 1897 (Fig. 14). Each of the eight failures was a tension break and in all cases the section of cable near the axis of the canyon was buried and had to be abandoned. The cable engineers recognized that the cable failures were related to the transport of sediment down the canyon from the Congo River. They realized that to avoid further failures they would have to loop the cable hundreds of miles out to sea. Instead they diverted the cable into the river mouth where in relatively shallow depths cable repairs could be more easily effected. Along

TABLE IV. BIOLOGICAL TRAWLS

Trawl No.	At Core Sta.	Surface Sediment	Orders or Classes of Animals	Number of Specimens / Number of Species/Trawl	Approx. No. Species/Trawl	Depth (Fms.)
22	V12-73	Diatomite	12	8.3	25	1635
23	V12-75	Red lutite	8	9.1	14	2145
24	V12-76	Sand	5	3.8	9	2137

this third route the cable failed 15 times between 1897 and 1937 (Fig. 15). Three failures occurred in a tributary canyon leading northwest from the sandspit terminating at Padrao Point. Two failures occurred in another tributary a short distance west. These failures were also all tension breaks and cable recovery was prevented by burial in most instances. The cable ships attempted to lay the cable across the canyon at about a 45° angle in an effort to decrease the drag on the cable. Repairs of the cable at the river mouth may have taken less time than the deepwater repairs but they were not without difficulties. The high currents at the Congo mouth carried away mark buoys, made grappling runs more difficult, and swamped the boats used in tending buoys. In 1914 it was decided to insert a "T" piece linking Banana with Saō Thomé and Luanda so when the cable failed in the canyon Saō Thomé would be at least linked to Banana. The breaks were not distributed uniformly through the years but occurred at definite seasons during two periods: 1893–1904 and 1924–1928. These cable breaks are associated with some highly significant characteristics of the Congo River itself.

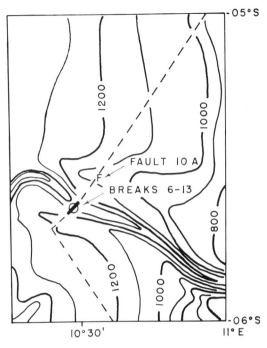

Fig. 14.—Cable failures in Congo Submarine Canyon; Cable Route 2 (deep water) 1894–1897. Circle indicates location of 8 breaks. Letter F indicates location of fault.

[*Editor's Note:* Material has been omitted at this point.]

DISCUSSION

The exact mechanism by which the river generates turbidity currents at peak-discharge months during periods of river channel excavation is not known. We may only speculate on possible relationships. It might be imagined that at periods of peak discharge enormous quantities of sand and silt are carried in the stream past Malela to the head of the submarine canyon and that appreciable quantities of sand- and silt size material are thus carried over the underlying salt water. At periods of peak discharge conceivably enough sand would in this manner settle into the underlying sea water to generate turbidity flows.

As another possibility, the bed load carried by traction and saltation along the bottom at times of peak discharge might separate at the head of the canyon and simply continue as a turbidity flow beneath the salt-water wedge which fills the canyon head within the river.

As a third possibility it might be imagined that in addition to one of the above mechanisms, sand bars carried away during peak discharge near the end of the braided stream and near Padrao Point create avalanches of sand which transform into turbidity currents.

A buried "submarine canyon" has been discovered beneath the Mississippi delta (Osterhoudt, 1946). This buried canyon leads to the head of the Mississippi cone distributary system. There has been considerable discussion of the rates of subsidence and rates of compaction, in relation to eustatic events in the Mississippi delta region (Fisk and McFarlan, 1955). Of critical importance to these discussions is the precise relation to sea-level of the braided-stream deposits and the eroded floor of the Pleistocene Mississippi River "trench".

If the submarine canyon of the Pleistocene Mississippi cut into the mouth of the Mississippi and formed a feature similar to the Congo Submarine Canyon, then the floor of the lower Mississippi trench could have been excavated hundreds of feet below sea-level. Buried canyons similar to the modern Congo Canyon may exist below the deltas of other great rivers (Heezen, 1963).

BIBLIOGRAPHY

Anonymous, 1922, The West Coast of Africa: British Admiralty Africa Pilot Pt. II, 276 p.
Bruun, Anton F., 1957, General introduction to the reports and list of deep-sea stations, p. 7–12, in GALATHEA Report v. 1, Scientific results of the Danish Deep-Sea Expedition round the world 1950–52: 260 p., Copenhagen.
Buchanan, J. Y., 1887, On the land slopes separating continents and ocean basins, especially those on the West Coast of Africa: Scot. Geograph. Mag., v. III, p. 217–238.
———— 1888, The exploration of the Gulf of Guinea: Scot. Geograph. Mag., v. IV, p. 177–200, 233–251.
Cahen, L., 1954, Geologie due Congobelge: 577 p., Liege, Hvaillant-Carmanne.
Devroey, E. J., 1946, La Vallee Sous-Marine de Fleure Congo Inst. Royal Colonial Belge: Bull. de Seances XVII, p. 1040–1074.
———— 1956, Annuaire Hydrologique du Congo Belge et du Ruanda-Urundi: Ministere des Colonies, Belgique, Pub. 11, 467 p. (for many years such a volume appeared each year).

———— and Vanderlinden, R., 1951, Le Bas-Congo, Artere Vital de Notre Colonie: 350 p., Bruxelles, Goemaere.
Elmendorf, C. H., and Heezen, B. C., 1957, Oceanographic information for engineering submarine cable systems: Bell System Tech. Jour., v. XXXVI, p. 1047–1093.
Fisk, H. N., and McFarlan, Jr., 1955, Late Quaternary deltaic deposits of the Mississippi River: Geol. Soc. America Special Paper 62, p. 279–302.
Folk, R. L., 1961, Petrology of sedimentary rocks: 154 p. Austin, Texas, Hemphill's Bookstore.
Furon, R., 1963, Geology of Africa: 377 p., Edinburgh, Oliver & Boyd.
Heezen, B. C., 1956, Corrientes de Turbidez del Rio Magdalena: Bol. Soc. Geog. Colombia, v. 51 and 52, p. 135–143.
———— 1963, Turbidity currents: Chapter 27, p. 742–775 in Hill, N. M., ed., The sea: 963 p., New York, John Wiley.
———— Ewing, M., and Menzies, R. J., 1955, The influence of submarine turbidity currents on abyssal productivity: Oikos, v. 6, p. 170–182.
Hull, E., 1900, The suboceanic river valleys of the West African Continent: Trans. Victoria Inst., v. XXXI.
———— 1912, Monograph on the sub-oceanic physiography of the North Atlantic Ocean: 41 p., London, Edward Stanford.
King, Lester, 1962, Morphology of the earth: 699 p., Edinburgh, Oliver & Boyd.
Kolbe, R. W., 1957, Fresh-water diatoms from Atlantic deep-sea sediments: Science, v. 126, no. 3282, p. 1053–1056.
Locher, F. W., 1954, Ein Beitrag zum Problem der Tiefseesande im westlichen Teil des äquatorialen Atlantiks: Heidelberger Beiträge zu Mineralogie u. Petrographie, v. 4, p. 135–150.
Mouta, F., and O'Donnell, H., 1933, Carte Geologique de l'Angola: Ministerio dans Colonis Republica Portugesa, 87 p.
NEDECO, 1959, River Studies, Niger and Beume: 100 p., Amsterdam, North Holland Publishing Company.
Osterhoudt, W. J., 1946, The seismograph discovery of an ancient Mississippi River channel (abs.): Geophys., v. 9, p. 417.
Schott, G., and Schulz, B., 1914, Die Forschugsreise S.M.S. MÖWE im Jahre 1911: Archiv der Deutschen Seewarte, v. XXVII, no. 1, 104 p., Hamburg.
Shepard, F. P., 1948, Submarine geology: 348 p., New York, Harper Bros.
———— 1951, Mass movements in submarine canyon heads: Trans. Am. Geophys. Union, v. 32, p. 405–418.
Spencer, J. W., 1903, Submarine valleys off the American coast and in the North Atlantic: Geol. Soc. America Bull., v. 14, p. 207–226.
Spronck, R., 1941, Mesures Hydrographiques Effectuees dans la Region Divagante du Bief Maritime du Fleure Congo: Mem. Inst. Roy. Colonial Belge, 156 p.
Trask, P. D., 1955, Movement of sand around Southern California promontories: Beach Erosion Board Tech. Memo 76, 66 p.
Veatch, A. C., and Smith, P. A., 1939, Atlantic submarine valleys of the United States and the Congo submarine valley: Geol. Soc. America Special Paper 7, 101 p.

5

Copyright © 1973 by the American Association of Petroleum Geologists

Reprinted from *Amer. Assoc. Petrol. Geol. Bull.*, **57**(9), 1679–1691 (1973)

Congo Submarine Canyon and Fan Valley[1]

F. P. SHEPARD[2] and **K. O. EMERY**[3]

La Jolla, California 92037, and Woods Hole, Massachusetts 02543

Abstract   Seventeen transverse profiles of the inner 460 km of the Congo Canyon and the Congo Fan Valley were made during a 4-day study in June 1972. These profiles show that the canyon is V-shaped with side slopes 400–1,400 m high between the coast and a point 240 km seaward, where the axial depth is about 2,700 m. Farther seaward, the continuation as a fan valley narrows and is bordered by levees a few tens of meters high, and distributaries are found. The latter continue beyond the 4,600-m limits of this study for at least an additional 320 km to depths of 4,900 m. Seismic profiles show that the canyon has been cut through a belt of diapirs—probably derived from evaporites of Early Cretaceous age, and still rising through the several kilometers of subsequent sediments. Highest side slopes and a major bend in the canyon are present within the belt of diapirs or in the thick sediments that are dammed by the belt. The absence of a broad delta at the mouth of the Congo River, the presence of a possible temporary fill at the head, a steep axial slope near the head, a submerged fan bordering the seaward side of the diapir belt, and levees at depth support the concept of origin of the canyon–fan-valley system largely through erosion and deposition by turbidity currents. Tidal-current scour probably helps to limit the amount of fill in the canyon head.

Introduction

Congo Submarine Canyon is unique among the valleys of the ocean floor in that it penetrates as a deep estuary into the continent at the mouth of an enormous river; in fact, the river has the largest water discharge in Africa, although much of its sediment is deposited in an interior delta above Kinshasa (Veatch and Smith, 1939). Submarine canyons are present off other large rivers, including the Columbia, the Eel of northern California, the Hudson, the São Francisco of Brazil, the Tagus, the Rhone, and the Var of southern France. Other deep trough-shaped valleys are off the deltas of the Ganges, the Indus, and the Mississippi Rivers. Trincomalee Canyon in Ceylon, Tokyo Canyon, and a group of canyons along the west coast of Corsica head into relatively deep estuaries of small rivers. However, only the Congo Canyon penetrates the land at the mouth of a large river. The significance of the deep mouth of the Congo is that it exists despite the enormous amount of sediment that is introduced by the river.

The Congo Canyon was first reported by Buchanan (1887) in a survey prior to the laying of a submarine telegraph cable. Considerable interest was aroused because this and other cables laid across the canyon had to undergo frequent repairs (Milne, 1897). Heezen *et al.* (1964) discussed the history of these cable breaks and made a topographic survey of the canyon and its seaward continuation as a fan valley,[4] and obtained cores having characteristics that suggest the movement of turbidity currents along the outer fan valley.

In June 1972, an opportunity for further study of Congo Canyon came during Leg VII of the Eastern Atlantic Continental Margin Program, International Decade of Ocean Exploration (IDOE), when R/V *Atlantis II* was operating in the area. Four days were devoted to seismic profiling of the canyon and fan valley between axial depths of 400 and 4,500 m. In addition, echo profiles were run along the margin of the valley, and a Savonius current meter obtained a 4-day record in the canyon head.

Topography

Many submarine canyons can be traced seaward to where they merge into leveed valleys

© 1973. The American Association of Petroleum Geologists. All rights reserved.

[1] Manuscript received, January 25, 1973; accepted, February 25, 1973. Contribution No. 3097 of the Woods Hole Oceanographic Institution.

[2] Geological Research Division, Scripps Institution of Oceanography, University of California.

[3] Woods Hole Oceanographic Institution.

This study was made possible by funding for the Eastern Atlantic Continental Margin, a program of the International Decade of Ocean Exploration, through National Science Foundation Grant No. 28193. Appreciation is expressed to the many shipboard participants for their help, especially D. E. Koelsch and E. M. Young (chief technicians) and R. C. Groman and D. E. Woods (computer center). We also acknowledge the support of National Science Foundation contract GA-19492 and Office of Naval Research contract Nonr-2216(23). Processing of the soundings and current meter data was conducted by Neil Marshall, Gary Sullivan, and Pat McLoughlin.

[4] Heezen *et al.* referred to the entire valley as a canyon although their profiles and contours show that the floor of the outer valley is only a few meters below its surroundings.

that wander across the sediment fans at the base of the continental slope. This change in character is well illustrated by the valley off the Congo River, where about 200 km of the V-shaped, deeply incised canyon merges seaward into 500 km or more of leveed valley having distributaries and low side slopes (Fig. 1).

According to available soundings from a 1933 Portuguese chart and other surveys, probably all made by wire rather than by echo sounding, the canyon head has a V-shaped gorge with steep sides and a twisting axis that can be traced at least 25 km up the estuary (Fig. 2). Our operation did not extend landward of Ponto do Padrão, but our profile (Fig. 2 inset) showed a flat floor at 370 m with a slight depression or channel reaching 399 m. This is definitely shoaler than the 500–550-m soundings indicated by the 1933 Portuguese chart and by all other sources. However, only about 2 km west of this point, depths were 580 m, almost as deep as those of the Portuguese chart. It would appear that the head of the canyon may have had an extensive recent fill. Beyond a section of the axis having a seaward slope of about 10 percent, the V shape returns and characterizes all but one profile seaward of the steep area (Fig. 3).

Most of our profiles were made nearly at right angles to the axial trend of the outer valley (Fig. 1), but a few are at a more acute angle, as can be seen from the distinct difference in depth at the top of each side slope. These transverse profiles differ in each of three main structural regions that are crossed by the canyon–fan valley (Figs. 1, 3). Profiles 1–5, in sediment fill landward of a belt of diapirs, have progressively greater axial depths (776–1,582 m) and higher side slopes (400–1,400 m); most of the side slopes also are relatively smooth. Profiles 6–12, within the belt of diapirs, reveal progressive deepening of the axis (1,787–3,045 m) and decrease in height of side slopes (1,300–300 m). All of the side slopes are irregular, partly because of the presence of tributaries (profile 7) or possible distributaries (profiles 11, 12), but probably mostly because of slump or disturbances caused by local diapiric uplifts. Profiles 13–17, seaward of the diapiric belt, exhibit a low axial gradient (3,380–4,587 in 345 km) and low side slopes that diminish seaward (260–40 m) and consist of natural levees that border the fan valley.

The longitudinal profile (Fig. 4) of the axis of the canyon-fan valley was obtained by connecting the greatest depth along each of our 17 transverse profiles (Fig. 3); it also includes the soundings near the current-meter station, 20 km landward of profile 1. The gradient decreases in steepness seaward from perhaps 10 percent near the current-meter station to an average of 0.8 percent out to profile 12. This is an unusually gentle slope for a submarine canyon. Seaward of profile 12, the slope is still further reduced to 0.45 percent, more or less typical for the gradient of a fan valley.

The earlier surveys depict definite tributaries entering the main canyon along its south side. We were able to run a few profiles along both the north and south sides of the canyon head (Fig. 5). These show that tributaries exist on both sides where the canyon crosses the continental shelf. Evidence for tributaries along the outer part of the canyon is incomplete, but a few short lines suggest that tributaries are rare, if present at all, where the canyon crosses the continental slope.

Seaward of profile 11, the valley has at least eight distributaries (Fig. 1). From the profiles, we can say tentatively that the change from a canyon to a typical fan valley (Shepard, 1965) occurs near profiles 10 and 11, and certainly by profile 12. There is some gradation in this change, because profiles 8–10 show possible levees on their south side. The change between canyon and fan valley is clearly exhibited in the profiles made by Heezen et al. (1964). Their profiles extend farther seaward than ours (to long. 5°E), and so far as one can judge from the contours, their outermost crossing shows only a few meters of side slope, approximately equal to the height of the levees.

SEISMIC PROFILES

Our principal addition to previous knowledge relative to the valley off the Congo comes from our seismic profiles (Figs. 6, 7). These profiles support previous studies (Baumgartner and van Andel, 1971; Leyden et al., 1972; Emery, 1972) in showing a diapiric belt that extends along the continental slope off Angola, Zaire, Congo, Gabon, and probably Cameroun. The diapirs are believed to consist of salt of Early Cretaceous age, the same as for the diapirs beneath the adjacent land. Upward movement of the salt at the southern end of the belt continues at present, as shown by the many unconformities in overlying sediments, the protrusion of the diapirs well above their surroundings, and the absence of thick horizontally bedded sediments in ponds between the diapirs. Diapirs off the Congo River are similar except that few of them project above their immediate surroundings; this is interpreted as meaning that sediment is contributed by the Congo and deposited atop the diapir belt at a rate faster

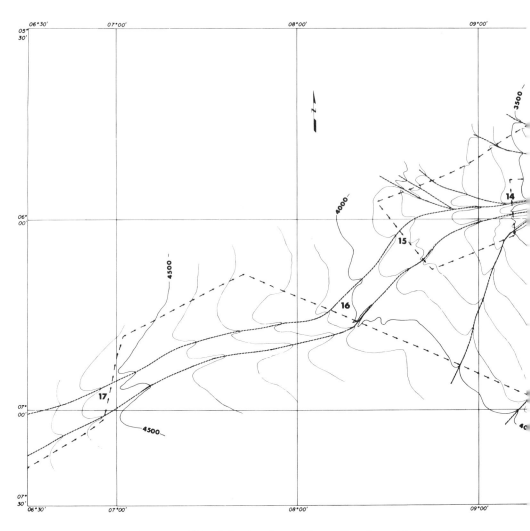

Fig. 1—General relief of Congo Canyon and Fan Valley based on 1972 soundings from *Atlantis II*. Approximate axes
4, and 7 are indicat

than the rate of diapir protrusion. Correspond-ingly, a scarp (similar to the Sigsbee escarp-ment in the Gulf of Mexico) borders the diapir belt except off the Congo River, where its place is taken by an apron of sediments. Moreover, the total thickness of sediments above acoustic basement (probably oceanic basement) is much greater off the Congo River than farther south. Reflection times exceed 4 seconds, meaning that the Congo Canyon is underlain by as much as 6 km of sediments, into the top of which it has been incised.

In interpreting details of the seismic profiles, one must realize that horizontal layers trun-cated by a canyon appear as a downbend next to the canyon side slope. The profiles clearly indicate that the canyon is cut into sedimentary formations, which are stratified except at the tops of the diapiric intrusions, as in profiles 7 and 8 and possibly profile 5 of Figure 7. Profile 8 has a U shape, in contrast to the V shapes of the other profiles. Apparently, erosion at this point started in the softer sediments and the valley became incised into the top of the diapir near one side. Clear evidence of slumping along the canyon side slopes is shown in the seismic profiles (Fig. 7), notably in profiles 3–7 and 9–12. The terraces for the most part seem to

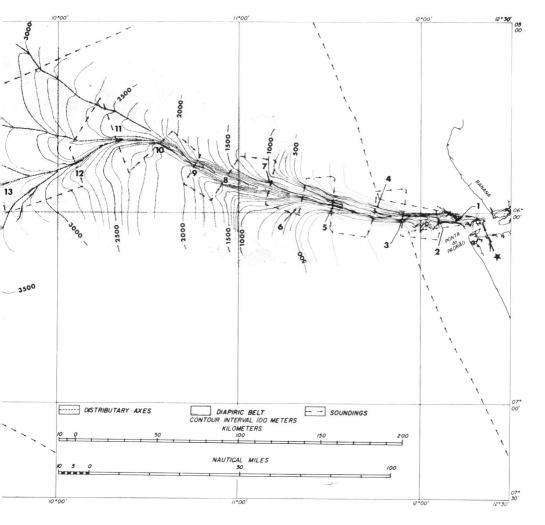

ibutaries shown by dashed lines. Locations of soundings (arrows), current-meter station (star), and profiles of Figures 3, piric belt is shaded.

be slump blocks. Only in profile 10 does the layer forming the block appear to continue beneath the adjacent intercanyon surface. In profile 8, where the canyon is cut into a diapir, there are no terraces or other indications of slumping.

The seismic profiles contain many indications of folding in the strata. Some of this folding is related to the diapiric intrusions, as in profile 11, but other folds could be the result of deposition in old valleys and the compaction of the fine-grained sediments. Also, slumping no doubt has contributed to deforming the beds, particularly along the sides of the canyon.

The seismic profiles that cross the fan valley (Fig. 8) indicate that, although erosion may have played a part, deposition on the flanks has been the predominant process of valley formation. Lenses are apparent in profiles 13 and 15, where the fan valley was built by deposition of the levees along the sides.

CANYON-FLOOR CURRENTS

Emplacement of an Isaacs–Schick-type Savonius rotor current meter (Isaacs et al., 1966) in the head of Congo Canyon provided a record of the currents on the floor for approximately 4 days. The current meter was dropped

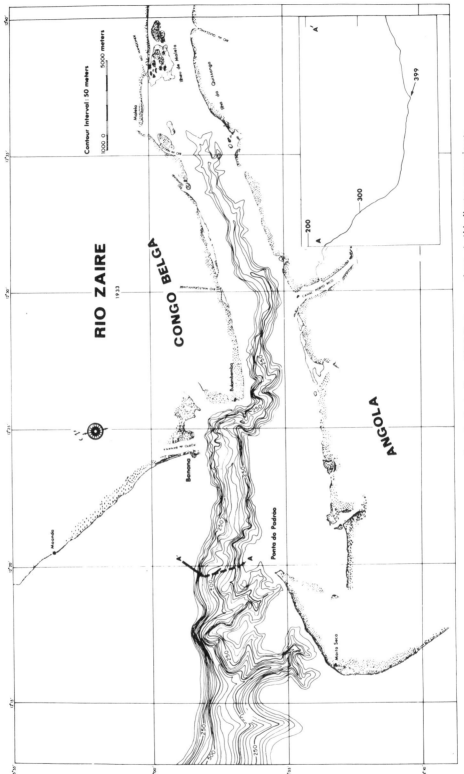

Fig. 2—Contours of head of Congo Canyon based principally on 1933 Portuguese chart. *A-A'* indicates approximate location of 1972 profile with its indications of fill in canyon head.

64

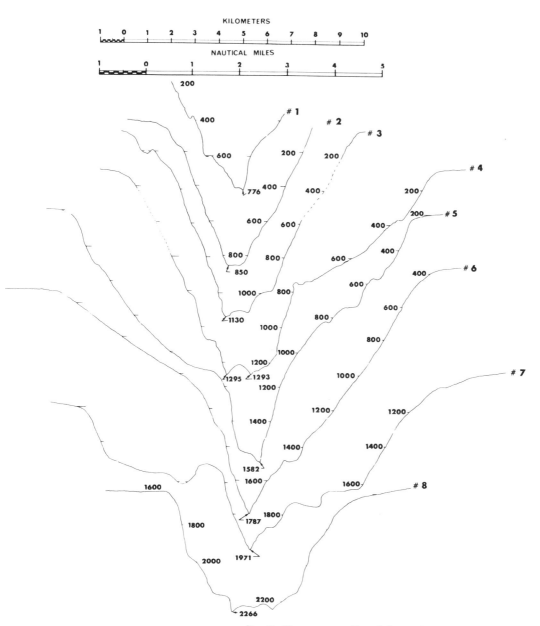

FIG. 3—Transverse profiles of Congo Canyon and Fan Valley.

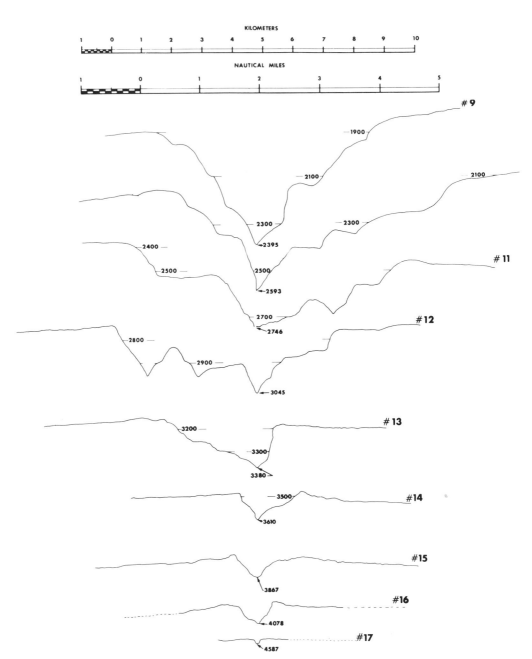

Profiles oriented to face downcanyon. For locations, see Figure 1.

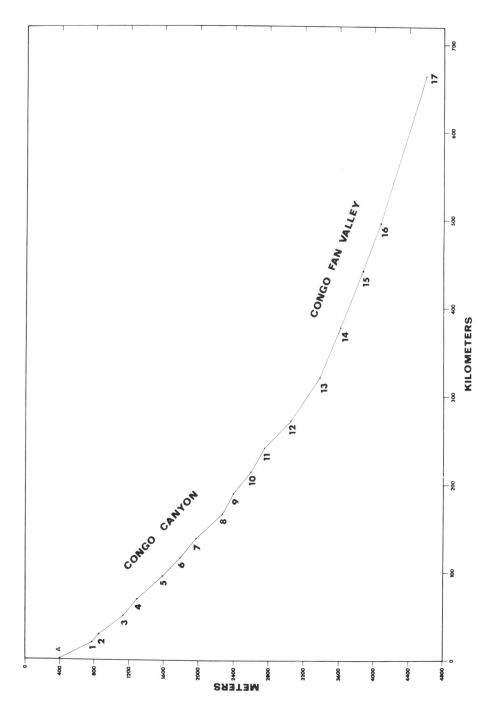

Fig. 4—Longitudinal profile down axis of Congo Canyon and Fan Valley. Note decrease in gradient at approximately point 12 where canyon changes into fan valley.

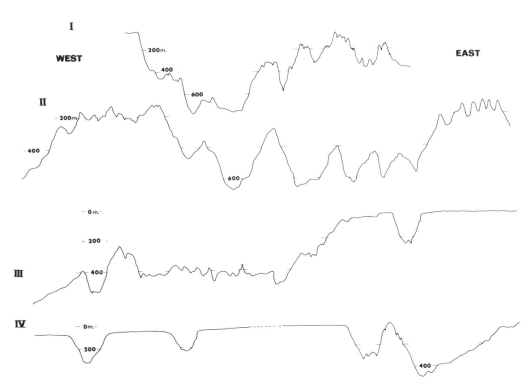

FIG. 5—Longitudinal profiles at head of Congo Canyon. Profiles I and II are on north side and III and IV on south; II and III are not located accurately.

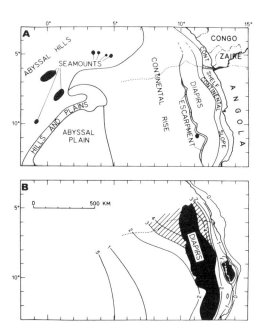

FIG. 6—**A,** Physiographic units of region showing Congo Canyon and Fan Valley (dots) crossing continental shelf, continental slope, diapiric belt, and continental rise. **B,** Thickness of all sediments above basement (in seconds of reflection time). Black denotes areas of salt diapirs of early mid-Cretaceous (Albian-Aptian) age. Diagonal hatching accents positions occupied by sediments thicker than 3 seconds reflection time.

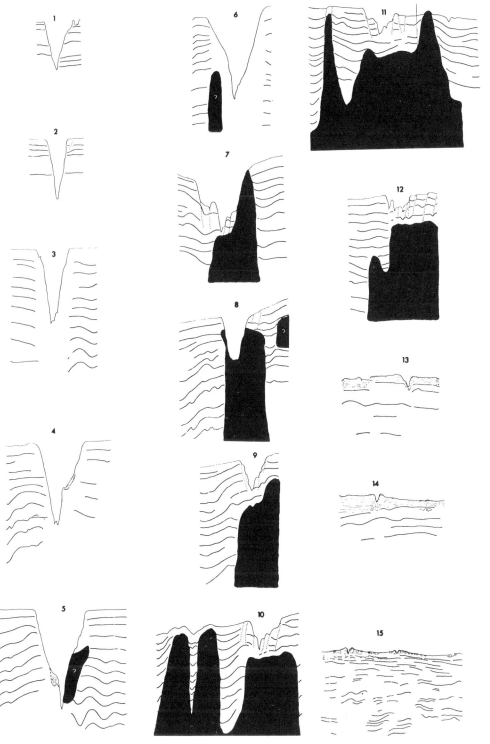

FIG. 7—Diagram from seismic profiles across axis of Congo Canyon and Fan Valley along same lines as in Figure 3. See also Figure 8. Solid zones represent diapirs.

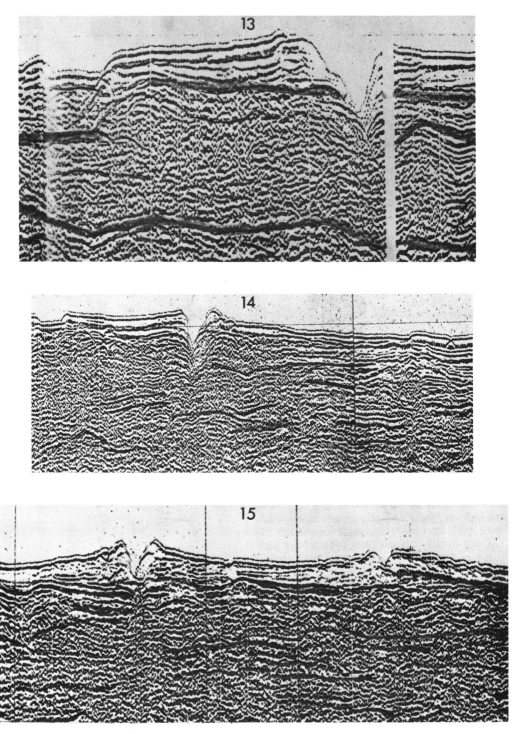

FIG. 8—Photographs of parts of seismic profiles from sections 13–15 showing contrast between recent sediments of levees and underlying beds.

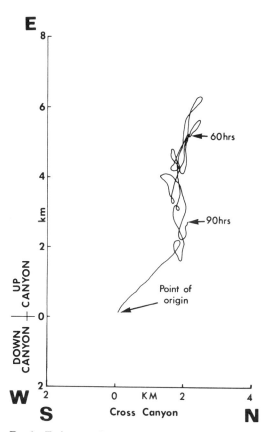

FIG. 9—Trajectory of movement of water particles along axis of Congo Canyon, showing net movement during 4 days of records.

into a 399-m channel that was bordered by a fairly flat floor of about 370 m in a section of the canyon where the axis trends nearly east-west (Fig. 2 inset). Principal results of the analysis of the currents in the record are indicated in Figures 9 and 10. The initial current was southeast (Fig. 9), which was across the general trend and at 180° from the predominantly northwest surface current. However, after the first 5 hours, the current was dominantly upcanyon and downcanyon with almost no net change during the last 84 hours (Fig. 10). Typical canyon currents illustrated by similar measurements at Scripps Institution (Shepard and Marshall, 1973) are downcanyon; apparently, the Congo does not follow this general rule and may well have predominant upcanyon motion. This is not surprising in view of the great seaward flow of water at the surface, presumably overlying a salt wedge (Buchanan, 1887), like that at the mouth of the Mississippi and of other large rivers that carry seawater upvalley.

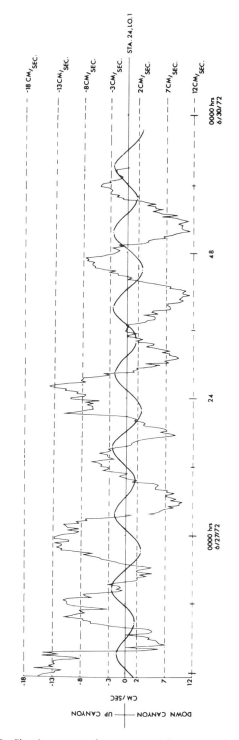

FIG. 10—Showing vectors of upcanyon and downcanyon currents along floor of Congo Canyon at depth of about 400 m. Note general upcanyon currents during rising tides and downcanyon during ebbing tides. Tides are for nearby Ponto do Padrao.

The most interesting feature of the current-meter record is that it seems to follow closely the tidal pattern (Fig. 10). Such a relation was rare in the currents measured off California and Baja California. The exact times of the tides at the mouth of the Congo are known only approximately, but apparently upcanyon flow accompanies the rising tide, and down-canyon flow the ebbing tide.

DISCUSSION

Our recent work on the Congo Canyon provides some new information but fails to explain completely the enigma of a canyon at the mouth of a large river where one should expect a delta. We know from previous information of cable breaking that active slides or strong turbidity currents occur along this canyon, mostly during the high stages of the Congo River (Heezen *et al.,* 1964). Our findings at the canyon head of a flat floor and of much shoaler depths than in 1933 are indicative of filling, or possibly that the previous soundings were made with a lead line dropped from a rapidly drifting boat. Our results here were obtained during the period of relatively low runoff. We need to make further soundings toward the end of a high runoff period to look for seasonal flushing of sediment accumulated in the canyon head.

The canyon appears to have been cut through a belt of diapiric salt masses that has risen as the canyon was eroded. The canyon cuts through only a few of the diapirs, but has cut deeply into the bedded sediments between the diapirs and into a wide thick prism of sediments that have been dammed by the belt. The highest side slopes of 1,400 m occur near the landward side of the diapiric belt. These side slopes probably are locally as steep as 45° and contain many slump blocks near their bases, testifying to instability of the sedimentary strata. If erosion had been restricted to a remote period, one would expect the slumping to have reduced them to gentle slopes and the canyon floor would have been flattened by the fill. Therefore, it seems likely that the canyon headward of profile 12 is actively eroding at present.

REFERENCES CITED

Baumgartner, T. R., and Tj. H. van Andel, 1971, Diapirs of the continental margin of Angola, Africa: Geol. Soc. America Bull., v. 82, p. 793–802.

Buchanan, J. Y., 1887, On the land slopes separating continents and ocean basins, especially those of the west coast of Africa: Scottish Geog. Mag., v. 3, p. 217–228.

Emery, K. O., 1972, Eastern Atlantic Continental Margin Program of the International Decade of Ocean Exploration (GX-28193) report for period March 1971 to July 1972: Woods Hole Oceanog. Inst. 72-53, 13 p.

Heezen, B. C., R. J. Menzies, E. D. Schneider, W. M. Ewing, and N. C. L. Granelli, 1964, Congo Submarine Canyon: Am. Assoc. Petroleum Geologists Bull., v. 48, no. 7, p. 1126–1149.

Isaacs, J. D., J. L. Reid, G. B. Schick, and R. A. Schwartzlose, 1966, Near-bottom currents measured in 4-kilometers depth off Baja California coast: Jour. Geophys. Research, v. 71, no. 18, p. 4297–4303.

Leyden, R., G. Bryan, and M. Ewing, 1972, Geophysical reconnaissance on African shelf: 2. Margin sediments from Gulf of Guinea to Walvis Ridge: Am. Assoc. Petroleum Geologists Bull., v. 56, no. 4, p. 682–693.

Milne, J., 1897, Sub-oceanic changes: Geog. Jour., v. 10, no. 2, p. 129–146.

Reyre, D., ed., 1966, Bassins sedimentaires du littoral africain, symp., pt. 2, Littoral Atlantiquê (New Delhi, 1964): Paris, Union Internat. Sci. Geol., Assoc. Services Geol. Africains, 304 p.

Shepard, F. P., 1965, Types of submarine valleys: Am. Assoc. Petroleum Geologists Bull., v. 49, no. 3, p. 304–310.

—— and N. F. Marshall, 1973, Currents along the floors of submarine canyons: Am. Assoc. Petroleum Geologists Bull., v. 57, no. 2, p. 244–264.

Veatch, A. C., and P. A. Smith, 1939, Atlantic submarine valleys of the United States and the Congo Submarine Valley: Geol. Soc. America Spec. Paper 7, 101 p.

Editor's Comments
on Papers 6 and 7

   While the controversies over the origin of submarine canyons continued, work went on with increasing enthusiasm and improved equipment in an effort to gather more facts about canyon morphology, the nature of the steep walls, and the sedimentology of the canyon fills. An important step forward was made when the old-fashioned heavy diving suit was replaced by Scuba-diving equipment, and geologists could move freely within the canyon heads, observing, measuring, and taking photographs. One of the pioneer divers in the Californian submarine canyon heads was Robert F. Dill, whose researches were presented in an unpublished Ph.D. thesis (Dill, 1964a). Some of these results, with excellent underwater photographs, were published in the volume dedicated to Francis P. Shepard (Miller, 1964) and this paper is reprinted here (6). Dill sought especially to find out the processes operating in submarine canyon heads, and emphasized for the first time the importance of slow sand creep, "rivers of sand," and localized slumps in moving the fill and eroding the canyon walls. He was able to observe Scripps Canyon Head before and after flushing out on two occasions, and measured slower movements by using marker stakes placed in the fill. Dill could prove that active submarine erosion was taking place by observing the truncation of pholad borings on the walls of the canyon head. Dill, Shepard, and their colleagues at Scripps Institution of Oceanography and the U.S. Navy Electronics Laboratory (now the U.S. Naval Undersea Research and Development Center) have since worked at greater depths using submersibles, increasing significantly our knowledge of the Californian and Baja Californian submarine canyons.

   Paper 7 takes us from the narrow confines of a canyon head to the vast chasms of the world's largest canyons. It is a sequel to those by Gershanovich (1968), who reported on Soviet oceanographic research in the north Pacific between 1958 and 1964 and presented new data on the geomorphology and recent sediments of the Bering Sea, and by Scholl and others (1968), who discussed the geologic history of the continental margin of North America in the Bering Sea. Both these papers draw attention to the very large canyons scarring the 3400-m-high escarpment which separates one of the earth's widest epicontinental seas (the shallow Bering Sea Shelf) from the Aleutian Basin.

The paper reprinted here (7) concentrates on Bering, Pribilof, and Zhemchug Canyons, three of the largest and longest submarine canyons in the world. Bering Canyon, 400 km long, is the world's longest known canyon, with a volume of 4300 km^3, while Zhemchug is the largest with nearly twice this volume, some 16 times bigger than the largest canyons outside the Bering region. They were studied using continuous seismic reflection profiles that crossed the valley floors.

Two of the three canyons described in this paper are unusual in having headward bifurcation of their slope axes to form elongated trough-shaped basins behind the regionally projected position of the shelf edge: these basins are structurally controlled, as are the main canyons. The authors attribute canyon cutting to turbidity currents, slumps, or sediment creep, feeding gently sloping submarine fans with up to 500 m thicknesses of turbidites. Bering Canyon was periodically cut and filled by axial sedimentation during late Tertiary and Quaternary time. Pribilof and Zhemchug Canyons were cut entirely in the Pleistocene.

In this paper, the excellent physiographic diagrams by Tau Rho Alpha give a vivid impression of the scale of these great canyons of the Bering Sea.

An error should be noted on profiles A-B and C-D, Figure 4, and profile E-F, Figure 5. The depth scales for these profiles are inaccurately calibrated to the reflection time scales: a depth of 3000 m should be coincident with a two-way reflection time of 4.0 sec. Figures 7 and 10, interchanged in the original paper, have been put back in the correct positions.

As an Addendum to this paper, David Scholl has kindly contributed a note on the lower of the two "unconformities." As a result of a DSDP drill hole along the eastern flank of Bering Canyon, this surface turns out to be a diagenetic boundary and not an erosional one, and this modifies the authors' views on the cutting of Bering Canyon.

Biographical Notes

Robert F. Dill was born in Denver, Colorado, in 1927. He received the B.S. degree in Geology in 1950 and the M.S. in Marine Geology in

1952 at the University of Southern California and the Ph.D. in 1962 in Oceanography (Geological) at Scripps Institution of Oceanography, University of California at San Diego. For twenty years, Dill worked as Oceanographer (Geological) at the Navy Electronics Laboratory (now the U.S. Naval Undersea Research and Development Center) at San Diego. During this period he was also cofounder of, and consultant in marine geology for, General Oceanographics, Inc., Newport Beach, California, and, since 1968, Research Associate on the faculty of the Department of Geology, State University at San Diego, In 1972, he became Associate Professor at the Department of Oceanography, George Washington University, Washington, D.C., and currently he is Oceanographer (Geological), Science Coordinator, with the Manned Undersea Science and Technology Office, National Oceanic and Atmospheric Administration, Rockville, Maryland. In its first year and a half of operations, Dill has provided program assistance and administrative guidance to over 240 scientists, utilizing eight research submersibles and three underwater laboratories to carry out its key scientific programs. He is also technical consultant on the mining of manganese and phosphorite nodules, offshore sand and gravel, and on the efficient harvesting of precious coral. He has participated in at least 34 deep-sea scientific expeditions in many parts of the world.

As a conservationist, Dill helped to establish the first underwater parks and marine reserves for the State of California and has served on State Boards and Commissions for management of marine and coastal resources. He received the California Resources Agency Golden Bear Award for contributing to the preservation of California's underwater recreational, scientific, and aesthetic resources.

During Scuba and submersible dives, and as a participant in several of Cousteau's *Calypso* operations, Bob Dill has made a number of well-known scientific films. He has published more than 60 scientific papers and is actively affiliated with many professional societies.

David W. Scholl was born in 1934. He received the B.S. in 1956 and the M.S. in 1958 at the University of Southern California and the Ph.D. in 1962 at Stanford University. He carried out research in marine sedimentation at the U.S. Navy Electronics Laboratory, San Diego, transferring in 1964 to the U.S. Naval Ordnance Test Station, where he directed assembly of several high-powered seismic reflection systems for research on the continental margin bordering southern California, the Bering Sea, and flanking the Peru-Chile trench. Scholl joined the U.S. Geological Survey in 1967 and has been involved in geological and geophysical studies in the Bering Sea, over the Aleutian Ridge and adjacent Aleutian Trench, off Kamchatka, the Kuril Islands, and the far northern Pacific. The National Science Foundation appointed Scholl to its JOIDES project Pacific and Antarctic Advisory Panels and he served in

1971 as Chief Scientist aboard the D/S *Glomar Challenger* during drilling operations of Leg 19 in the north Pacific and Bering Sea. In 1968 he received the A. I. Levorsen Memorial Award from the American Association of Petroleum Geologists and was selected to participate in its Distinguished Lecture Tour in 1970–1971. He is the author of more than 50 papers on eustatic changes of sea level, coastal sedimentation, structural and tectonic histories of continental margins, island arcs, and deep-sea trenches.

Edwin C. Buffington was born in Ontario, California, in 1920. He received the B.A. degree in Geology from Carleton College, Northfield, Minnesota, in 1941 and research at California Institute of Technology led to the M.S. degree and at the University of Southern California, where he gained the Ph.D. degree. He has spent 25 years with the U.S. Naval Undersea Center at San Diego and is Head of its Marine Geology Branch. During this period, he has dived as scientific observer in the bathyscaph *Trieste,* the Cousteau *Soucoupe Sous-marine* (diving saucer) and *Deepstar 4000,* and has been pilot-observer in 52 dives of the two-man submersible *Nekton,* owned and operated by the consulting firm of General Oceanographics, Inc., of which he is a director. A Scuba diver since 1953, Buffington has been the leader of several deep-sea oceanographic expeditions and has published numerous scientific articles on bathymetry, depositional and erosional processes on the continental shelf and slope, and the geology and tectonics of the Aleutian–Bering Sea region. During World War II he served in the U.S. Navy as a photo-intelligence officer in the central Pacific, and is presently a Commander in the Naval Reserve.

David M. Hopkins was born in 1921 and received his professional education at the University of New Hampshire (B.S. 1942) and Harvard University (M.S. 1948, Ph.D. 1955). He joined the U.S. Geological Survey in 1942 and has remained with the Survey ever since, except for a two-year period during which he served in the Army Air Force in Alaska and the Aleutian Islands (1944–1946), three months as Visiting Professor of Geomorphology at Stanford (1961), and a three-month Academy of Science Exchange Visit to the Soviet Union (1969). For two years, Hopkins was program supervisor of the Alaskan Terrain and Permafrost Section and for five years, organizer and head of the Geological Survey's marine program in the Bering Sea. He is currently a senior research geologist in the office of Marine Geology, U.S. Geological Survey, Menlo Park, California. Hopkins has served on a number of committees and commissions concerned with the evolution of slopes, periglacial morphology, Quaternary shorelines, the tundra biome, geologic names, and the Glacial Map of Alaska. He is the author of about 80 scientific papers on Quaternary chronology, biostratigraphy, paleogeography, paleoecology, paleoclimatology, and archaeology of northern regions. In 1968 he received the Kirk Bryan Award of the Geological Society of America for his article "Quaternary marine transgressions in Alaska," which appeared in *The Bering Land Bridge,* which he edited.

Tau Rho Alpha was born in 1939. He received his education at San Fernando Valley State College (B.A. 1962) and is engaged in independent graduate work at Hayward State College. Joining the U.S. Geological Survey in 1962, as a Science Technician, he has concentrated on the preparation of marine maps and charts. In 1969 he was appointed professional Cartographer. Tau Rho Alpha is experienced in maps and charts of all kinds but his main interest is in the representation of three-dimensional surfaces, especially the design and construction of physiographic diagrams of the ocean floor. He has supervised the building of an isometrograph which accurately displaces contours planimetrically, and has published several papers and 14 physiographic diagrams portraying ocean floor morphology and illustrating sub-sea geologic features.

6

Copyright © 1964 by Macmillan Publishing Co., Inc.

Reprinted with permission from *Papers in Marine Geology, Shepard
Commemorative Volume*, R. L. Miller, ed., Macmillan, New York, 1964, pp. 23–41

Sedimentation and Erosion in Scripps Submarine Canyon Head

by ROBERT F. DILL

U. S. NAVY ELECTRONICS LABORATORY, SAN DIEGO, CALIFORNIA

Introduction

THIS PAPER describes geological observations made in the Sumner Branch of the Scripps Submarine Canyon off La Jolla, California, which is in turn a branch of the large La Jolla Canyon emptying into the San Diego Trough at about 650 fathoms. The observations were visual and *in situ*, thanks to the use of SCUBA, which permits routine dives for scientific purposes down to 160 feet.

The observations reported here were mostly made on December 5, 1959, March 24, 1960, and during the concentrated period from January 1 to March 6, 1961. I was accompanied by U. S. Navy Electronics Laboratory diver–scientists J. A. Beagles, R. S. Dietz, J. Houchin, and W. Speidel, and by Earl Murray of the Scripps Institution of Oceanography. Without their help this work could not have been accomplished. Our primary interest was to examine the sedimentary and organic debris collected above and in the canyon head, determine if there is movement in the sediment mat they form after deposition, and look for evidence of submarine erosion that could be caused by such a movement. We hoped to throw light on the process whereby this material is emplaced and later flushed out, and the possible relation of this process to canyon erosion. For more than 20 years, Shepard (1949) has demonstrated the reality of this filling and flushing process at this canyon; sediment is trapped by the canyon head and gradually accumulates throughout the year due to the littoral drift of sand down the coast. Chamberlain (1960) has reported that on average about 1.8×10^5 cubic meters per year of sediments are lost into the head of the Scripps Canyon each year. This is roughly equivalent to the estimated littoral sand transport into the area. At the time of our December 1959 dive, the sedimentary accumulation had not been completely flushed for about one year. But Chamberlain (1960) subsequently found by an echo-sounding survey that the head did flush out seven days after our dive – i.e., on December 12-13, 1959. The removal occurred during a severe winter storm accompanied by unusually large storm waves. During the storm, large plumes of turbid surface water mixed with kelp and organic detritus extended at least a quarter-mile offshore above the canyon.

The dives in March 1960 confirmed the flushing out of the canyon. Exposed in the axis of the canyon were areas of rock that had been filled with sediment at the time of the December dives. In fact, the fill seemed to have been almost entirely flushed.

During the observations of January 1 to March 6, 1961, marker stakes were placed in one of the tributaries of Sumner Canyon and observed to move slowly down slope. During the two weeks prior to a mass removal of material from this tributary, this movement accelerated and was accompanied by a series of minor slumps. These resulted in as much as 1 foot of sediment being removed from the top of the sediment mat. The stakes, which were originally placed in an upright position, tilted down slope until in one spot they were 85 degrees from their original vertical position, thus showing a differential movement in the mat. The sedimentary mat attained almost vertical slopes just after slumping at the head of the slump scars. The intertwined nature of the grass-kelp sand mixture of the mat gave it the necessary strength to hold such slopes. Normally, slopes were of the order of 25 to 35 degrees from horizontal. On the week-end of March 4-5, 1961, after a two-week period of high swell and storm conditions, the canyon, as in 1960, completely flushed itself of all the sediment in the area of the marker stakes. The entire sediment fill (over 17 feet at the position of the marker stakes) was removed down to the bedrock over the two-day period. Other areas of the canyon were not affected by this slump and remained full of sediment up to April 13, 1961, when the last check dive was made in this area. These observations indicate that the slumps are localized and not of a catastrophic nature as indicated by Chamberlain (1960).

Techniques and Oceanographic Conditions

SEVERAL NEW techniques were used in the observations reported here. A portable tape recorder housed in an underwater case was used to immediately record pertinent observations made during the course of a dive. An underwater

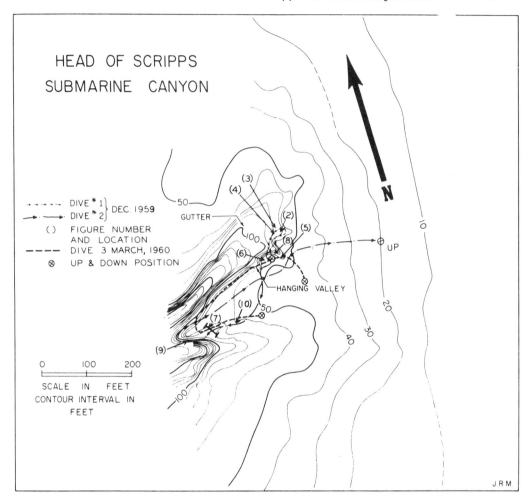

FIG. 3.1

Locations of the origins of accompanying illustrations, and designations of tracks taken by divers. Numbers correspond to illustrations; dashed lines distinguish individual dives. Base chart is from soundings by F. P. Shepard as drawn by J. P. Moriarty, Scripps Institute of Oceanography. The number (2) shows the origin of Figure 3-2, the number (3) shows the origin of Figure 3-3, etc.

navigation board, which contained a stopwatch, compass, depth gauge, and space for writing notes, was also useful for mapping and recording data. A Hasselblad camera with a "super-wide-angle" lens (90 degrees in air; 38-mm focal length) was used for photography. Photographs shown in this paper were obtained using Kodak Royal-X film, a lens opening of f 4.5, a shutter speed of 1/100, and natural lighting. Figure 3-1 shows the sites of the other illustrations used in this paper, and the routes covered by the dives of December 1959 and March 1960.

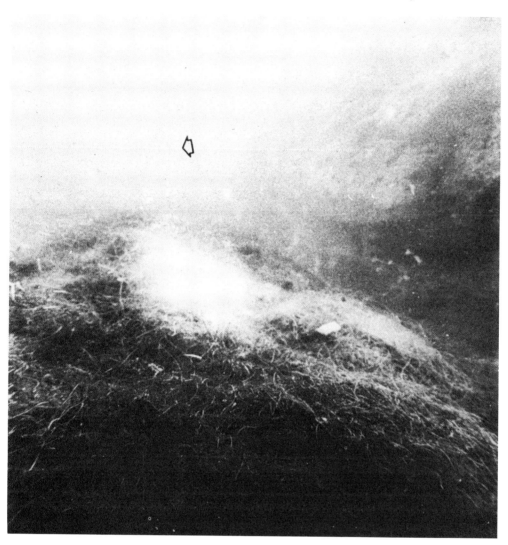

FIG. 3.2

Gas bubbling from the organic-sediment mat (see arrow) found in the axis of the canyon. Note the interwoven nature of the sea grass, which overlies a layer of loosely packed fine micaceous sand. The white area around the gas seep in a precipitate encrusting the organic portion of the mat. (Depth of photo 70 feet.)

U.S. NAVY PHOTO, by Dill

The oceanographic conditions prevailing during the December 1959 dives were extremely favorable for taking photographs. Horizontal visibility was in excess of 65 feet above the thermocline, which was at 80 feet. Below the thermocline the horizontal visibility decreased to about 20 feet. This deeper water was murky and whitish because of extremely fine suspended particles. Apparently, this was a colloidal suspension, possibly derived from the sediment mat. The currents, both along the bottom and in water away from the bottom, were practically nonexistent below 50 feet. Above this depth, a very slight oscillatory motion was set up by passing swells.

The visibility during the March 1960 and January to March 1961 dives was less than 10 feet everywhere. No useful photographs could be taken.

Sedimentary Mat

OVER THE past seven years, dives made by the late Conrad Limbaugh of the Scripps Institution of Oceanography and by me have shown that an accumulation of sedimentary and plant material forming a cohesive mat is usually found in the canyon head. This mat is composed primarily of fine sand and silt interbedded and mixed with broken pieces of sea grass and kelp. Its structure and composition has been described by Shepard (1949), Limbaugh and Shepard (1957), and Chamberlain (1960). During the time of the dives reported here, the organic portion was decomposing, forming gas which was actively bubbling to the surface (see Figure 3-2). Decomposition was associated with a strong hydrogen sulfide odor. Immediately around the gas seeps (Figure 3-3) are large white aureoles; upon close examination, the discoloration appears to be caused by a powdery deposit on organic material incorporated in the mat.

These white deposits are quite common in other places off southern California where gas seeps due to organic decay are found. The hydrogen sulfide and blackened sediment around the gas seeps indicate a reduced state. Small depressions are present along the contact between the rocky canyon and sediment mat. One of these (near the site of Figure 3-3) had a depth of about 2 feet, a width of about 1 foot, and a length of about 2 feet. In it the hydrogen sulfide odor was especially strong, and a yellowish discoloration was present on the rocks and the mat enclosing the depression.

In the area of active bubbling, the mat was loosely compacted so that a diver could push his arm deep into the material with very little effort. The material is, however, quite cohesive because of the interwinding of the long blades of surf grass and kelp (and it can be pulled away from the bottom in large chunks). In some places the mat gives the appearance of having shrunk in overall size subsequent to deposition. It seems thereby to have moved away from the canyon wall, leaving small depressions along the rock-to-mat contact. Additional evidence is the remnant mat fragments stranded on ledges higher than the general level of the existing mat (Figures 3-4 and 3-5). The formation of the sediment mat appears to take place only in the channel depressions at the extreme heads of the canyon, because except in these areas there

FIG. 3.3

Oblique view of the organic-sediment mat. Arrow points to an area of an active gas seep. Note that the mat is higher in the center than at the rock wall–mat contact and that the sediment layer underlies the area of predominate sea grass. (Depth of photo 75 feet.)

U.S. NAVY PHOTO, by Dill

is no mat-like material on the shallow shelf around the lip of the canyon. Therefore, it appears that the only way the canyon can be furnished with fill material is by the slow movement of the sedimentary mat from these areas of origin. Rapid movements and flushing would be indicated by large changes in bottom contours, a phenomenon that periodic echo-soundings indicated had

FIG. 3.4

Stranded organic-sediment-mat material on a ledge above the general level of the mat immediately adjacent to it. Mat material is not present on ledges above the one indicated by the arrow, nor on the lip of the canyon which is approximately 15 feet above this location. (Depth of photo 80 feet.)

U.S. NAVY PHOTO, by Dill

not taken place for at least one year prior to the dives made in 1959. A plausible explanation of the "nonfit" of the mat in the canyon is that during the compaction, which can be expected to take place in this type of mixed material as it decomposes and settles with time, the overall mat shrinks and is gradually pulled away and down (by gravity) from the rock wall which originally acted as its boundary.

FIG. 3.5

Stranded organic-sediment mat material in a tributary to the Sumner Branch of Scripps Canyon (arrow). The depressions at the contact between the mat and rock wall in some places contain yellowish water and gives off a strong odor of hydrogen sulphide. Note that the slope of the rock wall changes from almost vertical to approximately 25 degrees above the diver's hand. Below the sediment, shown here at its December 1959 level, the rock walls overhang and are much smoother. (Depth of photo 65 feet.)

U.S. NAVY PHOTO, by Dill

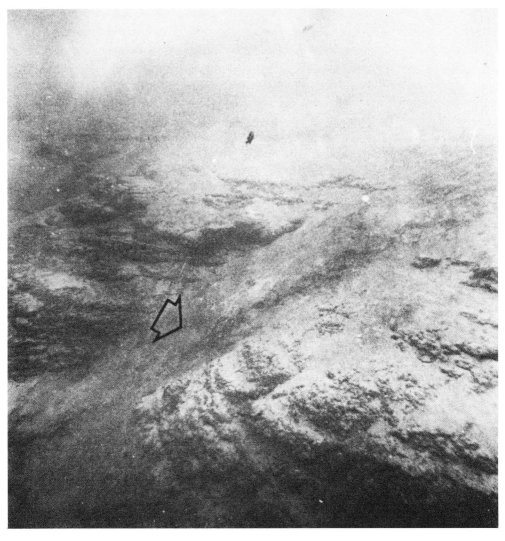

FIG. 3.6

Sand chute cut in the upper lip of the canyon. When the sediment is removed from these features, a smooth U-shaped rock channel is found. By following a rock bed into the area of the chute it was found that erosion is currently taking place, because the holes made by marine boring organisms are truncated whereas they are not eroded in the same bed on the sides of the chute. (Depth of photo 60 feet.)
U.S. NAVY PHOTO, by Dill

Proof of this slow and downward movement is the displacement of stakes placed in tributaries at the head of the canyon, and the movement of lobster-traps, man-made debris, etc. into deeper water when incorporated in the mat during its formation at the head of the canyon. As the mat material moves out of tributaries, it acts as a cohesive unit and is no longer able to fill the

86

entire bottom of the canyon. It can therefore be expected that depressions will develop, similar to the ones observed, at the rock-mat contact at the walls of the canyon.

Sand chutes of various sizes, ranging from very small rills a few inches across to some 8 to 10 feet across, cut into the rocky lip of the canyon and lead down to the sedimentary fill. They are filled with fine sand which contains very little broken kelp material (Figure 3-6). The well-developed sand chutes are U-shaped and smoothly worn. These clean-sided channels contrast markedly with the sea-growth-covered areas immediately adjacent to them.

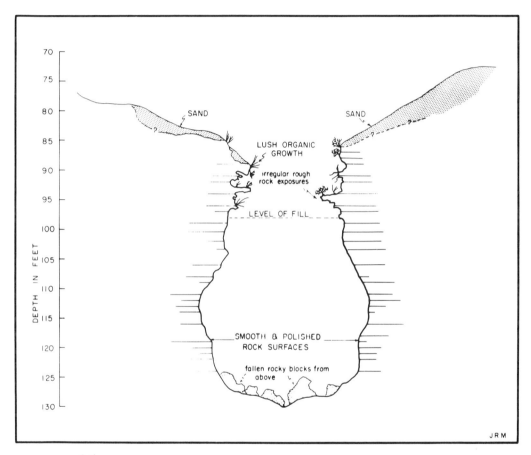

FIG. 3.7

Schematic cross-section of a typical canyon tributary at its seaward-most extremity, showing the undercutting and enlargement of the lower section that is normally filled with an organic-sediment mat. The blocks that periodically fall into these tributaries slowly move into deeper water. The notation "level of fill" indicates the approximate amount of sediment these features hold before the sediment becomes unstable and moves into deeper water. Drawing is by J. R. Moriarty, Scripps Institution of Oceanography.

Holes or burrows caused by boring organisms, such as the pholads, are truncated and smoothed, indicating active rock erosion has taken place within the period of existence of these animals. It seems, therefore, that these chutes are being presently eroded by flowing sand. Underwater movies, by W. North and C. Limbaugh (1960), and by North (1960), have caught such sand flows in action at Cape San Lucas Canyon on the tip of Baja California.

Further evidence of submarine erosion in the Scripps Canyon is found at the bottoms of the hour-glass-shaped tributaries that incise the rock walls along three branches of the canyon. These, in general, have wider cross-sectional areas at their bases than half-way up. Typically, above this half-way area the rock lip is covered with shelf sands, and here the typical such tributary expands in cross-section again (Figure 3-7). This shape suggests the existence of an erosional process capable of abrading rock from the wall of the tributary at levels where the tributary is usually filled with sediment. It is therefore necessary that such a process, if it does indeed exist, be an erosional process that can act when the tributary is full of sediment.

I do not wish to contend that this is the primary way in which the canyon heads were originally formed. There is too much evidence that they were initially cut subaerially during past periods of lower sea level (Shepard 1952). It does appear, however, that their original subaerial shapes are being modified considerably (by sand abrasion and sediment-mat movement) from the typical V-shape of subaerial canyons found on land shoreward of this submarine canyon. The polished rock walls (Figure 3-8) that have been observed by the writer over the past five years when the axis of the canyon was not filled with sediment, the smoothed irregularities, and the down-slope movements of large blocks of rock that have fallen into the tributaries (Murray, personal communication, 1960) tend to support this concept. Although one cannot entirely discount the possibility that the present topographic forms are preserved shore-zone features that originated during a late glacial lowering of sea level, the presence in minor erosion features of dissected borings of marine organisms that are today living in adjacent rock gives strong evidence for submarine origin of this erosion. In addition, I have not been able to find similar rock–sand chutes cut into the lips of the subaerial canyons of the region.

The sand chutes are small channels, through which sediment from the relatively flat sand areas surrounding the canyon is fed into tributaries or directly into the main canyon. Care must be taken not to confuse them with the tributaries, which are larger features. There are at least six tributaries in the Scripps Canyon head. These often appear as small hanging valleys more than 10 feet wide and 10 feet deep. In some instances, this lack of concordance with the main canyon appears to be due to differential rock hardness, permitting a more rapid erosion in the canyon than in the tributary. However, this is not always the case; quite possibly the construction is due to more material

FIG. 3.8
A tributary that leads into the main part of the canyon. The arrow points to the uppermost part of the beginning of the smoothed rock that underlies the sedimentary fill. One week from the time this photo was taken (Dec. 5, 1959), there was a removal of at least 4 feet of material from this location. (Depth of photo 70 feet.).
U.S. NAVY PHOTO, by Dill

moving down the canyon than down the tributaries, and greater loading of sediments on the bottom of the slow-moving mat that usually occupies the canyon axis.

There are at least four topographic forms that furnish routes for sediment to enter into the main canyon:

(1) Wide sand-depressions exist at the shallow shoreward end of the canyon. No bed rock is exposed in these areas, and their shape is controlled by the physical properties of the sediment. These lead directly into the tributaries and are areas where sea grass, algae, and sediment are mixed together to form the organic sediment mat.

(2) Small sand chutes, which are channels cut in the rock lip of the canyon or tributaries, permit the flow of sand moves into the chute areas by littoral drift parallel to the beach along the shelf and seaward of the mat-forming areas.

(3) Rock-walled tributaries are cut back into bedrock. In many places these tributaries have hour-glass-shaped cross-sectional profiles at their seaward extremities just before they enter the canyon. They are thus in many places widest – and therefore most highly eroded – in areas usually covered with sediment.

(4) The main channel, or the branch of the canyon that leads directly into the deep-sea, is, of course, a point where sediment can enter the canyon system.

Turbidity Currents

VERY LITTLE erosion of bottom sediment or rock slopes is caused by sediment in suspension (turbidity currents) at the head of the canyon. Instead, erosion is accomplished through abrasion and plucking by the sediment mat as it moves slowly down the canyon. This must be like glacier erosion. The smooth and polished bottoms observed in sand chutes and rock tributaries after the periodic removal of their sediment covers indicate that the movement of sand gradually grinds and erodes away the rock wall. This appears to me to be a more plausible explanation for the submarine erosion occurring in the upper parts of canyons than the concept of high-velocity turbidity currents, the latter of which have never been observed during hundreds of dives made in this area under many different conditions.

Differential Erosional Features

INTERBEDDED SHALES and sandstones form the walls of Scripps Canyon. Succeeding beds have different hardnesses, so when they are eroded, series of undercut ledges of hard sandstone or of silicious nodules crop out along the

FIG. 3.9

Overhanging ledges on the seaward wall of the junction of the two major branches of the canyon. Note the lush organic cover on the resistant sandstone that forms the protruding ledges and the clean appearance of the sand trapped on them. This wall is characteristic of the area and drops off to a depth of 225 feet before the organic-sediment mat is reached. (Depth of photo 110 feet.)

U.S. NAVY PHOTO, by Dill

steep sides of the canyon (Figure 3-9). In several areas, these ledges have broken off to form tali. Similar features called "blocs" have been observed in the submarine canyons off the southern coast of France (Nesteroff, 1958).

Organisms on the Rock Walls

WHERE THE rock walls are not cut by the sand chutes and are not periodically covered by the organic sediment mat, they are covered by abundant benthonic organisms (Figure 3-9), which continually dig in and weaken the rock walls. These organisms play an important role in making the rock easily erodable by other submarine processes. They also aid us in determining past levels of sedimentary fill in the canyon, because whenever the sedimentary fill is recently flushed (within a week or so), they leave a record of the level of past filling; the area formerly under the sedimentary fill is completely barren of organisms while that above contains a lush growth.

Organic Layers on the Surface of the Sediment

A FINE brown organic layer, a few mm thick, was present on the sediment found in the areas above the lip of the canyon during the December 1959 dives. E. W. Fager, D. L. Inman, and Earl Murray (personal communication) report this layer to be composed primarily of dead diatomaceous debris. This deposit disperses when disturbed by even the slightest current. Because of this extreme susceptibility to even the slightest motion of the overlying water, it follows that this layer must have been deposited by some means other than a slow particle-by-particle settling to the bottom; possibly flocculation has occurred. The deposition must also have been quite rapid, because as pointed out by the diver in Figure 3-10, the outline of a large ray is preserved in it. Presumably the ray was lying on the bottom when the material dropped out of suspension and formed the layer. Upon swimming away, the ray left its imprint on the bottom as a barren area not covered by the organic layer, although it is hard to see how he swam away without disturbing the layer. This organic material has been observed in many other areas in the southern California region during quiescent periods when no bottom currents were active. Its variant presence and absence, along with its patchy distribution, may be an important factor in explaining the extreme variability in organic carbon measurements of the surface sedimentary layers often encountered in coastal areas.

Conclusions

THE SCRIPPS Submarine Canyon has been studied for a number of years,

FIG. 3.10
R.S. Dietz points out the outline of a large ray in the oxidized organic layer that formed a thin film on the sand at the edge of the canyon. The small light patches. form when burrowing organisms cast material out of their holes from below, or when currents set up by the divers' fins hit the bottom. (Depth of photo 80 feet.)

Shepard and Emery (1941), Shepard (1949, 1952), and Chamberlain (1960). In view of the completeness of this prior work, it would be pretentious to go into the origin or the cause of submarine canyons in this short paper. The photographs taken at the head of Scripps Submarine Canyon during the dives discussed here are, however, believed to be valuable in pointing out to those who may not be divers the topographic forms and some of the environmental conditions that may prevail at the head of a canyon. Other canyon heads along the California coast present quite different pictures in that nearly all of them are, at diveable depths, cut in sediment.

Where found in a canyon, rock appears to be undergoing erosion by the movement of sediment down sand chutes and through tributaries leading into the main axis of the canyon. The hour-glass shapes of several tributaries, the polished and smoothed irregularities of the rock walls, the movement and smoothing of irregularities of fallen blocks of bedrock, the eroded and truncated holes of marine boring organisms, and the lack of organisms in areas that are only periodically covered with sediment are evidence of sediment movement; the writer believes that such sediment movement leads to erosion of the rock walls. The filling and rapid flushing by sand slides (one to two per year), and the slow movement of sedimentary material from the canyon heads in the form of an organic sediment mat are proposed as being the process that is now causing erosion of the rock walls of Scripps Canyon. The rapid refilling of the axes of tributaries by sand and organic material after flushings would seem to protect the underlying rock surfaces from turbidity-current erosion except during the brief period (one or two weeks) following flushing. Although not observed in the shallow portions of the Scripps Canyon, turbidity currents may be active further down the canyon where it empties upon the sedimentary fan in the San Diego Trough. Turbidity currents set up by the divers on the lips of the canyon and in the tributaries do not appreciably erode sedimentary material. This was shown by the inability of these currents to remove or even modify the small, fragile burrowing features formed in the sedimentary fill in the sand chutes and tributaries. The turbidity currents created by divers did not rapidly accelerate or erode the bottom sediment as the currents moved down the canyon, even when the currents were started on slopes of as high as 33 degrees; instead, the activity started by the divers gradually spread and dissipated (Figure 3-11). Buffington (1961) has had similar results with experimentally produced turbidity currents on the sea floor. The net result in such studies has been to make the water more turbid in the area of investigation, and by the removal of the fine-grained portion of the sediment to bring about a better sorting of the redeposited, coarser fraction of the original material.

The sedimentary slopes during the time of the first investigation reported here (December 1959) were stable and appeared to be at equilibrium with the prevailing conditions, yet only one week later a phenomenon took place that removed all the sedimentary fill of the canyon and exposed bare rock in all areas observed.

There appears to be submarine erosion at the present time in areas that are normally covered with sediment, and such an erosional process must take place while the canyon is filled with sediment. The grinding action of the sedi-

FIG. 3.11
Clouds of turbid sediment resulting from divers trying to start a turbidity-current-type flow down a steep-sloped sand channel that leads into a small tributary. Although the sediment could be made to slump for short distances (less than 20 feet), the clouds did not erode even the very fragile worm tube casts made by burrowing organisms or disturb the fine brown organic layer noted in Figure 3-10. The area here depicted is referred to as "Hanging Valley" in Figure 3-1. (Depth of photo 85 feet.)

U.S. NAVY PHOTO, by Dill

ment contained in the mat during its *slow* movement down slope as it consolidates is proposed by the author as such a process.

References

Buffington, E. C. (1961), "Experimental Turbidity Currents on the Sea Floor": *Bull. Am. Assoc. Petrol. Geol.,* **45**, 8.

Chamberlain, T. K. (1960), *Mechanism of Mass Sediment Transport in the Scripps Submarine Canyon, California:* Ph.D. thesis on file at the Scripps Institution of Oceanography, University of California, La Jolla, California.

Inman, D. (1953), *ONR Quarterly Progress Report:* Scripps Institution of Oceanography, University of California, La Jolla, California.

Limbaugh, C. and Shepard, F. P. (1957), "Submarine Canyons": *Geol. Soc. Am. Mem.* 67, **1**, pp. 632-639.

Nesteroff, W. D. (1958), *Recherches sur les sediments marine actuels de la region d'Antibes,* Ph.D. thesis, Faculte des Sciences de l'University de Paris, Paris, France, pp. 40-44.

North, W. J. (1960), "Fabulous Cape San Lucas": *Skin Diver Magazine,* **IX**, 5, pp. 24-26.

――― and Limbaugh, C. (1960), *Rivers of Sand:* Underwater motion picture report of submarine sand movement, on file at Department of Submarine Geology, Scripps Institution of Oceanography, University of California, La Jolla, California.

Shepard, F. P. (1949), "Terrestrial Topography of Submarine Canyons Revealed by Diving": *Bull. Geol. Soc. America,* 60, pp. 1597-1612.

――― (1952), "Composite Origin of Submarine Canyons": *Jour. Geol.,* 60, 1, pp. 84-96.

――― and Emery, K. O. (1951), "Submarine Topography off the California Coast: Canyons and Tectonic Interpretations": *Geol. Soc. America Special Paper 31,* 171 p.

7

Copyright © 1970 by Elsevier Publishing Co.

Reprinted from *Marine Geol.*, **8**(3/4), 187–210 (1970)

THE STRUCTURE AND ORIGIN OF THE LARGE SUBMARINE CANYONS OF THE BERING SEA*

DAVID W. SCHOLL[1], EDWIN C. BUFFINGTON[2], DAVID M. HOPKINS[1] AND TAU RHO ALPHA[1]

[1] *Office of Marine Geology, U.S. Geological Survey, Menlo Park, Calif. (U.S.A.)*
[2] *Seafloor Studies, Marine Geology Branch, Naval Undersea Research and Development Center, San Diego, Calif. (U.S.A.)*

(Received July 17, 1968)

SUMMARY

Three exceptionally large and long submarine canyons — Bering, Pribilof, and Zhemchug — incise the continental slope underlying the southeastern Bering Sea. Bering Canyon, the world's longest known slope valley, is approximately 400 km long and has a volume of 4,300 km³. The volume of Pribilof Canyon is 1,300 km³ and that of Zhemchug is 8,500 km³; Zhemchug Canyon may well be the world's largest slope valley; most other large submarine canyons have volumes less than 500 km³. Pribilof and Zhemchug canyons are further distinguished by the headward bifurcation of their slope axes to form elongated trough-shaped basins behind the regionally projected position of the shelf edge. These troughs are superimposed over structural depressions formed by down-faulted basement rocks of Mesozoic and older ages. Prior to canyon cutting these depressions were filled with as much as 2,600 m of shallow-water diatomaceous, tuffaceous, and detrital sediments largely of Tertiary age. Deposition of these sediments took place concurrently with general margin subsidence of at least 2,000 m.

The data and conclusions presented in this paper stress that the location, trend, and shape of the enormous submarine canyons cutting the Bering margin are structurally determined. However, axial cutting and headward erosion within the relatively unconsolidated Tertiary strata and the older, lithified basement rock is thought to have been caused by basinward-sliding masses of sediment; these unstable sediment bodies accumulated on the upper continental slope and outer shelf, probably near the mouths of major Alaskan rivers.

Bering Canyon was periodically cut and filled by axial sedimentation during Late Tertiary and Quaternary time. Pribilof and Zhemchug canyons, however,

---

* Publication authorized by the Director, U.S. Geological Survey.

are thought to have been excavated entirely during the Pleistocene. It is presumed that, during one or more periods of glacially lowered sea level, the Kuskokwim and Yukon rivers emptied into or near the heads of Pribilof and Zhemchug canyons. The enormous size and unusual shape of Zhemchug Canyon resulted from the breaching of the seaward wall of an outer-shelf basement depression and the subsequent removal of nearly 4,500 km^3 of Tertiary deposits filling it.

INTRODUCTION

The continental slope that marks the northeastern boundary of the deep Bering Sea or Aleutian Basin is scarred by numerous submarine canyons[1]. Although this is a common physiographic characteristic of many continental margins, the Bering slope is distinguished by the enormous size and length of several of its canyons (KOTENEV, 1965). The physiography, structure, and origin of three of these canyons — (1) Bering, perhaps the world's longest canyon; (2) Pribilof; and (3) Zhemchug, possibly the world's largest submarine canyon — are described in this paper.

Also presented are physiographic diagrams of these canyons. These were constructed from published bathymetric maps (NICHOLS and PERRY, 1966; SCHOLL et al., 1968) by setting up a two-point perspective grid. The bathymetric contour charts were plotted on this grid and then planimetrically displaced so that the vertical scale was 10 times the horizontal. On this displaced contour map the physiography is illustrated by the use of hachure lines. A three-dimensional effect is enhanced by varying the direction, thickness, and density of the hachure lines. The steeper the slope, the steeper the hachure line. The thickness and density of hachure lines corresponds to lights and shadows cast from a light in the upper left hand corner of the drawing. Where appropriate, the texture of the hachure lines was changed to illustrate outcrops of certain rock units.

GEOLOGIC SETTING

The continental margin (i.e., outer-shelf, continental slope, and continental rise) of the deep Bering Sea trends northwestward from Unimak Island, a short distance west of the tip of the Alaska Peninsula to near Cape Navarin, a promontory of northeastern Siberia (Fig.1). The margin is approximately 1,300 km in length, 3,400 m high, and lies as far as 550 km from the nearest continental shore. This escarpment separates two exceptionally broad and flat areas of the

---

[1] The expression "submarine canyon" is used in the sense suggested by SHEPARD (1965).

Fig.1. Physiographic setting of Bering, Pribilof, and Zhemchug canyons in relation to Aleutian Basin, Bering Sea shelf, Alaska Peninsula, and major drainage systems of western Alaska. Dashed lines are line of seismic reflection profiles shown on Fig.4, 5, 7 and 10.

earth, the Bering Sea shelf to the north and the Bering abyssal plain (Aleutian Basin) to the south.

Continuous seismic reflection records and rock samples from submerged outcrops show that the margin is constructed of three principal structural-stratigraphic units: (*1*) a basal unit of thoroughly lithified rock — *acoustic basement;* (*2*) an overlying sequence of stratified deposits — *main layered sequence;* and (*3*) a younger sequence of stratified sediments forming the continental rise at the base of the slope — *rise unit* (Fig.8, 11). Details of the structural and stratigraphic relations between these units have been given elsewhere (SCHOLL et al., 1966, 1968). The acoustic basement is in part constructed of folded and faulted mudstone and siltstone containing thin (1–2 cm) interbeds of graded sandstone that have the composition of a volcanic graywacke. These rocks contain a Late Cretaceous (Campanian) foraminiferal fauna and prisms of the pelecypod *Inoceramus.* Beneath the outer shelf and upper continental slope, the surface of this acoustic unit is thought to be an erosional unconformity that has been downwarped toward the Aleutian Basin.

Subsidence of the rocks of the acoustic basement evidently began in Early or Middle Tertiary time in conjunction with the deposition of the overlying main layered sequence, a sequence of poorly consolidated diatomaceous siltstone, tuffaceous and terrigenous siltstone, and sandstone strata. These strata are largely of Miocene, Pliocene, and Early Pleistocene age, but the basal part of the sequence may include rocks of Paleogene age (HOPKINS et al., 1969)[1]. Erosion, mainly in the form of canyon cutting, has removed much of the main layered sequence from the outer shelf and upper slope during later Quaternary time, exposing outcrops of the acoustic basement (SCHOLL et al., 1966). Sediment removed from these canyons and sediment transported along their axes by turbidity currents, slumps, or possibly by a slower-moving mechanism of sediment creep (SHEPARD and DILL, 1966, p.333), have in part accumulated at the base of the slope to form large, gently sloping, submarine fans. Deposits forming the fans constitute the rise unit, which is constructed of a sequence of turbidites as much as 500 m in thickness.

BERING CANYON

Physiography

Bering Canyon (or Aleutian Valley, KOTENEV, 1965) is nearly 400 km in length and may be the longest submarine canyon in the world (SHEPARD, 1948, p.216; SHEPARD and DILL, 1966, p.198). A physiographic diagram of this extraordinary submarine feature is shown in Fig.2.

---

[1] Older Tertiary rocks (Oligocene or possibly Eocene or Paleocene in age) have been recovered from the continental slope in and near Zhemchug Canyon, but the position of these rocks relative to the acoustic basement is uncertain; they may form part of the basement complex (HOPKINS et al., 1969).

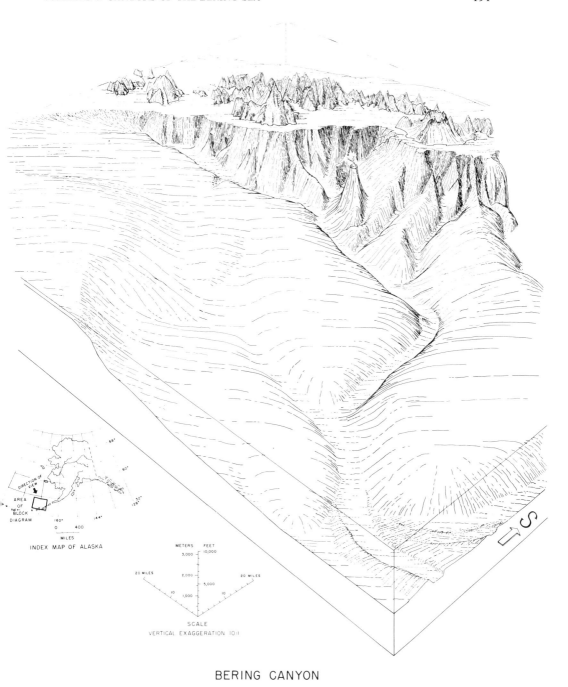

BERING CANYON

Fig.2. Physiographic diagram of Bering Canyon; view is to the southeast toward the Aleutian Ridge. Bathymetric control based upon charts published by NICHOLS and PERRY (1966) and SCHOLL et al. (1968); two-point perspective drawing by Tau Rho Alpha.

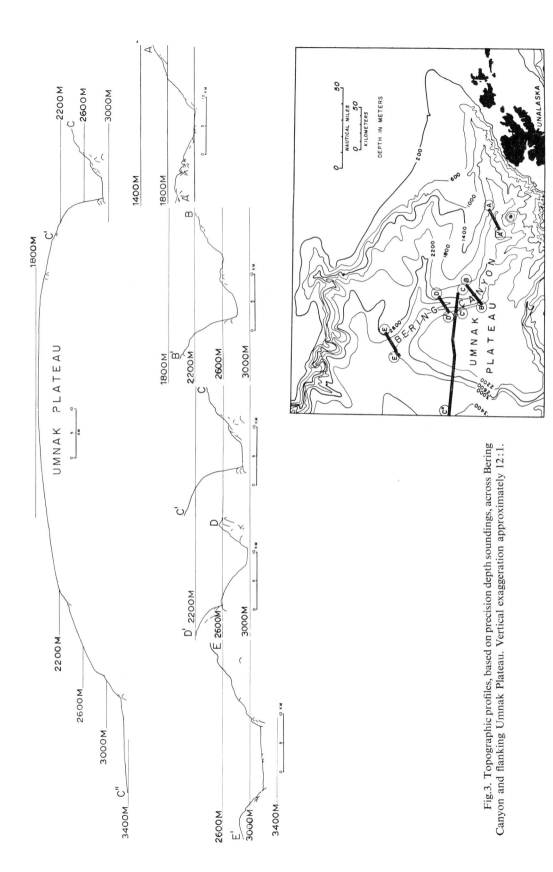

Fig.3. Topographic profiles, based on precision depth soundings, across Bering Canyon and flanking Umnak Plateau. Vertical exaggeration approximately 12:1.

102

Bering Canyon heads in a large, amphitheatre-shaped scar at the outer edge of the shelf near Unimak Island. The trend of the initial 200 km of the canyon is southwestward along the line of juncture formed by the intersection of the base of the Bering slope and the northern wall of the Aleutian Ridge. A number of tributary valleys enter the canyon from the ridge. At Bogoslof Island (Fig.1), which is an active volcano rising from the base of the ridge about 40 km north of Umnak Island (BYERS, 1959), Bering Canyon begins a gradual turn to the north-west. 100 km west of Bogoslof, at a depth of 2,800 m, the canyon turns abruptly northward and runs below the eastern rim of Umnak Plateau. This broad platform in the southeastern corner of the Bering Sea is underlain by at least 3,000 m of subhorizontal beds of the main layered sequence; the plateau has been geomorphically isolated from the base of the continental slope by the formation of Bering Canyon. The canyon's cross-sectional profile (Fig.3) is asymmetrical along its path adjacent to Umnak Plateau; the seaward, or plateau side, has slopes that locally may be as steep as 13°, while the eastern or continental flank is much gentler with 2–3° slopes. The floor of the canyon is generally flat (Fig.3) and from 3 to 4 km in width (KOTENEV, 1965). The northward swing of Bering Canyon persists for a distance of about 70 km and its thalweg drops from a depth of 2,800 to about 3,100 m — an average axial slope of 0°15', or 1/233. At this depth Bristol Canyon joins Bering Canyon, which then turns sharply to the northwest. Bering Canyon finally debouches on to the continental rise at a depth near 3,200 m. This point is only 30 km south of the mouth of Pribilof Canyon, which heads back of the shelf edge nearly 300 km northwest of the head of Bering Canyon (Fig.1). Near its mouth the floor of Bering Canyon is flat and approximately 11.5 km in width; it is flanked by a large natural levee (Fig.5, profile G–H). The volume of the Bering Canyon, computed as material missing from the continental margin, is approximately 4,300 km³. In comparison, most other large submarine canyons have volumes less than 500 km³.

Structure

The relationships between Bering Canyon and the structure of the rocks that form the canyon walls can be deduced from a study of continuous seismic reflection profiles that cross the valley floor. Some of these profiles have been presented in an earlier paper (SCHOLL et al., 1968); others, with added geological interpretation, are given in Fig.4 and 5.

Profile A–B (Fig.4), which crosses the canyon obliquely at the base of the Aleutian Ridge (Fig.1), indicates that the canyon axis coincides with a gentle synclinal depression in the lower strata of the main layered sequence at the base of the ridge. The lower unit of the main layered sequence consists of weakly reflecting horizons, whereas the upper unit, commonly 600–700 m thick, comprises strong and coherently reflecting horizons. The upper unit has evidently been removed from the sea floor in the vicinity of the canyon axis. Along the canyon's

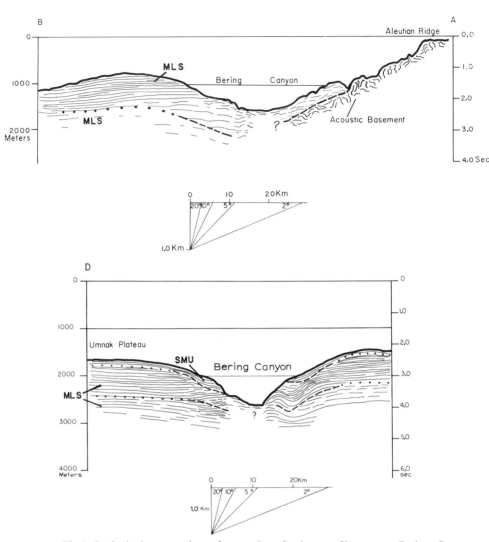

Fig.4. Geologic interpretation of acoustic reflection profiles across Bering Canyon; stratified units consist of surfacemantling unit (*SMU*) and main layered sequence (*MLS*). Acoustic basement is presumed to consist of Cenozoic volcanic units underlying Aleutian Ridge. Dashes separate unconformable units; dots separate conformable units that are acoustically distinguisable. Reflection time is given in seconds; depth in meters based on an acoustic velocity of 1,500 m/sec.

northern wall a slight divergence occurs in the dip between the lower and upper reflective units. This relationship was detected along most of the canyon's length.

Profile *C–D* (Fig.4) crosses Bering Canyon along its course below the eastern rim of Umnak Plateau. Examination of this profile, and those given by SCHOLL et al. (1968), reveals that the present canyon axis cuts through two unconformities within the stratified sequence forming its flanks. These unconformities dip toward the

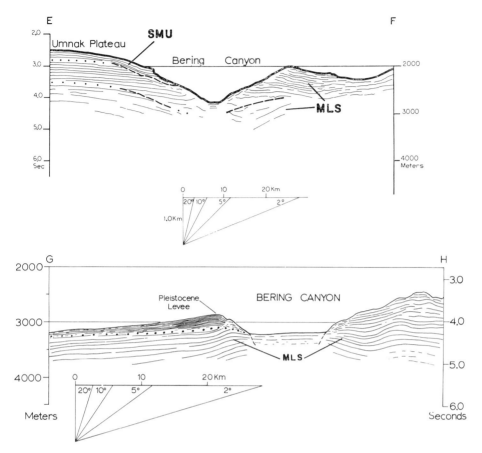

Fig.5. Geologic interpretation of acoustic reflection profiles *E–F* and *G–H* across Bering Canyon. See Fig.4 for explanation of symbols. Levee is constructed of deposits representing the continental rise unit. Right-hand levee is very large and may include beds generally recognized as belonging to the main layered sequence.

canyon axis. The upper unconformity lies only a few hundred meters below the sea floor (upper dashed and dotted line on profile *C–D*, Fig.4) and separates the upper reflective beds of the main layered sequence and an overlying *surface-mantling unit*, which was defined and described by SCHOLL et al. (1968). A second, and lower, unconformity occurs between the upper and lower divisions of the main layered sequence. Both unconformities can be projected across the canyon from the base of the continental slope to Umnak Plateau.

Profile *E–F* (Fig.5), which crosses Bering Canyon just below its confluence with Bristol Canyon (Fig.1), shows a cross-canyon structural-stratigraphic section that is similar to that shown along profile *C–D* (Fig.4). Both unconformities — the first separating the surface-mantling unit and the underlying main layered sequence, and the second lying between the divisions within this sequence — underlie Umnak

Plateau immediately adjacent to the canyon wall. Only the lower of the two discontinuities can be confidently projected across the canyon from the plateau to the stratified sequence underlying the continental slope.

Profile *G–H* (Fig.5) cuts across the outer-most reaches of Bering Canyon where it debouches onto the broad floor of the Aleutian Basin (Fig.1). A levee that rises 350 m above the valley floor flanks the left-hand (looking down-canyon) side of the canyon. At this point the canyon floor is flat, 11.5 km in width; where crossed along profile *G–H*, it is perhaps more properly termed a fan-valley (SHEPARD, 1965; SHEPARD and BUFFINGTON, 1968).

PRIBILOF CANYON

Physiography

Pribilof Canyon (KOTENEV, 1965; SHEPARD and DILL, 1966; SCHOLL et al., 1966, 1968) is approximately 150 km in length and has a volume of about 1,300 km^3. It is thus one of the world's largest and longest known submarine canyons (Fig.1, 6). The broad outer-shelf trough formed by the headward bifurcation of the main canyon axis behind the regionally projected position of the shelf edge is a further distinguishing characteristic. The trough trends northwestward approximately parallel to the strike of the continental margin. It is 90 km long, 30 km wide, and 1,500 m deep; this unusual outer-shelf trough was shown on early bathymetric charts of this region as a closed depression unconnected with a slope canyon.

Where it cuts through the regionally projected (restored) position of the shelf, Pribilof Canyon is approximately 45 km wide and 1,600 m deep. As it goes down the continental slope the canyon turns gradually to the west, broadens, and debouches onto the continental rise at a depth near 3,000 m. In Fig.6 a large leveed channel is shown extending from the mouth of the canyon; the leveed channel has been drawn without the benefit of bathymetric control, its assumed existence is based on the presence of levees adjoining the channels exiting from Bering and Zhemchug canyons.

Structure

Published seismic reflection profiles in the vicinity of Pribilof Canyon (SCHOLL et al., 1968) indicate that the outer-shelf trough associated with the lateral bifurcation of the canyon axis is related to a structural depression in the rocks of the acoustic basement. Profiles *I–J–K* and *K–L* (Fig.7), and structure contours on the acoustic basement (Fig.8), illustrate this morpho-structural superposition. Along profile *K–L* a normal fault is shown below the seaward wall of the canyon head, suggesting that the structural depression is a half-graben or rotated fault block. A similar fault underlies the shelf north of the canyon head; this fracture is presumed to strike westward and to pass beneath St. George Island, a Quaternary voclanic structure resting on a basement of crystalline rocks (BARTH, 1956; COX

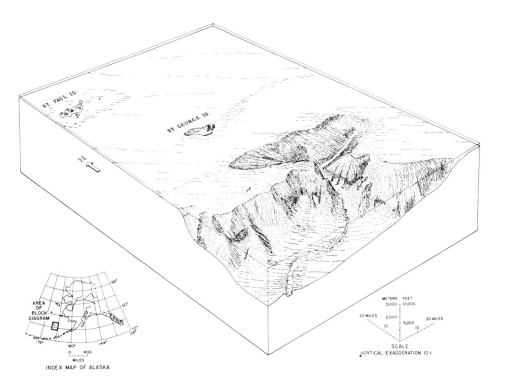

PRIBILOF CANYON

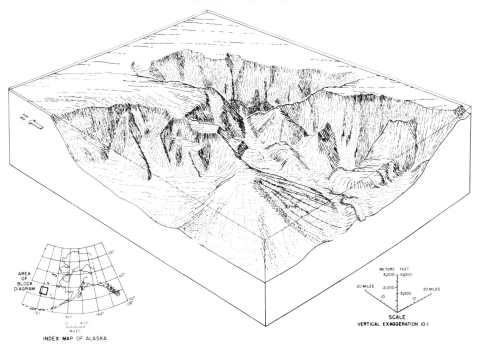

ZHEMCHUG CANYON

Fig.6. Physiographic diagram of Pribilof and Zhemchug canyons; views are toward Bering shelf. Bathymetric control based upon a chart published by SCHOLL et al. (1968); two-point perspective drawing by Tau Rho Alpha.

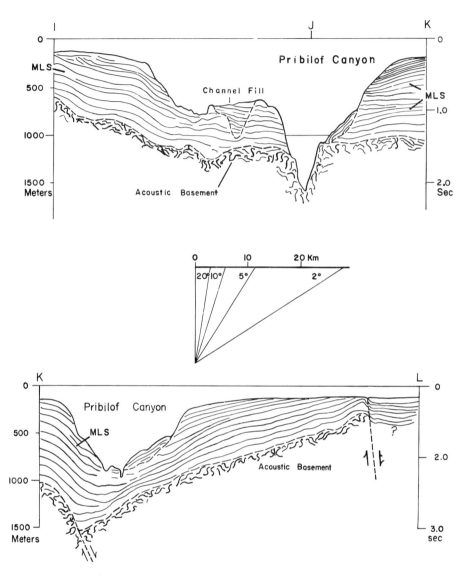

Fig.7. Geologically interpreted acoustic reflection profiles *I–J–K* and *K–L* across Pribilof Canyon. Acoustic basement is in part constructed of folded and faulted mudstone and siltstone beds of Late (Campanian) Cretaceous age (HOPKINS et al., 1969). See Fig.4 for explanation of symbols. Only the minimum dips of the fault planes are indicated; actual dips are probably greater than 45°.

et al., 1966). The acoustic basement exposed along the sides of the inner gorge of Pribilof Canyon along profile segment *J–K* (Fig.7) is constructed of lithified mudstone and siltstone of Late Cretaceous age (HOPKINS et al., 1969). The structure of the acoustic basement and overlying strata of the main layered sequence is shown by the pull-away geomorphic diagram of Pribilof Canyon in Fig.9.

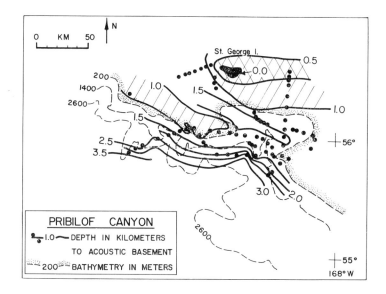

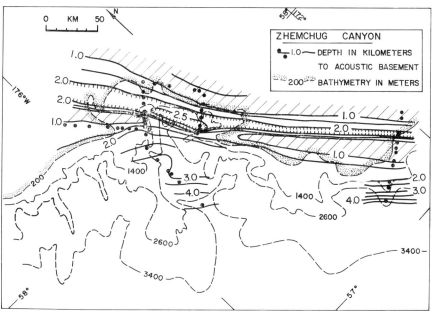

Fig.8. Structure contours, in kilometers, on surface of acoustic basement in vicinity of Pribilof and Zhemchug canyons. Basement rock — serpentinized peridotite intruded by aplitic dikes — is exposed on St. George Island, which is chiefly constructed of Pleistocene volcanic rocks (Cox et al., 1966); in central gorge of nearby Pribilof Canyon, acoustic basement is in part composed of indurated mudstone and siltstone of Late Cretaceous age (Fig.7). Lithology of the acoustic basement in Zhemchug Canyon area is unknown (HOPKINS et al., 1969). Structural contours have been reconstructed across canyon axes to shown probable pre-canyon configuration of acoustic basement.

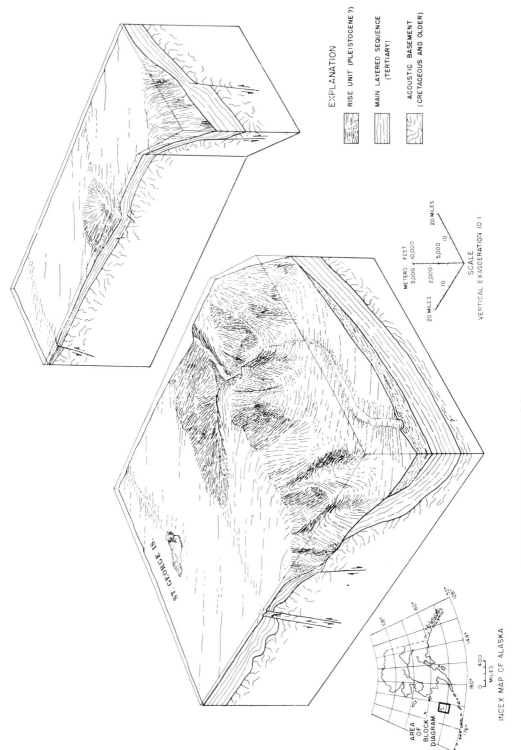

EXPLANATION

RISE UNIT (PLEISTOCENE ?)

MAIN LAYERED SEQUENCE
(TERTIARY)

ACOUSTIC BASEMENT
(CRETACEOUS AND OLDER)

SCALE

METERS   FEET
3,000   10,000

2,000   5,000

20 MILES   10   0   10   20 MILES

VERTICAL EXAGGERATION 10:1

ST. GEORGE IS.

INDEX MAP OF ALASKA

AREA
OF
BLOCK
DIAGRAM

MILES
0   400

PRIBILOF CANYON

ZHEMCHUG CANYON

Physiography

Zhemchug Canyon (i.e., "Pearl Canyon" in Russian) was described and named by KOTENEV (1965). Although geomorphically similar to Pribilof Canyon, Zhemchug is much larger (Fig.1, 6). Its volume is approximately 8,500 km³, which is enormous in comparison to other canyons that are described as "large". For example, Monterey Canyon off central California has a volume of only 450 km³; taken together, all central California submarine canyons have a combined volume of only 1,500 km³ (MENARD, 1960). Zhemchug may well be the world's largest submarine canyon.

Like Pribilof Canyon, the central axis of Zhemchug Canyon also bifurcates behind the face of the continental slope, and, in so doing, forms a northwestward elongated trough that is fully 160 km long, 25–30 km wide, and 2,600 m deep (Fig.1, 6). The seaward scarp of this trough is precipitous with slopes as steep as 20°. Relatively gentle slopes of 6–7° characterize the opposing landward flank of the trough.

Where it cuts the continental slope at a rim-to-rim depth of 200 m, the width of Zhemchug Canyon is 100 km. However, measured along this same crossing, the width of its central gorge (deeper than 1,500 m) is only 30 km. The canyon debouches on the continental rise at a depth near 3,400 m. Both our data and that of KOTENEV (1965) indicate that a leveed fan-valley extends southward from the mouth of the canyon (Fig.6). This channel is approximately 4 km wide and 100 m deeper than the adjacent sea floor. The right-hand (western) level is highest and rises approximately 300 m above the floor of the channel.

Structure

Profiles *M–N* and *O–P* (Fig.10) traverse the outer-shelf trough formed by the headward bifurcation of the canyon axis. These profiles show that the acoustic basement is down-warped and faulted below the trough; thus, like Pribilof Canyon, the outer-shelf trough is a bathymetric feature superimposed on a structural depression of the acoustic basement. Structure contours shown in Fig.8 and seismic profile *M–N* suggest that the depression is a rotated fault block (i.e., half-graben) with a hinge line located beneath the northeast wall of the trough and a shelfward dipping normal fault beneath its seaward or southwestern wall. This wall is a fault-line scarp and forms the landward flank of a basement high underlying the outer edge of the shelf (Fig.8); the vertical offset associated with this fault is at least 1,000 m. The physiographic relationship of the outer shelf trough of Zhemchug Canyon to the structure of the upper continental margin is illustrated by the exploded geomorphic diagram in Fig.11. It is apparent from this diagram that strata of the main layered sequence formerly extended across the trough and buried the outer-shelf basement high. Soviet researchers have collected deposits of

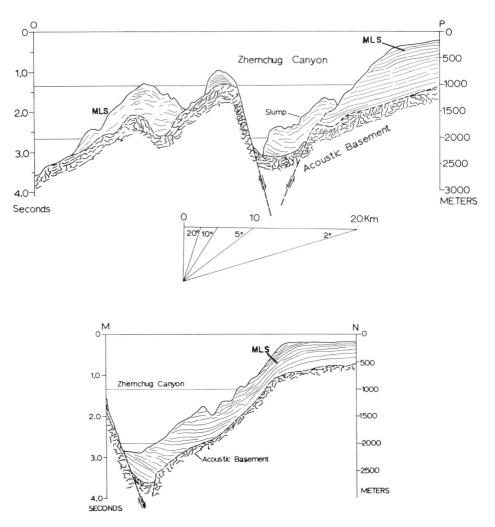

Fig.10. Geologic interpretation of acoustic reflection profiles across Zhemchug Canyon. See Fig.4 for explanation of symbols. Only the minimum dips of faults are indicated.

Neogene and Paleogene age in the vicinity of Zhemchug Canyon; but we do not know whether the Paleogene rocks were recovered above or below the acoustic basement (HOPKINS et al., 1969). Miocene and Pleistocene deposits were collected from the main layered sequence.

Buried outer-shelf basement highs, such as those associated with the inner troughs of Zhemchug and Pribilof canyons are common features of continental margins (DRAKE et al., 1959; CURRAY, 1965; ROSS and SHOR, 1965).

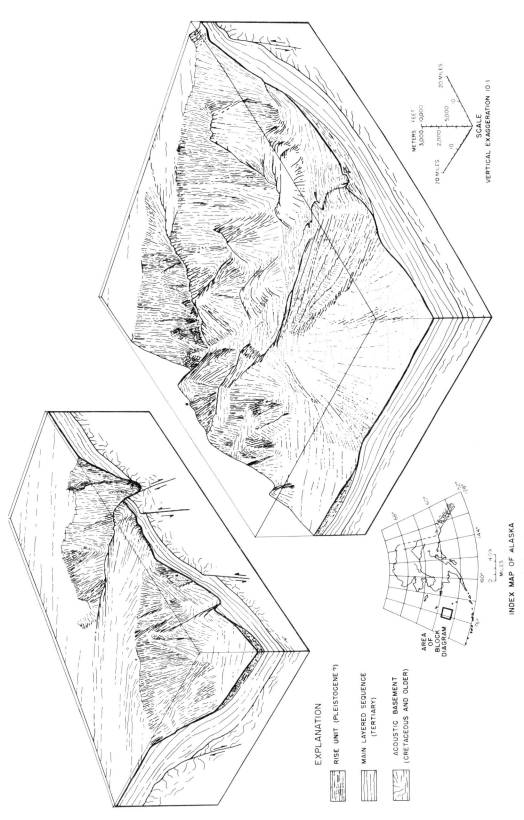

EXPLANATION

RISE UNIT (PLEISTOGENE ?)

MAIN LAYERED SEQUENCE
(TERTIARY)

ACOUSTIC BASEMENT
(CRETACEOUS AND OLDER)

INDEX MAP OF ALASKA

AREA
OF
BLOCK
DIAGRAM

SCALE

VERTICAL EXAGGERATION 10 1

Fig.11. Exploded geomorphic diagram of Zhemchug Canyon.

In this regard, it is of special interest to note that the structure of the continental margin in the vicinity of Zhemchug Canyon is similar to that associated
with Monterey Canyon, the largest submarine canyon cutting the California
continental slope (MARTIN and EMERY, 1967; STARKE and HOWARD, 1968). The
course of this canyon down the continental slope is fault controlled; we suspect,
as emphasized later, that the positions in which the main axes of Zhemchug and
Pribilof canyons cross the outer-shelf basement highs are also structurally determined.

GENESIS AND HISTORY OF THE CANYONS

Submarine erosion

Reflection profiles and regional geologic considerations imply that the
location, trend, and part of the relief of the large canyons cutting the continental
margin of the Bering Sea have been determined by structural features that affect
the geometry of the acoustic basement and the overlying main layered sequence.
However, seismic reflection records and submarine outcrops convince us that
extensive erosional deepening and widening of the canyons have taken place.
The thick sequence of Neogene (possibly older) strata cut by Pribilof and Zhemchug
canyons consists of shallow-water deposits that are unbroken by a major unconformity. This constitutes a record since the Miocene of essentially continuous
marginal subsidence and contemporaneous sedimentation at an average rate close
to 0.8 m/10,000 years. It is difficult, therefore, to consider that any significant
portion of these canyons was cut by subaerial processes. Although the exact
mechanism of submarine erosion is moot, it seems reasonable to suspect that it
took place in conjunction with mass sediment movements and turbidity flows
generated in areas of exceptionally intense sediment accumulation on the continental slope (MENARD, 1964, p.218; SHEPARD and DILL, 1966, p.332). The
sediment was most likely supplied by major streams and rivers draining Alaska
during periods of glacial-eustatically lowered sea level or during episodes of shelf
emergence resulting from regional uplift or coastal progradation.

Sediment sources

Alaskan stream systems that would provide adequate sediment sources
include the Yukon River, the Kuskokwim River, and the shorter system that
drains the rugged areas adjoining Bristol Bay (Fig.1). A well-defined system of
sea valleys leads from the present mouth of the Kuskokwim River and from the
head of Bristol Bay to the vicinity of the head of Bering Canyon. The highlands
adjoining Bristol Bay were repeatedly glaciated (COULTER et al., 1962) during the
Pleistocene; the effect of this must have been to augment the normal sediment
supplies from this area. It seems likely to us that the sediment contributed to the
head of Bering Canyon was chiefly derived from the highland surrounding Bristol

Bay, although some contributions might have come from the Kuskokwim.

A poorly defined sea valley extends from the position where the Yukon River entered the Bering Sea early in Holocene time to the east branch of Pribilof Canyon (HOARE and CONDON, 1968). The channel of the sea valley is presumed to have passed near St. Paul Island prior to possibly reaching the canyon head. This is supported by unpublished studies by Hopkins of a sheet of stabilized eolian sand that covers much of the northeastern part of the island. The sand is rich in quartz, muscovite, and potash feldspars that could not have been derived from the basaltic flows and volcanoclastic units that form the island. We reckon that the source was probably the flood plain of a major river draining western Alaska; however, the sand may also have been derived from a strand area well nourished with sediment contributed by a principal river. Although the positions that the Yukon and Kuskokwim Rivers occupied on the Bering shelf can be deduced for the Wisconsin interval of eustatically lowered sea level, these rivers may have traversed quite different and as yet unrecognized areas of the shelf during earlier low sea-level episodes (HOPKINS, 1967). We think it probable, however, that Zhemchug Canyon represents the position of debouchment of the Yukon River during one or more periods of low sea-level, and that the Kuskokwim River, as well as the Yukon, has at times entered the head of Pribilof Canyon.

Age and geomorphic development of canyons
Bering Canyon. That the position and course of Bering Canyon are structurally controlled has been emphasized previously (KOTENEV, 1965). Our seismic reflection records (SCHOLL et al., 1968; Fig.4, 5) confirm these earlier conclusions.

In an earlier discussion, while acknowledging that the position of the canyon is structurally controlled, the suggestion was made that the carving of the canyon might have been accomplished entirely during Pleistocene intervals of lowered sea level (SCHOLL et al., 1968). However, further study indicates that the erosional history is probably more complicated and that it covers a longer time span than was thought. The several unconformities that dip toward the canyon seem to imply several cycles of canyon cutting, separated by periods when the canyon was partially infilled. The earliest episode of canyon cutting is represented by the unconformity within the main layered sequence; and, because the main layered sequence seems to consist mostly of deposits of Neogene age (HOPKINS et al., 1969), we assume that the canyon first came into existence prior to the end of Tertiary time. The canyon was then partially filled by the deposition of the upper strata of the main layered sequence, probably during Late Neogene and Early Quaternary time. A second cycle of canyon cutting is represented by the unconformity between the main layered sequence and the surface-mantling unit, a second cycle of filling is represented by the sediments of the surface-mantling unit themselves, and a third cycle of canyon cutting is recorded by the erosion of the surface-mantling unit to form the present topography. The initial episode of

115

canyon cutting may have coincided with the strong Pliocene uplift of the Alaska Peninsula reported by BURK (1965); it seems reasonable to suppose that this uplift may have caused emergence of the nearby shelf, or coastal progradation, and initiated the development of turbidity flows that could deepen the topographic lows related to tectonic activity along the present axis of the Bering Canyon. The two subsequent canyon-cutting episodes probably took place during the Quaternary Epoch and may be related to sedimentation at the continental margin during periods of glacio-eustatically lowered sea level.

Pribilof Canyon. The configuration of the Pribilof Canyon is clearly structurally controlled. The topographic trough represented by the bifurcated head of the canyon has been formed by the removal of nearly 1,000 m of the main layered sequence in an east–west-trending structural basin, and the main stem of the canyon crosses the continental margin in a saddle, possibly a faulted area, in the basement ridge that bounds the basin to the south (Fig.8).

Pribilof Canyon probably originated later than Bering Canyon. The main layered sequence into which Pribilof Canyon is incised appears to record a rather simple history of slow subsidence and simultaneous prograding of the continental margin during Neogene and Early Pleistocene time. Because major unconformities are lacking within the main layered sequence in the Pribilof area, we suspect that the canyon was chiefly cut during Pleistocene time. At least two cycles of canyon cutting and one of partial refilling are recorded; the older cycle of canyon cutting is recorded by the deep filled channel that lies to the west of the present main stem of the canyon (Fig.7, section *I–J–K*). As previously emphasized, the most recent episode of canyon cutting seems to have been related to the debouchement at the continental margin of the Yukon River during Wisconsin time; the earlier episode may have involved either the Yukon or the Kuskokwim River or possibly both.

Zhemchug Canyon. The origin and history of the Zhemchug Canyon is generally similar to that of the Pribilof Canyon. Some of the relief of the Zhemchug Canyon can be ascribed to warping and faulting, but most was created by submarine erosion. Structure contours on Fig.8 suggest that the basement ridge underlying the outer shelf was breached by headward erosion along a structural saddle to form the slope axis or stem of the canyon. Subsequently, nearly 4,500 km³ of the main layered sequence in the faulted basin to the northeast were excavated to form the bifurcated head.

As we emphasized earlier, the structure of the continental margin in the vicinity of Monterey (MARTIN and EMERY, 1967; STARKE and HOWARD, 1968) and Zhemchug canyons is similar. In this regard, the stripping of the main layered sequence to reveal a basement depression is conceptually similar to the position taken by STARKE and HOWARD (1968) that the slope axis of the very large Monterey

Canyon off the central California coast originated by the fluming out of an ancient rock-walled canyon largely filled with non-marine deposits. Rocks dredged from the sides of Zhemchug Canyon (HOPKINS et al., 1969) and structures revealed by seismic reflection profiles (Fig.10) indicate that the canyon head is cut into marine sediments that probably accumulated contemporaneously with formation of the underlying graben. We are therefore of the opinion that the slope axis of Zhemchug Canyon is probably not the exhumed gorge of a subsided, subaerially carved canyon. However, there is insufficient geophysical information to be certain that a bedrock gorge exiting from beneath the shelf does not in fact cross the graben to connect with the presumed saddle in the outer basement ridge that structurally positioned the slope axis of Zhemchug Canyon (Fig.8).

The main layered sequence includes rocks as young as Early Pleistocene in the Zhemchug Canyon area, and we suspect that the major episode of canyon cutting was during the Pleistocene Epoch. A large river seems to be required to provide the sedimentary accumulations that could trigger erosive turbidity currents and debris flows capable of excavating this enormous canyon, and the Yukon River seems to be the only large stream that could have reached the continental margin in the Zhemchug Canyon area. It is of interest, therefore, to consider the size of Zhemchug Canyon in relation to the *maximum* quantity of Yukon-derived sediment that could have been funneled down its axis during the Pleistocene.

The Yukon is estimated to discharge annually about $100 \cdot 10^6$ metric tons of solid particulate matter (SAMOILOV, 1956; LISITZIN, 1966, p.78). During Glacial stages this discharge may have been somewhat greater, possibly by a factor of 1.5 (SCHOLL et al., 1968). The volume of Zhemchug Canyon is approximately 8,500 km^3, which is roughly equivalent to a mass of $9 \cdot 10^{12}$ metric tons if a porosity of 60% and a mean-grain density of 2.65 g/cm^3 is assumed. If it is further assumed that the Yukon emptied into the head of Zhemchug Canyon for $1.5 \cdot 10^6$ years (roughly half the length of Late Cenozoic glaciation, HOPKINS, 1967, p.460), then a probable maximum of $225 \cdot 10^{12}$ metric tons of Yukon-derived sediment could have sluiced down this canyon to the Aleutian Basin. The ratio of this figure to that of the mass of missing canyon rock is approximately 25. Interestingly this ratio of masses is comparable to the ratio of volumes (30) computed by MENARD (1960) for canyons cutting the central California continental slope. Menard's ratio is based on the volume of the canyons in comparison to the volume of their submarine fans, which presumably are constructed of the bulk of the sediment moved along their axes to the adjacent Pacific floor. Bathymetric data and seismic reflection profiles indicate that a sizeable submarine fan also lies at the base of Zhemchug Canyon; but the dimensions of this fan, as yet, are unknown.

MENARD (1960) also reckoned that lateral erosion or widening of canyons probably takes place by slumping and mass wasting of undercut and oversteepened walls. Thus sediment channeled down a continental slope need only erode vertically, and then sweep the canyon clean of debris, to account for both the relief and the width of an eventual canyon. For Monterey Canyon, the largest of the California slope canyons, MENARD (1960) chose an axial width of 100 m, from which he derived a canyon–fan volume ratio of about 5,000. If a 100-m width is considered for Zhemchug Canyon, the ratio of theoretically transported Yukon-sediment to eroded material is approximately 5,600; again a figure remarkably similar to that of the better-studied California canyon.

Although these figures do not have quantitative significance, their magnitude does reinforce the assumption that Zhemchug and other canyons incising the Bering continental slope could have been cut largely within the span of the Pleistocene ($3 \cdot 10^6$ years). This is no assurance that important cutting did not in fact begin in Late Tertiary time, when the outer region of the shelf was in part subaerially exposed (HOPKINS et al., 1969) and may have been reached by major rivers.

GENERAL CONCLUSIONS

The present form of the northeastern continental margin of the Bering Sea began to develop in Early to Middle Tertiary time as the result of subsidence and down-faulting of previously lithified and folded rocks of Cretaceous and older age. Subsidence proceeded differentially along the strike of the margin, and narrow depressions (rotated fault blocks or half-grabens) formed in the vicinity of the present shelf edge. These depressions and the seaward-dipping surface of the basement rock beneath the slope were filled and covered with 1,000–3,000 m of detrital, tuffaceous, and diatomaceous deposits of Tertiary age while faulting continued. During one or more cycles of intense submarine erosion; in Late Tertiary and Pleistocene time, these deposits were stripped from the continental slope and the outer-shelf basement depressions. Most of the exceptionally large submarine canyons that incise the Bering continental margin were formed at this time.

We believe that all of the large canyons cutting the continental margin were carved by the sliding of masses of sediment that had been deposited on the upper reaches of the continental slope by large Alaskan and Siberian rivers. The initial cutting of Bering Canyon is presumed to have begun in Tertiary time, perhaps in response to emergence of the shelf or to the coastal progradation associated with uplift of the nearby Alaska Peninsula during Pliocene time. The course of the canyon was determined by structural depressions or swales in the Tertiary strata of the main layered sequence. These depressions seem to be related to deeper structural offsets or flexures in a differentially downwarped basement surface (SCHOLL et al., 1968). At least two additional episodes of canyon erosion followed, separated by partial sediment infilling. These latter cycles of cutting probably took place in the Pleistocene after a eustatic emergence of the shelf and the dumping of large masses of sediment in the canyon head.

The Kuskokwim and Yukon rivers probably contributed the bulk of the sediment required for the cutting of Pribilof and Zhemchug canyons. Both canyons may have been excavated chiefly in the Pleistocene. The trough-shaped basins formed by their canyon heads are superimposed on structural depression in the rocks of the acoustic basement. The structural trend of the depressions — parallel to the continental margin — apparently accounts for the lateral bifurcation of the canyon axes to form the troughs.

REFERENCES

BARTH, T. F. W., 1956. Geology and petrology of the Pribilof Islands, Alaska. *U.S., Geol. Surv., Bull.*, 1028(F): 101–160.

BYERS JR., F. M., 1959. Geology of Umnak and Bogoslof Islands, Aleutian Islands, Alaska. *U.S., Geol. Surv., Bull.*, 1028(L): 267–369.

BURK, C. A., 1965. Geology of the Alaska Peninsula: island arc and continental margin, 1, 2. *Geol. Soc. Am., Mem.*, 90: 250 pp.

COULTER, H. W., HOPKINS, D. M., KARLSTROM, T. N. V., PÉWÉ, T. L., WAHRHAFTIG, C. and WILLIAMS, J. R., 1962. Map showing extent of glaciations in Alaska. *U.S., Geol. Surv., Miscellaneous Geol. Invest. Map*, I-415.

COX, A., HOPKINS, D. M. and DALRYMPLE, G. B., 1966. Geomagnetic polarity epochs: Pribilof Islands, Alaska. *Geol. Soc. Am., Bull.*, 77: 883–910.

CURRAY, J. R., 1965. Structure of the continental margin off central California. *Trans. N.Y. Acad. Sci., Ser. 2*, 27: 794–801.

DRAKE, C. L., EWING, M. and SUTTON, G. H., 1959. Continental margins and geosynclines: the east coast of North America north of Cape Hatteras. In: L. H. AHRENS, F. PRESS, K. RANKAMA and S. K. RUNCORN (Editors), *Physics and Chemistry of the Earth*. Pergamon, New York, N.Y., 3: 110–198.

HOARE, J. M. and CONDON, W. H., 1968. Geologic map of the Hooper Bay Quadrangle, Alaska. *U.S., Geol. Surv., Miscellaneous Geol. Invest. Map*, I-523.

HOPKINS, D. M., 1967. The Cenozoic history of Beringia — a synthesis. In: D M. HOPKINS (Editor), *The Bering Land Bridge*. Stanford University Press, Stanford, Calif., pp.451–484.

HOPKINS, D. M., SCHOLL, D. W., ADDICOTT, W. O., PIERCE, R. L., SMITH, P. B., WOLFE, J. A., GERSHANOVICH, D., KOTENEV, B. N., LOHMAN, K. E., OBRADOVICH, J. and LIPPS, J. H., 1969. Cretaceous, Tertiary, and Early Pleistocene rocks from the continental margin in the Bering Sea, *Geol. Soc. Am., Bull.*, in press.

KOTENEV, B. N., 1965. Submarine valleys in the zone of the continental slope in the Bering Sea. *Tr. Vses. Nauchn. Issled. Inst. Razved. Okeanol.*, 58: 35–44 (in Russian).

LISITZIN, A. P., 1966. *Processes of Recent Sedimentation in the Bering Sea*. Nauka, Moscow, 574 pp. (in Russian).

MARTIN, B. D. and EMERY, K. O., 1967. Geology of Monterey Canyon, California. *Bull. Am. Assoc. Petrol. Geologists*, 51: 2281–2304.

MENARD, H. W., 1960. Possible pre-Pleistocene deep-sea fans off central California. *Bull. Geol. Soc. Am.*, 71: 1271–1278.

MENARD, H. W., 1964. *Marine Geology of the Pacific*. McGraw-Hill, New York, N.Y., 271 pp.

NICHOLS, H. and PERRY, D., 1966. *Bathymetry of the Aleutian Arc, Alaska* (Scale 1:400,000). Dept. Commerce, Environmental Sci. Serv. Admin., Coast Geodetic Surv., Monograph 3: 6 maps.

ROSS, D. A. and SHOR JR., G. G., 1965. Reflection profiles across the Middle America Trench. *J. Geophys. Res.*, 70: 5551–5572.

SAMOILOV, I. V., 1956. *Die Flussmundungen*. VEB Hermann Haack, Gotha, DDR.

SCHOLL, D. W., BUFFINGTON, E. C. and HOPKINS, D. M., 1966. Exposure of basement rock on the continental slope of the Bering Sea. *Science*, 153: 992–994.

SCHOLL, D. W., BUFFINGTON, E. C. and HOPKINS, D. M., 1968. Geologic history of the continental margin of North America in the Bering Sea. *Marine Geol.*, 6: 297–330.

SHEPARD, F. P., 1948. *Submarine Geology*. Harpers and Brothers, New York, N.Y., 348 pp.

SHEPARD, F. P., 1965. Types of submarine valleys. *Bull. Am. Assoc. Petrol. Geologists*, 49: 304–310.

SHEPARD, F. P. and DILL, R. F., 1966. *Submarine Canyons and Other Sea Valleys*. Rand McNally, Chicago, Ill., 381 pp.

SHEPARD, F. P. and BUFFINGTON, E. C., 1968. La Jolla Submarine Fan Valley. *Marine Geol.*, 6: 107–143.

STARKE, G. W. and HOWARD, A. D., 1968. Polygenetic origin of Monterey Canyon. *Geol. Soc. Am., Bull.*, 79: 813–826.

ADDENDUM BY DAVID W. SCHOLL

After this paper was published in 1970, additional geophysical and geological data were gathered in the vicinity of several of the canyons, and a DSDP (Deep Sea Drilling Project) hole, Site 185, was drilled along the eastern flank of Bering Canyon (54.43°N, 169.24°W) at a depth of 2100 m. Although Site 185 was not located along profile C-D (Fig. 1), the drill penetrated a stratigraphic section equivalent to that revealed along this profile at a point approximately halfway between its eastern (right-hand) side and the axis of Bering Canyon (Fig. 4, lower profile). The upper of the two unconformities [i.e., the contact between the surface mantling unit (here about 160 m thick) and the main layered sequence] is overlain by silty and clayey diatom ooze of late Pliocene through Holocene age and underlain by similar but more indurated beds deposited prior to the late early Pliocene (Creager and others, 1973). The implication is that a major phase of canyon cutting took place in the middle Pliocene (i.e., between 3 and 4 m.y. ago).

A time hiatus was not coincident with the lower "unconformity." Instead, a transition from diatom ooze to mudstone, both of late Miocene age, was encountered; a transition possibly related to an upward-moving zone of diagenesis, and a transition that is recorded on seismic reflection records as a bottom-simulating reflector (BSR) (Scholl and Creager, 1973). Although it was not realized prior to the original publication of this paper, the BSR is a widespread acoustic phenomenon in the Bering Sea and not one restricted to the flanks of Bering Canyon. The lower "unconformity" is therefore not a surface of erosion but an acoustic recording of a diagenetic (?) boundary contained within a time-continuous sedimentary section.

These findings imply that Bering Canyon was initially cut during the middle Pliocene. Infilling probably took place in conjunction with the subsequent deposition of the surface mantling unit, but so did one or more cycles of erosional deepening. Ironically, in our paper we call for an initial episode of canyon cutting prior to the end of the Tertiary, an assertion that has turned out to be correct but for wrongly stated reasons.

We should perhaps mention that the oldest rocks of Tertiary age dredged from the walls of Zhemchug Canyon are middle Oligocene, and that they are part of the main layered sequence (see footnote p. 100).

REFERENCES

Creager, J. S., Scholl, D. W., and others (1973). *Initial Reports of the Deep Sea Drilling Project*, Vol. 19, Government Printing Office, Washington, D.C., 913 p.

Scholl, D. W., and Creager, J. S. (1973). Geologic synthesis of Leg 19 (DSDP) results; far north Pacific, and Aleutian Ridge, and Bering Sea. *In* Creager, J. S., Scholl D. W., and others, *Initial Reports of the Deep Sea Drilling Project*, Vol. 19, Government Printing Office, Washington, D.C., pp. 897-913.

Editor's Comments
on Paper 8

8   SHEPARD and DILL
     Excerpts from *Submarine Canyons and Other Sea Valleys*

The standard work on modern submarine canyons is the well-known
book *Submarine Canyons and Other Sea Valleys* by Shepard and Dill
(1966, Paper 8). In it they define the several types of submarine valleys,
give the history of submarine studies, and discuss the methods by which
they are explored. Following chapters deal in turn with the well-studied
canyons off La Jolla, California; other U.S. West Coast canyons; Baja
California (Mexico); Japanese examples (eastern Honshu); the many
canyons off the eastern U.S.; European canyons, including the huge
Cap Breton Canyon and those in the Mediterranean; and other areas,
including the Bahamas, Bering Sea, Ceylon, the Philippines, Hawaii,
Australia, New Zealand, and New Guinea; and the well-known Congo
Canyon.

Their next chapter, summarizing characteristics and dimensions of
submarine canyons, is a useful one to reprint here as the final contribution to Part I. In it, they deal with 93 canyons known up to 1966 and
classify them according to a number of distinctive features, which are
organized in the Appendix, and also reproduced in this volume as a convenient source of information on so many canyons. Shepard and Dill's
book also contains chapters dealing with various types of valleys other
than true submarine canyons; mass physical properties of canyon sediments; a valuable discussion of the origin of submarine canyons which
we would have included had space allowed; and a comprehensive list of
references which are reprinted here for those readers who wish to review early canyon studies (up to 1966). In the list of references cited by
the editor at the end of this volume, the post-1966 papers on modern
canyons highlight some of the better-known recent studies (see also pp.
2–5). Shepard kindly provided some of these references. On p. 147, the
reference to Kuenen 1937, 316–351, should read 327–351.

8

Copyright © 1966 by John Wiley & Sons, Inc.

Reprinted with permission from F. P. Shepard and R. F. Dill,
Submarine Canyons and Other Sea Valleys, Rand McNally, Chicago, 1966,
pp. 223–231, 343–351, 353–367

SUMMARY CHARACTERISTICS AND DIMENSIONS OF SUBMARINE CANYONS

F. P. Shepard and R. F. Dill

FROM THE FOREGOING discussion of submarine canyons, it will be seen that, although they do not all follow the same pattern, most of them have much in common. In compiling statistics from described canyons, it should be borne in mind that much more is known about canyons off the coasts of the United States and Baja California than elsewhere, and especially about those canyons with heads coming into the coast. We are actually determining the characteristics of the best surveyed canyons. However, this information should provide a moderately satisfactory picture of canyons in general. In compiling the statistics given in the Appendix, we are referring only to the canyons and not to the fan-valleys that form continuations of many of them. The location of most of the canyons included is shown in the frontispiece.

CANYON CHARACTERISTICS

In the Appendix, we have classified 93 of the canyons according to the following characteristics: (1) Length, (2) Depth of head, (3) Greatest known depth, (4) Character of coast, (5) Relation to points on downcurrent side, (6) Relation to river valleys, (7) Source of sediment for ca..yon heads, (8) Canyon gradient, (9) Nature of longitudinal profile, (10) Maximum wall height, (11) Channel curvature, (12) Abundance of tributaries, (13) Transverse profile character, (14) Nature of canyon wall material, (15) Core sediment found in axis, and (16) Relation to fan-valleys. In the table, the canyons from the same general area are given in somewhat of a geographic sequence, and at the end of the table are a few of the miscellaneous canyons discussed in Chapter IX. In determining some of the means, each group is averaged and treated as a unit and then these units are in turn averaged. The various parameters appear to be log-normal distributions, thus average values are meaningful descriptions of the group properties.

Lengths and Depths at Extremes of Canyons

The lengths of the submarine canyons have been measured along their axes from the canyon head to the point where the canyon disappears on the slope or enters a fan-valley. These lengths are subject to some uncertainty because many of the surveys do not extend far enough in either direction. Where this is the case, the length is given with a plus sign. For an average length, those with a plus sign are included since they often are the longest. Using this method, the average of the various groups is 30 nautical miles. Probably more complete knowledge would increase this by a few miles. By far the longest canyons are in the southern Bering Sea, where Bering Canyon is of the order of 230 miles in total length. Cap Breton Canyon, of western European canyons, is perhaps the next longest with a probable length of 135 miles. Great Bahama with 120 miles is tied with Congo Canyon for third. If the fan-valleys were included, much greater lengths would be found; for example, about 500 miles for the Congo Valley.

The shortest of the canyon groups are those of the Hawaiian Islands, which average only 6.5 miles although the outer limits are not too well established. Almost equally short are the canyons off southern California that terminate in the inner basins of the continental borderland.

The canyons head at depths of a few feet to more than a thousand feet. The average for the entire group is 350 ft (107 m), but this would be somewhat reduced if better surveys were available. A lower average is found in some of the best surveyed areas. Thus, California canyons average 110 ft (34 m) at their heads, and Baja California has three canyons coming in almost to the beach. By contrast, the East Coast canyons all head at depths of more than 300 ft (90 m).

The depth of outer canyon termination is well known off California and Baja California and for a few other canyons. Elsewhere, it is rather indefinite from the available soundings. However, if we used averages only for those that are well surveyed, the depths would be lower than if we include the estimated depths and the canyons marked with a plus sign. With all included, the average for the group is 6,946 ft (2,117 m). The average for the areas where terminations are well developed is lower, amounting to 5,958 ft (1,816 m). This is lowered particularly by the elimination of the deep Aleutian group and of the canyons of western Europe. The deepest canyon may be Aviles Canyon, off northern Spain, with 15,580 ft (4,750 m), a questionable depth; and the second is Nazare, off Portugal, with a sounding of 14,764 ft (4,500 m) in what appears to be the outer canyon. The deepest that is rather well-established is Great Bahama Canyon with a depth of at least 14,060 ft (4,280 m). It is notable that very few canyons in well-surveyed areas can be traced beyond axial depths of 7,000 to 8,000 ft (2,130 to 2,440 m). On the other hand, the canyons terminating at less than 5,000 ft (1,520 m) are found principally off southern California where they are stopped by the inner basins, and off eastern Honshu where they enter a fault trough. Most canyons continue down to the contact between the relatively steep continental slope and the gently sloping continental rise. There is some evidence that this virtually confines the canyons to the slopes that have rock outcrops or where the rock has only a thin cover of sediment. Such a relationship is well established by continuous reflection profiling off parts of California and Baja California. However, the fan-valleys beyond the canyons may have relatively deep cuts into the sediments of the continental rise and, at least in the Hudson Fan-valley, the cut has extended into Tertiary formations (p. 150).

Land Physiography and Coastal Configuration
Shoreward of Canyon Heads

Since the majority of the canyons extend in close to the coast, it is possible to determine whether their heads enter bays, lie directly off bays, off straight beaches, or off straight cliffed coasts. The tabulation for the entire 77 canyons for which such a classification was possible shows that 13 extend into estuaries, 25 lie directly off bays, 26 off straight beaches, including barrier islands, and 13 off relatively straight cliffed coasts. There are, therefore, about an equal number off embayed and straightened coasts inside the canyon heads. It is only off western Corsica that a series of adjacent canyons penetrate deeply into embayments. To date, only two canyons have been found directly off a point of land and at least one is related to recent volcanism.

Because there has been considerable discussion of a relationship of canyons to the upcurrent side of points of land (Davis, 1934), an analysis of the canyon heads was made to look for such a relationship. It was claimed by Davis that these points deflect longshore currents seaward. The result of the analysis was that of the 93 canyons, 73 showed no relationship to such a point, 11 were quite clearly located off the upcurrent side of the point, and nine were close enough to such a point so that a slightly lowered sea level would place them adjacent to it. The best indications of a relationship of this kind were found in southern California and along the coast of Portugal.

The juxtaposition of canyon heads to land valleys has been discussed by many authors. This is, of course, a criteria having significance in the interpretation of canyon origin, both because the drowning of valleys would leave the remaining land valleys shoreward of the drowned areas and, on the other hand, the excavation of submarine canyons by submarine processes should be dependent on a good nearby source of sediment, such as could be provided by an entering river. For this analysis, we have classified the canyon heads as occurring quite directly off river valleys, showing no relation to land valleys, or of an uncertain relationship because of heading far from shore. The result of the analysis is that 46 are located directly off river valleys and 11 have no valleys inside. The remainder head too far seaward to classify. Thus, there appears to be a common relationship to river valleys. In several recent surveys where it had been thought previously that there was no nearshore canyon, the heads of canyons have been traced into the coast at a point where there is a land valley. Perhaps, therefore, the relationship is even better than the figures indicate. However, it should be pointed out that some of the other types of marine valleys are located off river valleys, particularly off deltas. There is only one area where it seems quite clear that the canyons show no relationship to land valleys; that is off Georges Bank, east of New England. Here, it may well be that prior to the glacial excavation of the Gulf of Maine, Georges Bank constituted a coastal plain bordering the New England area.

Sources of Sediment for Canyon Heads

Of the 93 canyons that are included in the table, 34 appear to have a present-day source of sediment carried to them by longshore currents or contributed to them directly by river mouths. Of the remainder, 45 have heads far enough from shore and at sufficient depth so that neither longshore currents nor direct river-mouth supplies are now available. However, we know that considerable transport is taking place over much of the continental shelves, so sediment is probably provided from time to time, and during lowered sea-level stages of the Pleistocene these canyons may have had direct sources. The remaining 14 canyons probably head too deep to have had much supply, even during the lower sea-level stages. Nevertheless, their heads may now be filled, or they may have been depressed by warping or faulting.

Physical Characteristics of Canyons

The gradients of the submarine canyons have been discussed by several authors, most of whom point to the fact that the inclinations are far steeper than those found in adjoining land valleys. As in the case of land canyons, the gradients are closely related to the slope into which the canyons are cut. The average slope of the canyon floors is 58 m/km. This relatively high average gradient is considerably weighted by the preponderance of short canyons, virtually all of these having steeply sloping floors. For example, the Hawaiian Canyon group with an average length of 6.5 miles has an average slope of 144 m/km. Conversely, the long canyons all have low gradients. Thus, Bering Canyon has 7.8 m/km; Congo, 9.6 m/km; Cap Breton, 19 m/km, and Great Bahama, 13 m/km. These four canyons have an average length of 152 miles.

The variation along the longitudinal profiles has been classified in the table as concave upward, convex upward, relatively even, and having a profile with sharp increase in gradient. The analysis shows a great preponderance of concave upward profiles with 56[1] in that

---

[1]Where the profiles were concave but also showed a sharp step-like steepening, half credit was given to both types, so that actually a total of 66 profiles out of 93 are concave upward, and 8 convex upward.

category and only four with convex upward profiles. Twenty-three profiles showed a sharp step-like increase along the profile, most of these being otherwise concave. A comparatively even profile was shown in 10 cases. The concave profile with steep canyon heads is typical also of land canyons.

The great height of the walls of submarine canyons has always been a feature that has aroused interest among geologists. It is rather difficult to determine the maximum height of the walls of each canyon. Since one wall is ordinarily higher than the other, the two wall heights were averaged and the greatest height has been given to the nearest 1,000 ft (305 m). Averaging the various groups, we find that the average greatest height is just over 3,000 ft (915 m). This is, of course, far greater than one would find from averaging the same heights for land canyons. One factor that may make this comparison rather meaningless is that small submarine canyons in many areas may have been missed in the surveys, or were insufficiently sounded to be included in the present compilation. For example, there are many indications of short, relatively low-walled canyons along the coast of North Africa, shown in the maps of Rosfelder (1954), but these were not included here because of scarcity of data available to the writers. It is an exceptional land canyon that has walls as high as 3,000 ft (915 m). By far the highest walls of a sea canyon were found near the lower end of Great Bahama Canyon, where the walls slope up on both sides to the low islands from an axial depth of 14,060 ft (4,280 m). So far as we know, this exceeds the height of any canyon wall on the continents. Walls higher than the 5,500 ft (1,680 m) average for the Grand Canyon are found in two of the Bering Sea canyons, probably in Manila Canyon, and perhaps in Llanes Canyon off northern Spain.

The extent to which the canyon axes curve seemed worth determining since erosion valleys on land are much more sinuous than fault valleys. A model was set up as in Figure 107 to show four rather arbitrary classifications, between straight at one extreme and meandering on the other. In addition, a class with one large meander-like bend and a class with right-angle bends as in trellised drainage were included. The majority of the canyons, a total of 52, proved to be sinuous, and 36 of the rest were gently curving. Two canyons have a meandering course; two others showed right-angled bends. No canyon runs in a very straight course.

The statement has been made that the submarine canyons have few tributaries. This condition is certainly true of all the other types of marine valleys discussed in Chapter XI to XIV. To determine the situation for the canyons, four categories were selected: (1) canyons with tributaries as common as in typical land river valleys, (2) tributaries existent but less common than in typical land valleys, (3) tributaries

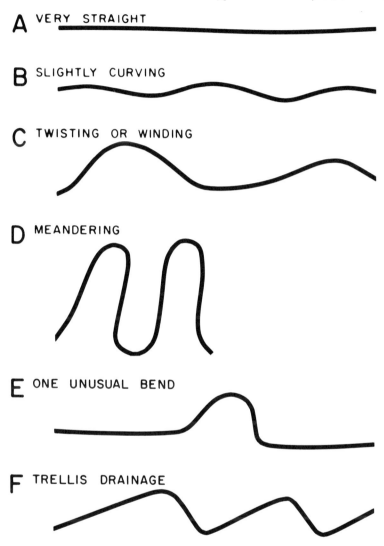

A VERY STRAIGHT

B SLIGHTLY CURVING

C TWISTING OR WINDING

D MEANDERING

E ONE UNUSUAL BEND

F TRELLIS DRAINAGE

Fig. 107. The patterns used for classifying the canyon channel trends used in the Appendix Table.

confined to the canyon heads, and (4) no tributaries. The result shows that the majority have tributaries comparable to land canyons with a total of 50, less common than land canyons with a total of 31, 10 apparently have tributaries confined to the canyon head, and only two do not have any tributaries shown. Thus, there is a very distinct difference between these canyons and the other types of marine valleys.

The character of the transverse profiles was classified in four categories also: (1) predominantly V-shaped with only a narrow floor (as in land canyons), (2) V-shaped in the upper half and trough-shaped (with a broad flat floor) in the lower half, (3) predominantly trough-shaped, and (4) of uncertain character. The analysis shows that 60 valleys have V-shaped transverse profiles, and only 13 have profiles that become trough-shaped in the lower portion, whereas only two appear to be trough-shaped along the entire length. The rest were not sounded well enough to make it possible to determine the nature of the profile.

The walls of the canyons have been dredged in a good many places. In the table, these walls are classified as (1) those having crystalline rock dredged at some place, (2) those with only sedimentary rock (3) those with only mud or other sediment, or (4) those with no information available. Forty-two of the canyons have no available information; virtually all of the rest have rock on their walls, with 22 yielding crystalline rock, mostly granite, and 30 having sedimentary rock alone. Four of the east coast canyons did not yield any rock, although we now know from continuous reflection profiles that rock is found near the surface in some of these canyons.

Many canyons have had cores taken in their axes. These are classified as (1) those with sand layers reported, (2) those with both sand and gravel in the cores, (3) with mud cores only, and (4) unknown. Of the 40 canyons that have core information from the axes, 16 show sand layers, 21 have both sand and gravel, and only four have mud alone. All of the last were found in the east coast canyons that head a long distance from shore.

The nature of termination of 42 canyons has been determined. It was found that 33 of these extend into fan-valleys and only nine terminated at the base of the continental slope without any visible fan-valley. Apparently, fan-valleys extending across the continental rise are ordinarily found outside the canyons. These usually terminate in flat basin floors.

COMPARISON WITH LAND CANYONS

In the preceding descriptions, land canyons have been compared to many of those of the sea floor. Some generalizations relative to this comparison are in order. The submarine canyons are remarkably similar in general character to land canyons cut into slopes of comparable steepness. They have courses that are approximately as winding as land canyons cut in similar slopes. Most of them have comparable V-shaped transverse profiles and a similar dendritic type of entering tributaries. Furthermore, the submarine canyons occur preponderantly outside the mouths of land valleys.

This resemblance to land canyons might suggest that they have the same origin. However, it is clearly evident that most of the submarine canyons are not simply drowned lower sections of land canyons. If they were, one would expect some continuity of the long profile between the land and the sea canyons, whereas, except in Corsica, there is no such relationship. Most of the submarine canyons show concave long profiles with decided steepness at their heads, compared to low gradients at the seaward end of land canyons. This major nickpoint, stressed by Woodford (1951), obviously requires more than a simple submergence and suggests that active marine erosion has taken place in the canyons. Furthermore, the vertical or overhanging walls in several gorges, observed by diving operations, form a remarkable contrast to the type of valley that exists in the landward continuations of the submarine canyons. This certainly suggests marine erosion. Finally, the observations in Scripps, La Jolla, San Lucas, and Los Frailes canyons all leave little doubt but that erosion is taking place in the canyon heads. It is clearly demonstrated that depth changes are also occurring in the canyons, and that sand and gravel are moving seaward along the canyon axes even out onto the fans beyond.

[*Editor's Note:* Material has been omitted at this point.]

Editor's Note: A key to the numbers and letters used in the Appendix will be found on page 140.

Canyon name and location	1 Canyon length	2 Depth at canyon head	3 Depth at canyon terminus	4 Coast character	5 Relation to points	6 Relation to river valleys	
California							
Coronado	8.0	240	5,580	C	C	C	
La Jolla	7.3	50	1,800	C	B	A	
Scripps (tributary).......	1.45	60	900	D	C	A	
Redondo	8.0	30	1,920	C	B	B	
Dume.................	3.0	120	1,860	C	A	A	
Mugu	8.0	40	2,400	C	B	C	
Sur Partington	49.0	300	10,200	D	C	C	
Carmel (tributary).......	15.0	30	6,600	A	A	A	
Monterey..............	60.0	50?	9,600?	C	C	B	
Delgada	55.0	90	8,400	D	C	B	
Mattole	16.0	60	5,720	B	B	A	
Eel	27.0	250	8,500	C	C	A	
Total or Ave. (12).......	21.5	110	5,290	1A1B 7C3D	2A4B 6C	6A 3B3C	
Oregon-Washington							
Columbia..............	37.0?	360	6,130?	B	B	A	
Willapa	60.0	500+	7,000	B	C	A	
Gray	30.0	500	6,440	B	C	A	
Quinault	25.0	500±	5,750	C	C	A	
Juan de Fuca	31.0+	800±	4,520+	B	B	C	
Total or Ave. (5)	36.6	532	5,968	4B 1C	2B 3C	4A 1C	
Bering Sea							
Umnak................	160.0	900	10,850	B	C	C	
Bering	220.0	600±	11,160	B	C	C	
Pribilof...............	86.0	500±	10,700	C?	C	C	
Total or Ave. (3)	155.3	667	10,903	2B1C	3C	3C	
U.S. East Coast							
Corsair................	14+	360	5,400	E	C	C	
Lydonia	16+	370	4,400+	E	C	C	
Gilbert	20+	480	7,680+	E	C	C	
Oceanographer	17+	600+	7,230+	E	C	C	
Welker................	27+	400	6,450+	E	C	C	

7	8	9	10	11	12	13	14	15	16
Sediment sources at head	Gradient in m/km	Nature of long profile	Maximum wall heights to nearest 1000 ft.	Channel curvature	Abundance of tributaries	Transverse profile character	Nature of canyon wall material	Sediment found in axial cores	Relation to fan-valleys
B	58	A	1,000	C	B	A	B	A	A
A	40	A	1,000	C	A	A	B	B	A
A	97	A	<1,000	B	C	A	B	B	B
A	39	A	1,000	C	B	C	B	A	A
A	97	A	1,000	C	B	A	A	D	C
A	49	A	<1,000	D	A	A	B	A	C
A	34	A	2,000	C	A	A	B	D	C
A	73	A	2,000	C	A	A	A	A	B
A	26.5	A+D	6,000	C or D	A	B	A	B	A
A	25	A	2,000	D	B	A	D	D	A
A	59	A	3,000	C	B	A	B	D	C
B	51	B	4,000	E	B	A	D	D	C
10A 2B	54	10A 1B1D	2,083	½B8½C 2½D1E	5A6B 1C	10A 1B1C	3A7B 2D	4A3B 5D	5A2B 5C
C	26	A	2,000	C	B	D	D	D	C
C	24	A	2,000	C	A	D	B	A	C
C	33	A	1,000	C	A	D	D	D	C
C	35	A	3,000	C	A	D	D	D	C
C	20	A	2,000	C	D	D	D	D	C
5C	27.6	5A	2,400	5C	3A1B 1D	5D	4D1B	4D1A	5C
C	10.4	A	4,000	C	A	D	D	D	C
C	8	A+D	6,000	C	A	A	B?	D	C
C	20	A	7,000	C	A	D	D	D	C
3C	12.8	2½A ½D	5,667	3C	3A	1A2D	1B2D	3D	3C
B	23	A	2,000	B	B	A	B	D	C
B	42	A	3,000	B	B	A	B	C	C
C	60	A+D	3,000	B	A	A	B	D	C
C	65	A	2,000	B	B	A	D	D	C
B	38	A+D	4,000	B	B	A	B	A	C

Canyon name and location	1 Canyon length	2 Depth at canyon head	3 Depth at canyon terminus	4 Coast character	5 Relation to points	6 Relation to river valleys
Hydrographer	27+	450+	6,600+	E	C	C
Hudson	50	300	7,000	B	C	A
Wilmington	23+	320	6,940+	E	C	C
Baltimore.............	28+	400	6,110+	B?	C	C
Washington	28+	360	6,740+	E	C	C
Norfolk	38	320	8,300	E	C	C
Total or Ave. (11).......	26.2	395	6,623	2B9E	11C	1A 10C
Hawaiian-Molokai						
Halawai	6.0+	300±	3,540+	B	C	A
Naiwa.................	7.5	380	4,880	D	C	A
Waikolu	9.0	<600	6,540	B	C	A
Pelekunu	10.0	<320	6,320	B	C	A
Hawaiian-Kauai						
Hanakapiai	6.0+	280	7,480	D	C	A
Hanakoa..............	3.7	600±	4,820	D	C	C
Hanopu	3.6	300	5,100	D	C	C
Total or Ave. (7)	6.5	397	5,526	3B 4D	7C	5A 2C
Western Europe						
Shamrock	30+	1200±	14,400	B?	C	C
Black Mud...........	30+	900±	12,200	E	C	C
Audierne	27	600±	10,500	B	C	C
Cap Ferret...........	50+	800	11,647	C	C	C
Cap Breton	135 or 70	400±	13,100	D	C	C
Aviles	65	60±	8,000	C	B	A (old river)
Llanes...............	38	450	13,300	D	B	C
Nazare	93	200±	14,764	C	A	A
Lisbon...............	21	400	6,450	B	B	A?
Setubal	33+	350	6,880	C	C	A
Total or Ave. (10)	52.2 or 45.7	536	11,224	3B4C 2D	1A 3B5C	4A5C
Mediterranean mainland						
Grand Rhone	15+	600±	5,550	C	C	A
Marseille	20+	600±	6,840±	B	C	C
Canon de la Cassidaigne	19+	360	6,630+	B	B	A
Toulon	12+	260	6,600	A borderline	C	C
Stoechades...........	17+	300	4,380+	A	C	C
St. Topez	25+	60?	5,750	A borderline	C	A
Cannes	17+	100?	6,600?	C or A	C	B
Var	15	160	6,550	D delta	A	A
Nice.................	12	150	5,840	D	C	B

7	8	9	10	11	12	13	14	15	16
Sediment sources at head	Gradient in m/km	Nature of long profile	Maximum wall heights to nearest 1000 ft.	Channel curvature	Abundance of tributaries	Transverse profile character	Nature of canyon wall material	Sediment found in axial cores	Relation to fan-valleys
B	37	A+D	3,000	B	B	A	C	A	A
B	25	A+D	4,000	B	A	A	B	A	A
B	48	A	3,000	C	A	A	C	C	C
B	34	A+D	3,000	B	B	A	C	C	C
B	38	A+D	2,000	B	A	A	C	C	C
B	35	A+D	3,000	B	B	A	A	A	C
9B	40	7½A	2,900	10B	4A	11A	6B4C	4A4C	2A9C
2C		3½D		1C	7B		1D	3D	
B	90	C	1,000	C	D	A	D	D	C
B	100	C	1,000	B	D	A	A	A	B
B	110	A	2,000	B	B	A	A	B	A
A	100	A	1,000	C	B	A	D	D	B
B	200	B+D	1,000	B	B?	A	A	D	C
C	190	B	1,000	B	B?	A	A	A	C
B	220	B+D	2,000	B	B?	A	B	A	A
1A5B	144	2A2B	1,286	5B	5B	7A	4A1B	3A1B	2A2B
1C		2C1D		2C	2D		2D	3D	3C
C	28	C+D	3,000	B	B?	D	D	D	C
B	57	D	3,000	B	B	B	B	A	A
C	60	D	4,000	B	D	D	D	D	C
C	31	A	3,000	C	A	D	D	D	A
B	58	A	6,000	B	B	D	D	L	C
A	20 or 16	C+D	5,000 or 6,000	B	A?	B	B	D	A?
B	45	C	5,000	C	A	D	D	D	C
A	36	D	5,000	F	B	D	A?	D	C
B	48	B+D	4,000	B	D	D	B	D	C
B	33	C?	2,000	C?	B	D	B?	D	C
2A	41.6 or	2A½B	4,000	5B3C	3A	1B	1A3B	9D	2A7C
4B3C	41.2	3C3½D		1F	4B2D	8D	5D		
C	55	C?	2,000	B	C	A	D	D	C
C	52	A+D	2,000	C	A	A	D	D	C
B	55	D	3,000	C	B	A	D	D	C
B	110	D?	4,000	B	A	A	D	D	C
A	40	A	4,000	B	A	A	B?	D	C
A	38	C?	3,000	C	A	A	D	D	C
A	65	A	3,000	C	A	A	D	D	C
A	71	A	3,000	C?	A	B	D	B	A
A	79	A+D	2,000	C?	A	B	B?	B	A

135

Canyon name and location	1 Canyon length	2 Depth at canyon head	3 Depth at canyon terminus	4 Coast character	5 Relation to points	6 Relation to river valleys
Cap d'Ail	14	320	6,870	B	C	C
Nervia	16	330	6,280	C	C	A
Taggia	12	300±	7,500	C	C	A
Mele	31	200	6,150	C curving	C	B
Noli	14	120	4,990	D curving	C	A?
Polcevera	49	300	8,830	C	C	A?
Genoa	20	260	6,260	B	C	A
Total or Ave. (16)	17.4	276.5	6,351	3A4B 5C3D	1A1B 14C	9A 3B4C
Mediterranean Islands						
Crete	4	<300	3,300	B?	C	A
West Corsica						
St. Florent	25	150±	7,850	A	A	A
Calvi	13	200±	7,800	B	C	A
Porto	20	150±	8,200	A	C	A
Sagone	29	150±	6,200	A	C	A
Ajaccio	34	150±	8,200	A	C	A
Valinco	35	150±	8,000	A	C	A
Total or Ave. (7)	22.8	178	7,078	5A2B	1A 6C	7A
Baja California						
San Pablo	20+	<400	8,400+	D	A	B
Cardonal	16+	<450	7,500+	C	B	B
Vigia	10	?	7,200	C+D	C	A or B
San Lucas—	19	30	6,900?	A-bay	C	A
Santa Maria	(24)		8,000			
San Jose	32	50	7,200?	C	C	A?
Vinorama— Salado	9	200	6,300	C	A	A
Los Frailes	9.5	10	5,200	A-bay	A (S. wind)	A
Saltito	6	1200	5,100	B?	C	B
Palmas— Pescadero	13 9.3	100?	5,300	C	A (S. wind)	C
Total or Ave. (9)	15.2	305	6,710	2A1B 4½C1½D	4A1B 4C	4½A 3½B1C
East Honshu						
Ninomiya	4.8	400	2,600	C	C	C
Sagami	5.0	310	3,300	C	C	C
Enoshima	6.7	450	3,250	E	C	C

7	8	9	10	11	12	13	14	15	16
Sediment sources at head	Gradient in m/km	Nature of long profile	Maximum wall heights to nearest 1000 ft.	Channel curvature	Abundance of tributaries	Transverse profile character	Nature of canyon wall material	Sediment found in axial cores	Relation to fan-valleys
B	78	D	1,000	B	B	B	D	D	A
B	62	D	2,000	B	B	B	D	B	A
B	100	A	2,000	C	B	A	D	D	A
B	32	A+D	1,000	C	B	B	D	D	B
A	58	A	2,000	C	B	B	D	D	B
B	29	A+D	3,000	B	B	B	D	D	C
B	50	D	2,000	B	B	C?	D	D	B
7A7B 2C	60.9	7A2C 7D	2,400	7B 9C	7A8B 1C	8A 7B1C	2B 15D	3B 13D	4A3B 8C
A	200	D	1,000	C	B	A	D	B	A
A	51	A	3,000	C	A	A	D	A	A?
B	97	A+D	3,000	C	Bor A	A	D	D	A
A	67	A	4,000	C	A	A	A?	D	C
A	35	A	3,000	C	A	A	D	D	A?
A	39	A	4,000	B	A	A	D	D	B?
A	37	A+D	4,000	B	A	A	D	D	C
6A1B	75	6A 1D	3,100	2B5C	5½A 1½B	7A	1A6D	1A1B 5D	4A1B 2C
A	67	B+D	3,000	C	B	A	B	D?	C
B	73	A+D	3,000	C	A	A	B	D	C
A	115	A	3,000	B	C	B	A	B	A
A	70 56	A+D	3,000	C	A	A	A	B	A
A	41	A+D	3,000	C	A	A	A (sed.)	B	A
B	113	A	1,000	C	A	A	A	B	A
A	91	A+D	2,000	C	C	A	A	B	B
C	108	A	1,000	C?	A	A	A	D	A
A	65 or 91	A	2,000	C	A	A	A	A	A
6A2B 1C	81	6A½B 2½D	2,333	1B 8C	6A1B 2C	8A 1B	7A 2B	1A5B 3D	6A1B 2C
B	77	A	<1,000	B	C	A	B	D	C
B	100	A	<1,000	C	A	B	B	D	C
B	70	A	<1,000	B	C	A	B	B	C

137

Canyon name and location	1 Canyon length	2 Depth at canyon head	3 Depth at canyon terminus	4 Coast character	5 Relation to points	6 Relation to river valleys
Hayama	13	300	4,600	E	C	C
Miura	15	173		E	C	C
Misaki	14.5	330	4,600	E	C	C
Jogashima	10	330		E	C	C
Tokyo	30	300	4,900	A	C	C
Mera	20	190	5,500	B (bay)	B	C
Kamogawa	25	200	9,100	B	C	A
Total or Ave. (10)	29.25	298	4,731	1A2B 2C5E	1B 9C	1A 9C
Miscellaneous						
Great Bahama	125	4800	14,060	B	C	C
Congo	120	80	7,000	A	C	A
Ceylon Trincomalee	20+	30+	9,500+	A	A?	A
Manila	31+	300	7,800+	B	C	A
Bacarra NW Luzon	15	300	6,000+	C+D	C	A
San Antonio, Chile	20+	<150	2,700+	C?	C	A?
Total or Ave. (6)	55	943	7,843	2A2B 1½C½D	1A 5C	5A 1C

7	8	9	10	11	12	13	14	15	16
Sediment sources at head	Gradient in m/km	Nature of long profile	Maximum wall heights to nearest 1000 ft.	Channel curvature	Abundance of tributaries	Transverse profile character	Nature of canyon wall material	Sediment found in axial cores	Relation to fan-valleys
B	54	A	2,000	C	B	A	A	D	C
B	48	A	2,000	C	B	A	B	B	C
B	49	A	2,000	C	A	A	B	A	C
B	71	D	1,000	C	A	A	A	D	C
B	26	A+D	3,000	C	A	B	A	B	A
B	44	C+D	2,000	C	A	A	B	B	C
B	59	A+D	5,000	B	B	B	B	A	C
10B	60	7A½C	2,000	3B	5A3B	7A	3A	2A4B	1A
		2½D		7C	2C	3B	7B	4D	9C
C	13	A	14,000	C	A	A	C	B	A
A	96	A	4,000	B	C	A?	D	A	A
								sand	
A	79	A	4,000	F	A	A	A?	D	C
B	40	A+D	6,000	C	A	D	D	D	C
B	63	C	3,000	C	A	B?	D	D	C
A	32	A	3,000	B+C	B	D	D	D	C
3A2B	54	4½A	5,666	1½B	4A1B	3A	1A1C	1A	4½A
1C		1C½D		3½C1F	1C	1B2D	4D	1B4D	1C½D

139

1. Length of canyon measured along axis (nautical miles)
2. Depth at canyon head (feet)
3. Depth at canyon terminus (feet)
4. Character of coast inside canyon head
   A. Heads in estuary
   B. Heads off embayment
   C. Heads off straight beach or barrier
   D. Heads off relatively straight cliff
   E. Uncertain
5. Relation of canyon head to points of land
   A. On upcurrent side of point
   B. Relatively near upcurrent side of point
   C. No relation to point
6. Relation of canyon head to river valleys
   A. Probable connection
   B. No connection
   C. Uncertain
7. Source of sediments to canyon head
   A. Receives good supply
   B. Supply restricted now, greater during lowered sea level stages
   C. Little known supply of sediment because of depth
8. Gradient of axis in meters per kilometer
9. Nature of longitudinal profile
   A. Generally concave upward
   B. Generally convex upward
   C. Relatively even slope
   D. Local step-like steepening along axis

10. Maximum height of walls in feet
11. Channel curvature (see Fig. 107 of Chapt. X)
    A. Straight
    B. Slightly curving
    C. Twisting or winding
    D. Meandering
    E. One meandering bend
    F. Right-angled bends
12. Abundance of tributaries
    A. As common as typical land valleys
    B. Less common than typical land valleys
    C. Confined to canyon head
    D. No known tributaries
13. Character of transverse profile
    A. Predominantly V-shaped
    B. V-shaped inner canyon, trough-shaped outer canyon
    C. Predominantly trough-shaped
    D. Uncertain
14. Nature of canyon wall material
    A. Crystalline rock dredged
    B. Rock dredged, but all sedimentary
    C. Mud only dredged on wall
    D. Unknown
15. Nature of core sediment from axis
    A. Includes sand layers
    B. Includes sand and gravel layers
    C. Mud cores only
    D. Unknown
16. Relation to fan-valleys
    A. Has fan-valley continuation
    B. No fan-valley continuation
    C. Unknown

Allen, J. R. L., 1964, The Nigerian continental margin: bottom sediments, submarine morphology, and geological evolution. *Mar. Geol.,* 1 (4), 289-332.

Athcarn, William D., 1963, Bathymetry of the Straits of Florida and the Bahama Islands. Pt. II, Bathymetry of the Tongue of the Ocean, Bahamas. *Bull. of Mar. Sci. of Gulf and Carib.,* 13 (3), 365-377.

Berthois, Leopold, 1962, Morphologie et géologie sous-marine. *Rev. Trav. Inst. Pêches Marit* II, 231-246.

Berthois, L. and R. Brenot, 1960, La morphologie sous-marine du Talus plateau continental entre le sud de l'Irlande et la Cap Ortegal (Espagne). *Jour. du Conseil Intl. Explor. de Mar.,* 25 (2), 111-114.

Bouma, A. H., 1964, Self-locking compass. *Mar. Geol.,* 1, 181-186.

Bouma, A. H., and F. P. Shepard, 1964, Large rectangular cores from submarine canyons and fan valleys. *Bull. Amer. Assoc. Petrol. Geol.,* 48 (2), 225-231.

Bourcart, Jacques, 1938, La marge continental. *Bull. Geol. Soc. France,* Ser. 5, v. VIII, 393-474.

———, 1950, Le socle continental de Toulon a la frontière Espagnole. *Conf. Centre Recherches et Études Oceanog.,* no. 3, 10 pp.

———, 1952, *Geographie du Fond des Mers.* Payot, Paris, 307 pp.

———, 1959, Morphologie du précontinent des Pyréneés a la Sardaigne. *Colloq. Intern. Centre Natl. Recherche Sci.,* LXXXIII, 33-50.

———, 1960, Carte topographique du fond de la Méditerranee occidentale. *Bull. Inst. Oceanogr., Monaco,* No. 1163, 3-20.

Bourcart, Jacques, François Ottmann, and Jeanne-Marie Ottmann-Richard, 1958, Premiers résultats de l'étude des carottes de la Baie des Anges, Nice. *Revue de Geographie Physique et Geol. Dynamique* (2), 1 (3), 167-173.

Bourcart, Jacques, Maurice Gennesseaux, Eloi Klimel, and Mme. Yolande le Calvez, 1960, Oceanographie. Les sédiments des vallées sous-marines au large dans le Golfe de Gênes. *Acad. Sci. (Paris) Comptes rendus,* 251, 1443-1445.

Brenot, Roger and Leopold Berthois, 1962, Bathymetrie de secteur Atlantique du Banc Porcupine (ouest de l'Irlande) au Cap Finisterre (Espagne). *Rev. Trav. Inst. Peches Marit.*, 26 (2), 219-246.

Bucher, W. H., 1940, Submarine valleys and related geologic problems of the North Atlantic. *Geol. Soc. Amer. Bull.*, 51, 489-512.

Buffington, Edwin C., 1952, Submarine "natural levees." *Jour. Geol.*, 60, 473-479.

———, 1961, Experimental turbidity currents on the sea floor. *Bull. Amer. Assoc. Petrol. Geol.*, 45 (8), 1392-1400.

———, 1964, Structural control and precision bathymetry of La Jolla submarine canyon, California. *Marine Geol.*, 1, 44-58.

Buffington, Edwin C. and David G. Moore, 1963, Geophysical evidence on the origin of gullied submarine slopes, San Clemente, California. *Jour. Geol.*, 71 (3), 356-370.

Casagrande, A., 1936, Characteristics of cohesionless soils affecting the stability of slopes and earth fills. *Jour. Boston Soc. Civil Eng.*, 23, 13.

———, 1938, Compaction tests and critical density investigations of cohesionless materials for Franklin Falls Dam. *Boston Dist., U.S. Army Corps. Eng.*

Chamberlain, T. K., 1960, Mechanics of mass sediment transport in Scripps submarine canyon, California. Ph.D. thesis, Univ. of Calif., Scripps Institution of Oceanography, 200 pp.

———, 1964, Mass transport of sediment in the heads of Scripps submarine canyon, California, in *Papers in Marine Geology* (Shepard Commem. Vol.) Macmillan, New York, pp. 42-64.

Collet, Leon W., 1925, *Les Lacs,* Paris, 320 pp.

Cooper, L. H. W. and David Vaux, 1949, Cascading over the continental slope of water from the Celtic Sea, *Mar. Biol. Assoc., U.K.*, 28, 719-750.

Coulomb, C. A., 1776, Essai sur une application des regles de maximis et minimis a quelques problemes de statique, relatifs à l'architecture; in *Mémoires de Mathematique et de Physique,* L'Academie Royale des Sci., Paris, pp. 343-382.

Crowell, J. C., 1952, Submarine canyons bordering central and southern California. *Jour. Geol.*, 60 (1), 58-83.

———, 1957, Origin of pebbly mudstones. *Bull. Geol. Soc. Amer.*, 68, 993-1009.

Curray, J. R., 1960, Sediments and history of Holocene transgression, continental shelf, northwest Gulf of Mexico, in *Recent Sediments, Northwest Gulf of Mexico,* F. P. Shepard, F. B. Phleger, and Tj. H. van Andel eds. Amer. Assoc. Petroleum Geol., pp. 221-266.

———, 1964, Shallow structure of the continental terrace, northern and central California. *Abs. Ann. Meeting Geol. Soc. Amer.,* pp. 37-38.

Curray, J. R. and D. G. Moore, 1964, Pleistocene deltaic prograda-
tion of continental terrace, Costa de Nayarit, Mexico, in *Marine
Geology of the Gulf of California—a Symposium*, Amer. Assoc. Petrol.
Geol. Mem. 3, pp. 193-215.

Daly, R. A., 1936, Origin of submarine canyons. *Amer. Jour. Sci.,* 31
(186), 401-420.

Dana, J. D., 1863, *A Manual of Geology*. Philadelphia, 798 pp.

————, 1890, Long Island Sound in the Quaternary era, with observations on the submarine Hudson River channel. *Amer. Jour. Sci.,*
ser. 3, 425-437.

Dangeard, L., 1961, A propos des phenomenes sous-marine profonds de glissement et de resedimentation. *Cahiers Oceanog.,* XIII
(2), 68-72; (31), 401-420.

————, 1962, Observations faites en "Soucoupe Plongeante" au large
de Banyuls. *Cahiers Oceanog.,* XIV (1), 19-24.

Dangeard, Louis, and Pierre Giresse, 1965, Photographie sous-
marine et geologie. *Cahiers Oceanog.,* 17 (4), 255-269.

Davidson, G., 1887, Submarine valleys on the Pacific Coast of the
United States. *Calif. Acad. Sci. Bull.,* 2, 265-268.

————, 1897, The submerged valleys of the coast of California,
U.S.A, and of Lower California, Mexico. *Calif. Acad. Sci. Proc.,*
ser. 3, 1, 73-103.

Davis, W. M., 1934, Submarine mock valleys. *Geog. Rev.,* 24, 297-308.

Day, A. A., 1959, The continental margin between Brittany and
Ireland. *Deep-Sea Res.,* 5, 249-265.

de Andrade, Carlos F., 1937. Os vales submarinos Portugueses e o
diastrofismo das Berlengas e da Estremadura (with English
summary). *Casa Portuguesa,* pp. 237-249.

Dietz, Robert S., 1953, Possible deep-sea turbidity-current channels in
the Indian Ocean. *Bull. Geol. Soc. Amer.,* 64, 375-378.

Dill, R. F., 1961, Geological features of La Jolla Canyon as revealed
by dive no. 83 of the bathyscaph TRIESTE. *Tech. mem.* no. TM-
516, U.S.N. Electronics Lab., 27 pp.

————, 1964a, Sedimentation and erosion in Scripps submarine
canyon head. In *Papers in Marine Geology* (Shepard Commem.
Vol.), Macmillan, New York, pp. 23-41.

————, 1964b, Contemporary submarine erosion in Scripps Submarine Canyon, Ph.D. Thesis, Univ. Calif., Scripps Inst. Oceanog.,
privately printed, 269 pp.

————, in press a, Submarine erosion in the head of La Jolla Canyon.
Bull. Geol. Soc. Amer.

————, in press b, Sand flows and sand falls. In *Reinholt Encyclopedia
of Earth Sciences*. R. W. Fairbridge, ed.

Dill, R. F., R. S. Dietz, and H. B. Stewart, Jr., 1954, Deep-sea channels and delta of the Monterey submarine canyon. *Bull. Geol. Soc. Amer.*, 65, 191-194.

Dill, R. F., and D. G. Moore, in press, A diver-held vane shear apparatus. *Marine Geol.*

Dill, R. F., and G. A. Shumway, 1954, Geologic use of self-contained diving apparatus. *Bull. Amer. Assoc. Petrol. Geol.*, 38 (1), 148-157.

Dott, R. H. Jr., 1963. Dymanics of subaqueous gravity depositional processes. *Bull. Amer. Assoc. Petrol. Geol.*, 47, 105-128.

Drake, C. L., M. Ewing, and G. H. Sutton, 1959, Continental margins and geosynclines: the east coast of North America north of Cape Hatteras. In *Physics and Chemistry of the Earth*, Pergamon, London, (3), 110-198.

Elmendorf, C. H., and B. C. Heezen, 1957, Oceanographic information for engineering submarine cable systems. *The Bell System Tech. Jour.*, 36 (5), 1047-1093.

Emery, K. O., 1960, *The Sea off Southern California.* John Wiley & Sons, New York, 366 pp.

———, 1965, Geology of the continental margin off eastern United States. In *Submarine Geology and Geophysics*, W. F. Whittard and R. Bradshaw, eds., Butterworths, London, pp. 1-20.

Emery, K. O., W. S. Butcher, H. R. Gould, and F. P. Shepard, 1952, Submarine geology off San Diego, California. *Jour. Geol.*, 60 (6), pp. 511-548.

Emery, K. O., and Jobst Hülsemann, 1963, Submarine canyons of southern California. Pt. I, Topography, Water and Sediments. *Allan Hancock Pacific Expeditions*, 27 (1), Univ. So. Calif. Press, Los Angeles, Calif., pp. 1-80.

Emery, K. O., and F. P. Shepard, 1945, Lithology of the sea floor off southern California. *Bull. Geol. Soc. Amer.*, 56, 431-478.

Emery, K. O., and R. D. Terry, 1956, A submarine slope of southern California. *Jour. Geol.*, 64 (3), 271-280.

Ericson, D. B., 1952, North Atlantic deep-sea sediments and submarine canyons. *Trans. N.Y. Acad. Sci.*, ser. II, 15 (2), 50-53.

Ericson, D. B., Maurice Ewing, Göesta Wollin, and B. C. Heezen, 1961, Atlantic deep-sea sediment cores. *Bull. Geol. Soc. Amer.*, 72, 193-286.

Ewing, John, Xavier Le Pichon, and Maurice Ewing, 1963, upper stratification of Hudson Apron region. *Jour. Geophys. Res.*, 68 (23), 6303-6316.

Ewing, Maurice, D. B. Ericson, and B. C. Heezen, 1958, Sediments and topography of the Gulf of Mexico. In *Habitat of Oil*, Amer. Assoc. Petrol. Geol., Tulsa, Okla., pp. 995-1053.

Fisk, H. N., and E. McFarlan, Jr., 1955, Late Quaternary deltaic deposits of the Mississippi River. In *Crust of the Earth, Geol. Soc. Amer. Spec. Paper 62*, pp. 279-302.

Francis, T. J. G., 1962, Black Mud Canyon. *Deep-Sea Res.*, 9, 457-464.

Fraser, G. D., and H. F. Barnett, 1959, Geology of the Delarof and westernmost Andreanof Islands, Aleutian Islands, Alaska. *U.S. Geol. Surv. Bull. 1028-1*, pp. 211-245.

Gates, Olcott, and William Gibson, 1956, Interpretation of the configuration of the Aleutian Ridge. *Bull. Geol. Soc. Amer.*, 67, 127-146.

Gealy, Betty Lee, 1956, Topography of the continental slope in northwest Gulf of Mexico. *Bull. Geol. Soc. Amer.*, 66, 203-227.

Gibson, William, and Haven Nichols, 1953, Configuration of the Aleutian Ridge Rat Islands — Semisopochnoi I. to west Buldir I. *Bull. Geol. Soc. Amer.*, 64, 1173-1181.

Gorsline, D. S., J. W. Vernon, and A. Shiffman, 1965, Processes of sand transport in the inner margins of the continental shelf. Abs. 39th Ann. Meeting, *Amer. Soc. Paleont. and Minerol*, New Orleans. p. 63.

Gorsline, D. S., and K. O. Emery, 1959, Turbidity-current deposits in San Pedro and Santa Monica basins off southern California. *Bull. Geol. Soc. Amer.*, 70, 279-290.

Gueze, E. C.: W. A. and T. K. Tan, 1954. The mechanical behavior of clays. In *Proc. Second Intern. Congr. Rheology*, Academic Press, New York, p. 451.

Hadley, M. L., 1964, The continental margin southwest of the English Channel. *Deep-Sea Res.*, 11, 767-779.

Hamilton, E. L., 1957, Marine geology of the southern Hawaiian Ridge. *Bull. Geol. Soc. Amer.*, 68, 1011-1026.

————, 1959, Thickness and consolidation of deep-sea sediments. *Bull. Geol. Soc. Amer.*, 70 (11), 1399-1424.

Hamilton, E. L., and H. W. Menard, 1956, Density and porosity of sea-floor surface sediments off San Diego, Calif. *Bull. Amer. Assoc. Petrol. Geol.*, 40 (4), 754-761.

Hayter, P. J. D., 1960, The Ganges and Indus Submarine Canyons. *Deep Sea Res.*, 6 (3), 184-186.

Heezen, B. C., 1956, Corrientes de turbidez del Rio Magdalena. *Bol. Soc. Geografica Colombia, Bogota,* nos. 51 and 52, pp. 135-143.

————, 1959, Note on progress in geophysics. Dynamic processes of abyssal sedimentation: erosion, transportation, and redeposition on the deep-sea floor. *Geophys. Jour. of Royal Astronom. Soc.*, 2 (2), 142-163.

Heezen, B. C., Roberta Coughlin, and W. C. Beckman, 1960, Equatorial Atlantic mid-ocean canyon. *Abs., Bull. Geol. Soc. Amer.*, 71, 1886.

Heezen, B. C., and C. L. Drake, 1964, Grand Banks slump. *Bull. Amer. Assoc. Petrol. Geol.*, 48 (2), 221-233.

Heezen, B. C., D. B. Ericson, and Maurice Ewing, 1954, Further evidence for a turbidity current following the 1929 Grands Banks earthquake. *Deep-Sea Res.*, 1, 193-202.

Heezen, B. C. and Maurice Ewing, 1952, Turbidity currents and submarine slumps, and the Grand Banks earthquake. *Amer. Jour. Sci.*, 250, 849-873.

Heezen, Bruce C., and C. Hollister, 1964, Deep-sea current evidence from abyssal sediments. *Marine Res.*, 1 (2), 141-174.

Heezen, B. C., and A. S. Laughton, 1963, Abyssal Plains. In *The Sea*, Vol. 3, M. N. Hill, ed., Interscience Pub., John Wiley & Sons, N. Y., pp. 312-364.

Heezen, B. C., R. J. Menzies, E. D. Schneider, W. M. Ewing, and C. L. Granelli, 1964, Congo submarine canyon. *Bull. Amer. Assoc. Petrol. Geol.*, 48 (7), 1126-1149.

Heezen, B. C., M. Tharp, and Maurice Ewing, 1959, The Floors of the Ocean. I, North Atlantic. *Geol. Soc. Amer. Spec. Paper 65*, 122 pp.

Heezen, B. C. and M. Tharp, 1964, physiographic diagram of the Indian Ocean. (Map) *Geol. Soc. Amer.*, New York.

Heim, Albert, 1888, Bergsturz und Menschenleben, *Vierteljahrschriften Naturforsch.*, Zurich.

Hess, H. H., 1932, Interpretation of gravity-anomalies and sounding-profiles obtained in the West Indies by the International expedition to the West Indies in 1932. *Trans. Amer. Geophys. Union*, 13th Ann. Meeting, pp. 26-32.

Hodgson, E. A., 1930, The Grand Banks earthquake. *Supp. Proc. Eastern Sect. Seis. Soc.*, pp. 72-79.

Holtedahl, Hans, 1958, Some remarks on geomorphology of continental shelves off Norway, Labrador, and southeast Alaska. *Jour. Geol.*, 66 (4), 461-471.

Holtedahl, Olaf, 1940, The submarine relief off the Norwegian coast. *Norske Videnskaps-Akad. Oslo*, 43 pp.

Hoshino, Michihei, and Takahiro Sato, 1960, On the topography and bottom sediment of Kamogawa submarine canyon, Boso Peninsula. *Quaternary Res.*, (6), 228-237.

Houtz, R. E., and H. W. Wellman, 1962, Turbidity currents at Kadavu Passage, Fiji. *Geol. Mag.*, 99 (1), 57-62.

Hoyt, W. V., 1959, Erosional channel in the Middle Wilcox near Yoakum, Lavaca County, Texas. *Trans. Gulf Coast Assn. Geol. Soc.*, 9, 41-50.

Hubert, John F., 1964, Textural evidence for deposition of many western North Atlantic deep-sea sands by ocean-bottom currents. *Jour. Geol.*, 72 (6), 757-785.

Hull, Edward, 1912, *The Sub-Oceanic Physiography of the North Atlantic Ocean*, Edward Stanford, London, 41 pp.

Hurley, Robert J., 1960, The geomorphology of abyssal plains in the northeast Pacific Ocean. Mimeo. Rept., Scripps Inst. Oceanog., ref. 60-7, 105 pp.

———, 1963, Analysis of bathymetric data from the search for U.S.S. THRESHER. *Program 76th Ann. Meeting, Geol. Soc. Amer.*, pp. 85a-86a.

Hurley, R. J., and F. P. Shepard, 1964, Submarine canyons in the Bahamas. *Abs. 77th Ann. Meeting, Geol. Soc. Amer.*, p. 99.

Inman, D. L., 1950, Submarine topography and sedimentation in the vicinity of Mugu submarine canyon, California. *Beach Erosion Board, Corps of Engrs. Tech. Memo No. 19*, 45 pp.

———, 1953, Areal and seasonal variations in beach and nearshore sediments at La Jolla, California, *Beach Erosion Board, Corps of Engrs. Tech. Memo. No. 39*, ii, 82 pp.

———, 1963, in Shepard, F. P. *Submarine Geology*, 2nd Edit., Harper & Row, New York, pp. 138-140.

Inman, D. L. and T. K. Chamberlain, 1960, Littoral sand budget along the southern California coast. *Rept. 21st Intl. Geol. Cong.*, Copenhagen, Vol., abs., pp. 245-246.

Inman, D. L., and Earl Murray, 1961, Mechanics of sedimentation. *ONR Progr. Rept.* Jan. 1—June 30, pp. 11-13.

Jennings, C. W. and R. G. Strand, 1965, Geol. map of California. Santa Cruz sheet. Div. of Mines, State of Calif.

Johnson, D. W., 1925, *New England—Acadian Shoreline*. John Wiley and Sons, New York, 608 pp.

———, 1939, *The Origin of Submarine Canyons*. Columbia Univ. Press, New York, 126 pp.

Kaplin, P. A., 1961, Diver studies of the heads of submarine canyons. *Okeanologiya,* 1 (6), 1034-1038.

Keith, Arthur, 1930, The Grand Banks earthquake. *Proc. Seismol. Soc. Amer.* Eastern Sec., Suppl. 5 pp.

Kiersch, George A., 1965, Vaiont reservoir disaster. *Geotimes,* 9 (9), 9-12.

Kiilerich, A., 1958, The Ganges Submarine Canyon. *Andhra Univ. Mem. Oceanog.* ser. 62 (11), 29-32.

Kindle, E. M., 1931, Sea-bottom samples from the Cabot Strait earthquake zone. *Bull. Geol. Soc. Amer.*, 42, 557-574.

Kjellman, W., 1955, Mechanics of large Swedish landslips. *Geotechnique,* 5, 74-78.

Kuenen, Ph. H., 1937, Experiments in connection with Daly's hypothesis on the formation of submarine canyons. *Leidsche Geologische Mededeelingen,* VIII, 316-351.

———, 1950, *Marine Geology.* John Wiley and Sons, New York, 568 pp.

————, 1953, Origin and classification of submarine canyons. *Bull. Geol. Soc. Amer.*, 64, 1295-1314.

Kuenen, Ph.H. and C. I. Migliorini, 1950, Turbidity currents as a cause of graded bedding. *Jour. Geol.*, 58, 91-127.

Lambe, T. W., 1958, The structure of compacted clay, *Jour. Soil Mechanics and Foundation Div., Amer. Soc. Civil Engineers,* 85 (SM2), 55.

Laughton, A. S., 1959, Photography of the ocean floor. *Endeavour*, 18 (72), 178-185.

Laughton, A. S., 1960, An interplain deep-sea channel system. *Deep-Sea Research,* 7 (2), 75-88.

Lawson, Andrew, 1893, The Geology of Carmelo Bay. *Univ. Calif., Dept. Geol., Bull.* 1, 1-59.

Le Conte, Joseph, 1891, Tertiary and post Tertiary changes of the Atlantic and Pacific coasts. *Bull. Geol. Soc. Amer.*, 2, 323-328.

Leonards, G. A., 1962, *Foundation Engineering.* McGraw-Hill, New York, 1136 pp.

Limbaugh, Conrad and F. P. Shepard, 1957, Submarine canyons. In *Marine Ecology,* Joel Hedgepeth, ed., *Geol. Soc. Amer. Mem. 67,* 1, 633-639.

Lindenkohl, A., 1885, Geology of the sea-bottom in the approaches to New York Bay. *Amer. Jour. Sci.*, ser. 3, 29, 475-480.

Lisitzin, A. P. and A. V. Zhivago, 1960, Marine geological work of the Soviet Antarctic Expedition, 1955-1957. *Deep-Sea Research,* 6 (2), 77-87.

Ludwick, J. C., 1950, Deep water sands off San Diego. Unpublished Ph.D. thesis, Univ. of Calif., Los Angeles, 55 pp.

Ma, Ting Ying H., 1947, Submarine valleys around the southern part of Taiwan and their geological significance. *Bull. Oceanog. Inst. Taiwan,* 2, 1-12.

Martin, B. D., 1963, Rosedale Channel—Evidence for Late Miocene submarine erosion in the Great Valley of California. *Bull. Amer. Asso. Petrol. Geol.*, 47 (3), 441-456.

————, 1964, Geology of Monterey Canyon. Unpublished Ph.D. Thesis, Univ. of So. Calif., Los Angeles.

Mathews, W. H., and F. P. Shepard, 1962, Sedimentation of the Fraser River Delta, British Columbia. *Bull. Amer. Assoc. Petrol. Geol.*, 46, 1416-1437.

Matthes, F. E., 1907, Quoted by A. C. Lawson in the California Earthquake of 1906, State earthquake investigation commission rept., *Carnegie Inst. of Washington, pub. 87,* I (1), 54-58.

Menard, H. W. Jr., 1955, Deep-sea channels, topography and sedimentation. *Bull. Amer. Assoc. Petrol. Geol.*, 39 (2), 236-255.

————, 1960, Possible pre-Pleistocene deep-sea fans off central California. *Bull. Geol. Soc. Amer.*, 71, 1271-1278.

————, 1964, *Marine Geology of the Pacific.* McGraw-Hill, New York, 271 pp.

Menard, H. W., E. C. Allison, and J. W. Durham, 1962, A drowned Miocene terrace in the Hawaiian Islands. *Science,* 138, 896-897.

Menard, H. W., and J. C. Ludwick, 1951, Applications of hydraulics to the study of marine turbidity currents. *Soc. Econ. Paleont. and Mineralog., Spec. Pub. No. 2,* 2-13.

Menard, H. W., S. M. Smith, and R. M. Pratt, 1965, The Rhone deep-sea fan. In *Submarine Geology and Geophysics,* W. F. Whittard and R. Bradshaw, eds., Butterworths, London, pp. 271-285.

Middleton, G. V., in press, Small-scale models of turbidity currents and the criterion for auto-suspension.

Milne, John, 1897, Sub-oceanic changes. *Geo. Jour.,* 10 (2), 129-146, 259-289.

Moore, D. G., 1956, Vane shear strength, porosity and permeability relationship of some sieved cohesionless sands. *XX Cong. Geol. Internat. Resum. de los Trabajos Presentados Mexico,* pp. 266-267.

————, 1960, Acoustic-reflection studies of the continental shelf and slope off southern California. *Bull. Geol. Soc. Amer.,* 71, 1121-1136.

————, 1961, Submarine slumps. *Jour. Sed. Petrol.,* 31 (3), 343-357.

————, 1965, The erosional channel wall in La Jolla sea-fan valley seen from bathyscaph TRIESTE II. *Bull. Geol. Soc. Amer.,* 76, 385-392.

Moore, D. G., and J. R. Curray, 1963a, Structural framework of the continental terrace, Northwest Gulf of Mexico. *Jour. Geophys. Res.,* 68, 1725-47.

————, 1963b, Sedimentary framework of continental terrace off Norfolk, Virginia and Newport, Rhode Is. *Bull. Amer. Assoc. Petrol. Geol.,* 47, 2051-2054.

Moore, D. G., and George Shumway, 1959, Sediment thickness and physical properties: Pigeon Point Shelf, California. *Jour. Geophys. Res.,* 64 (3), 367-374.

Morgan, J. P., and W. G. McIntire, 1956, Quaternary geology of the Bengal Basin. *Coastal Studies Inst., Louisiana State Univ., Tech. Rept. No. 9,* 56 pp.

Moriarty, J. R., 1964, The use of oceanography in the solution of problems in a submarine archaeological site. In *Papers in Marine Geology* (Shepard Commem. Vol.) R. L. Miller, ed. The Macmillan Company, New York, pp. 511-522.

Murray, G. E., 1960, Geologic framework of Gulf coastal province of United States. In *Recent Sediments, Northwest Gulf of Mexico,* F. P. Shepard, F. B. Phleger, and Tj. H. van Andel, edits., Amer. Assoc. Petrol. Geol., Tulsa, Oklahoma. pp. 5-33.

Nasu, Noriyuki, 1964, The provenance of the coarse sediments on the continental shelves and the trench slopes off the Japanese Pacific Coast. In *Papers in Marine Geology* (Shepard Commem. Vol.) R. L. Miller, ed., The Macmillan Company, New York, pp. 65-101.

Natland, M. L., and Ph. H. Kuenen, 1951, Sedimentary history of the Ventura Basin, California, and the action of turbidity currents. *Soc. Econ. Paleontol. & Mineralog. Spec. Publ. No. 2,* pp. 76-107.

Nesteroff, W. D., 1958, Recherches sur les sediments marins actuels de la region d'Antibes. Ph.D. thesis, Univ. of Paris, 347 pp.

Niino, Hiroshi, 1952, The bottom character of the banks and submarine valleys on and around the continental shelf of the Japanese Islands. *Jour. Tokyo Univ. Fisheries,* 38 (3), 391-410.

North, W. H., 1960, Fabulous Cape San Lucas. *Skin Diver Mag.,* May, pp. 24-26, 52.

Nota, D. J. G., and D. H. Loring, 1964, Recent depositional conditions in the St. Lawrence River and Gulf—a reconnaissance survey. *Mar. Geol.,* 2, 198-235.

Peck, R. B., W. E. Hanson, and T. H. Thornburn, 1953, *Foundation Engineering.* John Wiley and Sons, New York, 410 pp.

Peckham, V. O., and J. H. McLean, 1961, Biological exploration at the head of Carmel submarine canyon. *Ann. Rept. Amer. Malacolog. Union,* p. 43.

Pérès, J. M., J. Piccard, and M. Ruivo, 1957, Resultats de la Campagne de Recherches du Bathyscaphe F.N.R.S. III. *Bull. Inst. Oceanog. Monaco,* 1,092, 1-29.

Perry, R. B., and Haven Nichols, 1965, Bathymetry of Adak Canyon, Aleutian Arc, Alaska. *Geol. Soc. Amer. Bull.,* 76, 365-370.

Phipps, C. V. G., 1963, Topography & sedimentation of the continental shelf and slope between Sydney & Montague Island—N.S.W. *Austr. Oil and Gas Jour.,* Dec.

Phleger, F. B., 1942, Foraminifera of submarine cores from the continental slope, Pt. 2 *Bull. Geol. Soc. Amer.,* 53, 1073-1098.

———, 1956, Foraminiferal faunas in cores offshore from the Mississippi Delta, *Deep-Sea Res.,* 3 (Suppl.), 45-57.

Poland, J. F., A. A. Garrett, and A. Sinnott, 1948, Geology, hydrology, and chemical character of the ground waters in the Torrance-Santa Monica area, Los Angeles County, California. *U.S. Geol. Surv., Ground Water Branch,* 475 pp, (duplicated).

Powers, H. A., R. R. Coats, and W. H. Nelson, 1960, Geology and submarine physiography of Amchitka Island, Alaska. *U.S. Geol. Surv. Bull. 1028-P,* pp. 521-554.

Reineck, H.-E., 1963, Der Kastengreifer. *Natur und Museum,* 93 (2), 65-68.

Revelle, R. R., and F. P. Shepard, 1939, Sediments off the California coast. In *Recent Marine Sediments,* P. D. Trask, ed., Amer. Assoc. Petrol. Geol., Tulsa, Oklahoma, pp. 245-282.

Richards, A. F., 1961, Investigations of deep-sea sediment cores, I. Shear strength, bearing capacity, and consolidation. *U.S. Navy Hydrogr. Office Tech. Rept. No. 63,* 70 pp.

———, 1962, Investigations of deep-sea sediment cores, II. Mass physical properties. *U.S. Navy Hydrogr. Office Tech. Rept. No. 63,* 143 pp.

Richards, A. F., and G. H. Keller, 1962, Water content variability in a silty clay core from off Nova Scotia. *Limnol. and Oceanog.,* 7 (3), 426-427.

Richards, H. G., and J. L. Ruhle, 1955, Mollusks from a sediment core from the Hudson Submarine Canyon. *Proc. Penn. Acad. Sci.,* 29, 186-190.

Ritter, W. E., 1902, A summer of dredging on the coast of California. *Science,* n.s., 15, 55-65.

Roberson, M. I., 1964, Continuous seismic profiler survey of Oceanographer, Gilbert, and Lydonia Submarine Canyons, Georges Bank. *Jour. Geophys. Res.,* 69 (22), 4779-4789.

Rosfelder, André, 1954, Carte Provisoire au 1/500.000^{me} de la Marge Continentale Algerienne. *Bull. Geol. de l'Algerie,* n.s., 5, 57-106.

Royse, C. F., Jr., 1964, Sediments of Willapa Submarine Canyon. *Univ. of Washington Tech. Rept. no. 111,* 62 pp.

Schalk, Marshall, 1964, Submarine topography off Eleuthera Island, Bahamas. *Bull. Geol. Soc. Amer.,* 57, 1228.

Schoeffler, J., 1965, Le Gouf de Cap Breton, de l'Eocene inférieur à nos jours. In *Submarine Geology and Geophysics,* W. F. Whittard and R. Bradshaw, eds., Butterworths, London, pp. 265-270.

Scruton, P. C., 1960, Deltaic building and the delta sequence. In *Recent Sediments, Northwest Gulf of Mexico,* F. P. Shepard, F. B. Phleger, and Tj. H. van Andel, eds. Amer. Assoc. Petrol. Geol., pp. 82-107.

Seed, H. B., J. K. Mitchell, and C. K. Chan, 1960, The strength of compacted cohesive soils, Research Conf. Shear Strength of Cohesive Soils, *Proc. Amer. Soc. Civil Engineers,* pp. 877-964.

Shepard, F. P., 1934, Canyons off the New England Coast. *Amer. Jour. Sci.,* 27, 24-36.

———, 1937a, "Salt" domes related to Mississippi Submarine Trough. *Bull. Geol. Soc. Amer.,* 48, 1349-1362.

———, 1937b, Investigation of submarine topography during the past year. *Trans. Amer. Geophys. Union,* pp. 226-228.

———, 1948, *Submarine Geology.* Harper & Bros., New York, 338 pp.

———, 1949, Terriestrial topography of submarine canyons revealed by diving. *Bull. Geol. Soc. Amer.,* 60, 1597-1616.

————, 1951, Mass movements in submarine canyon heads. *Trans. Amer. Geophys. Union,* 32 (3), 405-418.

————, 1954, High-velocity turbidity currents, a discussion. *Proc. Roy. Soc.,* A, 222, 323-326.

————, 1955, Delta-front valleys bordering the Mississippi distributaries. *Bull. Geol. Soc. Amer.,* 66 (12), 1489-1498.

————, 1957, Northward continuation of the San Andreas Fault. *Bull. Seismol. Soc. Amer.,* 37 (3), 263-266.

————, 1961, Deep sea sands. *21st Intl. Geol. Congr. Repts.* pt. 23, pp. 26-42.

————, 1963, *Submarine Geology,* 2nd ed., Harper & Row, New York, 557 pp.

————, 1964, Sea-floor valleys of Gulf of California. In *Marine Geology of the Gulf of California—A Symposium,* Amer. Assoc. Petrol. Geol., Mem. 3, pp. 157-192.

————, 1965, Submarine canyons explored by Cousteau's Diving Saucer. In *Submarine Geology and Geophysics,* W. F. Whittard and R. Bradshaw, eds., Butterworths, London, pp. 303-311.

Shepard, F. P., and C. N. Beard, 1938, Submarine canyons: distribution and longitudinal profiles. *Geograph. Rev.,* 28 (3), 439-451.

Shepard, F. P., and G. V. Cohee, 1936, Continental shelf sediments off the Mid-Atlantic States. *Bull. Geol. Soc. Amer.,* 47, 441-458.

Shepard, F. P., J. R. Curray, D. L. Inman, E. A. Murray, E. L. Winterer, and R. F. Dill, 1964, Submarine geology by diving saucer. *Science,* 145, (3636), 1042-1046.

Shepard, F. P., and Gerhard Einsele, 1962, Sedimentation in San Diego Trough and contributing submarine canyons. *Sedimentology,* 1 (2), 81-133.

Shepard, F. P., and K. O. Emery, 1941, Submarine topography off the California Coast: Canyons and tectonic interpretation. *Geol. Soc. Amer. Spec. Paper No. 31,* 171 pp.

————, 1946, Submarine photography off the California Coast. *Jour. Geol.,* 44 (5), 306-321.

Shepard, F. P., and G. A. Macdonald, 1938, Sediments of Santa Monica Bay, California. *Bull. Amer. Assoc. Petrol. Geol.,* 22, 201-216.

Shepard, F. P., Hiroshi Niino, and T. K. Chamberlain, 1964, Submarine canyons and Sagami Trough, east-central Honshu, Japan. *Bull. Geol. Soc. Amer.,* 75, 1117-1130.

Shepard, F. P., R. R. Revelle, and R. S. Dietz, 1939, Ocean-bottom currents off the California Coast. *Science,* 89 (2317), 488-489.

Shipek, C. J., 1960, Photographic study of some deep-sea floor environments in the Eastern Pacific. *Bull. Geol. Soc. Amer.,* 71, 1067-1074.

Smith, E. H., F. M. Soule, and O. Mosby, 1937, MARION and GENERAL GREENE expeditions to Davis Strait and Labrador Sea under the direction of the U.S. Coast Guard, 1928-1931-1933-1934-1935. *U.S. Coast Guard Bull.*, 19, 1-259.

Smith, P. A., 1937, The submarine topography of Bogoslof. *Geograph. Rev.* 27 (4), 630-636.

Smith, W. S. R., 1902, The submarine valleys of the California coast. *Science*, XV (382), 670-672.

Spencer, J. W., 1895, Reconstruction of the Antillean continent. *Bull. Geol. Soc. Amer.* 6, 103-140.

————, 1898, On the continental elevation of the glacial epoch. *Geol. Mag.*, 4 (5), 32-38.

————, 1903, Submarine valleys off the American coast and in the North Atlantic. *Bull. Geol. Soc. Amer.*, 14, 207-226.

Sprigg, Reg. C., 1947, Submarine canyons of the New Guinea and South Australian Coasts. *Trans. Roy. Soc. S. Austr.*, 71 (2, Dec.), 296-310.

————, 1963, New structural discoveries off Australia's southern coast. *Australasian Oil and Gas Jour.* (Sept.), 9 (12), 32-33, 36, 40, 42.

Stearns, H. T., and G. A. Macdonald, 1947, Geology and ground water resources of the Island of Molokai, Hawaii. *Hawaii Div. of Hydrog. Bull.*, 11, 113 pp.

Stetson, H. C., 1936, Geology and paleontology of the Georges Bank Canyons, I. Geology. *Bull. Geol. Soc. Amer.*, 47, 339-366.

————, 1937, Current measurements in Georges Bank Canyons. *Trans. Amer. Geophys. Union*, pp. 216-219.

————, 1949, The sediments and stratigraphy of the East Coast continental margin. *Papers in Phys. Oceanog. and Meteorol.*, Mass. Inst. Tech. and Woods Hole Oceanog. Inst., 11 (2), 60 pp.

Stewart, H. B., Jr., R. S. Dietz, and F. P. Shepard, 1964, Submarine valleys off the Ganges Delta. *Abs., Ann. Meeting Geol. Soc. Amer.*, pp. 195-196.

Stewart, H. B., Jr., F. P. Shepard, and R. S. Dietz, 1964, Submarine canyons off Eastern Ceylon. *Abs., Ann. Meeting, Geol. Soc. Amer., p.* 197.

Stokes, W. L., and D. J. Varnes, 1955, Glossary of selected geologic terms with special reference to their use in engineering. *Proc. Colorado Sci. Soc.*, v. 16, 165 pp.

Stride, A. H., 1963, Current-swept sea floors near the southern half of Great Britain. *Quart. J. Geol. Soc.* London, 119, 175-199.

Suess, Edouard, 1900, *La Face de la Terre*, 3 vols., A. Colin, Paris.

Sundborg, Åke, 1956, The river Klarälven, a study of fluvial processes. *Meddelanden Frau Uppsala Univ. Geografiska Inst.*, no. 115, pp. 128-316.

Sverdrup, H. U., M. W. Johnson, and R. H. Fleming, 1942, *The Oceans, Their Physics, Chemistry, and General Biology*. Prentice Englewood Cliffs, N.J., 1087 pp.

Taylor, D. W., 1948, *Fundamentals of Soil Mechanics.* John Wiley & Sons, New York, 200 pp.

Terada, R., 1928, On the geophysical significance of the Kwanto earthquake. *Tokyo Imp. Acad. Proc.,* 4, 45-55.

Terzaghi, K., 1943, *Theoretical Soil Mechanics.* John Wiley & Sons, New York, 510 pp.

————, 1950, Mechanism of landslides. In *Application of Geology to Engineering Practice,* (Berkey Vol.), Geol. Soc. Amer., pp. 83-123.

————, 1953, Discussion. *Proc. Third Intl. Conf. Soil Mechanics and Found. Engr.,* 3, 158.

————, 1956, Varieties of submarine slope failures. *Proc. 8th Texas Conf. on Soil Mech. and Found. Engr., Spec. Publ. 29,* Bur. Engr. Res., Univ. Texas, Austin, Tex., 41 pp.

Terzaghi, Karl, 1962, Discussion. *Bull. Amer. Assoc. Petrol. Geol.,* 46, 1438-1443.

Terzaghi, K., and R. B. Peck, 1948, *Soil Mechanics in Engineering Practice.* John Wiley & Sons, New York, 566 pp.

Uchupi, Elazar, 1965, Maps showing relation of land and submarine topography, Nova Scotia to Florida. *U.S. Geol. Surv., Misc. Geol. Investig.,* Map 1-451.

Veatch, A. C., and P. A. Smith, 1939, Atlantic submarine valleys off the United States and the Congo submarine valley. *Geol. Soc. Amer. Spec. Paper 7,* 101 pp.

von Rad, U., F. P. Shepard, A. M. Rosfelder, and R. F. Dill, 1965, Origin of deepwater sands off La Jolla, California. *Abs. Geol. Soc. Amer. Ann. Meeting,* Kansas City, Mo., p. 177.

Wegener, Alfred, 1924, *The Origin of Continents and Oceans.* Eng. translation from German 3rd edit., E. P. Dutton & Co., New York, 212 pp.

Whitaker, J. H. McD., 1962, The geology of the area around Leintwardine, Herefordshire. *Quart. Jour. Geol. Soc. of London,* 118, 319-351.

Wilde, Pat, 1964, Sand-sized sediment from the Delgada and Monterey deep-sea fans. *Abs. Ann. Meeting Geol. Soc. Amer.,* pp. 224-225.

————, 1965, Estimates of bottom current velocities from grain size measurements for sediments from the Monterey deep-sea fan. *Trans. Mar. Tech. Soc.—Amer. Soc. Limnol. and Oceanog.,* 2, 718-727.

Wimberley, C. S., 1955, Marine sediments north of Scripps Submarine Canyon, La Jolla, California. *Jour. Sed. Petrology,* 25 (1), 24-37.

Winterer, E. L., and D. L. Durham, 1962, Geology of southeastern Ventura Basin, Los Angeles County, California. *U.S. Geol. Surv. Prof. Paper 334H,* pp. 275-366.

Woodford A. O., 1951, Stream gradients and Monterey Sea Valley. *Bull. Geol. Soc. Amer.,* 62, 799-852.

Yamsaki, Naomasa, 1926, Physiographical studies of the great earthquake of the Kwanto District, 1923. *Jour. Faculty Sci.,* Univ. of Tokyo, sec II, II (2), 77-119.

Part II
MODERN DEEP-SEA FANS

Editor's Comments
on Papers 9 Through 11

Our knowledge of submarine canyons results from nearly a century of soundings and, more recently, seismic profiles, coring, Scuba diving and submersible descents, and by other methods. Only since 1951 has knowledge in comparable detail become available on the great deep-sea fans or cones, thought of originally as submerged "deltas" at the mouths of submarine canyons.

To introduce this section on modern deep-sea fans, we have chosen the often-cited paper by Menard, published in 1955 (9). We are omitting the first part, dealing with the topography expected from various types of sedimentation (from suspension, turbidity currents, and slumps), with basins and troughs as sediment traps, and the nature of the dams that pond up the sediment, as these are topics marginal to the scope of this volume. In the second part there is a useful clarification of nomenclature, followed by a discussion on the flow of turbidity currents, in which the fact that deep-sea channels hook toward the left in the Northern Hemisphere is seen to be highly significant. Eliminating other possibilities, Menard shows that the Coriolis effect, causing a deviation to the right in the Northern Hemisphere, would encourage the surface of turbidity currents to tilt from right to left and build up right-side levees bordering the channels in which they flow. As a result of this, large turbidity currents would spill over the lower left-side levees and develop channels that hook left. This concept is supported quantitatively by the work of Komar (1969). The building of fans is attributed to sheet flow of large turbidity currents, with the coarser bottom layers following the channels but the more tenuous upper layers overflowing the levees and spreading radially over the fan.

We follow the previous paper, which deals with fans off the west coast of North America, with a more detailed study of a single fan, that of the Rhône in the Mediterranean Sea (10). It has large leveed channels and, as predicted for a Northern Hemisphere fan, these hook left. The authors, in passing, quote evidence for channels in the Southern Hemisphere which hook right.

The section of the paper on abyssal hills attributed to salt domes is omitted (see Mauffret and others, 1973, for a recent account), but it may be noted that these structures have been found in similar situations elsewhere, for example in the Gulf of Mexico (see the three papers by Ewing and others cited in the references to Paper 10) and associated with the Congo Canyon and Fan (Paper 5). Discussion of the area, thickness, and volume of the Rhône Fan and Balearic Abyssal Plain is also omitted, but the authors' conclusions on the age of the fan (Oligocene onward) and on the possible rate of denudation of the Rhône drainage basin to provide the sediments of the fan are included. The pre-Pleistocene age is in accord with Menard's (1960) calculations for the California fans, which he considers to date from the Tertiary: some authors, however, argue that many large fans could be wholly Quaternary in age (e. g., Moore, 1969).

During the 1960s, widespread interest in detailed work on deep-sea fans resulted in an increasing number of papers. One of the best of these is a well-illustrated paper by Shepard, Dill, and von Rad, published in 1969 and dealing with what is by now the most intensely studied canyon-fan system in the world, that of La Jolla, California (11). We have already seen something of Dill's diving researches in La Jolla Canyon Head (Paper 6), and now we go into deeper water to investigate the fan and its fan valley. The authors employed modern techniques such as closely spaced sounding lines, continuous reflection profiles, and deep-diving submersibles and brought back oriented box cores for sedimentological analysis, including X-ray studies and magnetic fabric analysis (Rees and others, 1968) and samples of kelp for ^{14}C dating. They found the fan valley to be winding, with levees more continuous on the north than on the south side and precipitous walls up to 50 m high on the outsides of bends. Slump blocks of clay were common on the valley floor. Contrary to expectations, they found some of the coarsest sediment, including gravel and mud balls in sand, farthest from the shore and at the greatest depths.

On the fan, the presence of shallow-water foraminifera and sand suggests material spilling over the levees onto the open fan as suggested in the two preceding papers, but the authors do not favor turbidity currents as transporting agents. They postulate a traction type of pulsating current as measured today by current-meter studies (e. g., Shepard and

159

Marshall, 1973a, 1973b) but believe that sedimentation may have been more rapid at times during the Pleistocene. Migration of the channel across the La Jolla Fan is suggested by discontinuous, inactive distributary channels that are parallel to the main fan valley but shallower: also cited is the possible evidence of buried older channels in the thick outer fan, a theme that recurs in the following paper by Normark (12). Another of our later themes touched on by the authors is a comparison with older sediments such as the deep-water Pliocene in the Ventura and Santa Paula Creek areas and the Miocene or Pliocene fan-valley filling at Doheny State Park, California.

In this paper, the caption for Figure 16B should read "thin section" instead of "X-radiograph"; Figures 18A and B have been interchanged so that the caption now reads correctly; and the caption for Figure 24 has been amended and is followed by a new figure, 24a, kindly supplied by Ulrich von Rad. On p. 214, second column, line 3, for Fig. 16 read Fig. 26.

Biographical Notes

Henry William Menard was born in Fresno, California, in 1920. He gained the B.S. degree in 1942 at California Institute of Technology and the M.S. in 1947 and Ph.D. in 1949 at Harvard University. Menard has been Geologist with Amerada Petroleum Corporation; associate to supervisory Oceanographer with the Navy Electronics Laboratory; Associate Professor of Marine Geology (1955–1961) and Professor (from 1961) at the Institute of Marine Research, University of California, San Diego; Guggenheim Memorial Foundation Fellow (1962–1963); and Churchill Foundation Fellow (1970–1971). He is a member of the National Academy of Sciences and has served on the President's Science Advisory Committee. From 1942 to 1948, he was with the U.S. Naval Reserve, rising from Ensign to Lt. Commander: he was decorated with the Bronze Star. Among his many publications are four books: *Marine Geology of the Pacific* (1964), *Anatomy of an Expedition* (1969), *Science—Growth and Change* (1971), and *Geology, Resources and Society* (1974).

Stuart M. Smith was born in Waterbury, Connecticut, in 1936. He studied Geology at Oberlin College, Ohio, gaining the B.A. degree in 1958 and at Harvard University where he gained the M.A. degree in 1960. Since that date he has been employed at Scripps Institution of Oceanography as Associate Specialist in Submarine Geology. His duties have included development of computer system and running data-processing group for digitizing, storage and display of navigation, bathymetric and magnetic data.

Richard M. Pratt obtained the B.S. degree in Mining Geology at the University of Idaho in 1951 and the M.S. in Geology (1954) and Ph.D. in Geology and Oceanography (1959) at the University of Washington. His varied experience in geology has included: working underground in Montana and as a petroleum geologist in California; participating in many Woods Hole Oceanographic Institution expeditions as Chief Scientist and with the U.S. Geological Survey in the Bering Sea; taking part in the Deep-Sea Drilling Project, Leg 18, and other marine programs, including geophysical research with the National Ocean Survey. He has also taught at Virginia Polytechnic Institute. After working as Senior Geologist with Fugro, Inc., in Puerto Rico, Pratt is now Senior Research Geologist with Western Geophysical Company of America in Houston, Texas.

Ulrich von Rad was born in Germany in 1935. He studied geology at Heidelberg, Tübingen, and the Technical University of Munich from 1954 to 1960, gaining the Master's degree at Munich in 1960. After graduate study at Princeton University, he gained the Ph.D. degree at Munich in 1964 with a dissertation on flysch sediments in the Bavarian Alps. Two years of postdoctoral studies on the Recent deep-water sands of La Jolla Fan with Shepard, Dill, and Winterer at Scripps Institution of Oceanography were followed by research work at the Marine Geology Branch of the Geology Department of the Technical University, Munich, in the Adriatic and Ionian Seas. Since 1970 von Rad has been Research Scientist at the Department of Marine Geology of the Federal Geological Survey, Hanover, Germany. He participated in Leg 14 of the Deep-Sea Drilling Project on the *Glomar Challenger* and in various cruises of RV *Meteor* and *Valdivia*. His main interests are in flysch sediments, recent terrigenous sediments from the continental margin, and deep-sea cherts.

9

Copyright © 1955 by the American Association of Petroleum Geologists

Reprinted from *Amer. Assoc. Petrol. Geol. Bull.*, **39**(2), 236, 246–255 (1955)

DEEP-SEA CHANNELS, TOPOGRAPHY, AND SEDIMENTATION[1]

HENRY W. MENARD, JR.[2]

San Diego, California

ABSTRACT

The type and distribution of much of the minor topography in the northeastern Pacific basin can be correlated with the accessibility of a given area to deposition from turbidity currents. Deep-sea areas separated from North America by basins or troughs, which act as sediment traps for turbidity currents, are characterized by a highly irregular relief of a few hundred feet. Other areas connected to the continent by a gradual continuous slope are characterized by very smooth plains like those of the North Atlantic sea floor. These plains slope out from the continent except in the vicinity of long ridges on the sea floor. The plains slope around the ridges because the slopes are formed by turbidity currents and the ridges act as dams to deflect the currents to one side. Rough-bottomed basins thousands of feet deep are found in the regions of smooth plains, but with one exception all are inaccessible to turbidity currents because they are surrounded by mountains, or lie on the "lee" side of ridges relative to the general direction of flow of turbidity currents. The exceptional basin is in a seismically active area, and it may have formed too recently to be filled by turbidity-current deposition.

Turbidity currents have formed deep-sea fans at the mouths of many submarine canyons, and deep-sea channels cross most, if not all, of the fans. All twelve of the channels which have been explored in any detail hook sharply to the left across the fans. This left hook can be explained as a secondary effect of the action of Coriolis force on the turbidity currents which formed the channels. Without channels, this type of flow would have no tendency to hook left. Unchannelized turbidity currents are required to form the fans.

[*Editor's Note:* Material has been omitted at this point.]

[1] Navy Electronics Laboratory Professional Contribution No. 34. Scripps Institution of Oceanography Contribution No. 750. Manuscript received, September 28, 1954.

[2] Navy Electronics Laboratory.

DEEP-SEA CHANNELS AND DEEP-SEA FANS

DIRECTION OF FLOW OF TURBIDITY CURRENTS

Regardless of the method of flow, it may be expected that a turbidity current debouching from a submarine canyon onto the relatively flat sea floor will be checked in velocity and will deposit some of its sediment in a deep-sea fan.[3] The available evidence for the existence of such deep-sea fans is fragmentary, but the best surveyed areas give the strongest support. "Deltas" or "delta-like fans" have been found associated with submarine canyons, notably the Hudson (Ericson, Ewing, and Heezen, 1951), La Jolla (Menard and Ludwick, 1951), and Monterey (Dill, Dietz, and Stewart, 1954) canyons. Similar fans appear to be associated with the Astoria Canyon and the Cascadia Deep-Sea Channel (Fig. 4), and the Arguello and Delgado canyons (Fig. 6), as well as many of the other canyons shown by Shepard and Emery (1941).

Deep-sea channels extend across the fans at the mouths of submarine canyons. The presence of these channels proves that part of the turbidity-current flow on the sea floor is channelized (Menard and Ludwick, 1951); therefore, it is unlike the flow observed in laboratory experiments, or the flow deduced from the shape of sedimentary beds supposedly deposited by turbidity currents (Kuenen and Migliorini, 1950), or the flow deduced from the time and position of cable-breaks (Kuenen, 1952). This inconsistency might not be significant if deep-

---

[3] The nomenclature for constructural landforms on the deep-sea floor is confused, as might be expected in a new field in which the evidence is fragmentary. Two of these landforms of particular interest are the "fans" or "deltas" shaped like gently sloping cones and found at the mouths of many submarine canyons, and the "channels" or "valleys" which lead across the "fans" from the mouths of the submarine canyons toward the deep sea. The earliest description of channels on the deep-sea floor known to the writer is in Shepard and Emery (1941) where they are called "valleys," "troughs," and "channels" because the soundings then available were not adequate to show that all the channels were the same landform. Subsequently, channels on the deep-sea floor have been called "canyons" or "mid-ocean canyons" (Ericson, Ewing, and Heezen, 1951, 1952; Ewing *et al.*, 1953); "channels" or "deep-sea channels" (Menard and Ludwick, 1951; Dietz, 1952; Dill, Dietz, and Stewart, 1954); and "valleys" or "outer valleys" (Shepard, 1951, 1952; Tolstoy, 1951). It is the writer's opinion that the term "canyon," with its connotation of high steep sides, should not be applied to landforms which typically are a few miles wide and only a few hundred feet deep; moreover, it would be unfortunate to link the genetically controversial term "submarine canyon" to these deep-sea channels for which few geologists would advocate a subaerial origin. The choice between "valley" and "channel" is more difficult, but "valley" has a connotation of erosion which may not be justified. The writer prefers the term "deep-sea channel."

The gently sloping cones at the mouths of submarine canyons were recognized by Shepard and Emery (1941) and called "deltas." This term has continued in use (Ericson, Ewing, and Heezen, 1951, 1952; Menard and Ludwick, 1951; Dill, Dietz, and Stewart, 1954) but as Kuenen (1952, p. 472) has observed, "so-called deltas are more like fans," and such terms as "delta-like fan" (Menard and Ludwick, 1951), "depositional fan" (Shepard, 1952), "delta-fan" and "delta-like apron" (Dill, Dietz, and Stewart, 1954), "coalescing alluvial plains" (Ericson, Ewing, and Heezen, 1951), and "apron" (Dietz, 1952) have also been used. Of the more conventional terms, the choice appears to be between "delta" and "fan" and depends on the assumed former position of sea-level. A delta is a deposit of sediment formed where a river enters a lake or the ocean, that is, in the vicinity of a shoreline. A fan is a deposit of sediment formed where the bed slope of a river markedly flattens. If, as Shepard (1952) argues, the shoreline was near the level of the mouths of submarine canyons when the sediment was deposited, the deposit should be called a delta. If, on the other hand, the sediment was deposited by turbidity currents because of a decrease in bed slope, the deposit should be called a fan. In the absence of any recognized term without genetic connotations, "deep-sea fan" is used in the present paper.

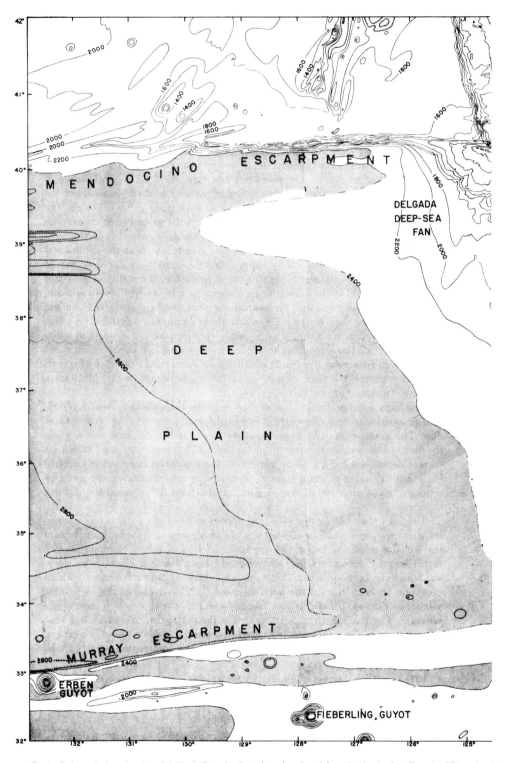

FIG. 6.—Bathymetric chart of sea floor off California. From shoreline to base of continental slope; chart is taken from Shepard and Emery (1941). Three great deep-sea fans lie at mouths of Delgada, Monterey, and Arguello canyons, and continental slope is notably more dissected around these three (and associated canyons or tributaries) than elsewhere. Deep-sea channels hook left from mouths of canyons. As ridge above Mendocino Escarpment acts as dam to turbidity currents, deep area on south can not be filled from north. Instead, both areas have smooth sea floor sloping west.

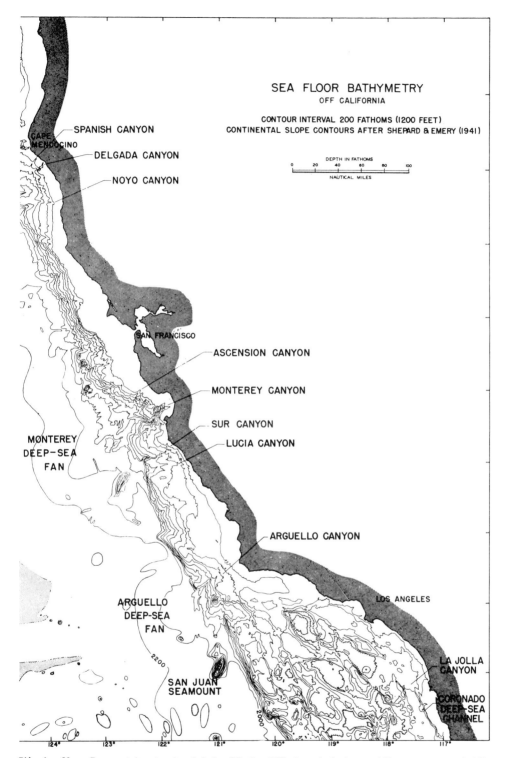

SEA FLOOR BATHYMETRY
OFF CALIFORNIA

CONTOUR INTERVAL 200 FATHOMS (1200 FEET)
CONTINENTAL SLOPE CONTOURS AFTER SHEPARD & EMERY (1941)

DEPTH IN FATHOMS

0   20   40   60   80   100

NAUTICAL MILES

SPANISH CANYON

CAPE MENDOCINO

DELGADA CANYON

NOYO CANYON

SAN FRANCISCO

ASCENSION CANYON

MONTEREY CANYON

SUR CANYON

LUCIA CANYON

MONTEREY DEEP-SEA FAN

ARGUELLO CANYON

LOS ANGELES

ARGUELLO DEEP-SEA FAN

2200

SAN JUAN SEAMOUNT

LA JOLLA CANYON

CORONADO DEEP-SEA CHANNEL

2000

124°   123°   122°   121°   120°   119°   118°   117°

Ridge above Murray Escarpment also acts as dam. As basins off Southern California are also barriers to turbidity currents, area south of Murray Escarpment (Baja California Seamount Province) is generally inaccessible to turbidity currents and is rough. Fan of Arguello Canyon seems to have surmounted easternmost part of Murray Escarpment, and small flood of sediment has covered irregular topography in this one small area of whole Baja California Seamount Province.

165

sea channels were few, but recent exploration with echo-sounders capable of showing such relatively small features suggests that they are very common.

Only the Cascadia, Monterey, and La Jolla deep-sea channels have been surveyed for any significant distance in the northeastern Pacific, but ten other channels in the area have been crossed in several places at random and are shown on at least one echogram. Approximately ten more doubtful channels can also be seen on echograms. Detailed contouring of the unpublished soundings of the Coast and Geodetic Survey indicates the presence of many other channels. The Cascadia and Dickens deep-sea channels appear to be at least 500 miles long, although the continuity of the channels is by no means proved. The slopes of the supposed channels are constant at about 0.001 except near the continental apron (Fig. 7). These slopes closely approximate the slope of the surveyed part of Cascadia Deep-Sea Channel; hence, they tend to confirm the reality of the outer part of the channels. The Cascadia Deep-Sea Channel reaches a very smooth, sloping plain which continues far toward the west. The channel itself can not be traced, but the slope of the plain is also about 0.001, so the gradient on which any turbidity currents would flow is relatively constant for more than 1,000 miles. The actual bed slope of a channel in this area would probably be somewhat less because of meander-like bends which would lengthen the distance of travel; it might approach the slope of 0.00045 of the Northwest Atlantic "Mid-Ocean Canyon" (Ewing *et al.*, 1953).

The large number and great length of deep-sea channels show that a significant proportion of turbidity-current flow in the oceans is not in the form of sheet flow. The question may be raised whether any of it is sheet flow and whether the mode of deposition deduced from the sheet-like form of supposed turbidity-current deposits is correct. Some answers to these questions and others concerning the distribution of grain sizes in turbidity-current deposition are given by considering the direction of flow of turbidity currents as indicated by deep-sea channels.

Deep-sea channels on deep-sea fans appear to have one outstanding characteristic: most, and possibly all that are not in structural troughs, hook toward the left. Inasmuch as important conclusions follow if this appearance is real, rather lengthy documentation is attempted.[4] Four deep-sea channels on deep-sea fans have been surveyed along the west coast of North America; without exception all hook toward the left compared with the trend of the connecting submarine canyon. ("Left" is also "south" and it is not clear which term is significant.) The Hudson Deep-Sea Channel is the only other one which has been surveyed where it crosses a fan. It appears that the channel is confined in a general way to a structural trough formed by a curve in the continental terrace; there-

---

[4] The general fact that deep-sea channels are located along the sides of fans was first noted by Shepard and Emery (1941, p. 84). "In all these cases these channels follow the side of the 'delta' rather than cutting the central portion. All the channels diverge from the general trend of the submarine canyons inside."

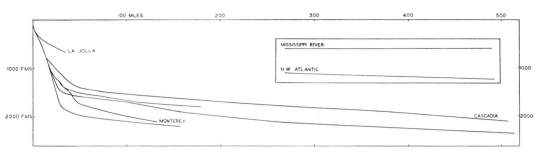

Fig. 7.—Longitudinal profiles of submarine canyons and associated deep-sea channels in northeastern Pacific Ocean, showing general similarities and compared with Mississippi River and Deep-Sea Channel of Northwest Atlantic which has not been traced to a submarine canyon.

fore, no prominent hook in the channel would be possible unless the turbidity currents could form a channel while flowing uphill.

Soundings are adequate in some places to permit tracing the course of deep-sea channels with reasonable assurance even though special surveys have not been made. Charts of the California coast (Shepard and Emery, 1941) show seven submarine canyons connected with what the writer would call deep-sea channels. One of these is the Monterey Deep-Sea Channel which unquestionably hooks toward the left (and south), but has already been counted as a specially surveyed channel. Of the remaining six, the Arguello, Noyo, and Delgada deep-sea channels hook toward the left; Lucia probably hooks toward the left but can not be traced with assurance; and the Pioneer and Sur deep-sea channels hook left for only a few miles and then bend right to the edge of the charts. The right bend in the Pioneer Channel occurs where it approaches the Guide Seamount from the north, and the bend in the Sur Channel is at the northern edge of the fan of the Lucia Deep-Sea Channel and of the Davidson Seamount. The right bends, therefore, may show merely that the tendency of turbidity currents to hook left is not so strong that they run uphill.

The unpublished soundings of the Coast and Geodetic Survey show several deep-sea channels off Oregon and in the Gulf of Alaska. Of these, the Cascadia Deep-Sea Channel has been surveyed and found to hook left. Three other nearby channels appear to hook left, and of them, the Astoria Deep-Sea Channel almost certainly does so. In the Gulf of Alaska the two most prominent channels hook left on their fans but the soundings are not adequate to show whether all others do so.

In summary, of the fifteen deep-sea channels which are surveyed or well sounded in the vicinity of their alluvial fans, twelve hook left, one probably hooks left, and two hook left for short distances but bend right where such a bend is required for the channel to slope downhill. This type of evidence for a left hook in the course of deep-sea channels may be convincing by itself, but other evidence may be adduced.

Many crossings along courses perpendicular to the continental apron of western North America show one to three deep-sea channels near the top of the apron. Crossings parallel with the general contours of the apron are relatively free of such channels. This relationship was particularly noteworthy when a line approximately along the 2,200-fathom contour was sounded for the whole length of central California. In this area, numerous well known submarine canyons dissect the continental slope and lead into deep-sea channels, and the continental apron grades smoothly out to sea into a sedimentary plain which extends at least to a depth of 2,600 fathoms. The topography shows clearly that sediment has been deposited from the continent outward across the area of the profile but only one channel was recognizable. It appears that the channels at the top of the apron must bend or hook rather than trend down the regional slope of the apron. The slope in this area, therefore, is not produced by turbidity currents flowing in channels.

ORIGIN OF THE LEFT HOOK

Twelve deep-sea channels off western North America certainly hook left (which is also south); none is known to hook right or trend straight out from the continental slope. The number of observations is not large, but it is enough to show that a real phenomenon is being observed. It is a reasonable approximation to consider that the channels are all in the most sinistral third of the available half-circle of arc of directions in which they might trend from the mouth of a submarine canyon at the base of the continental slope. Consequently, the odds are only about 3^{-12} that the twelve channels which hook left represent merely a chance sample from a group of channels which trend in all directions.

Only three plausible hypotheses of the origin of left (or south) hooks have been conceived.

1. Ocean currents exert a southward component and cause the turbidity currents to hook where they are slowed at the base of the continental slope.

2. The abyssal sea floor off western North America has been uniformly fractured by a series of step faults down on the south side. The hooks in the channels merely reflect adjustment to the slope of initial topography.

3. Some characteristics of the flow itself cause it to turn toward the left (south).

Quantitatively the first hypothesis appears improbable because it requires ocean currents with a significant southerly velocity at a depth of at least 3,000 meters and the California Current is negligible at a depth of only 500 meters (Sverdrup, Johnson, and Fleming, 1942).

The second hypothesis—control of slope by initial topography—might appear to apply to the basins of the Ridge and Trough Province and of the California Borderland, but the Deep Plain slopes about west-southwest and the deep-sea channels on the eastern margin of the plain trend slightly east of south. In other words, they do not trend down the regional slope; therefore, even if the slope were produced by block faulting, the faulting would not explain the channel trend.

The third hypothesis remains. Coriolis force—the tendency for any moving object to deviate from a straight path because of the deflecting force of the earth's rotation—acts on all currents in the ocean and must be considered as a possible cause of the left hook of turbidity-current channels. Unfortunately for simplicity of discussion, Coriolis force causes a deviation to the *right* in the northern hemisphere. Bates (1953) has considered the effect of Coriolis force on various types of flow of interest in the formation of deltas. He mentions that turbidity flows could turn to the right through an entire inertial circle because of Coriolis force; but, inasmuch as the evidence available to him indicated no such complete turn, he proposed that a tilting of the surface of the flow—so that it would be higher on the right than on the left—would be induced according to principles well known in oceanography. The tilting could cause the flow to trend toward the right parallel with depth contours instead of down the bed slope. It would further produce higher natural levees on the right side of the flow than on the left. The prediction regarding the height of natural levees is confirmed by surveys of the sea floor of western North America, and Bates' explanation appears satisfactory on this point. However, turbidity-current channels hook left and no direct reaction of the flow to Coriolis force is capable of causing it to bend toward the left in the northern hemisphere.

Coriolis force might produce a left hook of turbidity currents, however, according to the following reasoning—the wording of which applies only to the northern hemisphere.

1. A turbidity current flowing down a submarine canyon is confined so that it can not curve toward the right, despite Coriolis force. However, the top surface of the turbidity current is affected so that in transverse profile it tilts from right to left.

2. When the turbidity current debouches from the mouth of a submarine canyon onto the deep-sea floor, it forms natural levees along the sides of its channel, and the right-hand levee is higher because of the tilting of the turbidity-current surface.

3. The position of a subaerial stream channel on an alluvial fan at any given time is random, and a symmetrical deposit is formed. When one sector of the fan becomes higher than the others, some chance occurrence causes the channel to shift from its bed and it remains in a lower sector either at the right or at the left of its earlier channel.

4. The initial position of a turbidity-current channel on a deep-sea fan might also be random although the action of Coriolis force would tend to confine it to the right-hand quadrant.

5. As sedimentation continued, however, the asymmetry in the cross section of the channel would introduce a bias into the direction of migration of the channel. Unlike a subaerial stream channel, a turbidity current breaking out of an aggraded channel and migrating to some less elevated sector would always tend to move toward the left because the natural levee on that side is lower

than the one on the right. Regardless of the initial position of the turbidity-current channel, therefore, a series of migrations from elevated sectors would eventually bring it into a sector far on the left. In fact, the channels seem to have migrated as far toward the left as they can go without flowing uphill on the continental apron.

A general rule for the orientation of leveed turbidity-current channels may be proposed. "A turbidity current channel flanked by levees has a tendency to deviate from a straight line which varies directly with the amount of the deflecting force of the earth's rotation but is in the opposite direction." As the intensity of the deflecting force of the earth's rotation is directly proportional to the sine of the latitude, the tendency to hook will be correspondingly small in the equatorial regions. The deflecting force of the earth's rotation is toward the right in the northern hemisphere and toward the left in the southern hemisphere; the tendency to turn has the opposite sense.

MODE OF FLOW AND ORIGIN OF FANS

If a turbidity current debouching from a submarine canyon promptly hooks left, what forms the deep-sea fan which occupies roughly a semi-circle? Subaerial alluvial fans have a regular form like a half of a cone which is produced by the meandering of a stream across the area. Deposition occurs along a course until the sector becomes higher than others, and some chance occurrence shifts the course to a lower sector. This process of meandering can not produce the submarine alluvial fans, because the channels remain in the same sector. Some other process is required, a process that spreads sediment in a semi-circle around the mouth of a submarine canyon.

A suitable process would be the very sheet-flow of turbidity currents which the existence of turbidity current channels seemed to question. It is entirely possible that at times comparatively small turbidity currents flow down submarine canyons and follow the bends in deep-sea channels, and at other times large turbidity currents flow over the deep-sea channel and spread out in a semi-circle as a kind of vast sheet flow. However, the two types of flow need not be separate. A very pronounced vertical concentration gradient may be expected in turbidity currents, especially if coarse sand and gravel are carried in suspension (Menard and Ludwick, 1951). A large turbidity current with a vertical concentration gradient might split at the mouth of a submarine canyon. The dense bottom layer of coarse sediment would follow the channel but the more tenuous top layer might be higher than the channel levees and thus would flow in a random direction or toward the right instead of hooking left along the channel. The top layer would slowly spread radially because it would be denser than the surrounding water.

GEOLOGICAL IMPLICATIONS AND CONCLUSIONS

Turbidity currents are important agents, transporting and depositing sediment in the near-shore part of the abyssal northeastern Pacific. Witness the following facts.

1. Deep-sea channels ascribed to turbidity currents are widespread on the sea floor.

2. Areas which turbidity currents can reach, in general, are smooth plains. Areas which turbidity currents can not reach because basins form sediment traps between the deep-sea floor and the continents, in general, are areas with rough micro-relief.

3. The smooth sedimentary plains do not necessarily slope away from the continents. Some slope southward parallel with the continental margin in places where the westward flow of turbidity currents is obstructed by mountains which act as dams.

4. Deep basins with irregular bottom are common in places which are shielded from turbidity currents by ridges or troughs. Similar basins have not been found in places which are not so shielded except for one which is in a seismically active area.

The geological history of a submarine region apparently can be as complicated as that of a subaerial region. Consider the sediment deposited in the Ridge and Trough Province. One basin will be filled very rapidly because it lies on the course of a turbidity-current channel while another basin only a few miles away collects only pelagic organic remains and a thin layer of sediment moved by volume suspension. Later, after the turbidity current aggrades its channel, the ridge protecting the unfilled basin will be surmounted and sediment will pour into the basin. If the original channel is graded to some outlet by that time, the surface of the surrounding plain will become a disconformity, and little deposition will occur. The rate of deposition in the two areas will be reversed.

Deep-sea channels seem to show that turbidity currents flow at an angle to the slope instead of directly down slope. Most surveyed channels in the northeastern Pacific hook sharply toward the left across the alluvial fans at the mouths of submarine canyons. Because of the trend of the coastline, a hook toward the left is also a hook toward the south. It is not evident whether "left" or "south" is the direction related to the origin of the bend, but only a bend toward the left has been explained. For the present the most promising working hypothesis is that: (1) the surface of the flow becomes tilted in response to Coriolis force; (2) consequently, natural levees are higher on the right than on the left; and (3) the height difference biases "random" changes in the position of the channel so that all changes are toward the left.

Regardless of origin, some conclusions warrant discussion. Some turbidity currents hook upon emerging from the mouth of a submarine canyon (perhaps without exception if initial relief does not interfere). From the evidence now available it appears that sand bodies deposited on the deep-sea floor by turbidity currents should be confined to a narrow sector (left or south side) of the alluvial fan. The implications for paleogeography are apparent. Significant deductions may be drawn regarding the direction of ancient shorelines, and location of the mouths of ancient submarine canyons. With the knowledge that both channelled and sheet-flow modes of turbidity flow probably exist, some additional progress may also be made in attempting to explain the origin of minor sedimentary structures associated with graded bedding.

Inasmuch as turbidity currents are such controversial subjects, it may be discrete to close with a negative conclusion. Sea-floor topography, as shown by the echograms available to the writer, gives no proof that turbidity currents are able to erode the deep-sea floor. Deep-sea channels, for example, may be produced by non-deposition. On the other hand, many obviously erosional submarine canyons on both the east and west coasts of North America exhibit a peculiar curvature toward the left across the contours of the continental slope. If it develops that such curvature or hooking is a characteristic only of turbidity-current flow, sea-floor topography may give convincing evidence for submarine erosion.

BIBLIOGRAPHY

BATES, C. C., 1953, "Rational Theory of Delta Formation," *Bull. Amer. Assoc. Petrol. Geol.*, Vol. 37, No. 9 (September), pp. 2119–62.
BRAMLETTE, M. N., AND BRADLEY, W. H., 1940, "Geology and Biology of North Atlantic Deep-Sea Cores between Newfoundland and Ireland," *U. S. Geol. Survey Prof. Paper 196-A*, Pt. 1, pp. 1–34.
DIETZ, R. S., 1952, "Geomorphic Evolution of Continental Terrace (Continental Shelf and Slope)," *Bull. Amer. Assoc. Petrol. Geol.*, Vol. 36, No. 9 (September), pp. 1802–19.
DILL, R. F., DIETZ, R. S., AND STEWART, H., 1954, "Deep-Sea Channels and Delta of the Monterey Submarine Canyon," *Bull. Geol. Soc. America*, Vol. 65, pp. 191–94.
ERICSON, D. B., EWING, M., AND HEEZEN, B. C., 1951, "Deep-Sea Sands and Submarine Canyons," *Bull. Geol. Soc. America*, Vol. 62, pp. 961–65.
———, 1952, "Turbidity Currents and Sediments in North Atlantic," *Bull. Amer. Assoc. Petrol. Geol.*, Vol. 36, No. 3 (March), pp. 489–511.
EWING, M., HEEZEN, B. C., ERICSON, D. B., AND DORMAN, J., 1953, "Exploration of the Northwest Atlantic Mid-Ocean Canyon," *Bull. Geol. Soc. America*, Vol. 64, pp. 865–68.
KUENEN, PH. H., 1952, "Classification and Origin of Submarine Canyons," *Proc. Kon. Nederl. Akad. v. Wetenschappen-Amsterdam*, Ser. B, Vol. 55, pp. 464–73.
———, AND CAROZZI, A., 1953, "Turbidity Currents and Sliding in Geosynclinal Basins of the Alps," *Jour. Geol.*, Vol. 61, pp. 363–73.
———, AND MIGLIORINI, C. I., 1950, "Turbidity Currents as a Cause of Graded Bedding," *ibid.*, Vol. 58, pp. 91–127.
MENARD, H. W., in press, "Deformation of the Northeastern Pacific Basin and the West Coast of North America," *Bull. Geol. Soc. America*.
———, AND DIETZ, R. S., 1951, "Submarine Geology of the Gulf of Alaska," *ibid.*, Vol. 62, pp. 1263–85.
———, AND LUDWICK, J. C., 1951, "Applications of Hydraulics to the Study of Marine Turbidity Currents," *Soc. Econ. Paleon. Mineral. Spec. Pub. 2*, pp. 2–13.
NATLAND, M. L., AND KUENEN, PH. H., 1951, "Sedimentary History of the Ventura Basin, California, and the Action of Turbidity Currents," *ibid.*, pp. 78–107.
PHLEGER, F. B., 1951, "Displaced Foraminifera Faunas," *ibid.*, pp. 66–75.
SHEPARD, F. P., 1950, "Submarine Topography of the Gulf of California," *Geol. Soc. America Memoir 43*, Pt. 3, pp. 1–32.
———, 1951, "Composite Origin of Submarine Canyons," *Jour. Geol.*, Vol. 60, pp. 84–96.
———, 1952, "Transportation of Sand into Deep Water," *Soc. Econ. Paleon. Mineral. Spec. Pub. 2*, pp. 53–65.
———, AND EMERY, K. O., 1941, "Submarine Topography off the California Coast: Canyons and Tectonic Interpretation," *Geol. Soc. America Spec. Paper 31*, pp. 1–171.
SVERDRUP, H. U., JOHNSON, M. W., AND FLEMING, R. H., 1942, *The Oceans, Their Physics, Chemistry, and General Biology.* 1087 pp. Prentice-Hall, Inc., New York.
TOLSTOY, I., 1951, "Submarine Topography in the North Atlantic," *Bull. Geol. Soc. America*, Vol. 62 pp. 441–50.

10

Reprinted from *Submarine Geology and Geophysics*, W. F. Whittard and R. Bradshaw, eds., Butterworth, London, 1965, pp. 271–280, 283–284

The Rhône Deep-Sea Fan[1]

by

H. W. MENARD

Scripps Institution of Oceanography, University of California

S. M. SMITH

Scripps Institution of Oceanography, University of California

and R. M. PRATT

Woods Hole Oceanographic Institution, Massachusetts

Abstract. The River Rhône has been the major source of sediments making up the Balearic Abyssal Plain and the Rhône Fan, located in the northern end of the western Mediterranean Sea. The fan has an unusual system of large, disrupted, leveed channels possibly formed during the Pleistocene. The system is superimposed on what appears to be a typical asymmetrical fan, with channels on the left[2] side, produced by the combined effects of gravity and Coriolis force.

A band of abyssal hills, 40 by 150 km, runs parallel to the contours in the south-eastern quadrant of the fan. Isolated hills are 20–150 m high and 2–6 km across. The hills are tentatively identified as salt domes after examining alternative origins of vulcanism, mud extrusion and anticlinal folding.

Area of the fan is $7 \cdot 2 \times 10^4$ km² and of the abyssal plain $6 \cdot 3 \times 10^4$ km². The fan is underlain by an average of $2 \cdot 5$ km of unconsolidated and consolidated sediments. The plain has an average sedimentary thickness of $4 \cdot 2$ km. Total volume of sediments in the fan and plain is $4 \cdot 6 \times 10^5$ km³. Calculations using the present sedimentary load of the River Rhône show that this volume could be produced since the formation of the Alps in Oligocene times with an average post-Eocene rate of source-area denudation of 10 cm/10^3 years.

---

[1] Contribution, Scripps Institution of Oceanography, New Series, and Contribution No. 1594, Woods Hole Oceanographic Institution.

[2] That side on the left of the fan, looking down the slope.

1. INTRODUCTION

THE WESTERN Mediterranean basin has been intensively studied by Woods Hole Oceanographic Institution as well as by other laboratories, and precision sounding lines are numerous. Soundings were made available to Scripps Institution of Oceanography prior to the *Zephyrus* Expedition in 1962 and were used to plan a detailed survey of channels on the Rhône deep-sea fan. The survey was carried out in matchless weather on a flat, calm sea, and the resulting echograms show details of topography and sub-bottom structure never observed before by Scripps. The topography of part of the western Mediterranean is presented here because much of it has been unmapped, and because of the unusual quality of the records which are significant with regard to the origin of constructional relief of the sea-floor.

We are indebted to J. B. Hersey for permitting us to use unpublished information on sedimentary thicknesses, and for advice on planning the *Zephyrus* Expedition. J. Y. Cousteau kindly made oceanographical data collected by the Musée Océanographique available prior to, and during, the expedition. We wish also to thank Tj. H. van Andel for heavy mineral analyses of cores taken on the fan by the *Zephyrus* Expedition. Work on that expedition was supported by the National Science Foundation and a Guggenheim Fellowship.

2. REGIONAL SETTING

The deep basin of the western Mediterranean Sea is almost surrounded by land or shallow straits (fig. 105). Thus it may be regarded as an oceanic window in the continental crust. Relatively straight and steep continental slopes of 3°–6° border the basin along the French coast east of the River Rhône, and along Corsica, Sardinia and Africa. More irregular, steep slopes occur off Spain and the Balearic Islands. The deltas seaward of the Rivers Rhône and Ebro are associated with gentler continental slopes, averaging 1°, which bulge into the basin. Both steep and gentle slopes are cut by numerous submarine canyons (Bourcart, 1959; Shepard, 1963).

The basin is floored by an abyssal plain approximately bounded by the 2800 m contour at the base of the steep continental slope on the southern and eastern sides. The plain is several hundred metres less deep on the northern and western sides where it grades into the Rhône deep-sea fan. The deepest portion of the plain is a flat pond of sediment covering an area of about 5000 km² which lies at a depth of 2855 m. This flat area is located well south of the basin-centre. The location can be readily explained if most of the differential deposition is by turbidity currents. Thus volcanic ash or aeolian dust may fall uniformly over the whole basin, but turbidites are concentrated in deep-sea fans near the source-area. Judging by the topography, particularly of fans, only an insignificant amount of sediment is deposited by turbidity currents derived from Corsica, Sardinia and the Balearic Islands. Off Africa, some contribution from turbidity currents is indicated by the presence of minor fans as well as by the occurrence of sequential cable-breaks (Heezen & Ewing, 1955). However, by far the most important fans, and thus turbidites, are associated with the large rivers which empty into the northern end of the basin. Construction of these northern fans has gradually shifted the deepest part of the abyssal plain towards the south.

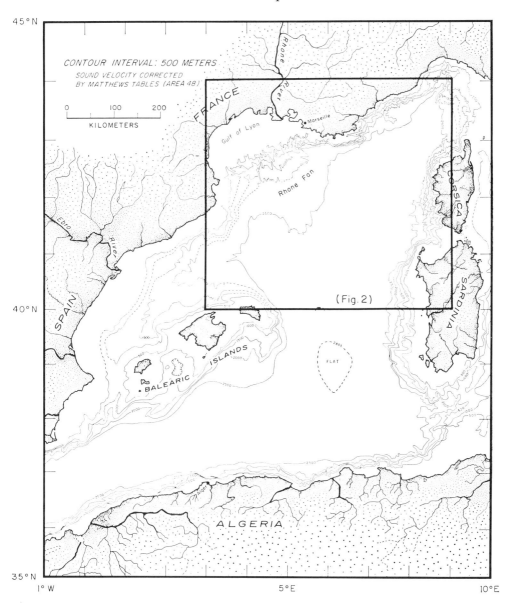

Figure 105. Regional setting of the Balearic Abyssal Plain and Rhône deep-sea fan. Contours compiled from various sources by Robert Nason.
For (Fig. 2) in inset read (Fig. 106).

3. CHARACTERISTICS OF DEEP-SEA FANS

Fan-shaped accumulations of sediment exist off the mouths of many submarine canyons and commonly they are crossed by channels bordered by levees. The great fans of the north-eastern Pacific, which are among the best known, have many characteristics in common (Menard, 1955; Hurley, 1960). They are notably asymmetrical particularly with regard to channels which are not randomly distributed but generally occur along the left edge of a fan. Channels hook sharply to the left as they emerge from submarine canyons. Moreover, the right levee in most places is markedly higher than the left. Finally, the whole shape of the fan is asymmetrical with a large lobe to the right of the channel, or channels, and a relatively minor lobe to the left.

All these characteristics are explained by the following hypothesis (Menard, 1955) which, as worded, refers only to the northern hemisphere. (i). The top surface of a turbidity current in a submarine canyon tilts up to the right because of the effect of Coriolis force on a fluid moving in a channel. (ii). Consequently, when a turbidity current emerges as a jet from a canyon and begins to construct natural levees, the right hand levee is generally higher than the left. (iii). The initial distribution of channels on a fan is probably random. However, the subsequent migration of aggrading channels is biased by the asymmetrical cross-section and most breakouts occur to the left because that levee is lower. Thus at any given time channels tend to be as far to the left on a fan as possible without flowing uphill. (iv). Small turbidity currents are confined to channels but large ones may commonly spill over the levees and spread across the fan as sheet-flows. Deposition from sheet-flows builds most of the fan, and the asymmetry of flow accounts for the construction of a large lobe to the right of a channel, and a small one to the left. Because of the size-gradient in a turbidity current, the sediment that spills over the levees tends to be finer than the sediment which is confined to the channel.

If this hypothesis is correct, the left hook of channels observed in the northern hemisphere should be reversed in the southern hemisphere, and the intensity of hooking should be directly proportional to the sine of the latitude. The hypothesis has proved difficult to test for a number of reasons. The shape of most fans and the course of most channels are clearly influenced by the initial relief of the surrounding region. Moreover, most of the relatively few abyssal plains and fans in the southern hemisphere are around the Antarctic and are little known.

A physiographical diagram of the South Atlantic (Heezen & Tharp, 1961) shows many channels in high latitudes off South America and Antarctica. A number of these hook sharply to the right at the base of the continental slope. The best examples are off central Argentina between latitudes 40° and 45°, and off Antarctica between longitudes 10° and 20° W. On the other hand, about the same number of channels in the South Atlantic are shown not to hook. The only apparent difference between the two classes of channels is that the ones that hook are in somewhat more level and open regions. Thus, accepting the accuracy of the diagram, it seems that many channels and particularly unconstrained ones, hook sharply right. This tends to confirm the hypothesis of the effect of Coriolis force on the flow of canalized turbidity currents. However, in some areas physiographical diagrams of the sea-floor are necessarily subjective and thus a more definite confirmation is desirable.

An elegant confirmation was obtained by Heezen (1963) in a seismic reflection profiler survey of a channel in the Hikurangi Trench off New Zealand. The channel "appears to have migrated several kilometres to the right during the deposition of the last kilometre of sediments. The consistently higher elevation of the left bank is apparently a reflection of the effect of the Coriolis force on turbidity currents which have built the left bank higher". Thus, in a relatively level region of the sea-floor, it appears that the Coriolis force has a pronounced effect on the flow of turbidity currents, causing a left hook in the northern hemisphere and a right hook in the southern. This in turn influences the distribution of sediment-sizes and minor constructional topography at the base of the continental slope.

4. RHÔNE FAN AND CHANNELS

The Rhône Fan may be examined in the light of the general background which has now been given. The fan, one of the great constructional landforms of the world, has a radius of about 300 km and forms approximately a semicircle filling most of the northern end of the western Mediterranean (fig. 106). The shape is asymmetrical, with relatively straight contours on the eastern side, and much more curved ones on the south and west. The head of the fan grades imperceptibly into the continental slope and the outer fringes of the fan merge with the Balearic Abyssal Plain. Slopes grading from 0·01 to 0·001 are similar to those found on Delgada and Monterey Fans off California (Menard, 1960). In short, the broad characteristics of the fan are normal.

A complex pattern of channels spreads across the fan (figs. 106, 107) and it is far from normal. The channels are in three systems: a group of large channels along the northern edge of the fan, another large group near the centre, and a group of smaller channels crossing the outer parts of the fan, particularly to the south. Let us first consider how the fan would look if the latter two groups did not exist. Two large submarine canyons cut into the continental shelf directly south of the mouth of the River Rhône (Bourcart, 1959). At the base of the continental slope these canyons grade into channels which hook sharply to the left until they trend along the base of the continental slope. As these channels aggraded, large turbidity currents would spill over the right banks and spread out to form an asymmetrical fan of the general shape observed. Thus, if only these two channels existed, the Rhône Fan would be a typical example of topography constructed by deposition from turbidity currents flowing under the combined effects of gravity and Coriolis force. It remains to be seen whether the other channels may be regarded as exceptional features which have been superimposed on an otherwise normal fan.

The second group of large channels consists of distributaries of a submarine canyon which lies south-west of the mouth of the Rhône. Several distributaries are flanked by conspicuous levees which rise 50–100 m above the surrounding fan. Confined between the levees, channels in some places have been aggraded some tens of metres above the fan. The composite picture closely resembles the birdfoot delta of the Mississippi except that the longitudinal slopes of the Rhône Fan are much steeper. This birdfoot pattern does not extend more than about 150 km from the mouth of the related canyon, or below a depth of about 2650 m. Moreover, it is partly chaotic rather than forming a

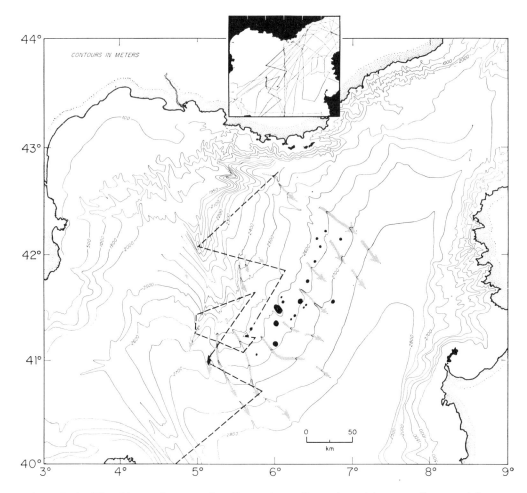

Figure 106. The Rhône deep-sea fan. Inset shows lines of precision soundings used for contouring. Pecked line indicates track of *Zephyrus* Expedition; black dots are abyssal hills and arrows represent paths of channelled turbidity currents.

completely integrated distributary system. In some places single crossings show a classical example of an elevated deep-sea channel bordered by high levees (fig. 107). These features were recognized during the *Zephyrus* survey, and the track of the ship was accordingly altered to trace individual channels as they were discovered. This technique has been highly successful in tracing continuous channels in the Pacific. In this survey, however, the technique proved that some leveed channels are discontinuous fragments.

The origin of the discontinuities is probably related to the exceptional height of the leveed channels. Sub-bottom echoes, with extraordinary resolution, were recorded in many places because the sea conditions permitted the use of a short sound pulse for echo-sounding. Some records show that natural levees have been built layer by layer

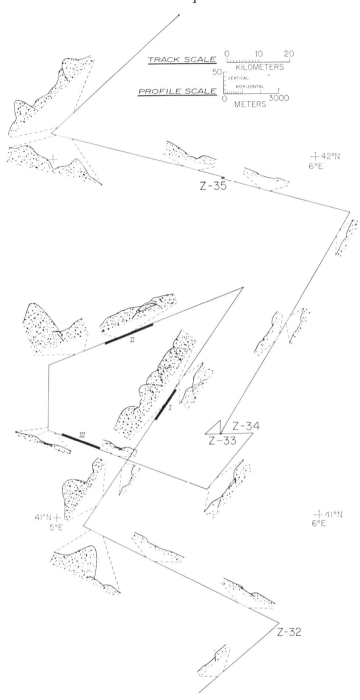

Figure 107. Profiles of channels crossed on *Zephyrus* Expedition and core-stations. Compare track in figure 106 for locations. Roman numerals refer to cross-sections shown in figure 108.

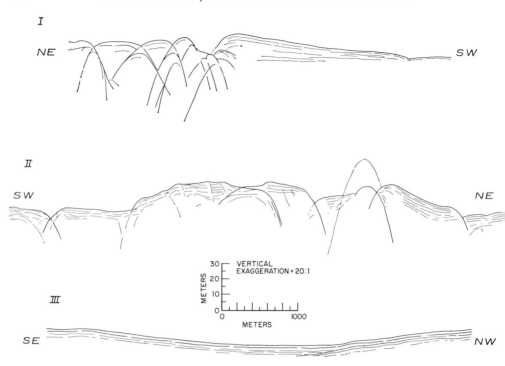

Figure 108.   Internal structure of leveed channels and of the fan traced from PDR records.
See figure 107 for locations.

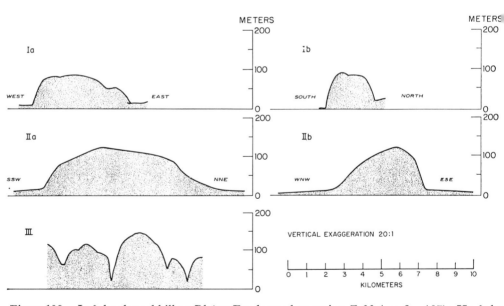

Figure 109.   Ia & b: abyssal hill on Rhône Fan located at station Z-33 (see fig. 107); IIa & b:
one of Sigsbee Knolls located at 23° 40′ N , 92° 40′ W. (Ewing, Ericson & Heezen, 1958); III:
abyssal hills in the north-eastern Pacific Ocean. Relief above arbitrary levels.

above an initially flat sea-floor (fig. 108). Consequently, the possibility that the levee-channel fragments are actually mud-lumps elevated from the surrounding fan can be discarded. By elimination, discontinuities are produced by removing parts of an initially continuous feature. It is not unreasonable that these ridges of unconsolidated sediment, as much as 100 m thick, should be unstable and subject to slumping. Indeed, some of the channel fragments appear disjointed and cut by faults or slump-planes (fig. 108).

The high and somewhat discontinuous leveed channels may be the record of an exceptional series of events which altered the relatively normal shape of the Rhône Fan. These events cannot have occurred in the distant past because the channels have not been buried by later sedimentation on a site of rapid deposition on a fan off a large river. Presumably they are Pleistocene events, when Alpine glaciation and de-glaciation introduced enormous amounts of relatively coarse sediment into the River Rhône. This sediment would have reached the basin at approximately the right time to build the birdfoot channels of the Rhône Fan. During a lowered stand of sea-level, the Rhône flowed across the continental shelf and may have dumped sediment into an unusual canyon—that leading to the present birdfoot channels instead of the usual ones. This change in course, coupled with an exceptional load in the river, would suddenly produce relatively large and frequent turbidity currents and build channels where previously the fan had been smooth. Deposition of the coarse fraction would aggrade the channels rapidly at the base of the continental slope and head of the fan. Rising sea-level would return the river to its normal terminus and might restore the earlier path of sedimentary transport to the basin. Several such events may have occurred during the wax and wane of Pleistocene glaciers. Slumping and disruption of the elevated channels would follow.

The smaller, outer channels form another group with many common characteristics. They are roughly parallel and relatively evenly spaced. Many curve gently to the left but this effect is slight. A few curve right and some even meander gently. These channels, a few tens of metres deep, a kilometre or so wide, and as much as 150 km long, appear to be secondary distributaries around the birdfoot channels, but this may be coincidental. It is not clear whether the channels are active at present or have become separated from the source of turbidity currents.

The Rhône Fan appears to be the site of complex turbidity currents following mutually dependent paths. Thus deposition around an active channel may partly fill an inactive adjacent channel or shift its axis. Moreover, sheet-flows may spread over each of the levees of two adjacent channels and join in a secondary canalized flow along the axis of the slight valley between the leveed channels. All these phenomena are suggested by echograms. Thus, small, perhaps secondary, channels occur between larger ones, and sub-bottom echoes show interfingering of turbidites which have spread laterally into the hollows between leveed channels (fig. 108).

Sub-bottom echoes were observed almost everywhere on the fan. The maximum depth of penetration was only about 20 m but half a dozen reflectors were common. Inasmuch as the reflectors were not sampled by coring, they cannot be positively identified. However, the local variations in thickness between reflectors, and the correlations of these variations with local topography, appear to rule out any possibility except that the reflectors are turbidites. The extent of individual reflectors, therefore, becomes of interest with regard to the problems of the dimensions of turbidity currents and the formation of extensive graded beds. Most modern turbidites cannot be traced from one core to the next, a fact which implies localized deposition. An exception is the "Madcap" area west of the Canary Islands where Cambridge University oceanographers found a single turbidite in many cores (Hill, pers. comm., 1962). On the other hand, ancient turbidites which are exposed on continents commonly can be traced for considerable distances. The turbidites of the Rhône Fan seem comparable. In several places they can be traced for 8–10 km and in one place for more than 16 km.

[*Editor's Note:* Material has been omitted at this point.]

7. AGE OF FAN AND DENUDATION OF SOURCE-REGION

The present sediment-load of the River Rhône is roughly $2 \cdot 6 \times 10^{-2}$ km^3 per year (van Andel, 1955). If this load is representative of the past, the total volume of fan and plain requires $1 \cdot 8 \times 10^7$ years of deposition. It is probable that erosion and sedimentation were much more rapid in Pleistocene times. If the unconsolidated sediment in fan and plain was deposited during Pleistocene times during 10^6 years, the sedimentation rate would have been about six times as fast as at present, thus leaving a minimum thickness for all the consolidated sediment of pre-Pleistocene time.

The Tertiary history of the Rhône drainage area is complex, because of the elevation of the Alps and of the marine transgressions and regressions in the present area of the Rhône Delta and Tertiary embayment (Gignoux, 1955). An elaborate exposition is unwarranted for the present purposes. Suffice it to say that erosional debris from the Alps has reached the western Mediterranean intermittently since the mountains were formed in Oligocene times, or during perhaps $4 \cdot 0 \times 10^7$ years. If the present sedimentary load of the Rhône can be used as a guide, $1 \cdot 2 \times 10^7$ years of deposition were required to deposit the consolidated sediment of fan and plain. The numbers are regrettably vague, but seem to warrant the conclusion that rivers comparable to the Rhône, acting through slightly more than a third of post-Eocene time, could deposit all the sediment in the Rhône Fan and Balearic Abyssal Plain by erosion of the Alps. Thus there is little indication of the products of pre-Tertiary continental erosion in the western Mediterranean.

Some indication of the total denudation and rate of denudation of the source-region can be derived from the volume of sedimentary deposits, the area of the source and the time the sedimentary system operates. The drainage area of the Rhône is about $1 \cdot 2 \times 10^5$ km^2 or slightly less than the area of sedimentation of the Mediterranean sea-floor. The total average denudation of the Rhône basin is $3 \cdot 8$ km. Denudation of $1 \cdot 3$ km is required to produce the unconsolidated sediment alone. For the whole of post-Eocene time, the average rate of denudation is about 10 cm/10^3 years, which compares with average rates of 12–21 cm/10^3 years for other mountains during comparable periods (Gilluly, 1949; Menard, 1961). On the other hand, if the unconsolidated sediment is Pleistocene and eroded during 10^6 years, the rate of denudation was 125 cm/10^3 years which compares with 100 cm/10^3 years at present for the Himalayas. During pre-Pleistocene time the rate was correspondingly lower.

8. REFERENCES

Bourcart, J. 1959 Morphologie du précontinent des Pyrenees à la Sardaigne. *Colloques int. Cent. natn. Rech. scient.* **83**, 33–50.

Ewing, J., Worzel, J. L. & Ewing, M. 1962 Sediments and oceanic structural history of the Gulf of Mexico. *J. geophys. Res.* **67**, 2509–2527.

Ewing, M., Ericson, D. B. & Heezen, B. C. 1958 Sediments and topography of the Gulf of Mexico, in *Habitat of Oil*, Weeks, L. (Ed.) *Am. Ass. Petrol. Geol.* 995–1053.

Ewing, M. & Ewing, J. 1962 Rate of salt dome growth. *Bull. Am. Ass. Petrol. Geol.* **46**, 708–709.

Fahlquist, D. A. 1963 Seismic refraction measurements in the western Mediterranean Sea. *Rapp. P.–v. Réun. Commn. int. Explor. scient. Mer Méditerr.* **17**, 963.

Gagel'gants, A. A., Gal'perin, E. I., Kosminskaia, I. P. & Krakshina, R. M. 1958 Structure of the earth's crust in the central part of the Caspian Sea determined by deep seismic data. *Dokl. Akad. Nauk SSSR.* **123**, 520–523.

Gignoux, M. 1955 *Stratigraphic Geology*, 682 pp. San Francisco.

Gilluly, J. 1949 The distribution of mountain building in geologic time. *Bull. geol. Soc. Am.* **60,** 561–590.

Harris, G. D. 1908 Rock salt. *Bull. geol. Surv. Louisiana,* **7,** 1–259.

Heezen, B. C. 1963 The Tonga-Kermadec and Hikurangi Trenches (Abstract). *Abstr. Pap. Int. Ass. Phys. Oceanography, XIII Assembly I.U.G.G.* **6,** 70.

Heezen, B. C. & Ewing, M. 1955 Orleansville earthquake and turbidity currents. *Bull. Am. Ass. Petrol. Geol.* **39,** 2505–2514.

Heezen, B. C. & Tharp, M. 1961 Physiographic diagram of the South Atlantic Ocean. *Geol. Soc. Am.*

Hurley, R. J. 1960 The geomorphology of abyssal plains in the northeast Pacific Ocean (unpubl. ms.). *Rep. Mar. Phys. Lab., Scripps Inst. Oceanogr.* Ref. 60–67.

Klemme, H. D. 1958 Regional geology of the Circum-Mediterranean Region. *Bull. Am. Ass. Petrol. Geol.* **42,** 477–512.

Mayev, Ye. G. 1960 Characteristics of sedimentation in the southern Caspian Sea. *Dokl. Akad. Nauk SSSR.* **130,** 154–157.

Mayev, Ye. G. 1961 Development of consedimentation folding on the floor of the southern Caspian. *Dokl. Akad. Nauk SSSR.* **137,** 146–149.

Menard, H. W. 1955 Deep-sea channels, topography, and sedimentation. *Bull. Am. Ass. Petrol. Geol.* **39,** 236–255.

Menard, H. W. 1960 Possible pre-Pleistocene deep sea fans off central California. *Bull. geol. Soc. Am.* **71,** 1271–1278.

Menard, H. W. 1961 Some rates of regional erosion. *J. Geol.* **69,** 154–161.

Menard, H. W. 1964 *Marine Geology of the Pacific*, 271 pp. New York.

Morgan, J. P. 1961 Genesis and paleontology of the Mississippi River mudlumps. *Bull. geol. Surv. Louisiana,* **35,** (1), 1–116.

Shepard, F. P. 1963 *Submarine Geology*, 2nd Ed., 557 pp. New York.

Solovyev, V. F., Kulakova, L. S. & Agapova, G. V. 1959 New data on the tectonic structure of the bottom of the south Caspian Sea. *Dokl. Akad. Nauk SSSR.* **129,** 1126–1129.

Solovyev, V. F., Kulakova, L. S. & Agapova, G. V. 1960 Tectonics of the south Caspian basin. *Izv. Akad. Nauk SSSR. Ser. Geol.* **4,** 1–8.

van Andel, Tj. H. 1955 Sediments of the Rhône Delta. II. Sources and deposition of heavy minerals. *K. Ned. geol.-mijn. Genoot. Geol. Ser.* **15,** (3), 515–547.

11

Copyright © 1969 by the American Association of Petroleum Geologists

Reprinted from *Amer. Assoc. Petrol. Geol. Bull.*, 53(2), 390–420 (1969)

Physiography and Sedimentary Processes of La Jolla Submarine Fan and Fan-Valley, California[1]

F. P. SHEPARD,[2] R. F. DILL,[3] and ULRICH VON RAD[4]

La Jolla, California 92037, San Diego, California 92132, and Munich, Germany

Abstract The depositional environments of La Jolla canyon, fan-valley, and fan are well known from closely spaced sounding lines, deep-diving vehicle observations, numerous undisturbed box cores, and continuous reflection profiles. The narrow rock-walled canyon changes seaward at 300 fm (549 m) to a wider valley cut into the compacted clayey sediments of a fan, and bordered by discontinuous leveelike embankments. The fan-valley merges gradually into the relatively flat floor of San Diego trough. Numerous dives into the fan-valley have shown precipitous walls along the outside of the bends of the winding channel. Slumping is taking place actively from these walls and large slump blocks of clay are common on the floor. Small scour depressions around isolated erratics suggest the erosive effect of relatively weak currents in some places but, for the most part, the muddy floor seems to have been little disturbed in recent years. Diagonal tension cracks cut the floor locally.

Box cores show that most of the sediment deposited on the valley floor in the past few thousand years consists of poorly sorted clayey silt, underlain by discontinuous layers of well-sorted fine-grained sand with a few coarse sand grains, gravel, and mud balls. Sand layers occur in 94 percent of the valley axis cores, of which 26 percent are graded; 59 percent have parallel laminations; and 41 percent have current-ripple cross-laminations. Sand layers are less common in the cores from levees and from the small discontinuous terraces along the sides of the fan-valley. Cores from the open fan have less and finer grained sand. In all these environments the sand shows no consistent or systematic grain-size variation with increasing water depths. Some of the coarsest sediments, including gravel and mud balls, are found in sand farthest from shore and at the greatest depths.

The character of the sand and the finding of shallow-water Foraminifera indicate the probability that sand has been carried from the coastal area along the valley axes and spilled over the levees onto the open fan. However, there is little evidence of recent high-velocity, high-density turbidity currents, because, in general, the covering mud layer is distinctly separated from the underlying sand deposits, and therefore does not suggest deposition at the terminus of a turbidity current. Also, the discontinuous character of the sands and series of laminae with heavy mineral concentrations indicate introduction by a traction type of pulsating current, such as has been seen during vehicle dives, and also has been measured in the few available current-meter records. The locally precipitous fan-valley walls and outcrops of gravel, and the sand layers on the levees and open fan, may be the product of stronger currents that moved down the valley during earlier more pluvial periods, when greater quantities of sediment entered the canyon heads. Possible confirmation of this idea comes from the available C-14 dates in plant layers, which suggest that deposition in the past few thousand years may have been considerably slower than that indicated for the Pleistocene. The finer sediments may be largely the

result of slow downslope movement of slightly higher density muddy waters coming from the coastal areas.

Continuous reflection profiles have shown that the inner La Jolla fan has only a thin cover of unconsolidated sediments overlying the folded and faulted Miocene-Pliocene rocks. The outer fan and adjacent San Diego trough contain a thick section (more than 1,000 m) of Quaternary sediments with probable buried older channels and possible thick lenses of sand sediments.

INTRODUCTION

Because of the convenient location directly outside Scripps Institution, the La Jolla submarine canyon, fan-valley and fan have been investigated more than any other comparable submarine environment in the world. The topography is known from closely spaced, well-located sounding lines (Shepard and Buffington, 1968). At depths too great for scuba dives, the general appearance of the canyon and fan-valley was observed from deep-diving vehicles. Sampling of the area has been very extensive and has led to several reports (summarized in Shepard and Einsele, 1962). Prior to 1963, coring along the canyon and fan-valley was rather unsatisfactory because of the

[1] Manuscript received, December 4, 1967; accepted, February 23, 1968.

[2] University of California, Scripps Institution of Oceanography.

[3] Naval Undersea Research Center.

[4] Institut für Geologie, Technische Hochschule München (Munich), Germany.

The writers acknowledge the help of the crews of various Scripps and Navy Electronics Laboratory ships on which the work was conducted. Also, numerous visiting scientists were helpful in the field work. Andre Rosfelder designed some of the instruments and accompanied us on several trips. Hugo Genser provided descriptions, photographs, and X-radiographs of several box cores. Neil F. Marshall was of great help in field work and in laboratory analyses. We are grateful for critical comments by D. G. Moore and M. L. Natland.

Von Rad gratefully acknowledges a NATO post-doctoral research fellowship granted by the Deutscher Akademischer Austauschdienst (Bad Godesberg, Germany) that enabled him to work for two years at Scripps Institution of Oceanography, The writers' work was supported in part by the Office of Naval Research under Contract Nonr-2216(23) and by the National Science Foundation Grant G.P.-3815.

Photos by Dill courtesy of U. S. Naval Undersea Research Center, San Diego, California 92132.

inadequacy of piston and gravity coring in penetrating compacted sands.

After the initiation of box coring (Reineck, 1963a), much better cores (20.3 × 30.5 cm wide, up to 60 cm long) were obtained, with well-preserved top layers and undisturbed sedimentary fabric. Many were oriented by a locked compass. Ninety-five successful box cores have been taken in this general area. All cores have been examined by the writers. Sediment characteristics were studied especially by von Rad; Foraminifera of many cores, by M. L. Natland; the magnetic fabric of 24 box cores, by A. I. Rees and von Rad (Rees *et al.,* in press); and size analyses were made largely by N. F. Marshall. Kelp layers found in a few box cores have been dated by the C-14 method, giving an estimate of the age of the layers and the relative rate of sediment accumulation. Continuous reflection profiles run by Moore (1966) and the writers along the fan and across the fan-valley were used to interpret the thickness of the fan sediments and the subbottom structure.

BATHYMETRY

The bathymetry of the area is discussed in more detail by Shepard and Buffington (1968). The regional chart with sample locations (Fig. 1) has been contoured mostly from their work.

La Jolla canyon and its concordant tributary, Scripps canyon, continue seaward beyond their 150-fm (274-m) juncture to a depth of about 280 fm (512 m), cutting through rock formations of Cretaceous and Eocene ages. Farther seaward, the channel continues as a fan-valley, cutting semiconsolidated Quaternary fan deposits that have filled San Diego trough. Where the rock walls terminate, the seaward slope of the rim and the valley axis decrease and the wall heights are lower. Also, the valley rim shows the first indication of marginal embankments, generally referred to as "natural levees" (Fig. 2). Between axial depths of 300 and 400 fm (549 and 732 m) these levees are virtually continuous along the north side of the fan-valley, but very intermittent on the south side. In this zone, the valley has relatively straight, steep walls about 40 fm (73 m) high. The valley is about 1 km wide but is cut by a narrow winding inner channel bordered by a low terrace (Fig. 3, secs. 6–8).

About 6 mi[5] (11 km) from its beginning, the fan-valley turns somewhat abruptly from its east-west course, and for about 6.5 mi (12 km) continues southeasterly in a relatively straight course. The walls form a distinct V-shaped profile with a relief of about 30 fm (55 m), and internal terraces of extremely variable heights are present, some near the rim and others near the floor (Fig. 3). The levees are better developed on the northwest wall than on the southeast.

In the next 4 mi (7 km), the V-shaped fan-valley continues to trend southwest in a more

[5] Nautical miles (6,080 ft) are used in the text.

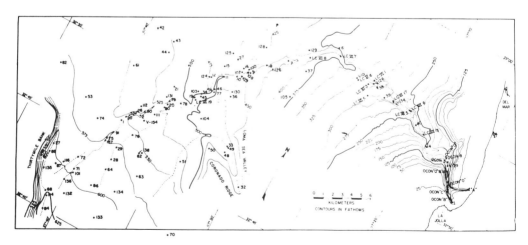

FIG. 1.—Submarine relief and locations of core samples along La Jolla fan and fan-valley. Contours largely from Shepard and Buffington (1968).

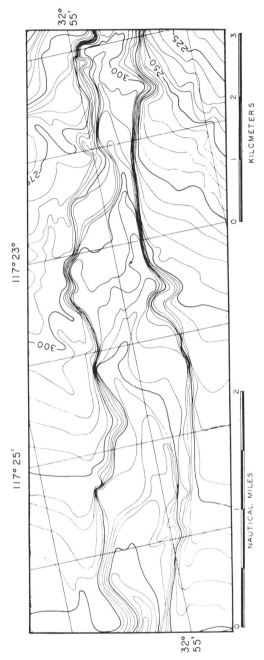

FIG. 2.—Inner La Jolla fan-valley showing winding inner channel and straight valley walls. CI = 5 fm. Original plotting scale 1:24,000. From Shepard and Buffington (1968).

twisting course, the wall relief decreases to about 25 fm (46 m), and the levees become better developed on the southeast side. In the outermost 7 mi (13 km), the fan-valley has a relatively straight course but gradually turns south (Fig. 4). Wall heights decrease until the valley merges into the floor of San Diego trough at 610 fm (1,116 m). In this section, continuous low levees are present on both sides. Small inactive distributary channels on the fan are essentially parallel with the main fan-valley. They are not at present connected with the main valley, and they have shallower depths. Another valley with low relief enters San Diego trough as a continuation of Loma sea valley, which borders the east side of Coronado ridge (Emery *et al.*, 1952, Fig. 1).

At the base of the eastern escarpment of Thirtymile Bank, there is a north-south-trending valley, referred to as a "rift valley" because of its location along a fault scarp. According to Moore (1966), this valley was formed by turbidity currents coming down from Newport and other submarine canyons north of San Diego. However, Dill's observations during *Deepstar-4000* dives in the Newport valley system suggest that the latter is not at present an active distributary system, even though it heads in shallow water.

DIRECT OBSERVATIONS FROM DEEP-DIVING VEHICLES

A total of 15 dives (mostly with Dill as observer) have been made into La Jolla fan-valley with the research submersibles *Trieste I* and *II* and *Deepstar-4000*. These have made it possible to gather detailed information on the topography and sediment cover of the valley floors and walls, not otherwise obtainable.

Perhaps the most surprising feature of the fan-valley observed in the deep dives was the extremely steep cliffs and vertical walls (as much as 50 m high) on the outside of some of the bends (Buffington, 1964; Dill, 1961; Moore, 1965). These walls have outcrops of horizontally bedded, semiconsolidated Quaternary sediment. Outcrops of pre-Pleistocene rocks are restricted to depths shallower than 300 fm (549 m), where they are overlain by unconsolidated or semiconsolidated fan deposits (Fig. 5).

Isolated cobbles and boulders (up to 1 m in

≫→

FIG. 3.—Profiles taken from fathograms of transverse crossings of La Jolla fan-valley. Includes 5 profiles from outer La Jolla canyon. Indicates character of terraces and levees along fan-valley. From Shepard and Buffington (1968).

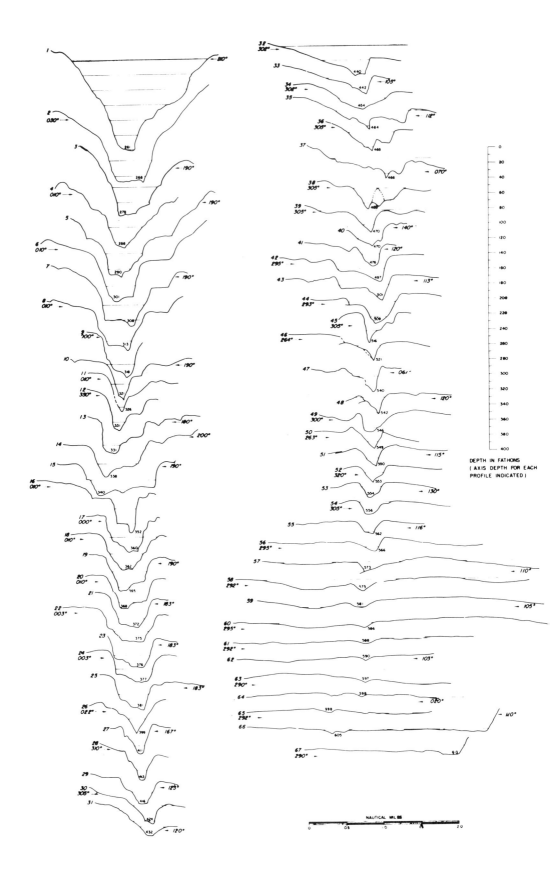

DEPTH IN FATHOMS
(AXIS DEPTH FOR EACH
PROFILE INDICATED)

NAUTICAL MILES

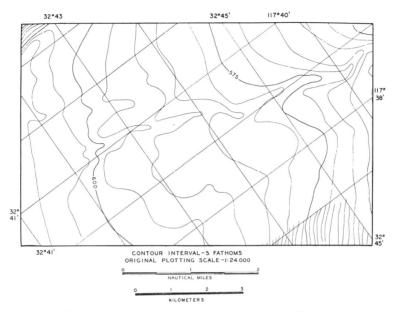

FIG. 4.—Outermost La Jolla fan-valley showing how it merges into San Diego trough and indicating two other fan-valleys and the "rift valley" at the base of Thirtymile Bank. From Shepard and Buffington (1968).

diameter) were first found by Dill in a Bathyscaph-*Trieste* dive near the landward end of the fan-valley at a depth of 325 fm (595 m). Subsequent dives in the *Deepstar-4000* showed that cobbles extend for a considerable distance along the floor of the valley (Fig. 6a). They also crop out in a terrace on the north side about 1–2 m above the valley floor at a constriction in the central axial channel at a depth of 320 fm (585 m) (Fig. 6b).

Rubble blocks of semiconsolidated clay, apparently broken off the steep walls, were found at many depths in the fan-valley. In the outer fan-valley, all stages of this process were observed (Fig. 7), from cracking clay on the walls to broken rubble at the base of steep slopes. Even the sides of the relatively low-walled outer fan-valley show exposures of horizontally bedded clay, locally slumping into the axis. Small vertical escarpments (80–150 cm high) can be seen (Fig. 8) where blocks have slumped away. Blocks of comparatively hard clay apparently are sliding gradually downslope toward the floor, and are seen standing 30 cm or more above the general surface (Fig. 9). This mechanism of slumping is important, because it produces concentrations of clay fragments along the valley axis that later can be picked up, transported, and redistributed, either

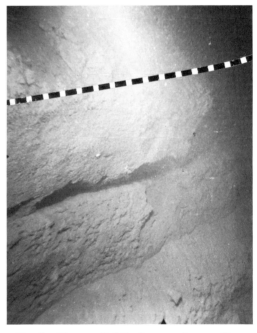

FIG. 5.—Sharp, well-defined contact between Eocene shale and poorly consolidated fan material (dotted line). Channel cuts through both sediment types, proving that downward erosion of canyon occurred following deposition of fan sediments. Photo by Dill.

by strong tractive currents or high-density suspension flows. Also, local spheroidal weathering was observed that produced rounded clay cobbles broken from the sides of the fan-valley (Fig. 10), instead of the more common angular blocks.

Dives also showed that the mud covering the channel floor is generally very thin (1–5 cm). Underlying sand was exposed when the mechanical arm of the submersible pushed aside the thin mud cover. Sand was also seen in

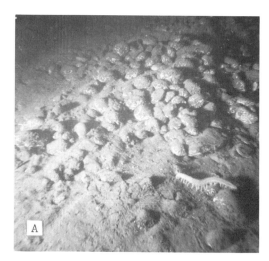

Fig. 6.—**A.** Rounded hard rock cobbles in central channel of fan-valley at depth of 328 fm (600 m). Average diameter is about 15 cm but large boulders were found with diameters up to 1 m. Photo by Dill. **B.** Cobble bed forms an internal wall terrace at base of outermost canyon at about 280 fm (512 m) depth. Wall slopes in area were between 50° and 60°. Photo by Dill.

Fig. 7.—**A.** Large blocks of angular and semiconsolidated fan material from walls of fan-valley form vertical cliffs about 1 m high at about 580 fm (1,061 m) depth. Photo by Dill. **B.** Blocks forming rubble pile at base of vertical walls. Blocks in photo are about 50 cm across. Photo by Dill.

mounds built up where crabs or other benthonic organisms had burrowed into the floor (Fig. 11). In traverses of the internal terraces,

Fig. 8.—Freshly exposed surfaces of semiconsolidated clay forming channel wall at 580 fm (1,061 m) depth. Note cone-shaped mounds formed by burrowing organisms at upper surface of side terrace. Vertical walls were caused by slumping of material into central channel of fan-valley. Photo by Dill.

Fig. 10.—Spheroidal weathering of clay blocks on north wall of La Jolla fan-valley at 445 fm (820 m) depth just above the channel floor. Photo by Dill.

however, no indications of underlying sand were encountered.

During a dive at 450 fm (823 m), an extraordinary number of small cracks and narrow ridges were seen in the channel fill near the south slope (Shepard and Dill, 1966, Fig. 30). Some of the cracks and ridges appear to extend

Fig. 9.—Relatively angular blocks of semiconsolidated wall sediment forming isolated group of erratics in predominantly sandy fill of fan-valley at 585 fm (1,070 m) depth. This group lacked scour depression. Large central block is about 60 cm across. Photo by Dill.

diagonally up-valley, and resemble tension crevasses on the sides of glaciers. Dill has observed similar cracks in the thin blanket of fine sediment that covers the fill in other parts of the valley at depths down to 610 fm (1,116 m). These cracks suggest tension on the floor, due perhaps to a slow downslope creeping motion of the entire channel fill. This is somewhat surprising because the regional slope of the channel axis in this area is slightly less than 1°.

In the outer fan-valley, many deep-sea holothurians (*Scotoplanes* sp.), gastropods (with tracks), and ophiuroids (brittle stars) covered the sea floor (10–20 brittle stars and 5–10 holothurians per m²) (Figs. 11, 12). Locally, these organisms were bowled over by the pressure wave, or water motion, created by the *Deepstar*. The surface layer of flocculated fine-grained mud also was readily set into suspension by the gentle current of the backwash from the craft's propellers. Both these observations suggest that at present the valley floor is not an area of high current velocities. This is indicated also by the lack of scour depressions around corroded and encrusted tin cans that lie on the channel floor (Fig. 13). If strong currents had swept this area recently, the upright can (Fig. 13b) would have been knocked over,

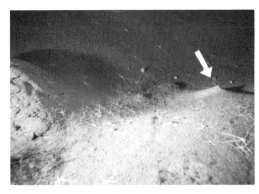

Fig. 11.—Mounds 15 cm high (see arrow), built by burrowing organisms, forming sand cones on fine-grained sediment in central channel at 590 fm (1,079 m) depth. Rounded erratic on left is semiconsolidated clay. Photo by Dill.

as well as the siliceous sponge that is seen in a vertical position (Fig. 13a) which indicates several years of undisturbed growth. However, many of the large clay blocks lying in the sandy fill had scour depressions surrounding them, suggesting the occurrence of local intermittent currents. It is notable that not all of the blocks are surrounded by scour depressions, and that the occurrence of depressions has no relation to depth or location within the valley.

Large sand waves with wave lengths of 15–30 m and heights of up to 1 m were observed in the outer fan-valley in depths of

Fig. 12.—Deep-sea holothurians (*Scotoplanes* sp.) at 582 fm (1,065 m) depth, bowled over by pressure wave of slowly moving *Deepstar*. Susceptibility to weak bottom current indicates that present bottom fauna has not developed in high-energy regime. Abundant bottom fauna along channel axis is actively reworking fine-grained flocculent layer that blankets walls and predominantly sandy fill of central channel of fan-valley. Photo by Dill.

Fig. 13.—Encrusted tin cans lying on sediment surface of central channel, indicate that area has not been subjected to strong bottom currents for considerable period of time. **A.** One-quart (0.8-liter) fruit can, at about 580 fm (1,061 m) depth, has siliceous sponge growing from upper rim. **B.** At nearly same depth, highly corroded and encrusted can stands upright surrounded by semiliquid flocculated rust deposit. Note lack of scour depression around base of can. Photo by Dill.

about 600 fm (1,097 m), where the predominantly sand-filled channel merges with the relatively flat floor of San Diego trough. The valley walls in this area are less than 5 fm (9 m) high. The sand waves are associated with large erratic blocks of semiconsolidated clay, which here are generally surrounded by scour depressions. Driftwood and shallow-water plant debris, consisting of waterlogged kelp, eelgrass, *etc*. (Fig. 14), were common in the scour depressions on the upslope side of large clay blocks that had fallen from the steep channel walls.

FIG. 14.—Gastropods with typical tracks concentrated in vicinity of relatively fresh-appearing clump of shallow-water debris, mostly kelp, 590 fm (1,079 m) depth. Note that partially buried block of semiconsolidated wall sediment is surrounded by shallow scour depression. Photo by Dill.

SURFACE SEDIMENTS

General Remarks

The numerous box cores obtained since 1962 have contributed much improved information, particularly because many of these cores are oriented (Rosfelder and Marshall, 1966), undisturbed, and of sufficient width to study the sedimentary fabric in great detail. The techniques that have been used for the fabric study of the box cores include core slicing, X-radiography of thin sediment slices (Bouma, 1964, 1965), obtaining small-scale lacquer peels, impregnation of small (5 × 5 cm) sediment samples with plastic resin (Reineck, 1963b), preparing "thin sections" (50–80 μ) for fabric study under the petrographic microscope, and measuring the anisotropy of magnetic susceptibility from oriented cylindrical specimens with a torque magnetometer (Rees, 1965).

In discussing the recent sediments of La Jolla fan, it is important to group the box cores according to their natural depositional environments (Table I). More than half the successful box cores (54 out of 95) were taken along the axes of the valleys. An additional 13 were obtained from or near the crest of natural levees, 7 from terraces within the main fan-valley, 16 from the open fan, and 5 from other environments. Table I gives average data for the 95 box cores and 16 gravity cores.[6]

* A much more complete table has been deposited as Document No. 9889 with the ADI Auxiliary Publications Project, Photoduplication Service, Library of Congress, Washington, D.C. 20540. A copy may be secured by citing the document number and by remitting $1.25 for photoprints, or $1.25 for 35 mm

Sedimentary Structures Along Canyon and Main Fan-Valley Axis

The sedimentary structures of 40 box and 5 gravity cores, and the longitudinal profile of the canyon and main fan-valley are schematically shown in Figure 15. Of the cores, 42 (93 percent) contain one or several thick layers of fine- to medium-grained terrigenous arkosic sand showing well-preserved sedimentary structures. Of the three mud cores of this group, two consist of still semiconsolidated clayey silt, and probably represent outcrops of the Pleistocene fan deposits. This exposure of older Pleistocene deposits indicates that the recent sediment has been eroded from, or was never deposited at, these locations.

A thin mud layer, rarely more than 5 cm thick, covers the sand layers in most of the cores. It is never greater than 20 cm, except in core 10 with 50 cm of mud and no sand layer. Nine cores of this group contain only sand, three of them are shorter than 10 cm, hence not representative. Predominantly sand cores are much more common than predominantly mud.

The most typical internal sedimentary structures observed in flysch beds and described as the "turbidity sequence" by Bouma (1962) are found in some of the sand layers. The a-b-c from Bouma's idealized sequence (a-e) for a turbidite has the following vertical succession, from top to bottom, for which some subdivisions have been added (see also Table I):

c_4 convolute laminated division
c_3 wavy laminated division
c_2 deformed cross-laminated division
c_1 current-ripple cross-laminated division
b_1 inclined parallel-laminated division
b   parallel horizontal-laminated division
a   massive graded division

Eighteen sandy cores of this group contain sand layers that are not laminated and show no size grading (a_1, Table I). This "massive ungraded division" is especially typical of the coarse-grained canyon axis cores, but occurs also at intervals all along the fan-valley.

Graded massive bedding was found in 29 percent of the sandy canyon-fan-valley cores,

microfilm. Advance payment is required. Make checks or money orders payable to: Chief, Photoduplication Service, Library of Congress. Samples are listed separately for each of the major depositional environments and given in order of increasing water depth; included are core lengths, the ratio of sand to mud layers, and a few important fabric and size parameters for each core.

Table I. Data from Cores, La Jolla Submarine Fan and Fan-Valley Area, California

	Canyon Axes	Depositional Environment of Group							
		Inner Fan-Valley	Middle Fan-Valley	Outer Fan-Valley	Total Canyon Fan-Valley Axes	Other Valley Axes	Terraces	Natural Levees	Open Fan
Depth ranges of groups	8–287 fm 15–525 m	310–400 fm 567–732 m	416–545 fm 761–997 m	546–607 fm 999–1,111 m	8–607 fm 15–1,111 m	454–645 fm 831–1,180 m	310–587 fm 567–1,074 m	316–610 fm 578–1,116 m	313–610 fm 571–1,116 m
Number of samples in group	10	7	17	11	45	15	10	15	21
Average ratio of sand layers to mud layers	70	50.7	52.1	63.5	64	44	15	9	12
Percent lacking all sand layers	0	0	17	0	6.6	26	50	26	57
Percent with massive ungraded sand (a_1)	70	28.6	17	45	37	33	0	40	9.5
Percent with massive graded sand (a)	10	0	41	35	26	13	20	7	4.8
Percent with horizontal parallel laminations (b)	80	57	53	54	58	20	30	20	19
Percent with inclined parallel laminations (b_1)	0	14	6	9	7	0	10	7	0
Percent with current-ripple cross-laminations (c_1)	30	43	47	45	42	33	20	47	9.5
Percent with deformed cross-laminations (c_2)	0	0	17	9	9	6.6	10	7	0
Percent with wavy laminations (c_3)	20	28	17	18	20	13	0	13	0
Percent with convolute laminations (c_4)	0	0	0	9	2	13	0	0	0
Percent with micro-erosion channels	10	0	17	18	13	0	0	27	0
Percent with appreciable coarse sand and/or gravel	30	14	6	18	13	33	10	0	4.8
Percent with mud balls in sand	20	14	12	45	22	33	10	0	4.8
Percent with plant-debris layers	30	14	17	18	20	0	0	0	4.8
Average median grain size ϕ of sand or coarse silt layer ϕ (No. of analyses)	2.2 (7)	1.5 (2)	2.5 (11)	1.9 (8)	2.2 (28)	2.5 (8)	2.8 (1)	2.9 (6)	3.17 (8)
Average maximum one- percentile ϕ (No. of analyses)	0.1 (6)	—	1.6 (11)	1.09 (6)	0.9 (23)	1.75 (8)	—	2.2 (2)	2.2 (3)
Standard deviation ϕ (No. of analyses)	.45 (2)	—	0.7 (14)	.08 (8)	0.7 (24)	1.18 (8)	0.7 (1)	0.48 (5)	1.18 (5)

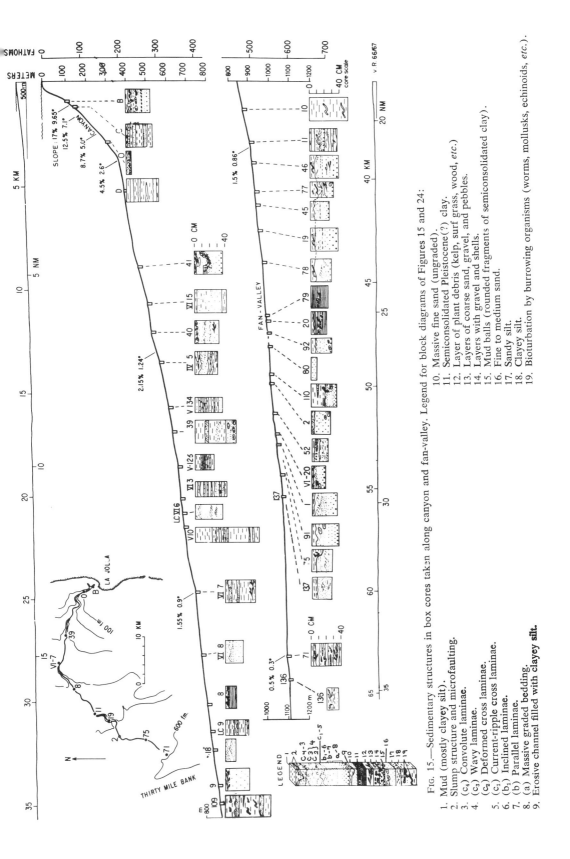

FIG. 15.—Sedimentary structures in box cores taken along canyon and fan-valley. Legend for block diagrams of Figures 15 and 24:

1. Mud (mostly clayey silt).
2. Slump structure and microfaulting.
3. (c_4) Convolute laminae.
4. Wavy laminae.
5. (c_3) Deformed cross laminae.
5. (c_1) Current-ripple cross laminae.
6. (b_1) Inclined laminae.
7. (b) Parallel laminae.
8. (a) Massive graded bedding.
9. Erosive channel filled with clayey silt.

10. Massive fine sand (ungraded).
11. Semiconsolidated Pleistocene(?) clay.
12. Layer of plant debris (kelp, surf grass, wood, *etc.*)
13. Layers of coarse sand, gravel, and pebbles.
14. Layers with gravel and shells.
15. Mud balls (rounded fragments of semiconsolidated clay).
16. Fine to medium sand.
17. Sandy silt.
18. Clayey silt.
19. Bioturbation by burrowing organisms (worms, mollusks, echinoids, *etc.*).

195

generally at the base of fully cored sand layers. Grading appears to be conspicuously absent in the canyon and inner fan-valley with the exception of core 41 at 254 fm (455 m). The intermediate and outer parts of the fan-valley yielded a series of cores with graded sand layers: six closely spaced cores (11, 46, 77, 45, 19, 78) between 498 and 531 fm (911 and 971 m), and three cores (LC-VI 20, 1, 9) between 562 and 575 fm (1,028 and 1,051 m). Most sand layers are well to moderately sorted at the base and generally better sorted toward the top. Some grade into overlying, very poorly sorted clayey and sandy silt. Inverse grading occurs rarely.

Parallel lamination (b) is the most abundant structure, and is found in about 60 percent (27 of 45) of all sandy canyon-fan-valley axis cores. It is produced in most cases by changes in composition, such as heavy mineral concentrations, biotite or fecal-pellet layers (Fig. 16b).

Small-scale (mostly less than 1 cm) current-ripple cross-lamination (c_1) (e.g., in core 1, Fig. 17) is the next most important structure, found in 43 percent of the sandy cores. This grouped cross-lamination with scoop-shaped sets of laminae clearly was caused by the migration of small current ripples. The stoss sides of the ripples generally are eroded, and wave lengths and ripple amplitudes are variable, with ripple indices (wave length/amplitude) ranging from 4 to 12 (average 8.5).

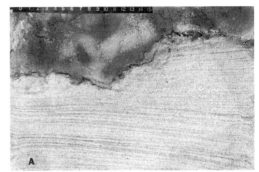

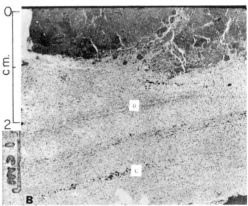

FIG. 16.—**A.** Parallel lamination in sand layer along fan-valley in box core 75 at 578 fm (1,057 m). Dark layers are concentrations of heavy minerals. Note erosional unconformity and overlying mud bed. **B.** Bands of dark minerals shown in X-radiograph of box core 36 along natural levee at depth of 472 fm (864 m).

FIG. 17.—X-radiograph of box core 74 taken from natural levee along fan-valley at depth of 556 fm (1,017 m), showing cross-lamination in sand layer.

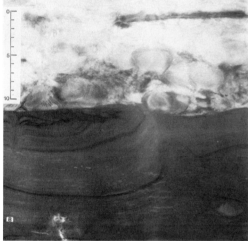

FIG. 18.—Convolute lamination and drag folds. **A.** From X-radiograph of box core 110 taken in fan-valley at depth of 552 fm (1,010 m). **B.** From Pliocene Pico Formation in Hall Canyon, Ventura basin. Laminated fine sands are overlain by coarse-sand formation. Photo by von Rad.

Heavy mineral concentrations (amphibole, epidote, magnetite, *etc.*) in distinct laminae may indicate that reworking by bottom currents is very common in most parts of graded and ungraded beds (Fig. 16). However, these features are rarely found at the base of sand layers. On the other hand, lamination, including alternation of dark and light layers, can be produced without current pulsation, as shown by the tank experiments of Kuenen (1966); but his experiments did not demonstrate that heavy mineral concentrations, like those shown in Figure 16b, can be produced without a distinct increase of current above that which deposited the intervening layers. Thin sections reveal micro-unconformities and erosional flutes filled by trapped coarse sand grains or cross-laminated foresets (scour-and-fill structures). The presence of these structures suggests that the sand layers probably have been deposited slowly, grain-by-grain, with intermittent times of nondeposition, erosion, and reworking.

Slightly deformed cross-lamination (c_2) has been produced by syn- or post-depositional deformation and oversteepening of ripple-drift cross-lamination foresets (Fig. 18, left). Wavy parallel laminations (c_3) in some places produced by a bending of laminae around mud pebbles, are relatively common. All transitions exist between the c_1-c_2-c_3 divisions of Table I and the convolute laminated division (c_4), characterized by sharp crested "anticlines" and broadly folded "synclines" (Fig. 18a, right).

Most thin sand layers are very fine grained and distinctly laminated throughout their thickness ("laminites," Lombard, 1963). This b-c sequence was found in about 20 percent of the canyon-fan-valley cores. Fully developed graded parallel and cross-laminated cycles (a-b-c) were present in about 28 percent of the sandy cores of this group (Fig. 19a). The thickness of fully cored sand layers along the canyon-fan-valley axis ranges from a few centimeters to more than 10 cm. Several sand layers more than 40 cm thick were not penetrated completely by the box corer. Hence, it is possible that in some cores, showing only the b-c sequence, the graded division (a) was not penetrated.

Small-scale slump structures with very fine sand layers (*e.g.*, Figs. 15, 20) showing down-slumped and micro-faulted cross-lamination were noted along (or across) the length of some box cores. Channels 5–15 cm wide, cut into parallel-laminated sand and later filled with homogeneous clayey silt (Fig. 16), indicate the activity of strong erosive bottom currents that did not subsequently deposit sandy material.

Semiconsolidated clayey pebbles (called "mud balls") are commonly incorporated in a matrix containing considerable medium- and coarse-grained sand, especially in the outermost fan-valley between 544 and 575 fm (995 and 1,051 m) (Fig. 21a). These inclusions have a diameter range of 6–13 cm. They are similar in texture and foraminiferal content to the scattered blocks of clay rubble, which apparently came from the erosion of the steep channel walls and terraces, and were transported

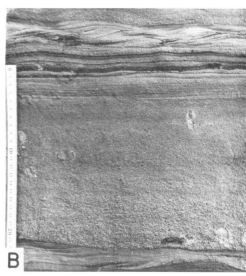

Fig. 19.—Unlaminated graded sand covered by laminated finer sand and, in turn, by cross-laminated sand. Typical of turbidite beds, according to Bouma (1962). **A.** From box core 1 at 568 fm (1,039 m) along fan-valley. **B.** From Pico Formation of Hall Canyon, Ventura basin. Photo by von Rad.

down-valley as float in a coarse-grained matrix.

Seven of the axis cores, especially those from the outermost part of the fan-valley, have distinct layers of plant debris. These layers consist of waterlogged kelp stipes and holdfasts, eelgrass, land-derived wood, and broken bits of Gorgonian coral fronds, interbedded with or overlying fine-grained laminated sand. Core 71 at 597 fm (1,094 m) (Fig. 29) contains four distinct plant layers, two of them 4–6 cm thick and two 0.5–1.5 cm thick. The plant layers have their closest possible source in the heads of the canyons, where similar mats are being deposited and moved downslope by slow glacierlike gravity creep (Dill, 1964a, p. 158).

Bioturbation and mottling, shown in X-radiographs, considerably affect the sedimentary fabric of the clayey sediments on the sea floor (Fig. 22). Most of the clayey-silt layers have been almost completely homogenized by the action of burrowing organisms. Only exceptionally faint parallel- or cross-lamination can be observed in X-radiographs of sandy silts. The term "mottling" describes a combination of burrow structures (Bouma, 1964, p. 307), such as lumps of silt caused by burrowing or homogeneously mottled "cloudy" structures. To ascribe typical burrow structures to definite types of burrowing animals is difficult. However, annelids and other worms certainly produce most of the small tubelike burrows. Echinoids (*e.g.,* the heart urchin *Echinocardium* sp.), pe-

lecypods and gastropods have been observed reworking the canyon fill and must be responsible for some of the larger burrow structures.

Grain size and thickness of sand layers.— Figure 23 and Table I show that the grain size and thickness of sand layers do not have a consistent decrease along the fan-valley in the

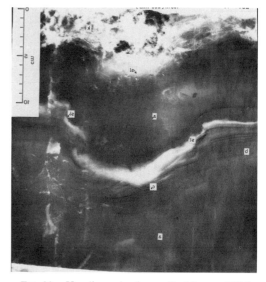

Fig. 20.—X-radiograph of core 41, taken at 250 fm (475 m), showing slumping effects, perhaps result of bioturbation.

198

downslope and offshore direction. This apparently is caused by irregularities in bottom topography and by the irregular and discontinuous processes involved in the initial deposition and reworking of sediment material. It is noteworthy that very coarse sand and gravel associated with reworked Pleistocene(?) mud balls are present in the intermediate and outermost parts of the fan-valley. The distribution of textural parameters along the longitudinal profile of La Jolla canyon and fan-valley is shown in Figure 23. The coarsest representative sand of the upper 20 cm of most cores has been analyzed. As the sand layers occur at different intervals in each core, the parameters shown in this graph cannot be correlated directly from core to core. Along the total length of the canyon-fan-valley axis, at least six samples contain medium- to coarse-grained sand (sometimes with gravel). The coarse sediments also have high medians and maximum one-percentile values, general poor sorting, and bimodality.

A thin layer of clayey silt lies on top of most fine-grained sand. In this top layer, the very poorly sorted mud, influenced by biological mixing, does not show much lower maximum grain sizes than does the underlying well-sorted sand. One exception to the general rule of complete randomness of sediment distribution along the fan-valley is the series of closely spaced cores (11–19) taken within a horizontal distance of about 2.5 mi (4 km) between depths of 498 and 521 fm (911–953 m). The thick sand layers in these cores can be compared, because they are compositionally similar and have several textural parameters that change in a consistent pattern; with one exception, median, maximum one-percentile, and mode-values decrease in the downvalley direction. The sediment from this stretch of the fan-valley may have been transported and deposited by a common mechanism of sedimentation, perhaps a turbidity current.

In two zones along the axis of the fan-valley, the recent sediment cover seems to be missing or very thin. One of these zones with hard semi-consolidated clay (core 8 at 432 fm (790 m)) is in an area where the continuous reflec-

Fig. 21.—Mud balls in sand formations. **A.** From box core 2 along fan-valley at 553 fm (1,012 m). **B.** From box core 72 taken in disconnected short fan-valley at depth of 593 fm (1,085 m). **C.** From Miocene at La Jolla, southern California. Scale center in cm. Photo by Neil Marshall.

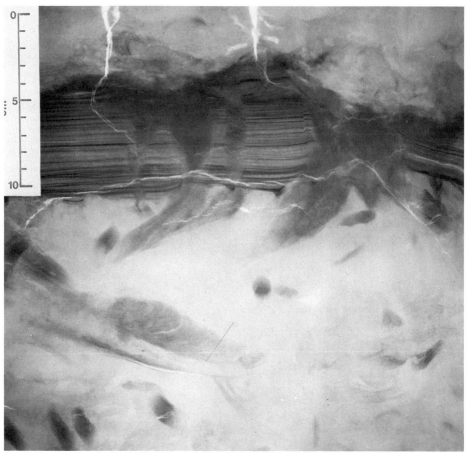

Fig. 22.—Laminated sand layer partly disrupted by bioturbation and considerably disturbed underlying mud formations. From X-radiograph of box core 125 at 453 fm (829 m) on natural levee.

tion profiles show bedrock close to the surface. The other zone (cores 79 and 20) is at a depth of 543 fm (994 m), where the Sparker records show no evidence of near-surface bedrock. The hard clay is apparently part of the fan deposit into which the valley has been cut. F. L. Parker's micropaleontologic examination of cores 8 and 20 suggests that these compact clays are of Pleistocene age (67 and 61 percent, respectively, left-coiling specimens of *Globigerina pachyderma,* including many small and large arctic-type specimens). Surprisingly, the other two samples of semiconsolidated clay (cores 76 and 79, the latter being very close to core 20) showed no indication of "cold Pleistocene" (having 93 percent right-coiling *Globigerina pachyderma*). Possibly the latter clays were deposited during an interglacial or interstadial period, when warmer temperatures similar to

those of the present day (89–98 percent right-coiling *Globigerina pachyderma*) prevailed (F. L. Parker, personal commun.).

Axes of Other Valleys

Examination of Figures 1 and 4 shows four small valleys that are independent of the main La Jolla fan-valley but in close proximity to its outer section. The slightly incised valley forming the continuation of Loma sea valley and extending around the outer end of Coronado ridge was cored in three localities. The cores consist of mud, except for core 49, which has a thin layer of silty sand. Core 76, at 556 fm (1,017 m), from the valley east of and parallel with the outer main fan-valley, contains semiconsolidated Pleistocene(?) clay overlain by soft clayey-silt, similar to that in nearby core

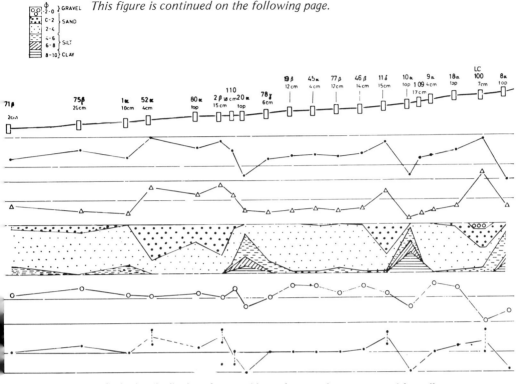

This figure is continued on the following page.

FIG. 23.—Grain-size distribution of top sand layers in cores along canyon and fan-valley.

0 in the main fan-valley. Three other cores 29, 28, 86) along this valley have layers of ne to very fine sand. Core 72, taken in the nort valley west of the outermost main fan-alley at 595 fm (1,088 m), has a thick lower ction of coarse sand and many large (up to 5 cm) rounded and partly imbricated mud alls (Fig. 21b). The matrix of medium- to parse-grained sand grades upward into a ross-laminated fine-grained sand.

Four of the box cores (27, 135, 114, 84) ken from the rift valley at the east foot of hirtymile Bank closely resemble core 72, hav-ng numerous large, rounded mud balls. Three f these cores contain moderately sorted, me-ium- to coarse-grained sand, and one has a aximum one-percentile of 2.8 mm (−1.4 φ). he virtual absence of glauconite and Forami-ifera in the coarse-grained sand of the rift val-y precludes the possibility that the mud lumps re due to slumping action from the nearby eep slope of Thirtymile Bank.

Natural Levees

The natural levees were sampled in 15 dif-ferent locations (13 box cores and 2 gravity cores; see Fig. 24). The main difference be-tween the fan-valley axis cores and the levee cores is in the ratio of sand to mud layers. Eleven of 15 levee cores contain sand layers, usually only 1–5 cm thick. Most of the sand is very fine grained and laminated throughout the layer. An exception is core 103 at 496 fm (907 m), which contained a 15-cm thick graded-sand layer with a complete a-b-c_1-c_2-c_3 cycle. The outermost levee sample, core 132 at 608 fm (1,112 m), has the thickest (22 cm) sand layer. The levees, like the channel, appear to have thicker sand layers along the outer part of the fan-valley. All the levee cores lacking sand layers are found at depths of less than 495 fm (905 m), where wall heights are greater than 30 fm (55 m). Eight of 12 levee cores with sand layers are current-rippled and cross-lami-nated, three are also parallel-laminated, and only three are massive ungraded. One of the levee cores shows an unusually high degree of bioturbation within a sand layer (core 125; see Fig. 22). Similar active reworking of the sea floor along the levees by burrowing organisms was observed by Dill during his deep dives.

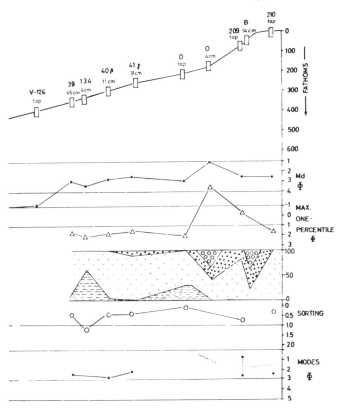

Shows that grain size varies irregularly, but with some increase in outer part.

Fan-Valley Terraces

Seven box cores and three gravity cores were obtained from the terraces along the sides of the fan-valley (Fig. 24). The terraces are probably slump blocks except in the inner fan-valley, where they may be due to rejuvenation after a period of deposition. Core 108 at 441 fm (807 m) is from the flat top of a small isolated hill in the fan-valley, and thus is comparable with the terrace samples. Five of the 10 samples contain sand as layers or lenses, generally well below the surface. Most of the samples lacking sand layers were from the upper part of the fan-valley at depths of less than 390 fm (713 m). Most sand layers are only 1–4 cm (maximum, 8 cm) thick, and consist of parallel- and cross-laminated, moderately well-sorted very fine- to fine-grained sand. Core 101 at 587 fm (1,074 m) contains a 7-cm thick sand layer with a complete a-b-b_1-c_1 sequence.

Open La Jolla Fan and Basin Slope

Twenty-one samples (16 box cores, a few of them undisturbed, 4 gravity cores, and 1 piston core) were taken from the open fan away from the fan-valleys and levees (see Fig. 24). Twelve consist of mud only, and most of the remaining 9 have only thin layers of very fine-grained terrigenous sand. However, core 134 at 595 fm (1,088 m) is primarily sand with a thin cover of mud and with large mud balls in the lower part. Core 116, at 598 fm (1,094 m), consists of relatively coarse-grained sand covered with a kelp layer. The proximity of this core to cores 71 (Fig. 30) and 136 in the outermost fan-valley is of interest because they all contain kelp layers. Again, in the open fan environments the deeper cores show an increase of thickness and grain size of sand layers.

Other Environments

Sample 50 from the northernmost tip of Coronado ridge has a poorly sorted, muddy, medium-grained sand layer overlying a very poorly sorted silty-sandy-gravel (maximum one-percentile: 6 mm) with angular rock fragments. Although this sample comes from near the edge of the fan, it does not belong in the open-fan group, because the rock fragments and

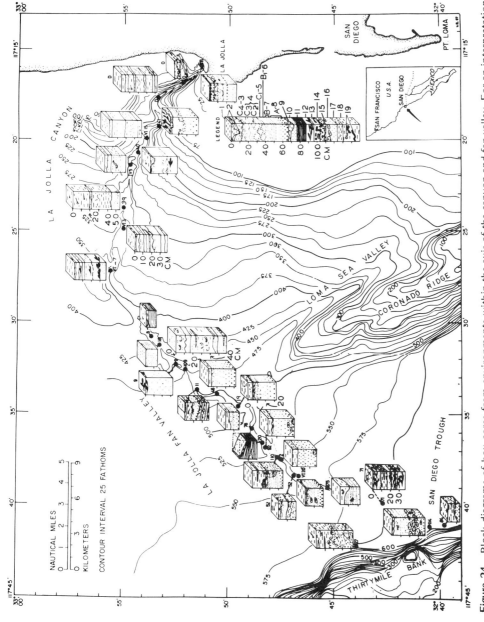

Figure 24. Block diagrams of box cores from environments within the axis of the canyon and fan-valley. For an interpretation of the symbols. see Figure 15.

sand grains consist of authigenic minerals (collophanite, glauconite, pyrite) and of volcanic, sedimentary, and metamorphic rock fragments, typical of the facies on top of Coronado bank (Emery *et al.*, 1952).

The short box cores from the top of Thirtymile Bank consist of sand coarser than most of the fan and fan-valley sands, and include considerable glauconite and an abundance of Foraminifera. The two samples taken on the fan-valley slopes are structureless mud with very little sand.

THICKNESS OF FAN DEPOSITS AND UNDERLYING STRUCTURE

Continuous reflection profiling has been used to determine the thickness of the fan sediments and the structure of the rock masses underlying La Jolla fan. The locations of a series of profiles taken with the high-energy, low-frequency Sparker system of the USNS *Davis* (see Curray and Moore, 1963) are shown in Figure 25 and the profiles in Figure 26. Six of the profiles (A-F) run essentially down the slope of the fan (northeast-southwest) and the other 11 (G-Q) cross the fan-valley (northwest-southeast). The profiles show a boundary between what Moore (1966) has called the deformed "preorogenic" bedrock and the "postorogenic" sediments.

The isopach map (Fig. 27), showing the approximate thickness of the postorogenic sediments, was constructed from the writers' seismic profiles and two from Moore (1966, Fig. 16). The approximate thickness of the postorogenic basin fill is calculated from the two-way travel time of sound, according to the values computed by Moore (1966, Fig. 19). These estimates take into consideration the velocity increase with increasing sediment thickness resulting from an increase of average sediment density by diagenetic compaction.

The inner part of the fan consists of a thin veneer (10–100 m) of younger unconsolidated sediments overlying an irregularly sloping bedrock surface. The records show that La Jolla fan is not a "wedge-shaped" cone of clastic sediments, generally considered to be typical of deep-sea fans (Gorsline and Emery, 1959; Menard, 1960; Walker, 1966). Because the Sparker method does not resolve sediment layers thinner than about 4 m, areas with the designation "no sediment cover" in Figure 27 may have a thin cover of sediments, *e.g.*, along the northeast side of profile A (Fig. 26). Profiles A and G also indicate that part of the canyon and

fan-valley may be located along vertical faults, as was suggested by Buffington (1964) in his interpretation of the inner fan topography.

The outer part of La Jolla fan (Fig. 26, profiles L-Q) is underlain by a thick sequence of semiconsolidated[7] Pleistocene(?) sediments (200–1,000 m). A 200–500-m thick sediment wedge comes from the southeast where the inactive Loma sea valley interfingers with the ponded sediments of La Jolla fan. The buried continuation of Coronado ridge, with its partly eroded northwest-plunging anticline, is indicated in Figure 26, and an older, much deeper Loma sea valley is covered by fan deposits. Traces of buried older channels indicate that the fan-valley has migrated considerably during Pleistocene and Holocene times.

The outermost part of La Jolla fan and the central part of San Diego trough have very thick sediment fills, up to 1,000 m, with a small bowl-shaped central "basin" containing up to 1,450 m of mainly Pleistocene and Holocene sediments[8] (Fig. 26, profiles A, B, D, and L-Q; Fig. 27). In some places the bedrock of the trough is not defined clearly by the reflection profiles. Near the eastern part of the trough, the sediment thickness decreases abruptly to less than a few hundred meters, and in the west, the pre-Pleistocene rocks crop out on Thirtymile Bank. Also, rock is found at the surface along much of Coronado bank and ridge. Near the contact of San Diego trough and Thirtymile Bank, the thick basin fill shows abnormal acoustical impedance characteristics, indicated by strong wavy reflection horizons near the escarpment (Fig. 26, profiles A, C, P, O). This may signify a thick sand body shown by weak subparallel and horizontal reflectors. This suggestion conforms with the distribution of surface sediments along the rift valley, as seen in the box cores.

The contact between the trough sediments and the escarpment of Thirtymile Bank (Miocene volcanic rocks) is probably a fault, but profiles A and C (Fig. 26) show no clear evidence of this fault, possibly because of the inability of the Sparker to interpret steep slopes. Coronado Bank and the basin slope on the east consist of folded and faulted Eocene through Pliocene(?) strata that are only partly covered by thin prograding fan deposits (Fig. 26, pro-

[7] Geophysicists commonly call sediments "unconsolidated" that have even higher velocities than these, but "semiconsolidated" seems more in keeping with geologic usage.

[8] Age based on probable rates of deposition.

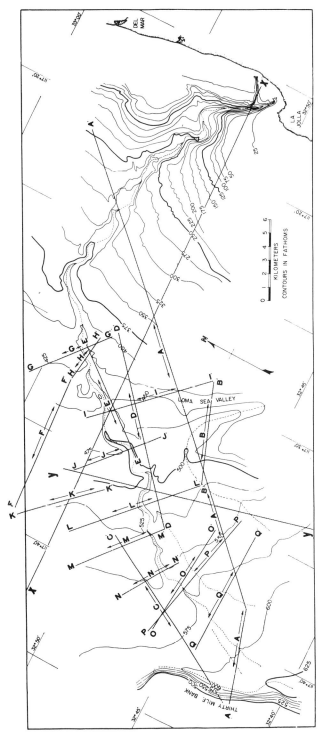

Fig. 25.—Location of continuous reflection profiles by writers (A-Q) and D. G. Moore (X-Y). Sediment thickness data from profiles were used to construct isopach map of sediment thickness for San Diego trough.

file A). Erosion of Miocene rocks is indicated from Foraminifera found in cores taken in the area (M. L. Natland, personal commun.).

SOURCES OF SANDY SEDIMENTS

The main source for sand on La Jolla fan is the sediment eroded from the batholithic and metamorphic rocks of the Peninsular Ranges, transported to the sea by rivers, redistributed along the coast of southern California by marine longshore currents, and funneled downslope by the canyon system. The mineralogic composition and texture of the sand deposits, the large amounts of plant debris, and the occurrence of displaced shallow-water Foraminifera (M. L. Natland, personal commun.) clearly indicate that most of the fan and fan-valley sand has been derived from the nearshore area.

A second, very minor source is from the active submarine erosion of the canyon walls (Dill, 1964b; Shepard and Dill, 1966, p. 38, 42, 58). All formations cropping out along the canyon walls have contributed some sediment material as talus. These include the conglomerate of the Rose Canyon Formation (Eocene), the siltstone and sandstone of the Rosario Formation (Late Cretaceous), and also the Pleistocene gravel beds. Possibly the coarse sediments come from the erosion of the fan-valley walls, although the deep-dive observers saw only clay walls, and a gravel bed in the low terrace near the inner end of the fan-valley (Fig. 6).

Shore-cliff erosion apparently contributes little to the total sand budget (Emery, 1960, p. 11–21). No significant similarities could be found between the size and composition of beach sands and of the sandstone beds exposed in the nearby cliffs (Rose Canyon Formation, La Jolla).

Evidently the submarine ridges and banks do not supply any appreciable amount of sand to the adjacent San Diego trough. Neither the heavy mineral assemblages nor the composition of the rock fragments in the coarse sands is indicative of the Thirtymile Bank sediment cover.

The localized patches of coarse-grained sand and fine gravel cored in the outer fan-valley can be explained as lag deposits or as the result of variation in the type of sediment entering the canyon at its head. This variation is mostly dependent on the intensity of winter storms, which move sediment from the beaches into the heads of the canyons. Coarse sand and cobbles up to 15 cm in diameter are common in the micaceous sands now moving down the head of Scripps canyon. Reworking of this mixed sediment by currents of variable velocity, like the ones observed and measured indirectly, would tend to concentrate the coarser fractions into patches of sediment with similar grain sizes.

Another possible source of coarse sediment is the submarine erosion of alluvium from the head of La Jolla canyon. The canyon is cutting shoreward locally at a rate as high as 0 6 m/year, through a sediment that was deposited in the canyon head during Pleistocene fluctuations of sea level. The erosion of La Jolla canyon head is most intense during winter storms, when sand from the beach spills over the headwall. On such occasions, coarse gravel and cobbles from the alluvium become part of the sediment fill that is moving downslope. The periodic incorporation of coarse material may result in the observed patchy distribution found in the outer fan-valley.

MECHANISMS OF SAND TRANSPORT

All investigations of the fans at the base of submarine canyons have led to the conclusion that sand and other coarse-grained fan sediments have been transported down along the canyons and their fan-valley continuations. The prevalent opinion has been that these sediments move seaward in turbidity currents. However, in recent years most investigators of these environments have acquired information that suggests a more complex explanation. The complications are discussed more fully elsewhere (see Shepard and Dill, 1966, p. 320–335), but will be considered here briefly.

Turbidity Currents

There are many indications that powerful currents or mass movements of some type transport all sizes of sediment along submarine canyons and prevent canyon filling. The precipitous, freshly eroded walls along the canyons and fan-valleys suggest the presence of powerful currents. Cobbles and even boulders found along many canyon axes imply current strength. Sand on the marginal levees 50 or 100 m above the channel floor shows that currents have had considerable thickness as well as power. All these facts appear to favor the existence of at least periodic high velocity currents.

However, many factors, especially in the area under discussion, are somewhat incompatible with the turbidity-current explanation. The failure of explosives set off in the canyon-head fill to start such currents confirms the work of

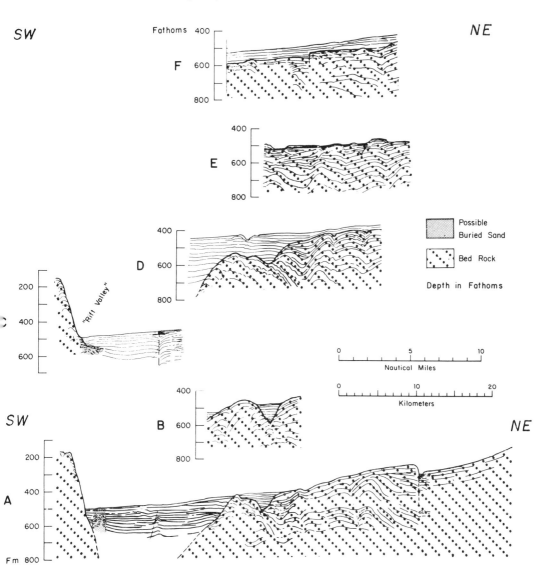

FIG. 26.—Continuous reflection profiles A-E roughly parallel La Jolla fan-valley axis (NE-SW), and F-Q are transverse to axis (NW-SE).

This figure is continued on the following page.

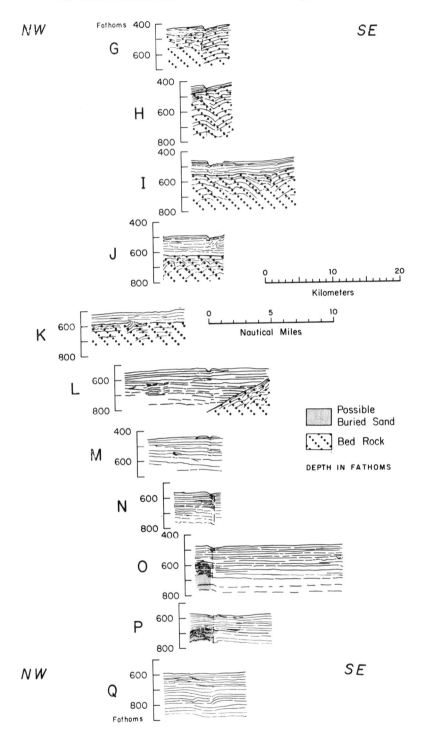

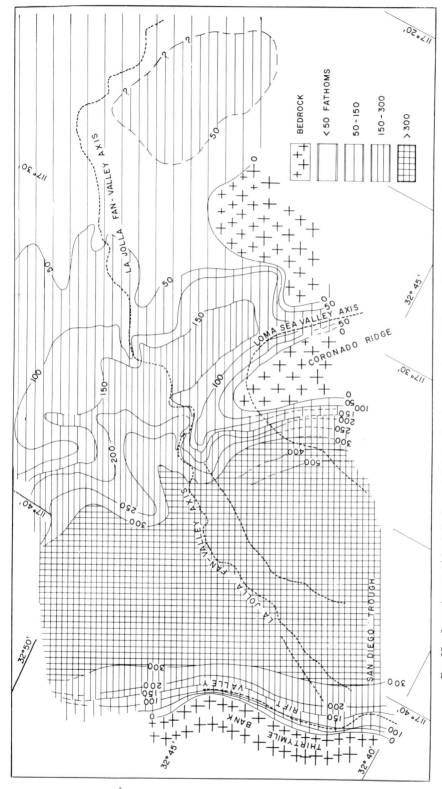

Fig. 27.—Isopach map of La Jolla fan and vicinity, showing approximate thickness of unconsolidated and semiconsolidated postorogenic fan deposits above bedrock boundary. CI = 50 fm.

Dill (*in* Shepard and Dill, 1966, p. 308), which indicated a lack of metastable conditions in the canyon floor fill. The nature of the sand layers along the valley axes also fails to confirm a turbidite origin. Graded sands are relatively rare, and the a-b-c sequence, said to be characteristic of turbidites, was found only in two small sections along the fan-valley axis. If unconformities like those which cut the sand laminae in core 75 (Fig. 16) were the result of a turbidity current, one would expect the tail of the eroding current to deposit a graded sand to mud bed above the eroded surface. Instead, there is only mud. Similarly, the mud layers covering most of the sand on the axis floor show little or no grading into the underlying sand. This fact suggests that the sands were deposited by other types of currents, presumably lacking the cloud of suspended sediment characteristic of turbidity currents. The cleanness of the sand contrasts with the poor sorting of most turbidites. The occurrence of laminae of heavy mineral concentrates (Fig. 16b) indicates alternate winnowing action during the deposition of the sand. Thus, the sand deposits appear to have been what Natland (1967) calls "tractionites" rather than "turbidites."

Observed and Measured
Ordinary Bottom Currents

As the result of vehicle dives down to 667 fm (1,219 m) in the canyons and fan-valleys, and of many remote-controlled bottom photographs to depths of 2,000 fm (3,700 m), it is known that tractive bottom currents are capable of transporting sand at all depths. Velocities up to 0.5 knots were recorded during several deep dives. At times, currents move up canyon.

Current-meter measurements confirming the existence of ordinary bottom currents were obtained from 320 fm (585 m) in La Jolla fan-valley by R. S. Schwartzlose and J. D. Isaacs (Fig. 28). Their meter (Isaacs *et al.,* 1966) makes current measurements at 30-minute intervals for a period up to four days. The records show pulsating currents with velocities up to what may be as much as 30 cm/sec (0.6 knots), although the calibration at the time of observation was not well established. Currents move both up and down the valley floor with a rather clear relation to the tidal cycle at Scripps pier. In general, they move up valley during rising tide and down valley during the ebb, although there is a lag from the coastal tide record. Mostly, velocities are lower when the tidal fluctuation is least. However, some of the highest velocities were observed during a period of small tidal fluctuation (midnight November 6–7, 1967).

Despite the fact that these actual observations show little difference between up- and down-valley currents, many dives have shown depressions scooped out on the up-valley side of obstacles, such as boulders, and dunelike deposits down valley. These features suggest that some type of current, probably neither tidal nor turbidity, is transporting material down the valley floors.

Gravity Creep, Slumps, and Sand Flows

Dill's scuba dives into the canyon heads have shown the importance of a glacial-like creep in moving the sediment fill down the steep floor of the inner gorges. It is possible that this process is also important as a transporting agent along the much reduced gradients of the outer canyon and fan valleys. Cracks and slump scars in the mud fill, noted in some places along La Jolla fan-valley, suggest such action. However, the sedimentary structure of the box cores reveals no distortional movement of this sort, and one could scarcely expect creep to be a major factor with gradients less than 1°.

Sand flows have been observed repeatedly in the 30° sand slopes along the walls of the south side of San Lucas canyon, Baja California, and rarely at the steep head of Scripps canyon. It seems unlikely that these flows would continue where the gradient is much reduced. Some other type of flow takes place in canyon heads causing a periodic deepening of many meters. Because there is little muddy material in these canyon heads to produce a turbidity current, the deepenings probably represent slumps. Presumably they do not travel far beyond the steep heads, but there is no positive information.

MECHANISMS OF FINE-GRAINED SEDIMENT TRANSPORT

The mud dominating the open fan cores and the relatively thin mud layers along the valley axes appear to have been transported from the land by several processes. In most submersible dives, fine-grained flocculent blobs of clay-sized material have been observed drifting as a slow gravity flow down the valleys, particularly during winter months. This material is stirred up by the waves along shore or introduced into the ocean by runoff from the occasional rains. The presence of suspended matter adds slightly to

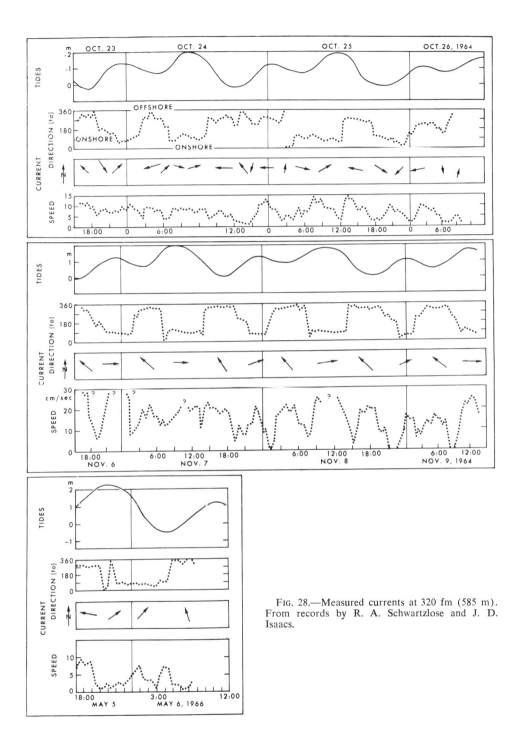

Fig. 28.—Measured currents at 320 fm (585 m). From records by R. A. Schwartzlose and J. D. Isaacs.

the density of the water, producing a slow gravity-controlled movement along the bottom, across the shelf, and down the slopes, until the mud clouds enter the valleys and are moved along them.

In study of the mud on the La Jolla fan, D. J. W. Piper, a student at Scripps Institution, found that the mud contains more silt near the fan-valley. This fact suggests that at least a significant amount of the fine sediment has come down the valleys and spilled over, depositing the coarser material near the valley margin. This is, of course, also implied by the existence of the levees. It is not known whether this process is the result of high-speed turbidity currents moving down the valleys or of some sort of broad density current of low velocity. The fine-grained sand layers found on the levees and less pronounced on the open fan suggest that at least locally the currents that spilled over onto the fan attained velocities of the order of 10 to 25 cm/sec, and probably much faster on the valley floor.

RATE OF SEDIMENT DEPOSITION

The only means of estimating the rate of deposition on the La Jolla fan and in the fan-valley comes from C-14 dates obtained from layers of matted kelp in three fan-valley cores and in one from the outer fan (Fig. 29). Dates from core 71, that has three kelp layers, provide the best information for the fan-valley, although they are inconsistent. The averages of dates at 7 and 30 cm by Isotopes Inc. and Scripps Institution laboratories, yield dates of 1,760 and 4,640 years. This difference of 2,880 gives an average rate for the 23 cm of 8 cm/ 1,000 years (15.2 mg/cm² for wet weight), which is equivalent to about 4 cm/1,000 years (8.5 mg/cm²) for compacted solid grains with zero porosity. Core 116 from the open fan had a kelp layer at a depth of 34-46 cm, with a date of 2,690 years. This gives an average rate of deposition of the overlying 34 cm of 12.6 cm/1,000 years, or for compacted material with zero porosity about 5.8 cm/1,000 years. These results suggest that deposition is faster on the open fan. However, it seems more likely that deposition has been very discontinuous along the fan-valley and the 2,690-year period probably includes long intervals of nondeposition and/or erosion.

Comparison of the rate of 12.6 cm/1,000 years on the open fan with a date from San Diego trough, obtained by Emery and Bray (1962), shows that their rate of 18 cm/1,000

years for the past 24,500 years is somewhat higher. Also, Inman and Goldberg (1963) found a much higher rate from the MOHOLE experimental drilling slightly north of the fan-valley in 510 fm (933 m). At a depth of 70 m below the surface, they obtained a sample that was dated as 34,000 years old. However, dates as old as this are now considered very suspect. If the rates were actually higher in periods extending back into the last glacial stage, an explanation may be that much more sediment was being introduced during the pluvial period that preceded the present. Possibly during such a pluvial period more powerful currents came down the valleys, building the natural levees and introducing the sand layers into the levees and the fans.

RELATION TO PALEO-ENVIRONMENTS

The sediments of the fan and fan-valley off La Jolla strongly resemble the Pliocene of equally deep-water origin found in the Ventura and Santa Paula Creek areas of southern California (Crowell et al., 1966; Natland, 1933; Natland and Kuenen, 1951; Winterer and Durham, 1962). The writers, with M. L. Natland and others, also have examined outcrops north of Doheny State Park, where an old Miocene or Pliocene channel filled with deep-water sediments is exposed in the sea cliff. Here, also, many features resembling the box cores were noted.

The resemblance between the deep-water Pliocene and the La Jolla fan deposits is illustrated by comparative photographs. Convolute lamination found in some cores in the present study resembles that found in Hall Canyon (Ventura area) Pico Formation (Fig. 18B). The sequence of massive, graded sand covered with parallel-laminated sand and, in turn, with current-rippled sand (the a-b-c sequence of Bouma for typical turbidites) is represented in core 46 and in a lacquer peel from Hall Canyon Pliocene (Fig. 19B). Mud balls commonly present in the outer La Jolla fan are equally common in the Miocene (Fig. 21C). Graded sand covered by a bed with abundant plant debris can be compared in a photograph of core 41 and a lacquer peel from the Ventura basin (Fig. 30). The Pliocene deep-water formations appear to have more graded beds than the recent near-surface fan deposits off La Jolla, but are otherwise very similar.

Longer cores, especially drill holes several hundred meters long, are needed to determine the three-dimensional shape of the sand bodies

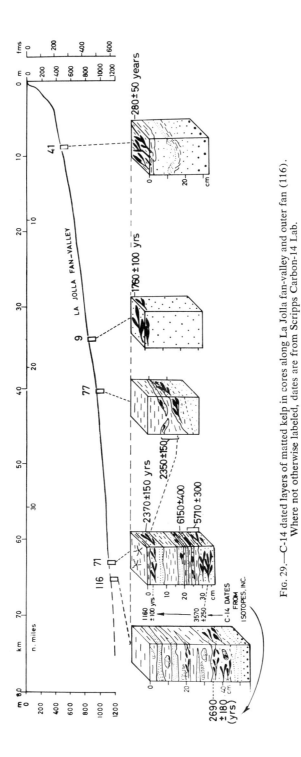

FIG. 29.—C-14 dated layers of matted kelp in cores along La Jolla fan-valley and outer fan (116). Where not otherwise labeled, dates are from Scripps Carbon-14 Lab.

Fig. 30.—Graded sand covered with bed high in plant material. (a) From box core 41 in fan-valley axis at 598 fm (1,094 m). (b) From Ventura basin Pliocene. Photo by von Rad.

in deep, present-day fans, and to find out if the sand layers that are so abundant in short cores have sufficient thickness to develop good stratigraphic traps for oil. The percentage of total organic matter is relatively high (1.5–3.4 percent organic carbon by weight) in the silt and very fine sand of La Jolla fan, so there would be a plentiful organic source for the formation of hydrocarbons during the diagenetic compaction of the basin sediments (Emery, 1960, p. 273–295). The discontinuous "shoestring sands" in direct contact with clay high in organic matter along the narrow fan-valley could become significant reservoirs for hydrocarbons. Sparker profiles indicate the presence of old buried channels, which very likely contain sands covered by muddy sediments.

The asymmetric filling of a basin, in which the coarser material is moved seaward and upward (in the column) with time, will form a wedge of clean sand within the basin fill. This sand will be surrounded by organic-rich clay. The Los Angeles basin, 100 mi north of San Diego trough, is a filled basin that probably received much of its sediment via ancient submarine valleys. The productive Wilmington oil field is on the seaward side of the basin, and production comes from sandstone intervals similar to those cored on the seaward side of San Diego trough.

A particularly intriguing possibility exists along the base of Thirtymile Bank, where the crinkly reflections from the unconsolidated sediments (Fig. 16, profiles A, C, O, and P) suggest the presence of a sand body of considerable thickness. If this proves to be true, it will be in agreement with the surficial sediment distribution patterns where sand becomes much more abundant and coarser as it approaches Thirtymile Bank, both in the channels and on the open fan. Here is an example of sediment becoming coarser seaward and away from the source of sediments. Imagine the difficulties of future geologists investigating the emerged La Jolla fan trying to find the old shoreline by the principle of increasing grain size, or by the use of ripple marks as a criterion for deposition in shallow water. The inclusion of shallow-water Foraminifera and land plants would add to the confusion.

REFERENCES CITED

Bouma, A. H., 1962, Sedimentology of some flysch deposits; a graphic approach to facies interpretation: Amsterdam-New York, Elsevier, 168 p.
———— 1964, Notes on X-ray interpretation of marine sediments: Marine Geology, v. 2, no. 4, p. 278–309.
———— 1965, Sedimentary characteristics of samples collected from some submarine canyons: Marine Geology, v. 3, no. 4, p. 291–320.
Buffington, E. C., 1964, Structural control and precision bathymetry of La Jolla submarine canyon, California: Marine Geology, v. 1, no. 1, p. 44–58.
Crowell, J. C., et al., 1966, Deep-water sedimentary structures, Pliocene Pico Formation, Santa Paula

Creek, Ventura basin, California: California Div. Mines and Geology Spec. Rept. 89, 40 p.

Curray, J. R., and D. G. Moore, 1963, Facies delineation by acoustic-reflections—northern Gulf of Mexico: Sedimentology, v. 2, p. 130–148.

Dill, R. F., 1961, Geological features of La Jolla Canyon as revealed by dive no. 83, of the Bathyscaph *Trieste:* San Diego, California, U.S. Navy Electronics Lab. Tech. Mem. No. TM-516, 27 p.

—— 1964a, Contemporary submarine erosion in Scripps submarine canyon: Ph.D. thesis, Calif. Univ., Scripps Inst. Oceanography, privately printed, 269 p.

—— 1964b, Sedimentation and erosion in Scripps submarine canyon head, *in* Papers in marine geology—Shepard Commemorative Volume: New York, Macmillan, p. 23–41.

Emery, K. O., 1960, The sea off southern California: New York, John Wiley & Sons, 366 p.

—— and E. E. Bray, 1962, Radiocarbon dating of California basin sediments: Am. Assoc. Petroleum Geologists Bull., v. 46, no. 10, p. 1839–1856.

—— *et al.,* 1952, Submarine geology off San Diego, California: Jour. Geology, v. 60, no. 6, p. 511–548.

Gorsline, D. S., and K. O. Emery, 1959, Turbidity-current deposits in San Pedro and Santa Monica basins off southern California: Geol. Soc. America Bull., v. 70, p. 279–290.

Inman, D. L., and E. D. Goldberg, 1963, Petrogenesis and depositional rates of sediment from the experimental Mohole drilling off La Jolla, California (abs.): Am. Geophys. Union Trans., v. 44, no. 1, p. 68.

Isaacs, J. D., *et al.,* 1966, Near-bottom currents measured in 4 kilometers depth off the Baja California coast: Jour. Geophys. Research, v. 71, no. 18, p. 4297–4303.

Kuenen, Ph. H., 1966, Experimental turbidite lamination in a circular flume: Jour. Geology, v. 74, no. 5, pt. 1, p. 523–545.

Lombard, A., 1963, Laminites: a structure of flysch-type sediments: Jour. Sed. Petrology, v. 33, p. 14–22.

Menard, H. W., Jr., 1960, Possible pre-Pleistocene deep-sea fans off central California: Geol. Soc. America Bull., v. 71, p. 1271–1278.

Moore, D. G., 1965, Erosional channel wall in La Jolla sea-fan valley seen from Bathyscaph *Trieste II:*

Geol. Soc. America Bull., v. 76, p. 385–392.

—— 1966, Structure, lith-orogenic units and post-orogenic basin fill by reflection profiling: California continental borderland: Ph.D. thesis, Rijksuniversiteit, Groningen, 151 p.

Natland, M. L., 1933, The temperature and depth distribution of some recent fossil Foraminifera in the southern California region: Scripps Inst. Oceanography Tech. Ser. Bull., v. 3, no. 10, p. 225–230.

—— 1967, New classification of water-laid clastic sediments: Am. Assoc. Petroleum Geologists Bull., v. 51, no. 3, pt. 1, p. 476.

—— and Ph. H. Kuenen, 1951, Sedimentary history of the Ventura basin, California, and the action of turbidity currents: Soc. Econ. Paleontologists and Mineralogists Spec. Pub., no. 2, p. 76–107.

Rees, A. I., 1965, The use of anisotropy of magnetic susceptibility in the estimation of sedimentary fabric: Sedimentology, v. 4, no. 4, p. 257–271.

—— U. von Rad, and F. P. Shepard, 1968, Magnetic fabric of sediments from the La Jolla submarine canyon and fan, California: Marine Geology, v. 6, p. 145–168.

Reineck, H. E., 1963a, Der Kastengreifer: Natur Museum, v. 93, no. 2, p. 65–68.

—— 1963b, Nasshaertung von ungestoerten Bodenproben im Format 5 × 5 cm für projizierbare Dickschliffe: Senckenbergiana Lethaea, v. 44, no. 4, p. 357–362.

Rosfelder, A. W., and N. F. Marshall, 1966, Oriented marine cores: a description of new locking compasses and triggering mechanisms: Jour. Marine Research, v. 24, no. 3, p. 353–364.

Shepard, F. P., and E. C. Buffington, 1968, La Jolla submarine fan-valley: Marine Geology, v. 6, p. 107–143.

—— and R. F. Dill, 1966, Submarine canyons and other sea valleys: Chicago, Rand McNally, 381 p.

—— and G. Einsele, 1962, Sedimentation in San Diego trough and contributing submarine canyons: Sedimentology, v. 1, no. 2, p. 81–133.

Walker, R. G., 1966, Deep channels in turbidite-bearing formations: Am. Assoc. Petroleum Geologists Bull., v. 50, no. 9, p. 1899–1917.

Winterer, E. L., and D. L. Durham, 1962, Geology of the southeastern Ventura basin, Los Angeles County, California: U.S. Geol. Survey Prof. Paper 334 H, p. 275–366.

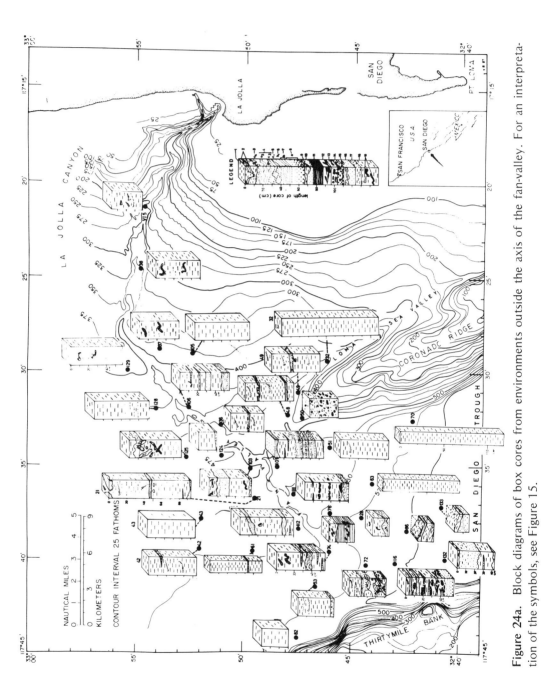

Figure 24a. Block diagrams of box cores from environments outside the axis of the fan-valley. For an interpretation of the symbols, see Figure 15.

Editor's Comments
on Papers 12 and 13

12   NORMARK
        Excerpts from *Growth Patterns of Deep-Sea Fans*

13   CURRAY and MOORE
        Growth of the Bengal Deep-Sea Fan and Denudation in the Himalayas

1970 saw the appearance of two more papers on La Jolla Fan, one by Piper on the Holocene sedimentation of the fan and the other by Normark, which is in part reprinted here (12), complementing the previous paper. Normark, surveying the La Jolla Fan and San Lucas Fan off the southern tip of Baja California, used a deep-tow vehicle kept close to the sea floor and equipped with a precision narrow-beam echo sounder and a 3.8-kHz reflection profiling system, side-looking sonars, stereo cameras, a proton magnetometer, and a precision positioning system, including bottom acoustic transponders for accurate horizontal positioning.

These techniques enabled Normark to obtain more detailed data than hitherto on levees, terraces, and slumps associated with the fan valleys and to differentiate the upper, middle, and lower fan environments and, in the case of the San Lucas Fan, a supra-fan area of convex-upward relief. He emphasized the relict and buried channels as representing former positions of the fan valley and its distributaries.

Normark's detailed analysis of these two modern fans and his consideration of published work on the Astoria, Monterey, Bengal, and other fans has proved valuable for erecting a general model for the growth of fans. This can be applied to other modern fans (e. g., the sublacustrine fan of the Rhône delta in Lake Geneva and modern alluvial fans) and used for comparison with supposed ancient fans (e. g., Normark and Piper, 1969, Miocene Doheny Channel and Fan; Jacka and others, 1968, Permian of the Delaware Basin; Nilsen and Simoni, 1973, Eocene Butano Fan; Nelson and Nilsen, Paper 26).

The author has just published a further paper on factors affecting growth processes on deep-sea fans (Normark, 1974) which carries on the work described in the paper that follows.

Having reprinted a paper on the world's largest canyons, it seemed appropriate to include also the world's largest fan, the Bengal Fan in the Indian Ocean. As Normark pointed out in the previous paper, this is nearly 100 times larger than other known fans and the leveed fan valleys on the upper fan are also much larger. Curray and Moore's short paper (13) outlines the main features of this tremendous mass of sediment, now known to be over 16 km in thickness, derived from the denudation of the Himalayas. The fan consists of mainly Tertiary and Quaternary sediments separated by two prominent unconformities, perhaps earliest Pleistocene above and late Miocene below: this dating of the unconformities is revised in a later paper (see below).

The authors are continuing their work on this fan. A recent report by Moore and others (1974) gives the results of further seismic profiling and of two drill holes made during the JOIDES program. The seismic profiling shows over 16 km of sediment and sedimentary rock in the northern part of the Bay of Bengal, consisting of undeformed turbidites above the upper unconformity, now dated from the drill cores as uppermost Miocene. The middle unit, again of turbidites in part, contains buried channels and levees, and is now deformed. The lowest unit, below the lower unconformity (proved now to be between the upper Paleocene and middle Eocene in holes 217 and 217A), rests on oceanic crust. The cores, from holes 773 m and 664 m in length, are of particular interest since few holes have been drilled into fan deposits: the latter hole reached upper Cretaceous. Another recent seismic study of the fan is reported by Naini and Leyden (1973). Curray and Moore (1974) describe modern and buried large sand-filled channels and levees on the proximal fan. On the central and distal fan, sheet-flow deposits show lateral continuity in section, with isolated channels and channel deposits.

Biographical Notes

Born in 1943, William R. Normark was raised in Wyoming and Utah. He received the B.S. degree in Geology from Stanford University and the Ph.D. in Oceanography from Scripps Institution of Oceanography, University of California, San Diego. He stayed on at Scripps for a year as Postgraduate Research Oceanographer, Geological Research Division, moving to the University of Minnesota in 1970 as Assistant Professor in the Department of Geology and Geophysics. Currently he is with the U.S. Geological Survey, Menlo Park, California. His interests within the field of marine geology include the structure of the continental margin, fine-scale sea-floor geophysical studies and deep-sea erosional processes: his other research interests include modern and ancient turbidites, especially deep-sea fans.

Joseph R. Curray was born in 1927 at Cedar Rapids, Iowa, and received the B.S. degree in Geology at the California Institute of Technology in 1949, the M.S. in Mineralogy at the Pennsylvania State University in 1951, and the Ph.D. in Oceanography at Scripps Institution of Oceanography, University of California in 1959 where he is currently Professor of Oceanography, Research Geologist, and Chairman of the Graduate Program. He is also a director of General Oceanographics, Inc., Nekton, Inc., and Desagon, Inc. His fields of specialization include the marine geology of the continental margin (sediments, structure, history, and origin), late Quaternary changes of sea level, and atolls and reefs, on which topics he has published about 50 papers. Curray's varied activities within the geological societies include serving on the editorial boards of *Sedimentology* and *Marine Geology:* he was editor of the latter journal from 1963 to 1966. In 1970 Curray was awarded the Francis P. Shepard Medal for Excellence in Marine Geology by the Society of Economic Paleontologists and Mineralogists.

David G. Moore was born at Long Beach, California, in 1925. He received the A.B. degree in Geology in 1950 and the M.S. in Marine Geology in 1952 at the University of Southern California. His postgraduate research was carried out at the University of Groningen in The Netherlands, where in 1966 he was awarded the Ph.D. (Cum Laude) in Marine Geology. From 1951 to 1955, Moore worked as Research Geologist and Field Party Chief at Scripps Institution of Oceanography, where he was involved with API Project 51 on a comparative study of nearshore recent sediments and environments in the northern Gulf of Mexico. He participated in 11 expeditions to the Gulf: on seven of these he was Chief Scientist. In 1955, Moore transferred to the U.S. Naval Undersea Center as Research Marine Geologist, where he is currently employed. He has made many cruises using deep submersibles and has taken part in 17 major deep-sea expeditions and has been leader or coleader on 14 of these. His main research interests are in the geology and geophysics of the continental margins, the strength of surficial sediments, slope stability, and sediment strength and sound velocity variation with depth of burial and rate of deposition: he has published over 50 papers on these topics. He has acted as consultant to a number of firms and to the cities of Los Angeles and San Diego on problems of marine geology and oceanography and patented a Free Corer for collecting sediment cores from the sea floor without the use of winch or wire. A member and officer of many geological societies, he was honored by the award of the Francis P. Shepard Medal for Excellence in Marine Geology by the Society of Economic Paleontologists and Mineralogists in 1967.

12

Copyright © 1970 by the American Association of Petroleum Geologists

Reprinted from *Amer. Assoc. Petrol. Geol. Bull.*,
54(11), 2170–2171, 2181–2192, 2194–2195 (1970)

Growth Patterns of Deep-Sea Fans[1]

WILLIAM R. NORMARK[2]

La Jolla, California 92037

Abstract The growth pattern of a deep-sea fan relates events in and around the fan-valleys to the structure and morphology of the open fan. The growth pattern cannot be determined without knowledge of the origin and recent history of the fan-valley system. The mapping of La Jolla and San Lucas deep-sea fans with the deep-towed instrument package developed at Marine Physical Laboratory of the Scripps Institution of Oceanography details the fine-scale morphology, structure, and internal fill of the fan-valleys and suggests the growth patterns of these fans.

The La Jolla fan, 20 km west of Scripps Institution, has one meandering fan-valley that extends across the entire fan. Except on the toe of the fan, the deeply incised valley has terraced walls with steeper walls on the outside of meanders. Very low-relief levees border the fan-valley in some localities. The present erosional valley bypasses the partly buried remnants of an older distributary system on the lower fan.

The San Lucas fan, off the southern tip of the peninsula of Baja California, shows a depositional lobe of sediment, or suprafan, below the short, leveed fan-valley extending from San Jose Canyon. The suprafan appears as a convex-upward bulge on a radial profile of the fan. The surface of the suprafan has a series of discontinuous depressions up to 55 m deep and 1 km wide. The depressions are generally asymmetric in cross section, commonly have terraced walls, and are underlain by coarse sand and gravel. They are interpreted to be channel remnants.

A model for deep-sea fan growth, based on this study, predicts that deposition on a fan will be localized in a suprafan at the end of large, leveed valleys commonly found on, and generally confined to, the upper reaches of deep-sea fans. The suprafan normally is on the midfan and is characterized by numerous smaller distributary channels. Rapid aggradation in the suprafan coupled with migration and meandering of the channels produces a surface marked by isolated depressions or channel remnants. Uniform deposition, producing a symmetrical half-cone morphology, results from the shifting through time of fan-valleys across the area of the fan.

[1] Manuscript received, January 8, 1970; accepted, April 27, 1970. Contribution of the Scripps Institution of Oceanography, new series.

[2] University of California, San Diego, Marine Physical Laboratory of the Scripps Institution of Oceanography. Present address: Department of Geology and Geophysics, University of Minnesota, Minneapolis, Minnesota 55455.

The success of the field work for this study is due in large part to J. R. Curray, F. N. Spiess, scientific leader for expeditions TOW MAS and TIPTOW, and the technicians and engineers of the deep-tow group, including M. S. McGehee, D. E. Boegeman, M. D. Benson, J. T. Donovan, F. W. Stone, and G. E. Forbes. The cooperation and help of the officers, crew, and scientific staff of the R/V *Thomas Washington* and ST-908 are greatly appreciated.

Fellow deep-tow enthusiasts, J. D. Mudie, R. L. Larson, B. P. Luyendyk, T. M. Atwater, and G. J. Miller, were of immense help in the manipulation and discussion of the geophysical records taken with the deep-tow instrument. E. L. Hamilton provided sound-velocity data from sediment cores collected during the surveys. F. J. Emmel discussed results of recent geophysical surveys on the Bengal fan, and G. B. Griggs, C. H. Nelson, and J. E. Andrews discussed their results from recent studies on deep-sea sedimentation.

D. J. W. Piper, a fellow student of deep-sea fans, and J. R. Curray provided many helpful suggestions during the early phases of this research. F. N. Spiess, J. R. Curray, P. Wilde, E. A. Silver, D. W. Scholl, and R. LeB. Hooke critically reviewed the manuscript.

The research was financed by the Office of Naval Research, the National Science Foundation, and the Deep Submergence Systems Project.

© 1970. The American Association of Petroleum Geologists. All rights reserved.

Introduction

Shepard suggested that "depositional fans" may represent turbidity-current material deposited as a result of a marked decrease in bed slope at the mouths of submarine canyons (Menard, 1955). Although most students of deep-sea fans (Menard, 1955) now accept this basic premise, there is little agreement as to the processes controlling fan morphology. Studies based on fan morphology and distribution of near-surface sediments are not in themselves adequate to determine the growth processes of the deep-sea fans. For this reason, the present study involves a comparative interpretation of the morphology, structure, and internal fill of deep-sea fan-valleys and their relation to the overall growth patterns of deep-sea fans. Major clues for constructing the growth patterns are gained from an understanding of (1) the origin of terraces, levees, and slump features associated with the fan-valleys, (2) the nature of the termination of the fan-valleys, and (3) the origin of fine-scale structures and the morphology of the fan surface outside the fan-valley. The deduced origin and recent history of the present fan-valleys on each fan then are related to the overall fan morphology to define the *growth pattern.*

A model for deep-sea fan growth combines the observations from the two fans described in this study with data on fans described in the literature. The model relates events in the fan-valleys to the development of the *shape* (lateral dimensions) and *profile* (radial surface profile) of deep-sea fans. If possible, the growth pattern for each fan is determined; otherwise, the reasons for the alteration of a growth pattern are evaluated. The model for fan growth is applied to the sublacustrine fan of the Rhone delta in Lake Geneva (Forel, 1885; Houbolt and Jonker, 1968). The processes governing deep-sea fan growth are shown to be somewhat analogous to those operating on subaerial alluvial fans.

EQUIPMENT

Although many investigators have concentrated on fan-valley systems of deep-sea fans, the morphology and internal structure of these features are not well known. With conventional echo-sounding devices it is possible only to determine the plan of fan-valleys: even this capability is handicapped in many surveys because of poor navigational systems. The only well-surveyed fan-valley to date is the La Jolla fan-valley (Shepard and Buffington, 1968); a precision shore-based radio navigation system (Buffington, 1960) provided accurate positioning for the survey.

Conventional wide-beam echo sounders provide no information from within narrow depressions on the sea floor (Krause, 1962), as shown in Figure 1. The upper trace of Figure 1A shows a typical profile over La Jolla fan-valley in 1,015 m of water taken with a conventional echo sounder operating on a 1-sec sweep. The lower trace (Fig. 1A) shows the same valley crossing as recorded by the narrow-beam sounding system on the deep-towed vehicle. All the terraces and other details in and near the fan-valley that are reflected by the narrow-beam echo sounder are obscured by the hyperbolic returns coming from near the upper edges of the valley on the conventional surface profile. The profiles of Figure 1A are from relatively shallow water. The problem of hyperbolic side echoes masking the true geometry becomes critical as the water depth increases, as illustrated in Figure 1B.

Study of the structure and morphology within fan-valleys obviously requires the use of a narrow-beam echo sounder. The deep-tow vehicle developed at Marine Physical Laboratory provides not only a precision narrow-beam echo sounder that operates near the sea floor, but also a 3.8-kHz reflection profiling system, side-looking sonars, stereo cameras, proton magnetometer, and a precision positioning system (Spiess *et al.*, 1967; Spiess and Mudie,

1970). The deep-tow instrument, or "fish," is maintained close to the sea floor on the end of an armored coaxial cable that provides power and commands to the fish and relays responses from the fish systems back to recorders on the towing ship (Spiess and Mudie, 1970).

FIELD WORK

The deep-tow data presented herein represent five separate field operations. The La Jolla fan, off southern California (Fig. 2), was surveyed during three separate short cruises: November 1967 and May 1968, test trips for the instrument, and May to June 1968, expedition TIPTOW TWO. During TIPTOW TWO, three bottom acoustic transponders (McGehee and Boegeman, 1966) provided accurate horizontal positioning for both the deep tow and the towing vessel, R/V *Thomas Washington*. The geographic positions of the transponders were determined by radar fixes on the San Diego coastline. The data from La Jolla fan are presented in two parts, the upper fan survey and the lower fan survey.

The deep-tow investigation of the San Lucas fan at the tip of Baja California (Fig. 2) included two separate surveys. Expedition TOW MAS, in November 1967, was terminated unexpectedly by loss of the deep-tow fish after six days of operation. In May 1968, expedition TIPTOW completed the work on the fan with a new instrument. Bottom acoustic transponders were used in both surveys although there was no common transponder in the two operations. Nevertheless, the relative positioning of the two transponder nets was accurate enough to allow recovery of the fish lost during the TOW MAS expedition by means of transponders dropped during expedition TIPTOW. Transponder navigation provides a very accurate relative positioning system with rms range errors (from three or more transponders to the fish) on the order of 4 m. (See Lowenstein and Mudie, 1967, and Spiess *et al.*, 1969, for more detailed discussions concerning positioning with bottom transponders.)

[*Editor's Note:* Part of the paper dealing with La Jolla Fan is omitted.]

Summary

The available evidence suggests that La Jolla fan is not growing at present. Throughout most of the construction of the fan, the channels or fan-valleys were shorter and less entrenched than the present valley. Deposition was confined to the fan itself, and the growth of the low cone of sediment was aided by a system of shifting distributary channels on the middle and lower fan. Low-relief levees probably flanked many of the channels. The incision of the present fan-valley modified a system of smaller distributary channels that had formed on the lower fan (Fig. 12). This incision is the most recent development of the fan-valley system and may reflect an attempt to develop a new equilibrium profile (Normark and Piper, 1969). The valley meandered during incision, leaving a varied series of terraces along the walls of the valley. Remnants of the older distributary system are preserved because of slow and spotty deposition on the lower fan. Apparently, many currents moving down the fan-valley still have a high enough velocity to bypass the lower fan with little deposition of sediment (Piper, 1970).

SAN LUCAS FAN

The San Lucas deep-sea fan is a composite of four small coalescing fans in 3 km of water off the tip of the peninsula of Baja California, Mexico (Fig. 2). Sediment moves onto the fan through four submarine canyons—Cardonal, Vigia, San Lucas, and San Jose (Fig. 13). A deep-sea channel extending south from Tinaja Trough, a structural rift along the west side of the tip of the peninsula (western border of Fig. 13 lies along axis of trough), marks the western limit of the fan. Cabrillo Seamount marks the approximate eastern edge, but sediments of the lower fan extend onto low abyssal hills flanking the East Pacific Rise 60 km south and east of the mouth of San Jose Canyon.

The canyons feeding the fan include the "San Lucas Group" of submarine canyons that has been studied extensively by Shepard (1964; Shepard and Dill, 1966). The bathymetry of the fan itself is poorly known except for the apex of the cone of sediment below San Jose Canyon. The structure of the continental margin of the tip of Baja California, including parts of the San Lucas fan, was studied by means of continuous reflection profiling (Normark and Curray, 1968). At the base of the continental slope, the sediments of the fan are

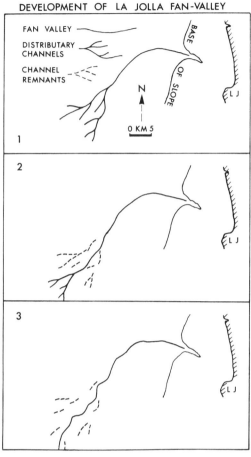

FIG. 12.—Recent history of La Jolla fan-valley. Incision of main valley distributary isolates upstream branches, and subsequent overbank deposition tends to smooth fan surface (2). Single valley cuts across entire fan as incision continues (3). Remnants of old distributaries are partly filled and smoothed, leaving discontinuous depressions on fan surface.

800–1,000 m thick and lie directly on the low-relief "basement," the seismic second layer of the oceanic crust (Phillips, 1964).

The San Lucas fan does not have a prominent fan-valley extending from any of the four submarine canyons. Short fan-valleys extend only 15–20 km below the mouths of the three eastern canyons. Shepard's bathymetric chart (1964, p. 180) shows a meandering valley south of San Jose Canyon between two irregular, hilly ridges. Bottom samples from the eastern ridge consist of sand. Seismic-reflection profiles (Fig. 14) show that these ridges are probably natural levees bordering the fan-valley. The section of sediment below the ridge

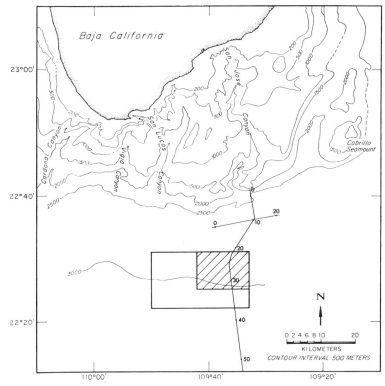

FIG. 13.—Bathymetry of tip of Baja California peninsula, after Shepard (1964). Location of deep-tow study areas on San Lucas fan are shown. Box encloses TOW MAS survey; hatched area denotes TIPTOW survey. Locations of seismic-reflection profiles of Figure 14 are shown with distance along profile in kilometers.

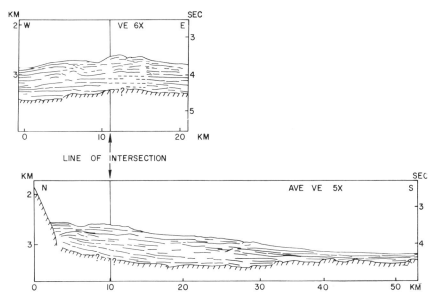

FIG. 14.—Seismic reflection profiles from San Lucas fan below San Jose Canyon (taken from Normark and Curray, 1968). Profile locations are shown in Figure 13. Study area lies between km 21 and 37 on north-south profile. See also Figure 23 for original record for north-south profile.

east of the valley exceeds 1 km and is the thickest anywhere on the San Lucas fan.

The San Lucas fan was selected for study of the area below the termination of the fan-valley associated with San Jose Canyon, and to relate events there to the growth of the fan. A north-south seismic-reflection profile across the fan (Fig. 14) shows irregular, hummocky topography with a generally convex-upward profile in the eastern part of the study area. The TOW MAS survey explored the western half of the fan (Figs. 13, 15) and the following TIPTOW survey (Fig. 16) concentrated on the area of hummocky relief. The steep gradient (22 m/km) through this region of hummocky relief is exceeded only at the apex of the fan, but the topography of the uppermost fan is modified by the hilly levees bordering the short fan-valley. The structure just beneath the hummocky topography is not decipherable from the reflection record. There is no clear fan-valley or channel morphology, and the reflection data suggest that the deeper sediments are horizontally bedded. The surface of the fan has a concave-upward profile on either side of the area of hummocky relief. The lower half of the fan has a smooth sea floor with very low gradients; near the toe of the fan, sediments lap onto low abyssal hills.

Results

TOW MAS survey.—Most of the area covered by TOW MAS survey (Fig. 17) has very low relief and was contoured at a 20-m interval because of relatively sparse data. Seen in profile, several shallow, channel-like trends suggested by the contours show none of the characteristic features of deep-sea fan valleys. Side-

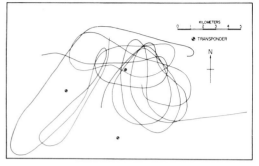

FIG. 16.—Fish track for expedition TIPTOW. Three transponders provided excellent positioning for deep-tow fish with rms error of 4 m for approximately 450 three-range fixes. TIPTOW and TOW MAS transponder nets were positioned with respect to each other by best fitting radar positions for surface, ship deep-tow topography, and deep-tow magnetics (Normark, 1969).

looking sonar records show that the bottom is smooth over these features with no scarps or irregularities indicative of channel walls. In the eastern end of the survey area, several steep-walled depressions were found. They could not be traced in any continuous system, and their lateral extent could not be determined.

TIPTOW survey.—The TIPTOW survey concentrated on the area of the depressions discovered during the earlier work. The closely spaced, accurately positioned traverses (Fig. 16) showed that the surface of the fan is indeed marked by a series of discontinuous depressions (Fig. 18). Although some of the depressions may appear to form a winding channel, good bathymetric control and side-looking sonar data confirm that many are unconnected features.

In the north-central section of the TIPTOW survey, contoured in Figure 18 at a 10-m interval, track coverage is extremely dense (Fig. 16). Side-looking sonar ranges overlap adjacent tracks across most of this area, and these data provided detailed morphology of several of the depressions. In this area, the depressions are somewhat irregular and trend nearly parallel with the gross contours of the fan. On the south, most of the depressions trend downfan and are arcuate in shape. Several are bordered by low leveelike relief. Profiles from five of them (Fig. 19) show terraced walls. The depressions generally are not symmetric either in relief or in development of the terraces.

Reflection profiles.—The information from the 3.8-kHz reflection profiling system is presented in Figure 20 as an isopach map of the

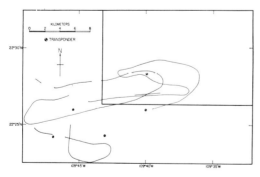

FIG. 15.—Fish track for TOW MAS survey. Rms position error for three or more range fixes was 6 m. Box outlines later TIPTOW survey area (Fig. 16). Two transponders along 109°40′W were dropped later in TOW MAS survey to extend study area eastward.

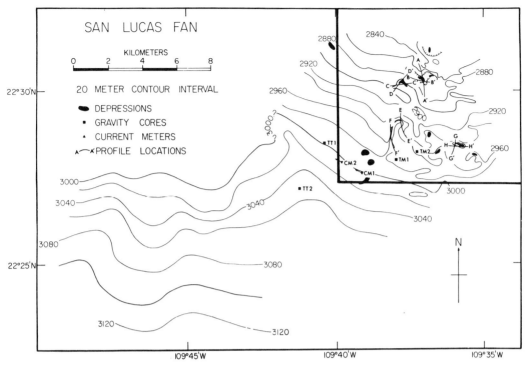

FIG. 17.—Generalized bathymetry of entire study area on San Lucas fan. Area enclosed in box is detailed in Figure 18. Location map for gravity core samples, current meters, and profiles is shown in Figure 19. Arrows on current-meter symbols indicate net current direction for bottom water (Normark, 1969).

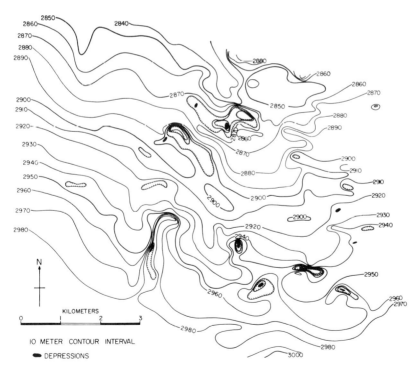

FIG. 18.—Detailed bathymetry of area of depressions on San Lucas fan. Closely spaced tracks across much of this area (Fig. 16) allowed many useful data on feature trends and additional soundings from side-looking sonar records to be used in constructing bathymetry (Normark, 1969, appendix I).

225

METERS

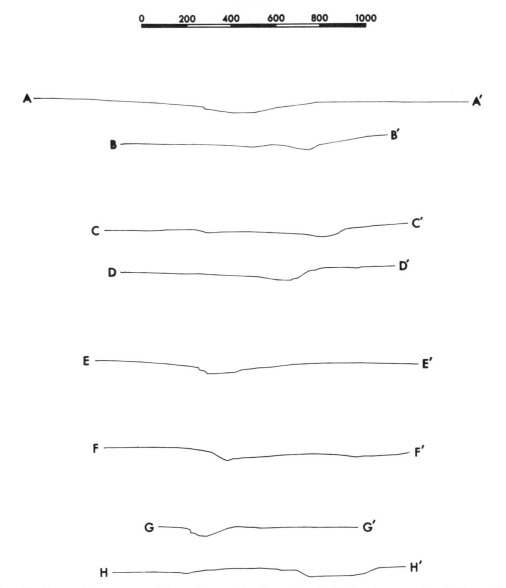

FIG. 19.—Line profiles over several depressions on San Lucas fan. Profile locations are given in Figure 17. No-vertical exaggeration. Profiles resemble in size and morphology, those from La Jolla fan valley (Figs. 5, 8).

depth of acoustic penetration, *i.e.,* the measured travel time to the deepest discernible reflector on the record. The fish height above the bottom ranged from 40 to 300 m, but had little effect on the depth of penetration. There was no penetration into the fan sediments at the bottoms of the depressions, but up to 13.4 m of penetration in the interdepression areas, with

deeper penetration farther from the depressions. Northwest of the depressions the penetration exceeds 13.4 m with a maximum penetration of 40 m. This plot is not continued for the TOW MAS area because of the limited operation of the reflection system during the earlier survey (Normark, 1969).

Only the area of penetration greater than

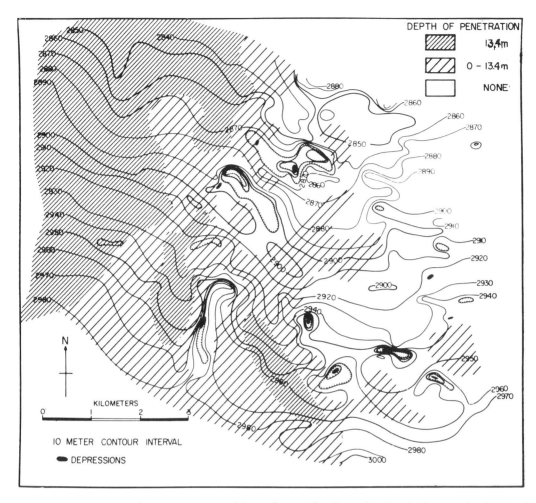

Fig. 20.—Sonar isopach map across area of depressions on San Lucas fan. Depth of penetration is plotted in meters; but arbitrary value of 13.4 m results from original plotting value in fathoms that was corrected to meters using sound velocity of sediments as measured in several gravity core samples (av. 1,550 m/sec) by E. L. Hamilton.

13.4 m shows reflecting horizons that can be traced for any distance. Little structure except for horizontally bedded sediments is suggested in this section of deepest penetration. The inter-depression areas are characterized by indistinct, discontinuous returns that give no structural information. Records over the bottoms of the depressions typically show no subbottom reflectors of any kind. Where some penetration is recorded, it is similar to that of interdepression areas with very shallow (< 7 m) returns.

Surface sediments.—The limited sampling of bottom sediments during the two deep-tow surveys (Normark, 1969) suggests a distribution of surface sediment similar to that of the La Jolla fan. Coarse sand and gravel are found within the depressions. Interbedded silty sands outside the depressions decrease in both thickness and grain size with increasing distance from the depressions. At greater distances, the predominant silty clays contain only a few thin (several centimeters) fine sand interbeds.

Discussion

Bathymetry.—The isolated depressions on the surface of the San Lucas fan are interpreted to be the remnants of a braiding and migrating channel system extending from the fan-valley near the apex of the fan. These depressions are negative-relief features and not part of a ridge and trough system. They are entrenchments in an area of otherwise smooth topography similar to the fan surface directly west. The hummocky topography described on the basis of the surface seismic-reflection profiles (Fig. 14) results from hyperbolic echoes from the edges of the depressions (Fig. 1B).

The cross-section profiles (Fig. 19) of the depressions resemble those of La Jolla fan-valley in size, relief, and morphology. Terraces are common on the walls of these depressions. Few profiles are symmetric. Steeper walls are common on the outside of arcuate depressions, and the terraces are wider on the opposite (or inside) wall. The fan surface away from the depressions is generally smooth, but low marginal ridges or mounds border several of the holes (Fig. 18).

Sediment distribution.—Little (< 7 m) or no subbottom penetration was achieved by the reflection system over the bottoms of the depressions (Fig. 20). The only available sample from a depression contained coarse sand and gravel similar to that described by Shepard (1964, p. 179) from the axis of San Jose Canyon. Intermediate penetration (< 15 m) characterized the interdepression areas and the two samples taken there consist of silty clay with up to 40 percent sand. West of the depressions, reflecting horizons as deep as 40 m below the sea floor were reached. One sample collected there consists of mostly mud with a few fine sand interbeds; the mud layers were 12–35 cm thick and the fine sand interbeds only 4–9 cm thick. The distribution of sediment types suggests a relation between the depth of penetration and the percentage of sand in the near-surface sediments. In general, the coarser the sediment, the poorer is the subbottom penetration.

Origin of Depressions

My observations suggest that the depressions are the result of erosion by channelized bottom currents into horizontally bedded fan sediments. This conclusion is supported by (1) the morphology of the depressions—arcuate, terraced walls, steep walls (narrower terraces) on the outside of curved depressions, and dimensions comparable to those of other fan-valleys, (2) the structure of the adjacent fan—truncation of flat-lying reflecting horizons near the depressions, and (3) the distribution of sediment —coarse sand and gravel inside the depression, becoming finer grained with distance from the channels. Initially a meandering fan-valley system formed, but a combination of circumstances transformed it into a series of disconnected depressions. Several mechanisms probably were involved. For example, the crescent-shaped depressions may be analogous to oxbows. Lateral migration of a fan-valley upstream might isolate short branches of the lower reaches of the system. A large slump mass generated by oversteepening could block the fan-valley and initiate deposition "upstream" from the slump, thus isolating the downstream segment of the valley. It is equally possible that not all turbidity currents would be large enough to traverse the entire length of the valley. Once sedimentation began within the valley, back-filling could have progressed (Andrews, 1967) until the upstream part was filled and the lower valley was abandoned. Depressions form in an area of rapid deposition; they cannot form by simple isolation of channel segments. In spite of uncertainties, I believe that the most reasonable interpretation of the data is that the depressions are the remnants of a former braiding and meandering fan-valley distributary system. A tentative conclusion from this reasoning is that small valleys, probably without levee development, are characteristic of the area of fan growth in the midfan (and perhaps upper fan) regions. This area of fan growth, which is characterized by numerous depressions, appears as a convex-upward bulge on the fan profile (Fig. 14).

Model for Deep-Sea Fan Growth

A constructional model for deep-sea fan development utilizes the topologic and structural data obtained with the deep tow, as well as available published data on other deep-sea fans and on current concepts for abyssal-plain sedimentation. To combine the data from many deep-sea fans into one coherent model, a growth pattern must be established for each fan.

Growth Pattern

To unravel the growth pattern of a deep-sea fan it is necessary to determine the history of the fan-valleys that have developed. It is of utmost importance to determine whether the

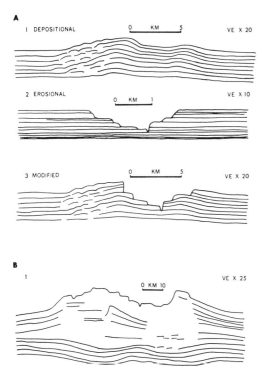

TYPES OF FAN-VALLEYS

A

1 DEPOSITIONAL   0   KM   5   VE X 20

2 EROSIONAL   VE X 10
                    0   KM   1

3 MODIFIED   0   KM   5   VE X 20

B

1   VE X 25
                 0 KM 10

2   VE X 25
               0   KM   5

FIG. 21.—**A.** Types of fan-valleys; cross-sections presented are taken from seismic reflection profiles of real fan-valleys wherever possible: *1,* after Tufts abyssal plain channels (Hamilton, 1967); *2,* after La Jolla fan-valley; and *3,* after Monterey fan-valley (Normark, in press). Observations on distribution of the surface sediments around these should support this classification; however, only examples of erosional (La Jolla fan) and compound (Monterey fan) fan-valleys have been sampled sufficiently to recognize any systematic distribution of sediments with respect to the valleys. **B.** Diagram illustrates two more configurations for leveed fan-valleys; both are based on reflection profiles from Bengal fan (Curray and Moore, 1968; F. Emmel, personal commun.). Large leveed valleys on fan apex (*1*) are clearly depositional features; weight of sediments in this feature has bowed down underlying layers. On the other hand, terraces are common on walls of valley, although this structure is more indicative of erosional valleys. Father south on Bengal fan, terraced wall morphology suggests erosional valley, but levees are present and levee structure is apparent from seismic reflection data (*2*). Fan-valley origin is not easy to determine. Samples shown in 1 and 2 are clearly only the extremes of a wide range of valley forms. A structural profile alone cannot uniquely identify a modified fan-valley (*3*).

present fan-valley system is in equilibrium with the present fan morphology. If the fan morphology represents deposition from a fan-valley pattern that predates the present fan-valley, the growth pattern of that fan cannot be deduced *a priori* by use of the present valley structure and morphology or the distribution of sediment within and around this valley. The growth patterns of deep-sea fans can be determined only if (1) the fans under consideration are growing at present or (2) the growth phase of the fan can be reconstructed after recognition of the alterations that have occurred since growth ceased.

The history of a fan-valley may be recorded in the internal structure of the fan sediments bordering the fan-valley. Accordingly, two fundamental types of fan-valleys are recognized (primarily by means of continuous reflection profiles) and are named on the basis of their inferred origins: (1) a *depositional* valley (Hamilton, 1967) commonly is bounded by levees with convex-upward bedding surfaces, and the valley floor may be built above the level of the surrounding sea floor; (2) an *erosional* valley (Laughton, 1968) cuts into previously deposited sediments. These definitions apply only on a large or generalized scale; for example, a depositional fan-valley may exhibit erosional sedimentary structures on the valley floor. Erosional and depositional fan-valleys represent the endpoints of a gradational series (Fig. 21). Changes from depositional to erosional conditions (or vice versa) can result in a *modified* valley form such as the present Monterey fan-valley (Normark, in press). None of these may be identified uniquely on the basis of a single structural profile, as emphasized in Figure 21B.

Constructional Model

The San Lucas fan appears to be the only fan described in this study that has preserved its depositional or growth pattern. This conclusion is based essentially on negative evidence in that there is no recognizable modification of the fan-valley pattern, *i.e.,* no evidence of fan erosion or extensive downcutting within the fan-valley. Whether the San Lucas fan is actively growing at present is a moot question, but certainly its formational morphology and structure have not been altered (Normark, 1969).

The morphology of the Astoria fan as described by Nelson (1968; personal commun.) compares very closely with that of San Lucas fan. The upper section of Astoria Canyon is filled, and the sediment load of the Columbia River no longer moves into and down the canyon. Instead, the sediment moves northward into Willipa Canyon, which feeds Cascadia Channel. The Astoria fan is essentially inactive (Nelson, 1968), compared to its development during times of lowered sea level, having lost its supply of sediment. Subsequent modification of the fan morphology is not recognized and the growth pattern of Astoria fan probably has been preserved.

Model description.—On the basis of my understanding of the San Lucas and Astoria fans, fan growth involves the formation of a leveed fan-valley (upper fan), a suprafan (usually on the mid-fan at the termination of the leveed valley), and a nearly flat-lying lower fan without channels. Under conditions of fan growth, the fan-valley (if one exists) extending from the mouth of the submarine canyon is depositional. The thalweg may follow an erosional channel within the valley, but the valley is characterized by natural levees and the valley floor may be built above the general level of the fan. The levees decrease in height and disappear down fan. Rapid radial deposition at the end of the leveed valley accelerates growth in this area, forming a depositional bulge or lobe on the fan surface. This depositional lobe is called a "suprafan" (Fig. 22). The suprafan is a small delta or fanlike deposit probably formed as the turbidity currents dissipate after leaving the confinement of the leveed valley. Deposition caused by spreading of turbidity currents was discussed by Buffington (1952), Johnson (1962), and Wilde (1965a, b).

On a radial profile of the fan, the suprafan appears as a low, convex-upward segment of an otherwise concave-upward profile. Channels are numerous on the suprafan. They are much smaller than the leveed fan-valley of the upper fan; their terracing is similar to that of erosional channels; they lack persistent levees; and they form a braiding and rapidly migrating system. Rapid shifting of these channels coupled with rapid aggradation of the suprafan may leave isolated channel remnants (San Lucas fan). If channel remnants are preserved as a series of depressions, deposition on the suprafan appears to be greatest in the vicinity of the channels. Deposition below the suprafan apparently results from broader, thinner flows, because not even small channels are developed on the lower fans. Figure 22 illustrates the morphology of this fan model; Figure 23 shows a longitudinal section of San Lucas fan for com-

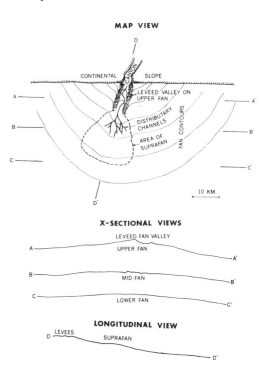

FIG. 22.—Model for deep-sea fan growth showing suprafan and simple morphology of fan related to growth area. Upper fan is characterized by large leveed fan-valley (*A-A'*), whereas midfan (or suprafan) shows many smaller distributary channels and some channel remnants or depressions (*B-B'*). Channels are absent from lower fan (*C-C'*). Suprafan appears as convex-upward bulge on longitudinal profile.

parison with the model profile, and representative surface echo-sounding profiles corresponding to the cross-sectional views of Figure 22.

Sands are found on the floor of the leveed fan-valley of the upper fan and on the suprafan. The coarsest fraction (coarse sand and gravel) is in the thalweg of the leveed valley and in the channels and depressions of the suprafan. Deposition of fine sand, silt, and silty clay occurs on the levees and the margin of the suprafan. Overflow deposition outside the levees and below the suprafan will be mostly silty clay with a few silt interbeds. Mud (silty clay) is found interbedded in all environments, but is thicker and more extensive with distance from the fan-valley floor and suprafan. This sediment distribution can be recognized on both San Lucas and Astoria fans (Nelson, 1968).

This model can be compared with other deep-sea fans only in a general way; published data on eight other fans (Tables 1, 2) are inade-

Table 1. Bathymetric Comparison of Deep-Sea Fans Ranked by Increasing Longitudinal Gradient

Name of Fan	Fan Dimensions (km)		Fan Gradient (m/km)	Depth (m)		Levees on Valleys	Apparent No. of Distributaries
	Length	Width		Apex	Base		
1. Bengal	2,600	1,100	1.1	1,500	4,400	yes	>20
2. Congo	>520	185	1.8	1,790	>2,700	yes	6
3. Rhone	300	250	3.0	1,900	2,800	yes	12
4. Astor	165	100	4.6	2,080	2,840	yes	40
5. Monterey	320	280	4.8	3,050	4,600	yes	4–6
6. Hudson	150	150	6.7	2,100	3,100	yes	7
7. Delgada	300	330	8.0	2,200	4,000	—	1+?
8. Mississippi	220	150	8.3	1,280	2,950	yes	7
9. San Lucas	65	45	8.6	2,500	3,060	yes	at least 11 remnants
10. La Jolla	36	30	18.0	470	1,100	yes	1
11. Redondo	7.5	11	22.6	590	760	yes	4
Sublacustrine fan of Rhone delta, Lake Geneva	15	5.5	10.5	150	308	yes	7

Sources of data:

1. F. Emmel (personal commun.); Curray and Moore (1969).
2. Heezen et al. (1964).
3. Menard et al. (1965).
4. Taken from Nelson et al. (in press).
5. Wilde (1965a); Normark (in press).
6. Taken from Nelson et al. (in press).
7. Winterer et al. (1968); Wilde (1965a).
8. Taken from Nelson et al. (in press).
9. This study.
10. This study; Shepard and Buffington (1968).
11. Taken from Nelson et al. (in press). Rhone delta fan, Lake Geneva, from Houbolt and Jonker (1968).

quate. The listings in Table 2 show that many deep-sea fans have undergone modification to the extent that no growth pattern can be discerned. It should be emphasized that features such as symmetry of outline, number of distributaries, and fan gradients, *per se*, are not defined rigidly by this model.

Control of fan shape.—A model for fan growth involving leveed fan-valleys and coalescing suprafans suggests that deposition may be localized to some extent. Deposition over the levees on the fan-valley, on the suprafan, and even on the lower fan where the turbidity currents spread out may include only a part of the entire fan area. A true fan morphology—a low, symmetric half-cone—will not form unless depositional sites through geologic time are random, leading to statistically uniform deposition on every sector of the fan. The position and length of the leveed fan-valley on the upper fan control the area of fan growth. To build a symmetrical fan, changes must occur in the leveed fan-valley on the upper fan; a new leveed valley must form near the apex and initiate growth on a new sector of the fan. Diversion to a new sector can occur if the existing fan-valley is blocked or if an extremely large turbidity current disrupts the existing levees. In

Table 2. Characteristics of Deep-Sea Fan Model Compared with Data Available for Deep-Sea Fans

Name of Fan	Growth Pattern Known	Leveed Valleys		Distributary Channel System	Suprafan	Outline of Fan
		Upper Fan	Lower Fan			
Bengal	no (insuff. data)	yes	yes?	yes	yes?	sym.
Monterey	no	yes	no	yes	no	asym.
Delgada	no	n.d.	n.d.	yes?	n.d.	asym.
Congo	no	yes	yes	yes	n.d.	sym.
Astoria[1]	yes	yes	no	yes	yes	asym.
Hudson	no	yes	n.d.	yes	n.d.	asym.
Rhone	no	yes	some	yes	yes?	asym.
Mississippi	no	yes	n.d.	yes	n.d.	n.d.
San Lucas[1]	yes	yes	no	yes	yes	sym.
La Jolla[1]	no	yes	no	yes (an older set)	no	asym.
Redondo	no	yes?	n.d.	yes	n.d.	n.d.
Hueneme-Mugu	no	yes	n.d.	yes	n.d.	n.d.
Lake Geneva, Rhone Delta Fan	yes	yes	no	yes	yes	asym.

[1] Model developed from studies in these fans.
n.d. = Not demonstrable with available data.

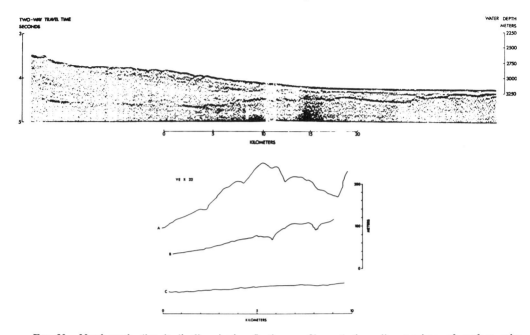

FIG. 23.—North-south (longitudinal) seismic-reflection profile and three line tracings of surface echo-sounding profiles from San Lucas fan illustrate relief and morphology envisioned in model for fan growth. Sections are comparable with those suggested in Figure 22. Suprafan appears on seismic reflection profile as convex-upward, hummocky area on upper midfan (see also Fig. 14). Only part of fan surface is shown in lower three echo-soundings tracings; these profiles emphasize type of relief comparable to sections of Figure 22.

either case, uniform deposition over the fan depends, more or less, on some catastrophic event to change the area of deposition.

In addition to building upward and laterally, levees tend to grow downstream because deposition of an unconfined (unchannelized) flow begins at the margins of the flow (Buffington, 1952, p. 477). Forward building (downstream) of the levees may proceed fast enough to prevent the formation of the suprafan morphology. Fans built from a source of well-sorted sediment may not develop the suprafan morphology because most of the deposition would be of the overflow type and levee growth (both upward and downfan) would be favored. If the levees did not build forward more rapidly than growth of the suprafan, the suprafan would build up and backfilling of the fan-valley could occur. However, as the valley floor and suprafan aggrade, the overflowing currents that formed the leveed valley would continue to increase levee height, resulting in elevation of the entire system. The longer growth continues, the more unstable the system

may become in terms of relief above the surrounding fan surface.

Menard et al. (1965) concluded that a typical deep-sea fan has an asymmetric outline. In some places, basement and/or basin morphology cause the asymmetry; my remarks are restricted to depositional processes that affect fan shape. Nelson (1968) recognized the asymmetry of the Astoria fan and tentatively accepted this in setting up his depositional model for deep-sea fans. The asymmetry is thought to reflect the tendency of turbidity currents in the northern hemisphere to hook left (looking downstream). Higher levees are formed on the right side of channels (looking downstream) in response to tilting of the surface of the channelized turbidity-current flow (Menard, 1955). Menard argued that through time the channels migrate leftward until confined at a point where leftward migration is impossible. The net result is an asymmetric fan.

Depositional channels on the Tufts abyssal plain have higher right-side levees, and leftward migration of these channels is common (Hamilton, 1967). The right-side levees on the Monterey fan are higher, but migration of the channel is not consistent in direction (Normark, 1969). In both cases, the lateral migration was only a few kilometers during deposition of the entire section of sediment, and thus this process is probably not important in developing fan morphology. Whether new channels preferentially break through the lower left-side levee and lead to asymmetric fan growth is not established. The old Monterey fan-valley broke through its right-side levee to capture the Ascension fan-valley (Normark, in press). On the Astoria fan, the main fan-valley is said to have shifted consistently to the left through time (Nelson *et al.,* in press); however, the present Astoria fan-valley is the youngest and is at the *right* of the next youngest fan-valley. Nelson *et al.* (in press) also observed that the left-side levee commonly is higher.

The asymmetry of outline of some deep-sea fans merely may reflect their youth. Many of these fans may be Pleistocene features (Menard, 1960; Menard *et al.,* 1965; Nelson *et al.,* in press) and, given more time, the outlines may develop more symmetry. In this sense, older and/or more rapidly growing fans should be more symmetric. In any case, it has not been established that leftward migration of the fan-valleys is the dominant factor determining fan shape.

Alternatively, the shape of deep-sea fans may reflect the importance of bottom currents in redistributing sediment brought to the fan by channelized turbidity currents. Asymmetric fans may be formed if bottom currents are strong enough to deflect sediment from its movement downslope (Wilde, 1965b). The bottom currents may deflect the turbidity currents that overflow or emerge from the fan-valleys. Erosion and redeposition of the fan sediments require more powerful bottom currents, and this effect should be less common. Heezen *et al.* (1966) concluded that deep geostrophic "contour" currents are capable of smoothing and shaping the continental rise along the western margin of the Atlantic Ocean.

Changes in Growth Patterns

La Jolla fan.—Piper (1970) suggested that 90 percent of the Holocene sediment contributed to La Jolla Canyon has bypassed the fan. It seems likely that sand-carrying turbidity currents now moving down the fan-valley are confined to the valley until they reach the lower fan (Normark, 1969). The fan is not being built up significantly at present. The small levees along the valley are a minor part of the relief in comparison to that caused by erosion.

The growth pattern of La Jolla fan cannot be reconstructed completely. The small levees bordering the valley may not have been deposited from currents moving through this fan-valley, although there is no evidence of an earlier depositional fan-valley. Small erosion channels on the lower fan below the 1,000-m contour suggest an earlier distributary system.

Monterey fan.—The Monterey fan has undergone a marked change since the growth of the large levees bordering the fan-valley (Normark, in press). The Monterey fan-valley has captured the more westerly Ascension fan-valley and subsequent erosion has deepened the valley —perhaps as much as 300 m. Deposition on Monterey fan now is confined to a deeply incised fan-valley and/or to the toe of the fan far south of the leveed valley. The growth pattern of Monterey fan cannot be reconstructed and the changes leading to the erosion of the present valley are unknown.

Bengal fan.—The tremendous Bengal fan (Curray and Moore, 1969; F. Emmel, personal commun.) is nearly 100 times larger than other known fans. The leveed fan-valleys on the upper fan are likewise much larger. The rate of deposition on the Bengal fan is relatively high, and a braiding system of leveed fan-valleys exists. Abandoned and buried fan-valleys also are common. Down fan the valleys become smaller and levee relief is much less, but no suprafan has been recognized definitely.

[*Editor's Note:* The section on sublacustrine and alluvial fans is omitted.]

Conclusion

A model for deep-sea fan growth based on available data on the morphology, structure, and sediment distribution of fan-valleys is proposed. Under conditions of fan growth, deposition is probably localized in time and space. A single leveed fan-valley on the upper fan feeds a distributary system of numerous smaller, ephemeral channels on the middle fan. Rapid deposition at the end of the leveed valley may build a small suprafan. Deposition on the lower fan occurs from sheetlike overflow (unchannelized currents). Aggradation within this system will continue until some catastrophic event leads to the development of a new leveed fan-valley on the upper fan. Those fans with strongly asymmetric shapes are probably morphologically young or are growing slowly; however, redistribution of sediment on the fan by bottom currents may lead to some asymmetry of shape. Given sufficient time, no tectonic disturbance, and a simple basement physiography, uniform deposition should occur within all sectors of the fan.

References Cited

Andrews, J. E., 1967, The Great Bahama Canyon and structure and sedimentary development of the outer channel of the Great Bahama Canyon: Ph.D. thesis, Miami Univ., 104 p.

Bartow, J. A., 1964, Stratigraphy and sedimentation of the Capistrano Formation, Dana Point area, Orange County, California: M.A. thesis, California Univ., Los Angeles, 102 p.

———— 1966, Deep submarine channel in upper Miocene, Orange County, California: Jour. Sed. Petrology, v. 36, p. 700–705.

Buffington, E. C., 1952, Submarine "natural levees": Jour. Geology, v. 60, p. 473–479.

———— 1960, Loran electronic surveying equipment (model 60)—operational appraisal: U.S. Navy Electron. Lab. Research Devel. Rept., v. 964, 32 p.

Bull, W. B., 1964, Geomorphology of segmented alluvial fans in western Fresno County, California: U.S. Geol. Survey Prof. Paper 352-E, p. 79–129.

Curray, J. R., and D. G. Moore, 1969, The Bengal deep-sea fan (abs.), in Abstracts for 1968: Geol. Soc. America Spec. Paper 121, p. 67.

Eckis, R., 1928, Alluvial fans of the Cucamonga district, southern California: Jour. Geology, v. 36, p. 224–247.

Forel, F. A., 1885, Les ravins sous-lacustres des fleuves glaciaires: Acad. Sci. Comptes Rendus, v. 101, p. 725–728.

Hamilton, E. L., 1967, Marine geology of abyssal plains in the Gulf of Alaska: Jour. Geophys. Research, v. 72, p. 4189–4213.

Heezen, B. C., C. D. Hollister, and W. F. Ruddiman, 1966, Shaping of the continental rise by deep geostrophic contour currents: Science, v. 152, p. 502–508.

———— R. J. Menzies, E. D. Schneider, W. M. Ewing, and N. C. L. Granelli, 1964, Congo submarine canyon: Am. Assoc. Petroleum Geologists Bull., v. 48, p. 1126–1149.

Hooke, R. L., 1967, Processes on arid region alluvial fans: Jour. Geology, v. 75, p. 438–460.

Houbolt, J. J. H. C., and J. B. M. Jonker, 1968, Recent sediments in the eastern part of the Lake of Geneva (Lac Leman): Geologie en Mijnbouw, v. 47, p. 131–148.

Johnson, M. A., 1962, Turbidity currents: Sci. Progress, v. 50, p. 257–273.

Krause, D. C., 1962, Interpretation of echo sounding profiles: Internat. Hydrog. Rev., v. 39, p. 65–122.

Laughton, A. S., 1968, New evidence of erosion on the deep ocean floor: Deep-Sea Research, v. 15, p. 21–29.

Lowenstein, C. D., and J. D. Mudie, 1967, On the optimization of transponder spacing for range-range navigation: Jour. Ocean Tech., v. 1, p. 29–31.

MeGehee, M. S., and D. E. Boegeman, 1966, MPL acoustic transponder: Rev. Sci. Instruments, v. 37, p. 1450–1455.

Menard, H. W., 1955, Deep-sea channels, topography, and sedimentation: Am. Assoc. Petroleum Geologists Bull., v. 39, p. 236–255.

———— 1960, Possible pre-Pleistocene deep-sea fans off central California: Geol. Soc. America Bull., v. 71, p. 1271–1278.

———— S. M. Smith, and R. M. Pratt, 1965, The Rhône deep-sea fan, in Submarine geology and geophysics: London, Butterworth, p. 271–286.

Moore, D. G., 1965, The erosional channel wall in La Jolla sea-fan valley seen from bathyscaph Trieste II: Geol. Soc. America Bull., v. 76, p. 385–392.

Nelson, C. H., 1968, Marine geology of Astoria deep-sea fan: Ph.D. thesis, Oregon State Univ., 287 p.

———— et al. (in press), Physiography of the Astoria canyon-fan system.

Normark, W. R., 1969, Growth patterns of deep-sea fans: Ph.D. thesis, California Univ., San Diego, 165 p.

———— (in press), Channel piracy on Monterey deep-sea: Deep-Sea Research.

———— and J. R. Curray, 1968, Geology and structure of the tip of Baja California, Mexico: Geol. Soc. America Bull., v. 79, p. 1589–1600.

———— and D. J. W. Piper, 1969, Deep-sea fan-valleys, past and present: Geol. Soc. America Bull., v. 80, p. 1859–1866.

———— et al., 1968, Detailed channel morphology of the La Jolla submarine fan-valley (abs.): Am. Geophys. Union Trans., v. 44, p. 212.

Phillips, R. P., 1964, Seismic refraction studies in Gulf

of California, *in* Tj. H. van Andel and G. G. Shor, Jr., eds., Marine geology of the Gulf of California: Am. Assoc. Petroleum Geologists Mem. 3, p. 90–121.

Piper, D. J. W., 1970, Transport and deposition of Holocene sediment on La Jolla deep-sea fan, California: Marine Geology, v. 8, p. 211–228.

Shepard, F. P., 1964, Sea-floor valleys of Gulf of California, *in* Tj. H. van Andel and G. G. Shor, Jr., eds., Marine geology of the Gulf of California: Am. Assoc. Petroleum Geologists Mem. 3, p. 157–192.

———— 1966, Meander in valley crossing a deep-ocean fan: Science, v. 154, p. 385–386.

———— and E. C. Buffington, 1968, La Jolla submarine fan-valley: Marine Geology, v. 6, p. 107–143.

———— and R. F. Dill, 1966, Submarine canyons and other sea valleys: Chicago, Rand McNally, 381 p.

———— ———— and U. von Rad, 1969, Physiography and sedimentary process of La Jolla submarine fan and fan-valley, California: Am. Assoc. Petroleum Geologists Bull., v. 53, p. 390–420.

———— and G. Einsele, 1962, Sedimentation in San Diego Trough and contributing submarine canyons: Sedimentology, v. 1, p. 81–133.

Spiess, F. N., and J. D. Mudie, 1970, Small-scale topographic and magnetic features, *in* A. E. Maxwell, ed., The seas, V. IV: New York, Interscience.

———— *et al.*, 1967, Deeply-towed marine geophysical observation system (abs.): Am. Geophys. Union Trans., v. 48, p. 133.

———— *et al.*, 1969, Detailed geophysical studies on the northern Hawaiian arch using a deeply towed instrument package: Marine Geology, v. 7, p. 501–527.

Sullwold, H. H., Jr., 1960, Tarzana fan, deep submarine fan of late Miocene age, Los Angeles County, California: Am. Assoc. Petroleum Geologists Bull., v. 44, p. 433–457.

Tolman, C. F., 1909, Erosion and deposition in the southern Arizona bolson region: Jour. Geology, v. 17, p. 136–163.

Wilde, P., 1965a, Recent sediments of the Monterey deep-sea fan: Ph.D. thesis, Harvard Univ., 153 p.

———— 1965b, Estimates of bottom current velocities from grain size measurements for sediments from the Monterey deep-sea fan: Marine Tech. Soc. Am. Litt. Soc. Proc., v. 2, p. 718–722.

Winterer, E. L., J. R. Curray, and M. N. A. Peterson, 1968, Geologic history of the Pioneer fracture zone with the Delgada deep-sea fan northeast Pacific: Deep-Sea Research, v. 15, no. 5, p. 509–520.

13

Copyright © 1971 by the Geological Society of America

Reprinted from *Geol. Soc. America Bull.*, **82**, 563–572 (Mar. 1971)

JOSEPH R. CURRAY   *Scripps Institution of Oceanography, La Jolla, California 92037*
DAVID G. MOORE   *Naval Undersea Research and Development Center, San Diego, California 92132*

Growth of the Bengal Deep-Sea Fan and Denudation in the Himalayas

ABSTRACT

A geological and geophysical survey in 1968 has shown that the Bengal Deep-Sea Fan is almost 3000 km long, and 1000 km wide. We estimate that it may exceed 12 km in thickness. The sediments of the fan have been transported by turbidity currents from the Ganges-Brahmaputra River delta, through the "Swatch of No Ground" submarine canyon and into an extensive, complex, meandering, and braided net of fan valleys. Present rate of sediment influx suggests a regional rate of denudation in the Himalayan source area of over 70 cm/10^3 years. The sediment section in reflection profiles of the fan has been subdivided into three units separated by prominent unconformities. Volumes of the upper two units compared with the sediment influx rate extrapolated into the past suggest that the unconformities may be late Miocene and earliest Pleistocene. These times correspond to periods of orogeny in the Himalayas and suggest contemporaneity between plate-edge orogeny and mid-plate tectonic activity.

INTRODUCTION

The modern Himalayan Mountains are well known as the loftiest of Earth's mountain ranges, and they may represent the greatest terrestrial relief generated on Earth throughout geologic time. This great mountain chain is unique also for its youth, having been elevated most recently during the Quaternary. Very late vertical movement in the Himalayas is demonstrated by Pleistocene lake and river deposits uplifted 1525 to 1825 m in the Kashmir Valley and by overthrusting of Pleistocene alluvial gravels by older Himalayan rocks (King, 1962, p. 516). This vast and rapid upthrusting of rock masses is being countered by tremendous erosional activity.

Sinking of the Ganges alluvial plain, for example, has been concomitant with uplift of the Himalayas and several kilometers of alluvium have accumulated (Krishnan, 1960, p. 573–574).

Although the total volume of alluvium deposited on the south flank of the Himalayas is impressive, it is apparent that the great bulk of erosional products is carried to the sea. In terms of total discharge and watershed, the confluent Ganges-Brahmaputra Rivers do not match the greatest world rivers, but on the basis of published estimates, they surpass all others in total amounts of sediment carried to the sea (Table 1). The Ganges River drains much of the south slopes of the Himalayas, and similarly, the Brahmaputra drains most of the north slopes. These two great rivers are confluent above the head of the Bay of Bengal where they have formed the world's largest deltaic plain (Coleman, 1969). Most of the detritus, however, passes through the delta to form an enormous submarine sedimentary fan (Fig. 1), hereafter referred to as the Bengal Fan, which floors the entire Bay of Bengal and slopes south from its head at lat 20° N. to 10° S. of the equator, a total length of about 3000 km. The fan is about 1000 km wide, as it is contained on its flanks by continental margins of India, East Pakistan, and Burma, and the submarine mountains of the Ninetyeast Ridge.

In May 1968, as part of the CIRCE Expedition, we used Scripps Institution of Oceanography's R. V. *Argo* to survey the Bay of Bengal in the first of two cruises planned specifically to determine the volume, structure, and origin of this great sedimentary wedge. The 1968 survey of the fan included coring, dredging, interval velocity determinations by wide-angle reflection, and 12,000 km of seismic reflection, magnetic and 3.5 kHz bottom-penetrating echo-sounding pro-

TABLE 1.   SELECTED RIVERS OF THE WORLD RANKED BY SEDIMENT YIELD*

Name	Location	Drainage Area $10^3 km^2$	Average Annual Suspended Load 10^3 metric tons	Average Discharge at Mouth $10^3 m^3/sec$
Yellow	China	673	1,890,000	1.5
Ganges	India	†1116	1,452,000	†11.7
Ganges and Brahmaputra		†2048	2,179,000	†31.5
Brahmaputra	E. Pakistan	†935	726,000	†19.8
Yangtze	China	1942	499,000	21.8
Indus	W. Pakistan	968	436,000	5.5
Ching (Yellow Trib.)	China	57	409,000	.06
Amazon	Brazil	5773	363,000	181.1
Mississippi	U.S.A.	3220	312,000	17.8
Irrawaddy	Burma	430	300,000	13.6
Missouri (Miss. Trib.)	U.S.A.	1370	218,000	2.0

*After Holeman (1968) except as noted
†Coleman (1969)

files (Fig. 2). Sound source for the reflection profiling was two to four triggered air guns of 164 cm³ to 328 cm³ displacement each. The velocity traverses utilized sonobuoys in the method described by Le Pichon and others (1968). The second cruise planned for spring 1971 will provide seismic refraction as well as further dredging and profiling data. The purpose of this paper is to present a preliminary analysis of data from the 1968 cruise as it relates to the rate of growth of the Bengal Fan and, correlatively, to compare this growth to the rate of denudation in the Himalayas.

PREVIOUS STUDIES

During the International Indian Ocean Expedition of 1960–1965, thousands of kilometers of bathymetric profiles were run by ships of several nations. Heezen and Tharp (1964) have used these data to construct bathymetric and physiographic charts of the entire Indian Ocean, including the Bay of Bengal. Dietz (1953) had suggested that the very large sea floor channels trending out of the Bay of Bengal at the latitude of southern Ceylon were probably of turbidity current origin. Construction of the more recent topographic charts has enhanced the plausibility of this origin and shown that large numbers of the sea floor channels were present, some with considerable continuity. The chart of Heezen and Tharp connects many of these channels together to suggest that most of the features originate from the huge "Swatch of No Ground." This submarine canyon lying off the Ganges and Brahmaputra River Delta was described by La Fond (1957) as a source of turbidity currents.

Very little reflection profiling had been done prior to the present studies within the Bay of Bengal. The U.S. Coast and Geodetic Ship *Pioneer* had collected a good profile crossing the Swatch of No Ground where it cuts the continental shelf, but no traverses down the slope or into the deep bay were available. Soviet geologists (Neprochnov and others, 1964) reported on a series of single station conventional explosive seismic reflection measurements trending roughly northward up the bay. They reported sediment thicknesses averaging somewhat less than 2.4 km, and a maximum thickness of 3.0 km off the Indian margin near lat 18° N.

FAN VALLEYS

The 3.5 kHz bottom-penetrating echo sounder utilized in this survey was effective in recording bathymetry and resolving details of structure within the upper 100 m beneath the sea floor. Contours of this and previous bathymetric data confirm that the surface morphology of the fan is controlled by a system of turbidity current channels, most of which issue from the Swatch of No Ground canyon at the head of the bay. Channels shown on the new bathymetric chart presented here (Fig. 1) are based on all available data.

In the northern end of the Bay, off the Swatch of No Ground, the profiles show the remarkable development of very large fan valleys which sit on the crests of huge elevated expressways or ridges made up of the natural levee deposits of the channels (Fig. 3a). The Bengal Fan valleys are also unusual in that even on the grand scale of their development, they have been periodically

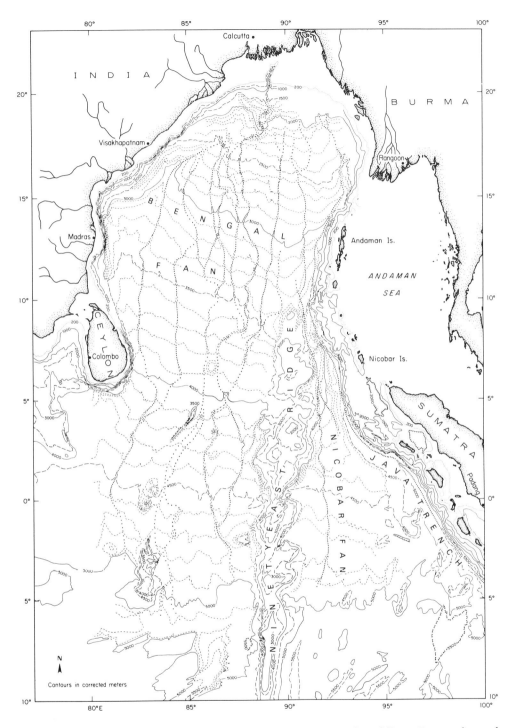

Figure 1. Bathymetric chart of Bengal Fan based on all available soundings. Fan valleys are indicated by dotted lines. Contour interval is variable.

238

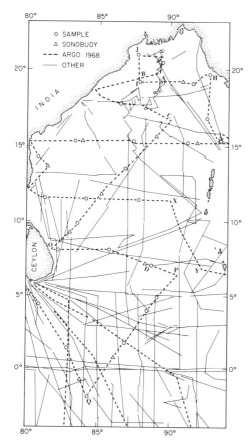

Figure 2.   Sounding lines on which bathymetry is based, plus ARGO 1968 tracks, stations, and sonobuoy runs. Location of photographs of Figure 3 and line drawings of Figure 4 are indicated.

abandoned while a new valley was formed on either flank (Fig. 3b). The present complex main channel (Fig. 3a) has a width of about 27 km between levee crests and is more than 180 m deep with the levees and channel floor well above the flanking fan surface. Yet even this impressive constructional feature is but one of several which can be seen in section one above the other and offset laterally by over a hundred km (Fig. 4, line J-I-H). These great submarine fan valleys are also remarkable for their internal complexity. Within the confines of the major levees there commonly exist several apparently active subchannels with ample evidence of terracing and probable meandering of channels.

Details of shallow sub-bottom structure show a great variety of channel types. In some parts of the fan surface, channels are partly or completely filled (Fig. 3c). Elsewhere the channels appear to be braided or are incised for over 100 m into sedimentary fill within formerly much deeper valleys. The most significant feature brought out by these records is the unmistakable evidence for pronounced migration of the channels by cut-and-fill processes analogous to those of subaerial streams. The turbidity current channels, in fact, show all of the depositional and erosional capabilities of subaerial streams for adjusting to variable base levels, stream loads, and discharge volumes.

STRATIGRAPHY AND STRUCTURE

Although four air guns of up to 328 cm^3 each were simultaneously used for much of the profiling, the sediment column of most of the fan was too thick to allow delineation of the base of the sediments. Well over 3 km of sediment were measured over most of the fan, and only near the margins of the fan, over isolated basement highs, and near the distal end of the fan, could the base of the sediments be observed. Internal structure of at least the younger sediments of the fan slopes upward on a regional scale to the north, as does the surface of the fan, and little effect can be detected from flanking sediment sources.

In the southern end of the Bay, and particularly rising up over the flanks of the Ninetyeast Ridge, are older strata underlying the younger sequence of sediments comprising the modern fan (Figs. 3d and 4, lines O-P and Q-P). The surface of these older folded and faulted strata can be followed in depth as very clearly defined unconformities in the reflection profiles. By tracing the unconformities the sediments and rocks of the fan have been divided into four units (Fig. 4). The uppermost sequence, arbitrarily termed "W" sediments, are the younger sediments of the modern Bengal Fan. These are deformed in only two places in the entire region: in an apparent subduction zone at the northern extremity of the Java Trench on the east side of the Bay of Bengal (Fig. 4, line X–Y), and on the upper part of the continental rise, where some recent slumping is evident. Underlying these "W," or younger, sediments are the "Y" sediments which are exposed in outcrops locally in the southern part of the fan and over the Ninetyeast Ridge (Figs. 3d and 4, lines O-P and Q-P). These

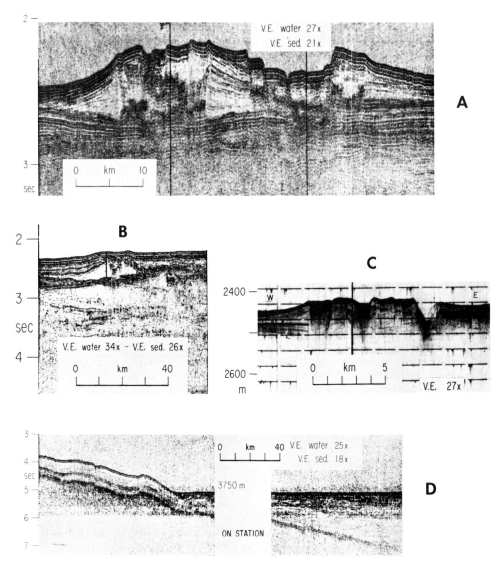

Figure 3. See Figure 2 for locations. All travel times are two way. A. Reflection profile of a large fan valley and natural levee system near northern apex of fan. Note the more nearly acoustically transparent sediments of the levees and the old surface of the fan depressed beneath the fan valley system. B. Reflection profile of an old abandoned and buried fan valley near northern apex of the fan. C. 3.5 kHz record of a fan valley in the central part of the fan. Note the fan sediment stratification, natural levees, active channel and two abandoned filled channels. The filled channels may represent an abandoned cut-off meander like an ox bow. D. Reflection profile of west flank of the Ninetyeast Ridge, showing "Y" sediments overlying "O" sediments folded and faulted over the ridge, with the younger "W" fan sediments ponded against the surface of the "Y" sediments.

clearly older sediments have been deformed, folded, and faulted locally; they will be the target of dredging in our forthcoming second phase of this study. The unconformity between the "W" and "Y" sediments can be traced with some degree of confidence throughout the entire Bengal Fan.

Underlying the "Y" sediments are older sedimentary rocks (Fig. 4, "O" sediments) and basement rocks, probably volcanics,

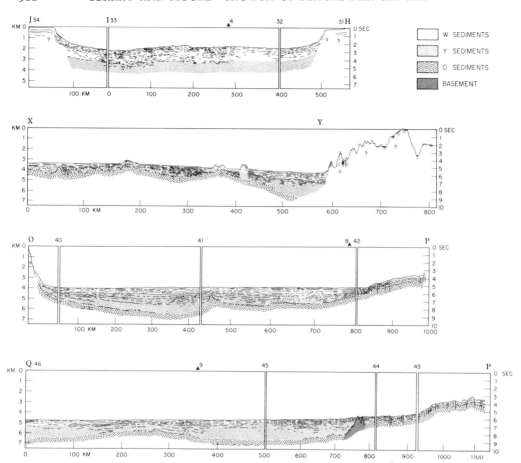

Figure 4.   Line drawings prepared from reflection profiles, with stratigraphic subdivision into three sediment units and acoustical basement rock, probably volcanic. Triangles represent locations of wide-angle velocity determinations and breaks in records represent coring locations. All travel times are two way. See Figure 2 for locations.

which appear locally as hills, particularly in the Ninetyeast Ridge. The "O" rocks can be identified clearly as sediments in many of the reflection profiles. Where they have been sampled on the Ninetyeast Ridge farther to the south (E. C. Allison, written commun., 1970), the older rocks are early to middle Tertiary. It appears that most of the "basement" reflections of the published Soviet reflection shooting (Neprochnov and others, 1964) may have been from the top of these older sedimentary rocks underlying the fan. As a result, the true volume of sediment beneath the fan is probably at least an order of magnitude greater than previously published figures have suggested, and also the total volume of sediment in the Indian Ocean may have been underestimated by an order of magnitude (Ewing and others, 1969).

On the basis of the geology of the Himalayas as described by Gansser (1964), we have assumed as a working hypothesis that the "W" sediments are of Quaternary age and the "Y" sediments are of late Miocene to Pliocene age. These then would be the sediments derived from the two late Cenozoic orogenies of the Himalayas. A major orogeny occurred in middle Miocene time, probably an important episode in the collision between the continents of India and Asia and the closing of the Tethys seaway. The next major orogeny was latest Pliocene and Pleistocene, and has resulted in the modern Himalayan Mountains. By extrapolating these unconformities throughout the entire fan, isopach maps of the "W" and "Y" sediment units have been constructed (Figs. 5a and 5b).

The thickest "W" sediments (Fig. 5a) occur in the northern part of the fan, closer to the river sources, but are noticeably thinned over the Ninetyeast Ridge and over a broad north-south swell in the northern part of the fan at about 85° E. The "Y" sediments (Fig. 5b) show greatest thickness along the eastern margin in the zone of apparent subduction (Fig. 4, line X–Y) and in the central and western parts of the fan. The sediments thin markedly over the Ninetyeast Ridge, but the structural north-south swell at 85° E. was not active during the period of deposition of these sediments. Particularly in the southern part of the fan, local uplifts have produced

folding and faulting of the "Y" sediments. Throughout the whole fan, the "Y" sediments appear to be thicker than the younger "W" sediments. In the filled northern end of the Java Trench, all three of the stratigraphic units "W," "Y," and "O" sediments are deformed, dipping easterly and apparently being subducted beneath the Andaman-Nicobar Island arc (Fig. 4, line X–Y).

A longitudinal section approximately down the center of the Bengal Fan (Fig. 6), was constructed utilizing our reflection data and published seismic data. Refraction work has been done thus far only at the very southern or distal end of the Fan (Gaskell and others, 1959). Velocities in the equivalent of our "O" unit are indicated as 4.4 to 5.1 km/sec. Velocities in the underlying basement rocks

Figure 5. Isopach maps prepared from reflection profiles of the sediment units. Two-way travel times are converted to approximate thicknesses by velocities measured by wide-angle reflection.

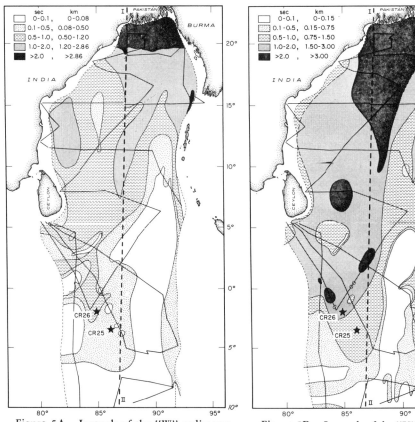

Figure 5A. Isopach of the "W" sediments, or uppermost unit.

Figure 5B. Isopach of the "Y" sediments.

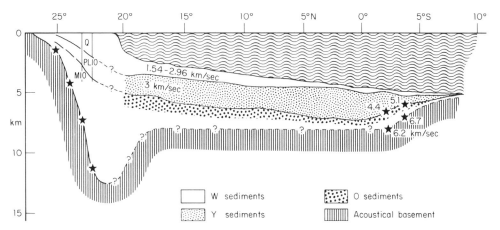

Figure 6.   Hypothetical longitudinal section from the Bengal Basin, Ganges-Brahmaputra delta, across the continental terrace, and down the Bengal Fan. See Figure 5 for line of section. Well data and land seismic results from Sengupta (1966); refraction stations at sea from Gaskell (1959). The "W," "Y," and surface of "O" sediments are interpreted from reflection profiles and extrapolation of unconformities. Basement surface between data points is conservatively interpolated. Basement reflections of Neprochnov and others (1964) generally approximate the surface of the "O" sediments.

are 6.2 and 6.7 km/sec. On shore in the Bengal Basin, drilling has been carried to depths of approximately 4 km, and deep seismic reflection studies have been carried out by several oil companies (Sengupta, 1966). Where last detected, under the Bengal Basin, the basement rocks lie at nearly 12 km below sea level and still dip steeply toward the south. In drawing the section (Fig. 6) we have taken the most conservative possible interpretation by assuming an essentially flat basement extending from the northernmost of the offshore refraction stations, south of the fan, northward to the descending basement reflections from the onshore seismic results. Drilling on land indicates an exceptionally thick Quaternary section, and a thick Pliocene section underlain by Miocene and older rocks. The coincidence shown as an interpolation between our unconformities in the Bengal Fan and the drilling results lends some credence to our working hypothesis of the ages of the units as "W" representing Quaternary, "Y" representing Pliocene and late Miocene, and "O" representing pre-late Miocene.

Thus the "W" sediments represent turbidite deposition of the most recent period of growth of the Bengal Fan. The "Y" sediments also thicken toward the northern source and show evidence of channel structures, suggesting that they too are primarily prod-ucts of turbidity currents. The "O" sediments, on the other hand, are of unknown depositional origin. The longitudinal section (Fig. 6) suggests that they may also thicken toward the north, but no obvious turbidite structures or channels have been observed in the profiles of "O" sediments, and they could be pelagic.

DENUDATION IN THE HIMALAYAS AND SIGNIFICANCE OF SEDIMENT VOLUME IN THE BENGAL FAN

The present rate of influx of sediments into the Bay of Bengal from the Ganges and Brahmaputra Rivers may be taken from published estimates (Table 1). Holeman (1968) estimates that 2.2×10^9 metric tons of suspended sediment are carried into the Bay of Bengal per year by the confluent Ganges and Brahmaputra Rivers. We have assumed that 25 percent of the total sediment load is bed load, giving a total sediment influx into the Bay of 2.9×10^9 metric tons per year. Although this proportion of bed load is very high, the rivers are known to be very sandy, and J. M. Coleman (written commun., 1970) estimates that the proportion of bed load may even be as high as 50 percent. Drainage area of the confluent rivers is 2.05×10^6 km². Of this we have used only the area above 200 m elevation as the assumed erosional area (1.54 × 10⁶ km²), thereby eliminating the flood plains or regions of deposition from the drainage area. We have ignored weathering products carried in solution and have assumed an average source rock density of 2.6 g/cm³. Thus we obtain a present rate of denudation of 72 cm/10³ years for the erosional area of the confluent rivers. This is somewhat lower than the 100 cm/10³ years estimated by Menard (1961).

If we now make the assumption of a uniform rate of erosion and sediment influx in the past, we can use our seismically measured fan sediment volumes to calculate the time necessary for deposition of the "W" sediments and "Y" sediments. This assumption of uniform rate of influx may be in error because of unknown effects of the Pleistocene glaciation and changes in erosional rates in the Himalayas. In balance, we are ignoring the very thick known alluvial deposits of the basins fronting the Himalayas and the Bengal Basin at the head of the Bay of Bengal. We also are ignoring contributions to the Bengal Fan from other sources such as rivers from India and Burma. Loss of sediment by subduction beneath the Andaman-Nicobar arc is also negligible.

Sonobuoy wide-angle velocity measurements made during the expedition enabled us to determine velocity gradients and velocities in the "W" sediments and a mean velocity in the "Y" sediments. By use of these interval velocities the areas of the isopachs (Fig. 5a and 5b) were converted to sediment volumes for the portion of the fan covered in the isopach maps and reflection profiling. Significant volumes of sediment may have been neglected outside of this coverage, such as under the continental shelf on the north and the delta of the Ganges and Brahmaputra. The volumes obtained are 3.1 × 10⁶ km³ for the "W" sediments and 7.0 × 10⁶ km³ for the "Y" sediments. Assuming a compacted density of 2.0 g/cc, 1.8 m.y. would be required for deposition of the "W" sediments. Assuming a sediment density of 2.2 g/cc for the "Y" sediments, 7 m.y. would be required for their deposition. These estimates of deposition time tend to corroborate further our working hypothesis of predominantly Quaternary age for the "W" sediments and late Miocene and Pliocene age for the "Y" sediments.

DISCUSSION

If our age assignments are indeed valid, the mid-plate tectonic activity which resulted in the unconformities in the fan section appear to be contemporaneous with orogeny in the Himalayas at the plate edge. Elsewhere on this plate edge along the filled northern continuation of the Java Trench, subduction appears to be occurring, resulting in scraping off, crumpling, and uplifting of at least the upper part of the section (Fig. 4, line X–Y). The validity of the age assignment for the "Y" sediments would further indicate that the northern part of the Ninetyeast Ridge has been uplifted as recently as post-Pliocene time.

ACKNOWLEDGMENTS

This study was supported by budgets of the Naval Undersea Research and Development Center and the Office of Naval Research. We wish to thank the officers, crew, and other scientific party of R/V *Argo* for their assistance. F. J. Emmel has given invaluable aid in reduction and interpretation of the data.

REFERENCES CITED

Coleman, J. M. Brahmaputra River: channel processes and sedimentation: Sediment. Geol., Vol. 3, p. 129–239, 1969.

Dietz, R. S. Possible deep-sea turbidity-current channels in the Indian Ocean: Geol. Soc. Amer., Bull., Vol. 64, p. 375–378, 1953.

Ewing, M.; Eittreim, S.; Truchan, M.; and Ewing, J. Sediment distribution in the Indian Ocean: Deep-Sea Res., Vol. 16, p. 231–248, 1969.

Gansser, A. Geology of the Himalayas: Interscience Publishers, 289 p., London, 1964.

Gaskell, T. F.; Hill, M. N.; and Swallow, J. C. Seismic measurements made by H.M.S. *Challenger* in the Atlantic, Pacific and Indian Oceans and in the Mediterranean Sea: Roy. Soc. London, Phil. Trans. Ser. A, Vol. 251, p. 23–84, 1959.

Heezen, B. C.; and Tharp, M. Physiographic diagram of the Indian Ocean, the Red Sea, the South China Sea, the Sulu Sea, and the Celebes Sea: Geol. Soc. Amer., scale 1:11,000,000, 1964.

Holeman, J. N. The sediment yield of major rivers of the world: Water Resources Res., Vol. 4, p. 787–797, 1968.

King, L. Morphology of the Earth: Oliver and Boid, 699 p., Edinburgh, 1962.

Krishnan, M. S. Geology of India and Burma: Higginbothams (Private) Ltd., 604, p., Madras, 1960.

La Fond, E. C. The Swatch of No Ground: Nat. Inst. Sci., India, Bull., No. 11, p. 84–89, 1957.

Le Pichon, X.; Ewing, J.; and Houtz, R. E. Deep-sea sediment velocity determination made while reflection profiling: J. Geophys. Res., Vol. 73, p. 2597–2614, 1968.

Menard, H. W. Some rates of regional erosion: J. Geol., Vol. 69, p. 154–161, 1961.

Neprochnov, I. P.; Korylin, V. M.; and Mikhno. The results of seismic research on the structure of the earth crust and sedimentary mass in the Indian Ocean: Int. Geol. Congr., 22nd Sess., Reports by Soviet geologists, p. 52–61, 1964.

Sengupta, S. Geological and geophysical studies in western part of Bengal Basin, India: Amer. Ass. Petrol. Geol., Bull., Vol. 50, p. 1001–1017, 1966.

MANUSCRIPT RECEIVED BY THE SOCIETY OCTOBER 1, 1970

CONTRIBUTION OF THE SCRIPPS INSTITUTION OF OCEANOGRAPHY

CONTRIBUTION OF THE NAVAL UNDERSEA RESEARCH AND DEVELOPMENT CENTER

Part III
ANCIENT SUBMARINE CANYONS

Editor's Comments
on Papers 14 and 15

14   SATO and KOIKE
      Excerpts from *A Fossil Submarine Canyon near the Southern
      Foot of Mt. Kano, Tiba Prefecture*

15   TANAKA and TERAOKA
      Excerpts from *Stratigraphy and Sedimentation of the Upper
      Cretaceous Himenoura Group in Koshiki-jima, Southwest Kyushu,
      Japan*

Japan has several well-developed ancient submarine canyons, but owing to language barriers these are not well known by Western geologists. A Royal Society Study Grant enabled the Editor to visit Japan in 1973, and there he met Japanese geologists who had worked on, or knew of, buried canyons. He was shown Tertiary examples on the Miura Peninsula by Iijima and Kanie (Kanie, 1969) and an older channel fill, the Permian Wadano Conglomerate, 50 km north of Nagoya, by Mizutani. Figure 5 in Mizutani (1964) shows the channel form by isopachs: there is a distinct left hook. The clast sizes are also plotted and the larger clasts appear to be concentrated nearer the head and at the base of the channel. Deposition of the conglomerate is attributed to turbidity flow.

Matsumoto and Okada (1973) deal with the Saku Formation of the Yezo Geosyncline in Hokkaido. These authors describe channel-fill sediments from the western marginal parts of the Cretaceous Yezo Geosyncline, characterized almost exclusively by conglomerates, consisting of irregular-shaped fragments of sandstone and mudstone derived from the Saku Formation and well-rounded clasts from farther away. These conglomerates are mainly matrix-free and the authors suggest that they are grain- and debris-flow deposits. The largest channel is shown in Figure 6: it is more than 100 m wide and over 20 m deep and has, near the top, two smaller channels containing turbidite sandstone and shale. The channels obviously acted as corridors for gravity sediment flows passing from western shallower areas to eastern deeper regions, a characteristic and important function for both modern and ancient submarine canyons.

The two papers that follow have been briefly summarized in English through the kind help of Hakuyu Okada, who also translated the figure captions of Paper 14. In reprinting this paper, we have omitted the list of references, which are mainly in Japanese. This is one of the earliest records of a "fossil" canyon; the paper by Tanaka and Teraoka (15) is one of the most recent.

Biographical Notes

Takahiro Sato was born in Tokyo in 1932. He studied at the Geological Institute of Tokyo University until 1954 and took the Graduate Course there, completing this in 1956. He obtained the Doctor of Science degree in 1964. Sato joined the Japanese Hydrographic Department in 1956 and he is at present Deputy Director of the Chart Division.

Kiyoshi Koike was born in 1926, also in Tokyo. He obtained his degree in 1948 from the Geological Institute of Tokyo University and stayed on at this Institute as Assistant. He was, regrettably, killed in an accident in 1957.

Keisaku Tanaka was born in Kobe, Japan, in 1924. He studied at the Geological Institute, University of Tokyo, and gained the B.Sc. degree in 1946. In 1961 he was awarded the D.Sc. at Kyushu University. In 1946 he joined the Geological Survey of Japan, where he is currently Senior Research Official. His publications have dealt with the Cretaceous System, stratigraphy, sedimentation, and paleontology (especially of Echinoidea and trace fossils).

Yoji Teraoka was born in Hiroshima, Japan, in 1934, gaining the B.Sc. degree in 1958 and the D.Sc. in 1965 at the Geological Institute, Hiroshima University. Since 1958, he has worked at the Geological Survey of Japan, where he is now Senior Research Official. He has published on Mesozoic stratigraphy, sedimentation, and tectonics.

14

Copyright © 1957 by the Geological Society of Japan

Reprinted from *Jour. Geol. Soc. Japan,* **63**(737), 100–101, 104, 107 (1957)

A Fossil Submarine Canyon near the Southern Foot of
Mt. Kano, Tiba Prefecture

Takahiro Sato and Kiyosi Koike

(Abstract)

Various discussions are presented with regard to the stratigraphical relation of the Higasi-higasa formation to the sub- and superjacent strata distributed around Minegami Village, Kimitu-gun, Tiba Prefecture, Japan.

To clarify this relation, the mode of deposition of this formation has been investigated. As the result, it was found that the sediments had been deposited in a paleo-submarine canyon eroded by a subaqueous mudflow: this is here designated as the Higasihigasa Fossil Submarine Canyon. Its dimensions are deemed to be 1.5 km.wide, 7.5 km. long, 150 m. deep, and showing probably a V-shaped transverse profile. The Higasihigasa Fossil Submarine Canyon might expose the base rocks in its upper stream, and its deposits interfinger with the alternation of sand and mud of the deep sea type in the Umegase formation in its down stream.

Since other examples of this kind of fossil submarine canyons have been reported among the Tertiary System in southern Kantô, it can be suggested here that subaqueous mudflows might form many submarine canyons in this district.

In addition, it was observed in the record of sediments that the regression of the sea level did not exceed more than 150 m.during the Quaternary Period in southern Kantô.

[*Editor's Note:* Translation of the Japanese text (with the exception of a single paragraph) is omitted, owing to limitations of space.]

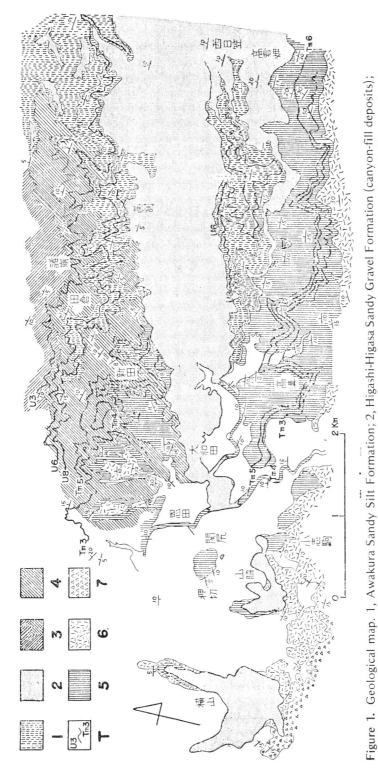

Figure 1. Geological map. 1, Awakura Sandy Silt Formation; 2, Higashi-Higasa Sandy Gravel Formation (canyon-fill deposits); 3, Morokue Fine Sand Formation; 4, Iwasaka Fine Sand Formation; 5, Takamizo Silt Formation; 6, Tomiya Tuffaceous Sandstone Formation; 7, Takeoka Tuff-Breccia Formation; T, key bed of tuff.

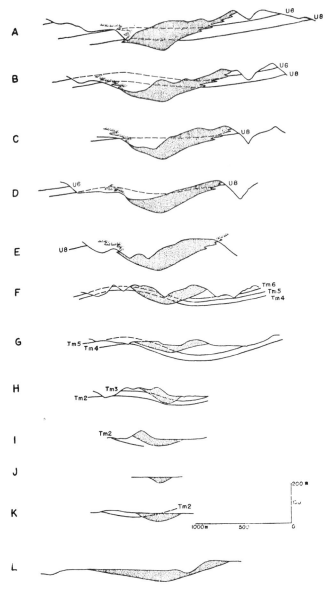

At the time of deposition of the Higashi-Higasa Formation (canyon-fill deposits), the basement movement tended to form northeasterly dipping submarine slopes. Molluscan fossils show that some consist of shallow-sea forms, fragmentary and abraded, and others of deeper-sea forms showing good preservation. Sediments show size variation from coarse to fine in a west to east direction. Supply of vast amounts of sediment by means of subaqueous mudflow coupled with synclinal structure of the basement caused submarine erosion and, during the interval Tm 5 to U8, formation of a submarine canyon.

Figure 7. Geological cross sections. A-L are shown in Figure 9. Dotted parts indicate the Higashi-Higasa Formation.

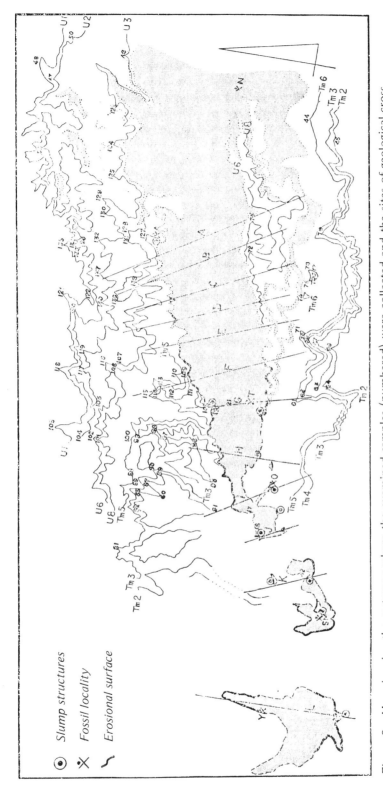

Slump structures

Fossil locality

Erosional surface

Figure 9. Map showing the outcrops where the examined samples (numbered) were collected and the sites of geological cross sections (A–L) in Figure 7.

15

Copyright © 1973 by the Geological Survey of Japan

Reprinted from *Geol. Survey Japan Bull.*, **24**, 157, 161, 168–169 (1973)

STRATIGRAPHY AND SEDIMENTATION OF THE UPPER CRETACEOUS HIMENOURA GROUP IN KOSHIKI-JIMA, SOUTHWEST KYUSHU, JAPAN

Keisaku Tanaka and Yoji Teraoka

Abstract

The Upper Cretaceous Himenoura Group of the Koshiki Islands, Kagoshima Prefecture, is unconformably overlain by Paleogene strata, forming a northeast-plunging synclinorium. It is more than 3,600 m thick and consists of sandstone, siltstone and mudstone. The sandstones are often cross-bedded, with occasional oyster beds and shallow sea shell ones. The group is stratigraphically divisible into six units, provisionally named A to F (Fig. 3).

Division A consists chiefly of mudstone and seems to show an open shelf facies. Division B, over 1,200 m thick, is dominated by sandstone of shallow sea deposition. Division C is represented by 150–200 m thick siltstone largely of open shelf facies, which contains ammonoids, inocerami and other molluscan fossils abundantly. In the succession of the upper part of division C to division D, over 600 m thick, sandstone sequences in combination with subordinate mudstone sequences show repeated cycles of upward-coarsening grain size, which are probably attributable to deltaic deposition. The upper part of division D carries a thin seam of coaly shale. Division E is occupied by a thickness of at least 850 m of mudstone often intercalated with turbidite sandstones and slump beds, showing a deep basinal facies on the whole. Noteworthy is the occasional occurrence of submarine channels up to more than 20 m deep in this division. Division F, over 750 m thick, as a whole marks a regressive phase, its lowest part containing fining-upward sequences probably of fluviatile origin.

The group is correlated partly to the Uppermost Urakawan to the Lower Hetonaian, the Campanian of the international scale, because of the occurrence of *Texanites* (*Pleisotexanites*) cf. *shiloensis*, *Inoceramus* (*Endocostea*) *balticus toyajoanus*, *I. orientalis orientalis* and *I.* cf. *schmidti*.

The Himenoura Group in the study area is similar in lithology and thickness to the group in the eastern part of the Amakusa-Shimojima area, but it has a much wider stratigraphic range than does the latter. The Himenoura Group of the area, on the other hand, differs from the group in Amakusa-Kamishima (the type area), in that the Lower Hetonaian is considerably thick and the sequence lower than that is composed mainly of sandstone.

From the mineral composition of sandstones, the kinds of pebbles and the paleocurrent directions, in conjunction with other available data, it is suggested that the coarse clastics were derived mainly from the western source areas where acid volcanics and granitic rocks occurred extensively.

[*Editor's Note:* Translation of the Japanese text (with the exception of a single paragraph) is omitted, owing to limitations of space.]

Stratigraphic division			Columnar section	Diagnostic sedimentary features *: not common	Fossils	Correlation		
Himenoura Group	F	F₃ 90		Slump structures Cross-bedding	Echinoids Pelecypods			Hetonaian Series
		F₂ 160		Cross-bedding				
		F₁ 500+		Cross-bedding				
	E	E₂ 350+		Graded bedding Directional sole markings Slump structures Submarine channels	Nuculanids *Inoceramus* Echinoids *Tosaloboris*	Lower Stage		
		E₁ 500+		Graded bedding Directional sole markings Slump structures Submarine channels	*Inoceramus (Endocostea) balticus* Ammonoids Echinoids			
	D	D₄ 200+		Cross-bedding Ripple marks	Oyster beds Trigonians *Thalassinoides*			
		D₃ 70+		Cross-bedding Slump structures*	Shell beds *Thalassinoides*			
		D₂ 90+		Cross-bedding Slump structures*	Oyster bed			
		D₁ 230+		Cross-bedding Ripple marks Slump structures	Oyster beds Trigonians *I. (E.) balticus toyajoanus*			
	C	150–200			*Inoceramus* cf. *schmidti* *I. orientalis orientalis* *I. (E.) balticus toyajoanus* *Texanites (Pleisotexanites)* cf. *shiloensis* *Glyptoxoceras indicum*	Uppermost Substage	Urakawan Series	
	B	B₄ 200+		Cross-bedding	Trigonians *Inoceramus*			
		B₃ 300+		Cross-bedding Flaser bedding	Oyster beds Shell beds			
		B₂ 200+		Cross-bedding				
		B₁ 500		Cross-bedding				
	A	50+		Slump structures*	Echinoids			

(mudstone/siltstone symbol)	Mudstone or siltstone
(symbol)	Mudstone interlaminated with sandstone
(symbol)	Sandstone and mudstone in thin-bedded alternation
(symbol)	Sandstone and mudstone in thick-bedded alternation (mudstone being predominant)
(symbol)	Sandstone and mudstone in thick-bedded alternation (sandstone being predominant)
(symbol)	Cross-bedded sandstone
(symbol)	Flat-bedded sandstone
(symbol)	Massive sandstone
(symbol)	Conglomerate
(symbol)	Pebbly mudstone
(symbol)	Coaly shale
(symbol)	Tuff or tuffaceous rock
-----	Fault relation

Stratigraphic summary of the Himenoura Group, Koshiki-jima, Kyushu.
Arabic figures indicate the approximate thickness in meters.

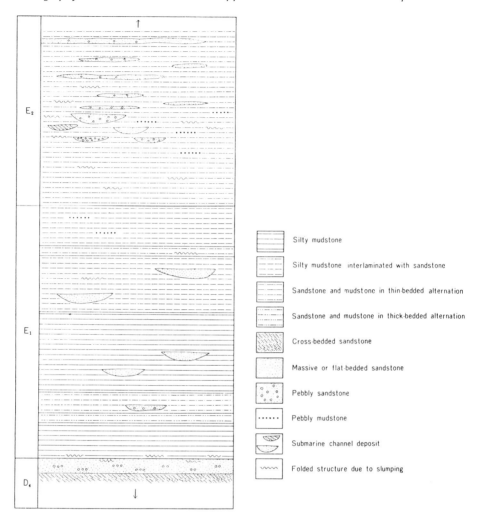

Diagrammatic section showing the facies distribution of division E, Himenoura Group, Nakakoshiki-jima.

Submarine channel structures are common in Division E of the Upper Cretaceous Himenoura Group. Each of these channels is several meters wide and one to a few meters deep, frequently associated with pebbly mudstone and slumped beds. The largest channel has an apparent width of about 100 m and a depth of some 20 m. Channel-fill deposits show a fining-upward sedimentation, accompanied by large flute casts (about 20 cm wide and 70 cm long). Channel-fill sandstones are medium to coarse grained, sometimes conglomeratic, and attain a thickness of more than 20 m. Large mudstone fragments are common in the sandstones.

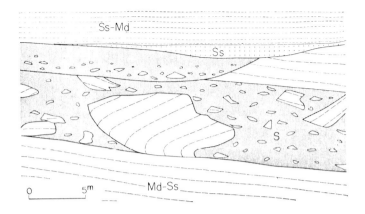

Detached masses originating from slumping. E$_2$, Himenoura Group, north of Umano-ri-zaki, Nakakoshiki-jima. From a photograph. S, slump bed (chaotically mixed silt and sand); Ss, sandstone; Ss–Md, sandstone and mudstone in thin-bedded alternation; Md–Ss, mudstone interlaminated with sandstone.

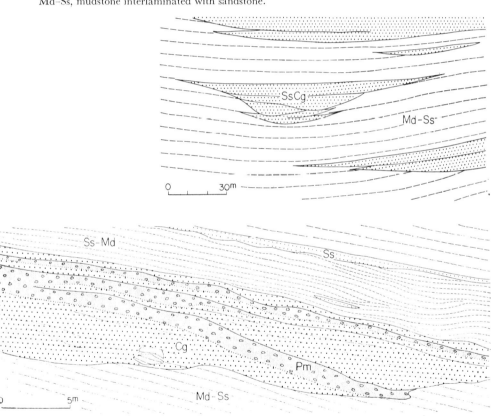

Large scale submarine channels. Above: E$_2$, Himenoura Group, north of Umanori-zaki, Nakakoshiki-jima. From a photograph. SsCg, sandstone and conglomerate; Md-Ss, mudstone interlaminated with sandstone. Below: E$_2$, Himenoura Group, Umanori-zaki, Nakakoshiki-jima. From a photograph. Cg, conglomerate; Pm, pebbly mud-stone; Ss, sandstone; Ss–Md, sandstone and mudstone in thin-bedded alternation; Md-Ss, mudstone interlaminated with sandstone.

Large scale submarine channel cutting into mudstone interlaminated with sandstone, and filled with sandstone. E_1, Himenoura Group, Oshika-zaki, Nakakoshiki-jima.

Large scale submarine channels cutting into interbedded sandstone and mudstone, and filled with conglomerate abounding in mudstone fragments. E_2, Himenoura Group, Umanori-zaki, Nakakoshiki-jima.

Editor's Comments
on Papers 16 and 17

16   HOYT
   Erosional Channel in the Middle Wilcox near Yoakum, Lavaca County, Texas

17   DICKAS and PAYNE
   Upper Paleocene Buried Channel in Sacramento Valley, California

These two papers highlight the contributions made to the discovery of ancient canyons by petroleum geologists. By geophysical means and by deep drilling, information from the subsurface has enabled oil geologists to reconstruct many canyons of the past which would never have been located by surface methods. This has been accomplished most successfully in the U.S. Gulf Coast area (notably in Texas and Louisiana) and in California. In these areas, sandy canyon fills occurring as shoestring bodies are potential reservoirs for oil and gas: equally, shale-filled canyons may trap hydrocarbons in sands below them. In the latter case, reservoir sands may occasionally be lost by erosion at the time of canyon cutting (e.g., Hoyt, Paper 16).

Since Bornhauser's pioneering paper of 1948 on a possible ancient canyon in southwest Louisiana, the U.S. Gulf Coast has been a fruitful source of buried submarine canyons, located during the search for oil. One of the largest and best delimited is the mid-Eocene channel near Yoakum, Texas, which is the subject of William V. Hoyt's paper (16). Data from a number of boreholes enabled him to draw transverse and longitudinal sections of the shale-filled channel and he used isopachs to demonstrate a long, sinuous canyon structure more than 50 miles (80 km) long, 10 miles (16 km) wide, 3000 ft (914 m) deep, comparable to many large modern canyons. Hoyt's clear text and diagrams caught the imagination not only of other Gulf Coast geologists but also foreign stratigraphers and sedimentologists, so that ancient canyons began to be sought, and found, in increasing numbers. Hoyt postulated rapid cutting and filling for his canyon and attributed erosion to faulting triggering off slumps and slides and setting up turbidity currents. The fill is of such tremendous volume that it shows up as a positive anomaly on the gravity map of the area. He cites evidence for considerable compaction of the fill after deposition, a theme taken up in the following paper. Note that the reference to Bornhauser 1959 should read 1948.

In California, several mainly subsurface Tertiary submarine canyons have been discovered and mapped during the search for oil and gas, especially in the Sacramento Valley. Details of these Californian ancient canyons are given by Whitaker (1974, Paper 21). The well-known Meganos Channel of upper Paleocene age is selected here for study. This channel was first mentioned in print by Silcox (1962) and then received detailed treatment by Edmondson (1965) in a publication of the San Joaquin Geological Society. Edmondson noted that the Meganos Gorge is a large fossil channel, 44 miles (70 km) long, 2500ft (770 m) thick, and 25 cubic miles (104 km^3) in volume, cut into Paleocene and upper Cretaceous sediments and filled during the late Paleocene. He also noted that the channel was filled by shale under quiet and stable conditions and that the fill is a factor of major importance in providing entrapment of gas and oil in several fields in the Sacramento Valley.

The following short paper (17) was given before the San Joaquin Geological Society in 1966 and later published by the American Association of Petroleum Geologists. In it, Dickas and Payne developed the ideas of Edmondson and modified his channel dimensions to 50 miles (80 km) long, up to 6 miles (10 km) wide, and filled with at least 2015 ft (614 m) of sediments, more than 95 percent of which are shales, now compacted to 35–60 percent of their original volume. They emphasized the rapidity of cutting and filling of the channel and the influence of downfaulting on its development.

In a recent paper, Edmondson (1972) considered in further detail the compaction of the Meganos fill. Compaction began before all the fill was deposited, and is considerable where there is all shale and negligible where there is all sand (as in the gorge axis at Walnut Grove). Dickas and Payne considered the sand facies to be an island in the gorge while Edmondson interpreted it as part of the fill: on either interpretation the effects and potential errors of interpretation induced by compaction would apply. Other major gorges in the Sacramento Valley, the Princeton Gorge (Redwine, 1972) and Markley Gorge (Almgren and Schlax, 1957), show little or no compaction because of the greater amount of coarse sediment and poorer sorting. Martinez Gorge, near Maine Prairie, may or may not be a gorge.

Biographical Notes

William V. Hoyt was born in 1890 at Junction City, Kansas. After graduating from the University of Kansas, he was engaged in petroleum exploration geology in Houston for many years before moving to Yoakum, Texas, where he was consultant to the petroleum industry. On retirement, he moved to Hamilton, Texas, where he died in 1968.

Albert B. Dickas holds graduate degrees from Miami University, Ohio (B.A. 1955 and M.S. 1956) in Geology and from Michigan State University (Ph.D. 1962) in Geology and Geophysics. Prior to 1966, Dickas was with Mobil Oil Company in the offshore area of the Gulf of Mexico and with Standard Oil Company of California on the West Coast. Between industrial positions he was a faculty member at Michigan State University. In 1966, he joined the University of Wisconsin, Superior, as Associate Professor in the Department of Geology and is responsible for curriculum in sedimentation, stratigraphy, oceanography, and geophysics, with research interests in sedimentary rock particle size statistics and environments of deposition. Currently, Dickas is also Director for the Center for Lake Superior Environmental Studies at the University of Wisconsin.

James L. Payne was born in 1937. He graduated from Colorado School of Mines in 1959 with the degree of Geophysical Engineer. He was employed by Standard Oil of California throughout the western states, including Texas, California, Washington, and Alaska, for the next eleven years as a geologist and geophysicist with increasing responsibilities. He transferred in 1971 to Chevron Overseas Petroleum, Inc., in San Francisco as Staff Exploration Geologist for Central America. In 1973–1974, he was Exploration Manager for Europe.

16

Copyright © 1959 by the Gulf Coast Association of Geological Societies

Reprinted from *Trans. Gulf Coast Assoc. Geol. Soc.,* 9, 41–50 (1959)

EROSIONAL CHANNEL IN THE MIDDLE WILCOX
NEAR YOAKUM, LAVACA COUNTY, TEXAS

William V. Hoyt*

ABSTRACT

A large mid-Eocene channel near Yoakum, Texas, is evident from a study of contour maps and cross sections assembled from the information provided by the electrical logs of more than 50 wells. The channel or canyon, clearly erosional, is filled largely with silty shale, which contrasts sharply on the well logs with the sandy character of the typical Wilcox strata through which it was gouged.

This canyon can be traced for more than 50 miles, from its mouth evidently near the southeast line of Lavaca County up-dip in a north-northwesterly direction to the outcrop of the Wilcox in Bastrop County. At its maximum known development near the town of Yoakum it has a width of 10 miles and a depth of 3,000 feet.

A completely satisfactory explanation of the origin is difficult. The following factors are indicated:

A. The presence of a major stream.

B. The great differential in thickness of the Wilcox now known to exist, attaining as much as 8,000 feet at the outer edge of the former continental shelf, which placed an unstable mass of sediments adjacent to deep water and a subsiding sea floor.

C. Slumps and slides at the mouth of the large stream, triggered in part by the fault action known to have been prevalent during Wilcox time, stoped inland guided by the stream channel and provided a source of turbidity currents powerful enough to cut a gentle gradient to the sea bottom.

D. An abrupt transgression of the sea led to rapid filling of the gorge and deposition of a thin blanket of shale over a large area outside the channel. This was followed by regression during which the thick and extensive sands of the Carrizo were deposited. These clean massive sands of the uppermost Wilcox entirely obliterated any evidence of the great channel below.

INTRODUCTION

Very unusual erosional and depositional phenomena which involve a major portion of the middle Wilcox formation are indicated by evidence from over 50 electrical logs of tests in the Wilcox producing belt of the central Gulf Coast of Texas. This evidence points plainly to the existence of a very large and deep fossil channel or canyon, evidently erosional, which was rapidly gouged and then completely filled with a sedimentation largely of silty shale which is in contrast with the predominantly sandy sedimentation of the normal middle Wilcox.

This entire erosional and redepositional cycle occurred just prior to the laying down of the upper massive sands of the Carrizo, the uppermost formation of the Wilcox.

Because of the magnitude of depth of the channel, the words canyon and gorge will be used synonymously with channel as being appropriately descriptive of the occurrence.

*Consulting Geologist, Yoakum, Texas.

Grateful acknowledgment is made to DeWitt C. Van Siclen, James A. Wheeler, and others for their helpful suggestions and stimulating discussions concerning the channel occurrence. Since there may be long continued controversy concerning the factors and mechanics involved, responsibility for the statements herein must rest entirely with the author.

DISCOVERY AND EXTENT

The first evidence of the existence of this channel was disclosed shortly after the discovery of the Yoakum gas field by the Pure Oil Company in the year 1945. While attempting an extension of the field to the east, the Pure, #1 Vick unexpectedly encountered a section of 1,585 feet of almost continuous shale instead of the normal Wilcox deposition and the zone of sands found in the nearby producing wells. In 1947, the Pure, #1 Reese, located a short distance northwest of the well on the Vick farm and a little closer to the gas field, encountered 1,750 feet of shale in the same section as the Vick well.

Many futile speculations were ventured in attempts to explain the situation by lithologic change, faulting, etc., and the occurrence was popularly dubbed the "shale bank."

From the discovery area at the Yoakum gas field in southwestern Lavaca County, subsequent drilling now extends the channel northwestward, progressively diminishing in size in an upstream direction, then curving more to the north as it crosses the east tip of Caldwell County and enters Bastrop County where it evidently reaches the surface up-dip from the outcrop of the Carrizo sandstone.

Downstream and down-dip in a southeasterly direction

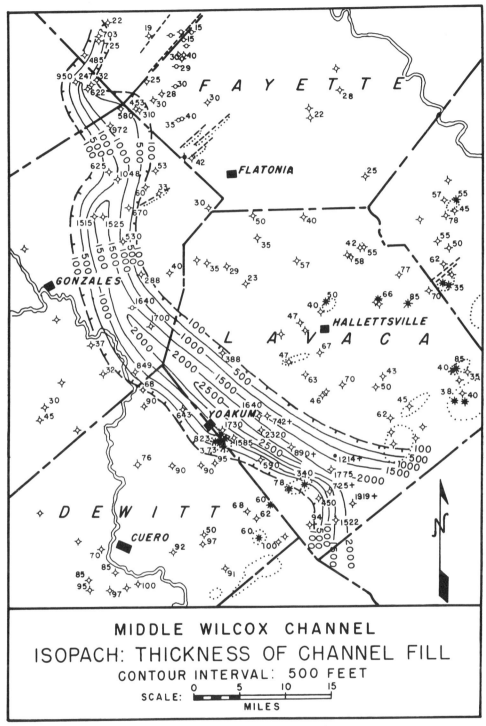

MIDDLE WILCOX CHANNEL
ISOPACH: THICKNESS OF CHANNEL FILL
CONTOUR INTERVAL: 500 FEET
SCALE: 0 5 10 15
MILES

Figure 1. Isopachous contours draw a canyon picture.

263

from the discovery area, very few wells have penetrated the entire thickness of the shale fill of the channel because of the increasingly excessive depth. However, from the existing data there is a suggestion that the channel widens, which might indicate that the mouth of the channel is being approached.

ISOPACHOUS CONTOURS DRAW A CANYON PICTURE

With the lack of fossil evidence, if it were not for the contrast in lithology between the normal Wilcox sandy sediments and the almost uniform silty shale of the channel fill it would be difficult to detect the feature.

An isopachous contour map (fig. 1) built upon the variations in thickness of the shale fill as shown in the various wells, pictures a gorge of unexpected magnitude which extends over 50 miles in length and 10 miles in width. From a contour thickness of 2,500 feet at the maximum depth there is a regular thinning to the canyon rims and an even gradient to shallower depths in an upstream direction.

The base of the upper massive sands of the Wilcox rests directly upon the shale of the channel fill and also upon the blanket stratum of shale. This and the fact that the channel shale thins progressively to both the east and west rims of the channel to connect with this comparatively thin stratum of shale should be sufficient indication that the time of deposition of the shale fill was contemporaneous with the deposition of the blanket shale stratum.

The thickness of the shale stratum is indicated at the wells beyond the confines of the channel (fig. 1), but no attempt has been made to contour these lesser variations of thickness of this section which varies from a maximum of 90 feet in DeWitt County to an average of 50 feet in Lavaca County and a lesser thickness in an up-dip direction.

It is an interesting fact that the shale mass of the channel fill is of sufficient magnitude to be distinctly recorded by gravity instruments, as a regional gravity map of the area discloses a consistent maximum anomaly along the exact axis of the channel as shown by well logs.

CROSS SECTIONS

Four cross sections of the channel drawn from electrical well logs are presented. Figure 2 gives their location.

The first section, A - A' (fig. 3), located at the discovery locality and including two producing wells of the Yoakum gas field, presents a section near the maximum development of the channel and fill. The first and last wells, numbers 1 and 8, have a normal Wilcox section which has a similar zonation on both sides of the channel over quite a large area.

This continuity of sedimentation on both sides of the channel is a good argument in favor of the erosional origin of the channel. The rapid descent of the shale in the wells at the center of the section shows the great depth to which the channel was gouged. The Chavanne, #1 Carter, number 5 in the section, has a total thickness of 2,320 feet of shale, the greatest to be encountered in the wells drilled to date.

It is noticeable that there is a pronounced thickening of the upper massive sand over the center of the channel, which amounts to as much as 500 feet. This is a logical

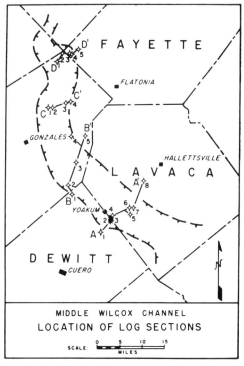

MIDDLE WILCOX CHANNEL
LOCATION OF LOG SECTIONS

SCALE: 0   5   10   15
                MILES

Figure 2. Location of channel sections and wells used.

result of differential compaction of the thick body of shale as loading of the sand section progressed and the overburden increased.

Section B - B' (fig. 4) is located a little farther upstream, in Gonzales County. It is very similar to the first section and the same explanations are applicable, although the size of the channel is a little smaller.

In the central well of the section, Gulf Coast Leaseholds, #1 Roznovsky, more sand occurs in the shale fill. This situation would be expected in an up-dip direction and closer to the source of sediments. The maximum shale thickness in this section is 1,890 feet. The blanket shale horizon is 68 feet thick on the west rim and 40 feet on the east rim. Total Wilcox formation thicknesses are shown, being 3,790 feet in the H. R. Smith, #1 Kruse (1), and 3,837 feet in the Tex-Penn, #1 Thompson (5).

Section C - C' (fig. 5), the next upstream and in central northeast Gonzales County, is confined to the eastern portion of the channel. This channel picture conforms to the preceding sections. It affords a study of the base of the Wilcox in relation to the presence of the channel above, as the wells penetrated to the Midway formation. It shows that the base of the Wilcox remained unchanged as originally deposited and was not in any way altered by the influences that resulted in the channel.

At this locality the maximum thickness of the shale fill is 1,518 feet at the center of the channel. The total Wilcox

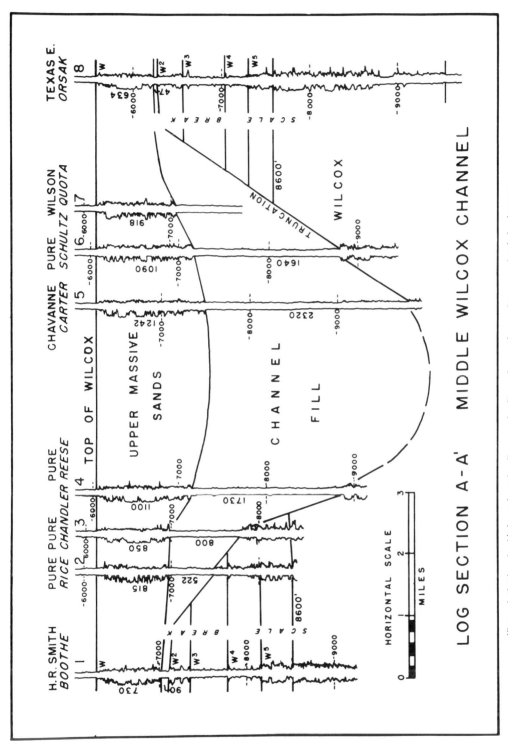

LOG SECTION A-A'   MIDDLE WILCOX CHANNEL

Figure 3. Section A-A', located at the discovery locality and near the maximum erosional development.

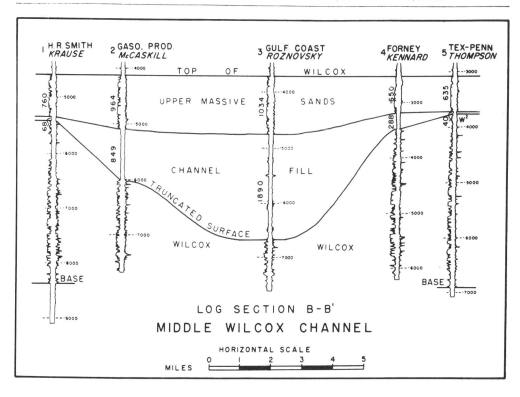

Figure 4. Section B - B', a little farther upstream in Gonzales County.

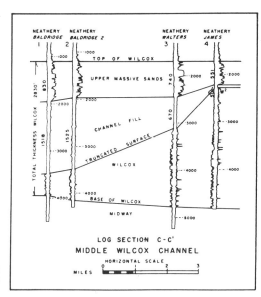

Figure 5. Section C - C', confined to the eastern portion of the canyon. This section affords a study of the base of the Wilcox below the axis of the channel.

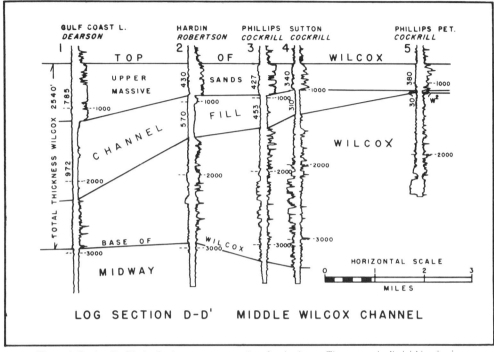

Figure 6. Section D - D', the farthest upstream section that is shown. The canyon is diminishing in size

thickness in the wells of the section averages 3,100 feet, with no thickening shown under the maximum development of the channel.

Section D - D' (fig. 6) is also confined to the eastern half of the channel. There is an indicated eastward widening of the channel due to the swing of the channel axis to a more westward direction at this point. The greatest thickness of shale fill in this section is 972 feet. The total thickness of Wilcox deposition now averages 2,500 feet. Two more sections nearer the channel source and likewise confined to the eastern portion of the channel could be shown but will be omitted for the sake of brevity as they show similar pictures and tendencies toward a playing out of the channel in an upstream direction.

YOAKUM GAS FIELD

Because of the intimate association of the Yoakum gas field to the channel both as to information of channel presence and relations of structure and closure of the producing reservoir, a cross section and contour maps of the field are presented. Figure 7 is a plat of the Yoakum gas field showing the location of the wells in the field cross section.

The field section (fig. 8) is important mainly because it shows the regularity of primary Wilcox deposition clear to the truncation at the channel. There is no tendency toward facies change or lateral gradation from sand to shale as the plane of truncation is approached. The sub-sea alignment of the logs of the section shows a regular structural rise of the sand horizons until the point of truncation.

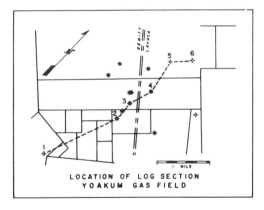

Figure 7. Plat of the Yoakum gas field showing location of wells in the field and those used in the field cross section, figure 8.

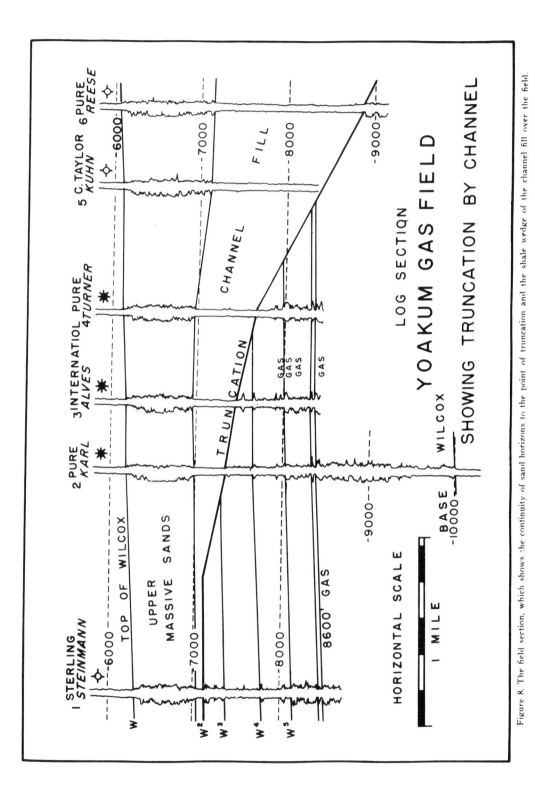

Figure 8. The field section, which shows the continuity of sand horizons to the point of truncation and the shale wedge of the channel fill over the field.

268

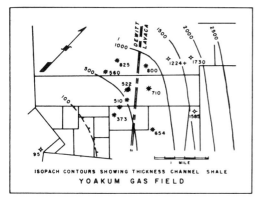

ISOPACH CONTOURS SHOWING THICKNESS CHANNEL SHALE
YOAKUM GAS FIELD

Figure 9. Isopachous contours showing thickness of channel fill over the field and into the depths of the channel.

From a normal rim thickness of shale in the Sterling, #1 Steinmann (1), the shale increases rapidly in thickness in an eastward direction over the field. The center wells of the section, Pure Oil Company, #1 Karl (2), #1 Alves (3), and #1 Turner (4) are gas distillate producers. The Taylor, #1 Kuhn (5), which did not penetrate through the channel fill, and the Pure, #1 Reece (6) with a shale section of 1,730 feet, did not produce because of the truncation of the zone of producing gas sands.

Figure 9, a plat of the Yoakum gas field with isopach

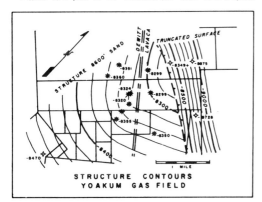

STRUCTURE CONTOURS
YOAKUM GAS FIELD

Figure 10. The subsurface structure of the lower producing sand of the field. Also shows the contoured surface of the truncated edge of the channel adjacent to the field.

contours of the channel fill over the field, shows the rapid thickening of the shale in an eastward direction into the depths of the channel. The swing of the contours westward at the north limit of the field suggests a possible tributary entering the channel from the west, which may have been a factor for structural closure and the accumulation of gas.

Figure 10 shows the present subsurface structure of the lower producing sand (8,600') of the field and its intersection with the erosional unconformity, the plane of

which is also contoured. The contours suggest a domal structure which might have been sufficient of itself to accomplish accumulation, and it is possible that the truncation of the structure destroyed a portion of the producing reservoir. However, the truncation and shale fill form a part of the structural closure as we find it today.

SUBSURFACE MAP OF THE CHANNEL

The regional subsurface setting of the channel is shown in figure 11, on the following page. Contours are drawn on the base of the blanket shale horizon and the base of the shale fill of the channel, using an interval of 500 feet. The picture is one of a deep gorge winding southward from the Bastrop-Caldwell county line into Gonzales County, then bending gradually to the southeast in Lavaca County and becoming larger and deeper as it progresses toward the outer shelf of the Wilcox where it must have opened into deep water. The swing to the westward in northern Gonzales County was probably caused by structural uplift and faulting in the vicinity of the Arnim and Muldoon oil fields near the town of Flatonia, which caused an early shift of the stream channel.

The contours from the western rim area could be connected directly across the channel with the contours to the east of it and would show little deviation from the regional structure as mapped on the top of the Wilcox formation. A connection of the —7,000 foot contour from the west rim to the east rim of the channel would pass at right angles over the —10,000 foot contour of the channel floor, indicating that the channel was 3,000 feet deep at this point.

COMPARISON WITH KNOWN CHANNELS

A study of the abundant literature dealing with submarine canyons and turbidity currents, with their capacity to transport and erode, and a look at the contour pictures of several present-day submarine canyons, leads one to accept the Yoakum channel or canyon as being very similar to these other known occurrences.

Ph. H. Kuenen, in his treatise on submarine canyons

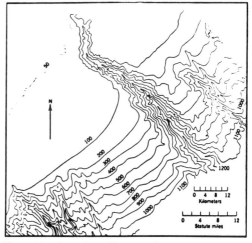

Figure 12. Chart of the Hudson submarine canyon, based on a chart by Veatch and Smith 1939. (Reproduced from Kuenen 1950, with permission of the publisher.)

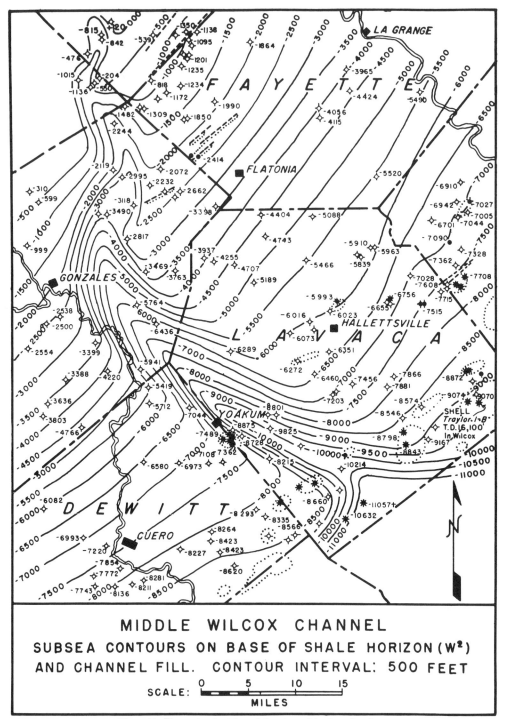

MIDDLE WILCOX CHANNEL
SUBSEA CONTOURS ON BASE OF SHALE HORIZON (W²)
AND CHANNEL FILL. CONTOUR INTERVAL: 500 FEET

SCALE: 0   5   10   15
MILES

Figure 11. The regional setting of the channel in the subsurface.

270

(1951), reproduced a chart of the well-known Hudson canyon (after Veatch & Smith 1939) which is so similar in form, depth, and shelf location to the subsurface map of the Yoakum channel (fig. 11) that it is reproduced here (fig. 12) for comparison. The Hudson canyon is not directly connected with the mouth of the Hudson River but is separated from it by an alluvial fan on the submerged shelf. The Hudson canyon must wait for the passage of geological time and the sedimentary filling and covering to be complete to become such an occurrence as that presented herein.

EXPLANATION OF OCCURRENCE

Before an entirely satisfactory explanation of all of the erosional and sedimentary features involved with the Yoakum channel occurrence is consummated there will have to be much controversy and investigation. Existing publications and opinions, and the research of scientists dedicated to the study of submarine problems and features, suggest common sense applications to the problems involved in this occurrence.

In making an attempt at a scientific explanation of the channel and fill as pictured in the foregoing maps and sections, several factors must be discussed:

A. The existence of a major stream is necessary. Ordinarily for a stream to maintain a channel and to gouge to canyon depths, a long period of structural growth for the area is necessary, (e.g., the Grand Canyon of Arizona, the Gunnison and the Royal Gorge of Colorado). In following this line of thought, a careful examination was made of the surface upon which the blanket shale stratum was deposited, and practically no evidence of disconformity was found to exist. In the absence of evidence of extensive erosion and the development of a well-formed tributary system to the channel, the existence of a period of structural uplift must be ruled out.

B. The few very deep wells near the southeast line of Lavaca County show that the Wilcox attains a very great thickness along what was then the outer continental shelf. The Shell Oil Company, #1-B Traylor, in the Providence City Field of eastern Lavaca County, at its total depth of 16,102 feet was still in Wilcox formation topped at 8,480 feet. This total of 7,622 feet of Wilcox, with an unknown amount yet to be drilled, is almost twice the average thickness of the entire Wilcox a short distance up-dip. This must have placed an unstable mass of unconsolidated sediments adjacent to deep water and a subsiding sea floor.

C. Natural slumps and slides occurring on the face of the steep, unstable mass of thick sediments at the outer shelf, triggered by faulting which was especially active in this general area in middle Wilcox time, would set up turbidity currents as the bottom of the gorge was reached. Waters of the stream would provide more fluidity and mobility to the current which would be necessary for the erosion and deepening of the channel.

Once started there must have been frequent and continued action in order to stope back through the shelf and up-dip to the extent recorded herein. Incredible as it may seem the entire erosion of the channel must have been accomplished near the close of the more marine phase of the middle Wilcox, which is present over a large area in southwest Texas.

D. A period of sharp transgression of the sea terminated the erosional forces at work on the channel and initiated a change which resulted in the rapid filling of the channel with shale and silt, and laid down the extensive shale stratum outside the channel.

A climatic change to a period of torrential rainfall may have aided the outflow of clay and silt which preceded the deposition of the well-washed sands of the massive Carrizo which followed and which indicates an extensive regression of the Gulf waters. A closer study of sedimentation, earth movements and climatic conditions during Wilcox time may throw more light on the situation.

CONCLUSIONS

Sufficient reliable information from the logs of many oil tests drilled in the general area is now at hand to prove conclusively the existence of the middle Wilcox channel or canyon as described herein. The evidence points conclusively to an erosional origin which is undoubtedly the same as that of the many other features of a similar nature located on the present-day continental shelves, which have received much greater attention elsewhere than in the Gulf Coast region.

With lack of fossil evidence the means of detection is limited to the sedimentary picture afforded by the study and correlation of electric well logs. While the Yoakum channel now stands out as an isolated occurrence for the Wilcox of the Gulf coastal region, it may not be the only one to exist. Should the depositional fill of a channel resemble the primary deposition of the formation gouged, as would be very likely in most occurences, it would be practically impossible to detect the presence of the channel.

It is to be hoped that this paper will lead to a more general investigation of the entire Gulf area for other channel occurrences and that the suggestions ventured herein as to the mechanics of origin will prompt constructive discussion and study which will allow a more conclusive explanation.

SELECTED BIBLIOGRAPHY

Bornhauser, Max, 1959, "Possible Ancient Canyon in Southwestern Louisiana," Bull. Amer. Assoc. Petrol. Geol., Vol. 32, pp. 2287-90.

Frick, J. D., Harding, T. P., and Marianos, A. W., 1958, "Eocene Gorge in Northern Sacramento Valley," (abs.) Bull. Amer. Assoc. Petrol. Geol., Vol. 43, No. 1, p. 255.

Kolbe, R. W., 1958, "Turbidity Currents and Displaced Fresh Water Diatoms," Science, p. 1505.

Kuenen, Ph. H., 1950, "Geomorphology of the Sea Floor," Marine Geology, John Wiley & Sons, Inc., New York, pp. 480-531.

—————, 1951, "Properties of turbidity currents of high density," Soc. Econ. Paleontologists and Mineralogists, Spec. Pub. No. 2, pp. 14-33.

Menard, Henry W., and Ludwick, John C., 1951, "Applications of hydraulics to the study of marine turbidity currents," Soc. of Econ. Paleontologists and Mineralogists, Spec. Pub. No. 2, pp. 2-13.

Natland, M. L., and Kuenen, Ph. H., 1951, "Sedimentary history of the Ventura Basin, California, and the action of turbidity currents," Soc. Econ. Paleontologists and Mineralogists, Spec. Pub. No. 2, pp. 76-107.

Shepard, Francis P., 1948, Submarine Geology, Harper & Brothers, New York.

—————, 1951, "Submarine Erosion, a discussion of recent papers," Bull. Geol. Soc. America, Vol. 62, pp. 1413-18.

—————, 1951-b, "Transportation of sand into deep water," Soc. Econ. Paleontologists and Mineralogists, Spec. Pub. No. 2, pp. 53-65.

Veatch, A. C., and Smith, P. A., 1939, "Atlantic Submarine Valleys of the United States and the Congo Submarine Valley," Geol. Soc. Am., Spec. Paper No. 7, 101 pp.

17

Copyright © 1967 by the American Association of Petroleum Geologists

Reprinted from *Amer. Assoc. Petrol. Geol. Bull.,* 51(6), 873–882 (1967)

UPPER PALEOCENE BURIED CHANNEL IN SACRAMENTO VALLEY, CALIFORNIA[1]

A. B. DICKAS[2] AND J. L. PAYNE[2]

Superior, Wisconsin, and Oildale, California

ABSTRACT

The Meganos channel, a fossil submarine channel, is in the central part of the Sacramento Valley of California. This channel was cut and filled in a marine environment during a relatively short interval of late Paleocene time. It has been traced in the subsurface more than 50 miles and has a maximum width of 6 miles. The greatest known thickness of sediments filling the channel is 2,015 feet. Formations truncated by the channel range in age from latest Paleocene through Late Cretaceous.

The rocks eroded by the channel are principally arenaceous; however, sediments filling the channel are composed of more than 95 per cent shale. Paleontologic studies indicate that the channel deposits accumulated in water depths ranging from neritic to upper bathyal.

A major factor contributing to the formation of the Meganos channel is thought to be regional faulting. Before the channel existed, the Midland fault, a major north-south-striking feature, began to form. It is the writers' opinion that the declivity created by this down-to-the-west normal fault set up conditions favorable for extensive slumping and turbidity currents which caused erosion of the sea floor and development of the channel. Subsequently, fine-grained terrigenous clastics filled this erosional feature and, at the same time, were deposited in a thin layer in the areas outside of the channel.

The channel shale contained an original large volume of interstitial water. Burial and overburden pressures compacted the shale to 35–60 per cent of its original volume.

Comparison among ancient subsurface and existing submarine canyons shows that features comparable with the Meganos channel have been formed in the past and are being eroded today, all under differing geologic settings.

Truncation of underlying formations by the channel shale combined with the local structure formed commercial accumulations of hydrocarbons at the Brentwood oil and gas field and the Dutch Slough, River Break, and West Thornton gas fields.

INTRODUCTION

An anomalous shale section, ranging in thickness from less than 100 feet to more than 2,000 feet, has been traced 50 miles in the central part of the Sacramento Valley of California. Subsurface data and isopachous studies indicate that this shale sequence occupies an ancient marine channel which was eroded and then filled with fine-grained clastics (Fig. 1).

Paleontologic and regional lithologic studies indicate that this channel was eroded in a marine environment into a predominantly arenaceous section. After erosion, sedimentary infilling took place in a very short period of geologic time during late Paleocene time. The sediments which filled the eroded channel have been assigned to the upper Paleocene Meganos C formation of Clark and Woodford (1927); this assignment is based on the Foraminifera found in the channel sediments.

[1] Manuscript received, March 10, 1966; accepted, July 22, 1966. Presented to the San Joaquin Geological Society by J. L. Payne on June 14, 1966.

[2] Geologists, Standard Oil Company of California, Western Operations, Inc., Northern Division, Oildale, California. A. B. Dickas' present address: Wisconsin State University. J. L. Payne's present address: Chevron Oil Company, Houston, Texas.

Sediments ranging in age from late Paleocene through Late Cretaceous were truncated during the creation of this feature. Where the regional dip is opposite to that of the slope of the channel walls, favorable local structure created hydrocarbon traps in the truncated sandstone beds abutting channel shale. To date, gas and oil production has been developed in four fields along the trend of this channel.

EARLY DEVELOPMENT OF CHANNEL CONCEPT

Although the anomalously thick shale section comprising the Meganos channel has been known for many years, a better understanding of the morphology of this shale section was not possible until detailed information became available through the discovery and development of the Brentwood field by the Shell Oil Company in 1962. In this field, the channel shale provides the updip closure on a northwest-dipping homocline. Lateral closure is created by a northwest-trending system of parallel faults (Sullivan, 1963).

As the Brentwood field was developed, isopachous studies of the Meganos shale made it possible to identify these sediments as a subsurface-filled erosion channel. However, in this area, only the northern flank of the channel remains intact;

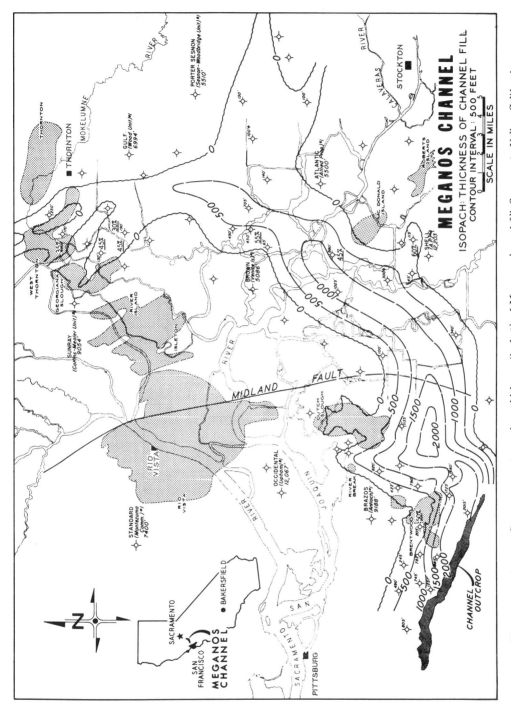

FIG. 1.—Location map. Isopacuous contours show thickness (in feet) of Meganos channel fill, Sacramento Valley, California.

273

the original extent and morphology of the southern flank have been destroyed by post-channel uplift and erosion along the Mount Diablo mountain range. Remnants of this southern flank are present at the surface on the north side of the Mount Diablo uplift (see Figs. 1, 5).

Once the existence of the buried Meganos channel was recognized and understood in the Brentwood area, other "anomalous shale" wells, some of which had been considered evidence for local channels (Silcox, 1962; Safonov, 1962), were for the first time fully comprehended. The Meganos channel was extended eastward from Brentwood to the position of the McDonald Island field and northward to what is now recognized as the "headward" area of the channel in the vicinity of the West Thornton field. The present extent of the Meganos channel was first defined publicly by W. F. Edmondson at a meeting of the San Joaquin Geological Society on October 13, 1964, and later in the selected papers presented to the San Joaquin Geological Society (Edmondson, 1965).

REGIONAL MORPHOLOGY OF CHANNEL

From the headward area of the channel north of the West Thornton gas field, the course of the Meganos channel has been traced 50 miles south and west to the Brentwood field as shown in Figure 1. At present, the downslope termination is not known because of the paucity of drilling in the area west of Brentwood and because of the extensive unconformities found in the general San Francisco Bay area. However, it is the writers' hypothesis that the effluence of the channel was situated in the Pacific Ocean west of San Francisco. If this is correct, the large volume of arenaceous sediments which was eroded as the channel was formed should be present in the offshore regions of the Pacific Ocean. To date, however, it is not known where the large volume of eroded sediments was redeposited.

The channel width is variable, ranging from a minimum of 2 miles to a maximum of 6 miles. This variation is probably a result of sea-floor topography at the time of channel erosion. Except in the headward region, tributary channels flow into the main channel course with an orientation of approximately 90° to the main axial trend and thus form a modified trellis drainage pattern.

In the headward area in the vicinity of the West Thornton gas field, a complex drainage pattern developed. From south to north in this region, the channel course divides into several distinct tributary channels. One very interesting feature associated with these tributary channels is the complete erosional isolation of a large island of the older sandstone sediments (Figs. 1, 3). Some geologists (Edmondson, 1965) have considered this sandstone island to be part of the channel fill, with channel sandstones resting directly on older Martinez sandstones. However, the writers would correlate these sandstones with the older Martinez sandstones in the manner shown in Figure 3. In view of the documented tributary channels in this region, it would not be unreasonable for an isolated island of older sandstone to develop as shown in Figure 1.

Isopachous studies, based on electric logs and paleontologic data, show that the deepest development of this channel is in the Brentwood area. Here, a maximum thickness of 2,015 feet of channel shale is mapped. Because of the characteristic electric-log pattern of the gorge shale and age dating by Foraminifera, there is no difficulty in establishing the base of the sediment fill even though in this area some shale-on-shale contacts are found.

Although the average plunge of the channel axis is southwesterly at approximately 2°, local areas exist where the axis plunges opposite the direction of the channel current flow. This is probably a result of later structural warping. Numerous measurements along the entire course of the gorge indicate that the slope of the walls ranges from 5° to 15°. This variation is random.

The regional morphology of the Meganos channel is outlined in Figure 1. The shale isopachous contours show that the channel depth increases in a downstream direction. The width also increases gradually in this direction, thus producing an overall gorge morphology very similar to present-day offshore channels. It was not possible to make a detailed interpretation of the channel configuration in the vicinity of the Midland fault because of the sparse well control in this region. However, as more subsurface information becomes available, it would be expected that some effect of the Midland fault on the gorge morphology will be noted.

Three electric-log cross sections (Figs. 3–5) have been constructed across the Meganos chan-

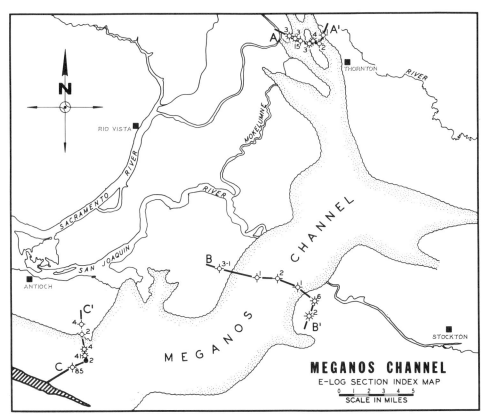

Fig. 2.—Location of electric-log sections and wells used, Meganos channel, Sacramento Valley, California. Cross section A-A′ = Figure 3; cross section B-B′ = Figure 4; and cross section C-C′ = Figure 5.

nel and their locations are shown in Figure 2. The erosional origin and later infilling of shale sediments is well demonstrated by these sections, especially Figures 4 and 5.

LITHOLOGY OF CHANNEL FILL

Silty shale and shaly siltstone comprise more than 95 per cent of the channel-fill sediments. In some places glauconitic units are found. Though most of the siltstone is confined to the basal sedimentary units, the overall fill sequence is remarkably uniform. This dominant fine-grained lithologic character is readily recognized by the nondescript spontaneous potential curve on electric logs.

In some areas, especially near the "headwaters," sandstone units of measurable thickness are recorded. These sandstone units appear to be restricted in their geographic development and, like the siltstone, are found generally in the older part

of the channel section. To date no production has been discovered in these sandstones.

The principal shale lithologic features contrast sharply with the underlying section into which the channel was eroded. Except for a minor sandstone and shale sequence west of the Midland fault, the channel has been eroded into massive sandstone units. This abutment of channel shale with regional sandstone makes recognition of the channel floor possible from electric logs alone. However, in the areas of deepest channel erosion, the channel shale is in juxtaposition with Upper Cretaceous shale. Here the base of the gorge can be recognized best by paleontologic studies.

GEOLOGIC HISTORY OF CHANNEL FORMATION

During the middle of the Paleocene Epoch, tectonic stresses caused initial movements along the Midland fault, a major structural feature of

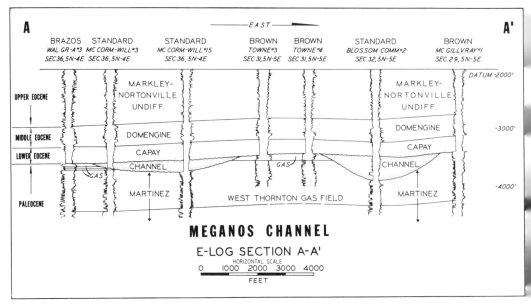

FIG. 3.—West-east cross section A-A', showing erosional isolation of island of older sandstone at West Thornton gas field. Location of section shown on Figure 2.

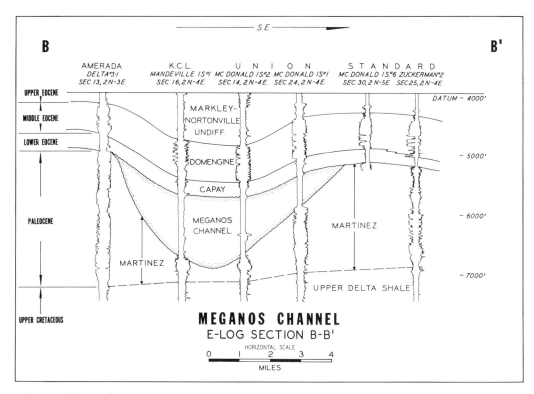

FIG. 4.—West-northwest—south-southeast cross section B-B'. Section indicates magnitude of differential compaction of channel shale beds. Location of section shown on Figure 2.

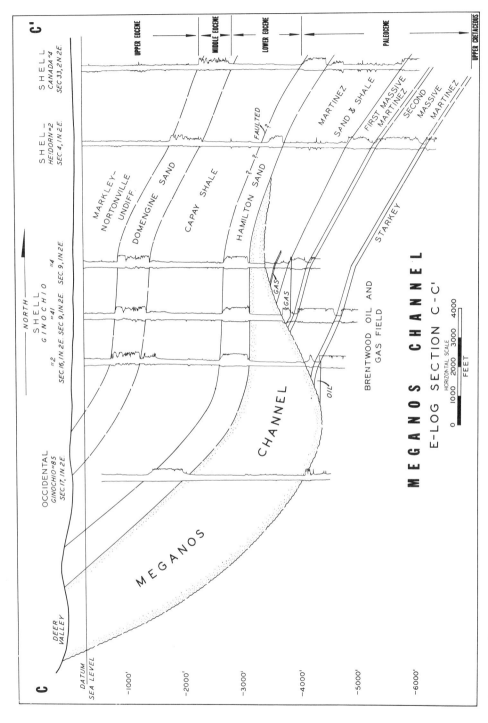

FIG. 5.—South-north cross section C-C'. Section shows updip entrapment of oil and gas by channel shale fill at Brentwood oil and gas field. Location of section shown on Figure 2.

the central Sacramento Valley (Fig. 1). This north-south-striking, down-to-the-west, normal fault was intermittently active through the Eocene Epoch.

Restored regional cross-section studies indicate that the first vertical movements along this fault took place just before the formation of the channel. Because of the relationship in the timing of these two events, it is believed that the faulting directly contributed to conditions favorable for channel formation.

The initial declivity caused by rupture along the Midland fault is believed to have increased local slope to the extent that turbidity currents were triggered. These turbidity currents probably were localized by ocean-bottom topography and a major onshore river system, such as an ancestral Sacramento River. The evidence of any river system has been destroyed by basin-edge erosion. The turbidity currents flowed westward down the footwall of the fault and initially eroded massively bedded sandstone. The relatively high velocity of these currents, and the abrasive nature of their suspended bed load quickly eroded a trough across the fault scarp. This trough was the beginning of the Meganos channel.

In a short period of time, lateral extension took place both landward toward the east and north, and seaward toward the west, by continued sporadic turbidity-current activity. As this development proceeded, the channel became a focal point for other turbid currents developing in the area. These added currents hastened vertical erosion. Tributary channels soon formed and flowed into the trunk channel, thus greatly altering its overall outline.

Eventually, a balance was established between discharge, on one hand, and gradient and current velocity, on the other. The resulting gorge profile established a state of erosive equilibrium and the further extension of this feature was retarded. A period of transgression of the area followed the erosional cycle and initiated a period of rapid infilling of the channel and the deposition of a thin shale section outside of the channel. Paleobathymetric analysis by M. Polugar[3] of the foraminiferal suite found within the channel fill suggests that the fill was deposited in a marine envi-

ronment ranging in depth from neritic to upper bathyal.

The shales of Meganos channel age which were deposited outside of the limits of the channel generally are 100 feet thick, a figure which is relatively constant. As a result, the channel edge was established arbitrarily at the 100-foot isopachous contour of the Meganos shale. Because of post-channel erosion, this blanket of Meganos shale is present only on the hanging-wall side of the Midland fault.

The presence of a thin layer of Meganos shale outside of the channel confines indicates that deposition was taking place throughout the area at the same time that the eroded channel was filling. The thinness of the shale section outside of the channel is explained by the very process of channel infilling. It is postulated that offshore currents were continually sweeping the sediments from the ocean floor and infilling the channel cavity. Therefore, only a very thin layer of sediments could build up outside of the channel as the channel itself was infilled from the sides. This method of infilling submarine channels might actually be the rule rather than the exception, because the phenomenon of a thin section of channel-age sediments outside of the channel confines has been noted elsewhere (Hoyt, 1959). After infilling of the erosional channel, normal sedimentation continued in the area. The overburden of the post-channel sediments caused differential compaction of the channel shale.

DIFFERENTIAL COMPACTION OF CHANNEL SHALE

The marine environment of the channel shale and the high initial porosity inherent to shale caused the fine-grained clastics which originally filled the eroded channel to have a very high volume of water. Weller (1960, p. 297) indicates that the initial porosity values for fine-grained marine clay, which constitutes most of the channel fill, can exceed 80 per cent, whereas the initial porosity of sandstone is estimated by Weller (1960, p. 292) to be approximately 37 per cent. Gradual elimination from the shale of the water and the closer packing of the sedimentary particles as a result of the pressure exerted by the accumulating overburden resulted in greater compaction of the channel shale relative to the arenaceous section. This is well displayed in Figure 4.

[3] Standard Oil Co. of California, Western Operations, Inc.

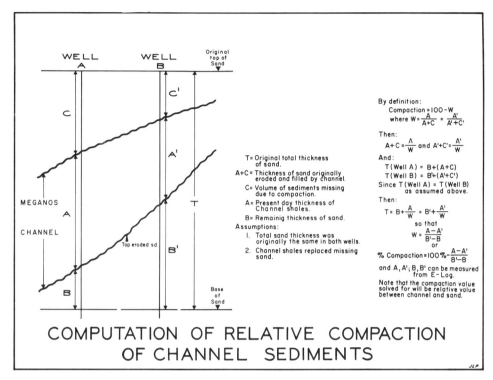

FIG. 6.—Method for computing relative compaction of channel shale beds.

The magnitude of the differential compaction of the channel sediments was computed along the length of the channel, and representative values are shown in Figure 1. In general, the differential-compaction values for the channel shale range from 60 to 35 per cent.

The simplest way to determine the differential compaction of the channel shale would be to compare a channel well with a non-channel well where the two are located close to each other and where both penetrated adequate section below the channel base to have reached good pre-channel correlation markers. Unfortunately, these conditions were not found in many places, so that it was necessary to devise the method outlined in Figure 6. The accuracy of this method should be as good as that of comparing a channel with a non-channel well.

COMPARISON OF MEGANOS CHANNEL WITH
OTHER SUBSURFACE AND SUBMARINE
CHANNELS

Much has been written on the origin and morphology of both ancient and modern submarine channels. It is not the writers' intent to present a detailed comparison between these many documented features and the Meganos channel. However, in reviewing the literature, certain aspects of the Meganos channel in relation to documented subsurface and modern channels should be noted.

One of the most outstanding examples found, and certainly the most interesting from the viewpoint of comparison with the Meganos channel, is the Yoakum channel (Hoyt, 1959). This mid-Eocene feature is in the central Gulf Coast of Texas. From the viewpoint of geometry, age of gorge erosion, thickness, lithologic character of the sedimentary fill, and the presence of a regional thin blanket of shale which is correlatable with the channel shale, the Meganos and Yoakum channels are unusually similar. In fact, the only apparent difference between these two features is the well-documented tributary system associated with the Meganos channel and the striking absence of such associated with the Yoakum channel.

In comparing the Meganos channel with existing submarine valleys, one must picture the Meganos channel before deposition of the shale fill. Just after complete erosion of the Meganos channel and before sedimentary infill, this gorge resembled many of the well-documented, present-day submarine canyons. One of the largest and best known of these is the Congo Canyon off the mouth of the Congo River on the western coast of Africa. With a main-branch length of approximately 200 miles, a maximum width of 5 miles, and a maximum depth below the canyon rim of 3,000 feet, the physical parameters of the Congo Canyon are similar to those of the Meganos channel. The erosion mechanism of the Congo Canyon is thought by Heezen *et al.* (1954) to be turbidity currents which were triggered in the confines of the mouth of the Congo River during months of maximum bed-load transport of that river.

ECONOMIC IMPORTANCE OF MEGANOS CHANNEL

Truncation of reservoir rock by channel shale has aided in the entrapment of economic quantities of gas and oil in four fields. These are the Brentwood oil and gas field and the Dutch Slough, River Break, and West Thornton gas fields (Fig. 1). Economically the Brentwood and Dutch Slough fields are classed as major discoveries in California. The Brentwood field, discovered in July, 1962, contains the only economic accumulation of oil discovered in the Sacramento Valley. The Dutch Slough field, a 1963 discovery, is presently ranked as the third largest gas field in California, with total recoverable gas reserves calculated at 300 billion cubic feet (Reedy, 1966).

To date, the thin but porous sandstone beds associated with the channel fill have not produced hydrocarbons.

CONCLUSIONS

There is sufficient subsurface evidence to prove conclusively the existence of the Meganos channel described here. The evidence indicates an erosional origin which is similar to that of the many submarine canyons which are known to exist on the present-day continental shelves. It is postulated that the method of channel infilling which is indicated for the Meganos channel occurred in the past in other areas and could be expected to take place again in the future.

The Meganos channel is not unique in that other fossil channels have been recorded previously in the literature. In fact, it is probable that submarine channeling is a much more common occurrence than generally is realized. This might be especially true if the possibility is considered that, should the depositional fill of a channel resemble the sediments of the formation eroded (as might be expected to occur in many areas), it would be difficult to detect the presence of a channel except through paleontological information.

It is hoped that this paper has added additional information to the study of submarine channels or canyons and possibly has raised some questions regarding the mechanisms of erosion and infilling of the channels. Certainly, the part of the Meganos channel discussed in this paper provides a good study of fossil channels because of the degree of documentation possible from the many wells drilled in the area.

SELECTED BIBLIOGRAPHY

Almgren, A. A., and W. N. Schlax, 1957, Post-Eocene age of "Markley Gorge" fill, Sacramento Valley, California: Am. Assoc. Petroleum Geologists Bull., v. 41, p. 326–330.
Bornhauser, Max, 1948, A possible ancient submarine canyon in southwestern Louisiana: Am. Assoc. Petroleum Geologists Bull., v. 32, p. 2287–2294.
Clark, B. L., and A. O. Woodford, 1927, The geology and paleontology of the type section of the Meganos Formation of California: Univ. Calif. Pub. Dept., Geol. Sci., Bull. 17, p. 63–142.
Edmondson, E. F., 1965, The Meganos gorge: San Joaquin Geol. Soc., Selected Papers, v. 3, p. 36–51.
Frick, J. D., T. P. Harding, and A. W. Marianos, 1950, Eocene gorge in northern Sacramento Valley (abs.): Am. Assoc. Petroleum Geologists Bull., v. 43, p. 255.
Goudkoff, P. P., 1945, Stratigraphic relations of Upper Cretaceous in Great Valley, California: Am. Assoc. Petroleum Geologists Bull., v. 29, p. 956–1007.
Heezen, B. C., R. J. Menzies, E. D. Schneider, W. M. Ewing, and N. C. L. Gionelli, 1964, Congo submarine canyon: Am. Assoc. Petroleum Geologists Bull., v. 48, p. 1126–1149.
Hoyt, N. W., 1959, Erosional channel in the middle Wilcox near Yoakum, Lavaca County, Texas: Gulf Coast Assoc. Geol. Soc. Trans., v. 9, p. 40–50.
Hunter, W. J., 1964, Dutch Slough gas field, *in* Summary of operations—California oil fields: Calif. Div. Oil & Gas, v. 50, no. 2, p. 63–69.
Kuenen, P. H., Marine geology: New York, John Wiley & Sons, 568 p.
Martin, B. D., 1963, Rosedale channel—evidence for late Miocene submarine erosion in Great Valley of California: Am. Assoc. Petroleum Geologists Bull., v. 47, p. 441–456.

Reedy, R. D., 1966, Dutch Slough field typifies Sacramento Valley gas finds: World Oil, January, p. 85–88.

Safonov, Anatole, 1962, The challenge of the Sacramento Valley, California: Calif. Div. Mines and Geology Bull. 181, p. 89–97.

Shepard, F. P., and K. O. Emery, 1941, Submarine topography off the California coast—canyons and tectonic interpretation: Geol. Soc. America Spec. Paper 31, 171 p.

Silcox, J. H., 1962, West Thornton and Walnut Grove gas fields, California: Calif. Div. Mines and Geology Bull. 181, p. 140–148.

Sullivan, J. C., 1963, Brentwood oil field, in Summary of operations—California oil fields: Calif. Div. Oil & Gas, v. 49, no. 2, p. 5–15.

Veatch, A. C., and P. A. Smith, 1939, Atlantic submarine valleys of the United States and the Congo submarine valley: Geol. Soc. America Spec. Paper 7, 101 p.

Weller, J. Marvin, 1960, Stratigraphic principles and practice: New York, Harper and Brothers, 725 p.

Editor's Comments
on Papers 18 Through 21

Papers 18 to 20 deal respectively with examples of Paleozoic, Mesozoic, and Tertiary canyons, and Paper 21 summarizes the present state
of knowledge of ancient canyons and fan valleys.

During the 1950s, members of the Ludlow Research Group were actively mapping in detail areas of Ludlovian (Upper Silurian) rocks in
Wales, where basinal facies occurs, and in the Welsh Borderland, where
there is shelf facies. The area around Leintwardine is intermediate in position and represents mainly a shelf-edge region. It was here that detailed
mapping on 6 inches to the mile and 25 inches to the mile scale revealed
six parallel channels, trending and deepening rapidly from shelf to basin. These have many similarities with modern submarine canyon heads
off California and probably funneled sediment into a basin of Japan Sea
type (Mitchell and Reading, 1971) on the southeastern margin of the
closing proto-Atlantic or "Iapetus Ocean" (McKerrow and Ziegler, 1972;
Harland and Gayer, 1972) in late Silurian time. The Leintwardine channels were probably the first European submarine canyon heads to be described.

As Whitaker's paper (18) is mainly a stratigraphical and paleontological account of the Leintwardine area, much of this part is omitted,
although detailed subdivision of strata on their total fossil content was
important in helping to delimit the channels in this rather poorly exposed region. The parts of the paper reprinted here deal with the evidence on which the six canyon heads were reconstructed, especially
their geometry and the primary sedimentary structures of the fill. The

complexity of the channel-fill faunas, where four different assemblages may be intimately associated, is dealt with more fully in a later article by Whitaker (Paper 21). Paleoecological aspects of the channels and interchannel areas are discussed by Whitaker (1963) and Jones (1969). In the paper, eight-figure references are to the British National Grid system. The genus *Conchidium* should now read *Kirkidium* throughout.

Officers of the Institute of Geological Sciences have recognized middle Wenlock to lower Ludlow canyons in North Wales. These are directed toward the northeast and east, exactly opposite to those at Leintwardine. Slumping from north and south occurred, and turbidity flows carried sands from the canyon mouths during lower Ludlow times. The Wenlock canyons contain an unusual fauna comparable with that at Leintwardine (P. Warren, personal communication, 1975).

Important work by Italian geologists on Mesozoic and Tertiary canyons and fans is exemplified by a short paper on Middle Jurassic channels by A. Bosellini, published in Italian in 1967. A translation, kindly made by Hugh Jenkyns, is included here (19). The well-exposed, narrow, deeply cut gullies are illustrated in Plate III and their wall contacts and fills of breccia in Plate II. The shape of the canyon in Plate III, fig. 2, bears close resemblance to the cross section of a typical modern canyon tributary at its seaward extremity, as illustrated in Paper 6 by Dill (1964b, Paper 6, fig. 3.7).

Since 1967, further work has slightly modified Bosellini's conclusions (written communication, 1974). Only the Lower Jurassic breccias, slumps and slides (Plate I, fig. 1; Plate III, fig. 1) have a northwest provenance. Gullies at the base of the Middle Jurassic Vajont Limestone (Plate II, figs. 1, 2; Plate III, fig. 2) are cut into the Lower Jurassic by a very large submarine fan of oolitic limestones migrating from the southeast, as indicated by the five black arrows on Fig. 1, modified by Bosellini from the original figure: the dashed line marking the facies change has also been moved slightly. The Middle Jurassic fan of oolitic grains is described by Bosellini and Masetti (1972): it is perhaps the largest ancient carbonate fan known.

The discovery of oil in the shelf regions of Australia has led to intensive study of parts of these continental margins, especially around the southern and northwestern parts of the continent and around New Guinea. Many large modern submarine canyons have been located and studied (e. g., Hopkins, 1966; Conolly and Von der Borch, 1967; Conolly, 1968, Paper 20; Von der Borch, 1968). One interesting result of the seismic profiles has been the discovery of a number of buried canyons of various ages within the Tertiary strata, cut and filled and then entirely covered with later shelf deposits. Hopkins (1966) shows two excellent profiles about 25 miles (40 km) southwest of Portland, Victoria. His figure 4 (Otway Basin profile) reveals a channel 3 miles (4.8 km) wide and approximately 2800 ft (850 m) deep, cut and filled during the late Tertiary. His figure 5, a seismic line across the Gippsland Basin, shows a canyon eroding Latrobe Coal Measures and infilled by Oligocene shale. Houtz and others (1967) and Von der Borch and others (1970) also discuss possible buried canyons off New Zealand and southern Australia, respectively.

Owing to space limitations, we have had to select mainly those sections of Conolly's paper (20) that deal with ancient canyons, brought to notice by seismic profiling. The confidential files of many oil companies must have similar fascinating profiles, for filled and buried canyons are one type of shoestring body that may act as hydrocarbon reservoirs (see Papers 16 and 17). On page 315, the Von der Borch reference should read "Southern Australian submarine canyons: their distribution and ages."

As the scattered literature on ancient canyons grew, it became necessary to summarize their diagnostic features and the criteria by which they could be recognized. This was first done by Stanley (1967, 1969) and later by Stanley and Unrug (1972). The most recent attempt to review all known ancient submarine canyons and fan valleys and to list criteria for identifying them was published by Whitaker (1974) and is reproduced in part here (21). The large Table 1 does for ancient examples what Shepard and Dill's (1966, Paper 8) Appendix does for modern canyons, giving as full information as possible on the geometry, dimensions, age, and sedimentology of all well-described ancient submarine canyons and fan valleys. The comprehensive reference list (with recent additions highlighted on pp. 7–8) should be useful to readers wishing to pursue this topic. The article sums up the material dealt with Part III and, in the few fan valleys mentioned, looks ahead to Part IV.

Biographical Notes

John H. McD. Whitaker was born in 1921 in Cambridge, England. He received the B.A. and M.A. degrees at Cambridge University, the B.Sc. from London University, and the Ph.D. from Leicester University. After three years as Assistant Lecturer at the University of Manchester, he moved to the University College (later the University) of Leicester in 1951 to start a new Department of Geology: he is currently Senior Lecturer there. His main research interests are in the Silurian and lower Devonian stratigraphy and sedimentology of the Welsh Borderlands and southern Norway, where he worked for 14 months as a postdoctoral Fellow with the Royal Norwegian Council for Scientific and Industrial Research. He is also working on the lower Tertiary flysch of Japan, where he recently spent three months on a Royal Society Study Visit.

Finding ancient submarine canyon heads in the Silurian of the Welsh Borderlands, as outlined in Paper 18, led him to a general study of ancient canyons and fan valleys, and he has had the opportunity of visiting examples of these features in California and Japan. His review of pre-Pleistocene submarine canyons and fan valleys follows in Paper 21.

Born in 1934 at Mantua, Italy, Alfonso Bosellini became doctor in Geological Sciences at the University of Padua in 1959. He received the *Libera docenza* (Ph.D. equivalent) in Geology in 1966 and in Sedimentology in 1969. He is Chief Research Scientist of the National Research Council and Professor of Sedimentology at the University of Ferrara. His fields of specialization include ancient carbonate sedimentology, paleogeography, and paleotectonics of the Southern Alps. Currently he is working on deep-water carbonates of ancient continental margins. Bosellini held a NATO postdoctoral Fellowship at the Johns Hopkins University in 1968–1969, working with R. N. Ginsburg, and was recently awarded a senior NATO Fellowship for three months at Scripps Institution of Oceanography in 1974. Bosellini is currently on the Editorial Board of *Sedimentology.*

John R. Conolly was born in Sydney, Australia, in 1936. He gained the B.Sc. in 1958 at the University of Sydney and the M.Sc. (1960) and Ph.D. (1963) at the University of New South Wales. Since then he has traveled widely, participating in many oceanographic expeditions, first as Research Scientist at Lamont Geological Observatory, Columbia University, then as Visiting Professor of Geology at Louisiana State University and Research Associate at Scripps Institution of Oceanography, University of California at San Diego. In 1966, Conolly returned to Australia as Queen Elizabeth II Fellow at the University of Sydney, later becoming consultant geologist in that city. He returned to the United States in 1969 as Associate, later full, Professor in the University of South Carolina and consultant to the U.S. Navy. After acting as District Geologist with BP Alaska, Inc., Conolly has now become a consultant. He is a member of 11 professional societies. His many publications reflect an extensive knowledge of modern and ancient sediments from many parts of the world.

18

Copyright © 1962 by the Geological Society of London

Reprinted from *Quart. Jour. Geol. Soc. London*, **118**, Pt 3, 319–320, 327–334, 339–341, 346–347 (Sept. 1962)

THE GEOLOGY OF THE AREA AROUND LEINTWARDINE, HEREFORDSHIRE

BY JOHN HARRY McDONALD WHITAKER, M.A. B.SC. F.G.S.

Submitted 20 June 1960 ; revised manuscript received 12 January 1962 ; read 23 November 1960

[PLATE XIV]

CONTENTS

SUMMARY

Wenlockian, Ludlovian, and lowest Downtonian strata have been mapped over an area of about ten square miles east of Leintwardine, where the structure is an asymmetrical syncline plunging east-north-east. The southern limb dips northward at about 35°, the northern limb eastward at about 8°.

Wenlock Shale and Wenlock Limestone (together 1560 feet thick) are succeeded by up to 1900 feet of Ludlovian rocks, divided into Elton, Bringewood, Leintwardine, and Whitcliffe Beds (Holland, Lawson & Walmsley 1959). Where complete, the succession is similar to that at Ludlow, but is thicker and less calcareous and with more evidence of instability (crinkle-marks and slump-structures) reflecting deposition nearer to the edge of the shelf. Basin elements in the fauna support this view of the palaeogeography. The lowest Downtonian strata (140 feet mapped) compare closely with those of neighbouring areas.

There is one major break in the sequence, within the Lower Leintwardine Beds ; six parallel channels, trending and deepening towards the west-south-west (i.e. from shelf to basin), are interpreted as Ludlovian submarine canyon-heads and show many features in common with their modern counterparts. Later these tended to fill up, with occasional 'flushing out', until they were eliminated by Upper Leintwardine and later sedimentation.

[*Editor's Note:* Material has been omitted at this point.]

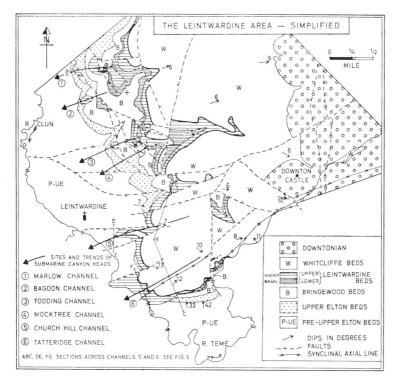

FIG. 1.—Simplified geological map of the Leintwardine area, showing the location of the submarine canyon-heads.

At several localities, lying in six belts trending ENE.–WSW., the normal succession of the Lower Leintwardine Beds is not developed and the higher members rest with a distinct and often striking break upon basal Lower Leintwardine Beds, Bringewood Beds, Upper Elton Beds, and Middle Elton Beds (Fig. 1). In each belt the break becomes more pronounced towards the west-south-west, and as much as 680 feet of the estimated normal succession may be cut out. These belts are interpreted as the sites of former submarine channels, cut and mainly filled during Lower Leintwardine times. They trend and deepen from the shelf area towards the basin, and are comparable with the heads of modern submarine canyons.

1. The Marlow channel. Fairly continuous exposures on both sides of the sunken lane from Marlow farm to Woodhead show normal strike-faults dividing the section into fault-blocks (Fig. 2a). The most westerly block shows slightly slumped and crinkle-marked

Lower Leintwardine Beds lying on basal Upper Elton Beds ; in the next block they lie on Lower Bringewood Beds ; in the next they are missing at the western end (where Upper Leintwardine Beds rest on Upper Bringewood Beds with phosphatic pebbles at the junction, which is otherwise not well defined) but at the eastern end of this block, and in the most easterly block, beyond Woodhead, they rest in normal succession upon the basal subdivision. There is therefore a double unconformity, below and above the upper subdivision. This is interpreted on p. 340. The gradient of the channel axis (Fig. 2b) is approximately 7° or 12 per cent.

2. The Bagdon channel. This is the least well-preserved channel ; most of its infilling has been eroded away, so that only a small oval outlier of higher Lower Leintwardine Beds appears on the map south of Hollow Bagdon ; its length runs ENE.–WSW., parallel with the other channels.

3. The Todding channel. This, like the Marlow channel, is complicated by faulting, but there are excellent exposures along the old Leintwardine–Ludlow road (Fig. 3). This interpretation differs from Alexander's (1936, pl. viii, inset ii) : the evidence for another channel here seems to the author to be particularly strong.

Small quarries at 4103 7537, north of the old road, show Upper Elton Beds overlain by a few feet of Lower Bringewood Beds and then, with discordance of dip and strike, very fossiliferous Lower Leintwardine Beds, which contain derived Upper Bringewood limestone cobbles and small boulders bearing *Conchidium knightii*. One boulder, 18 by 12 inches, rests at the bottom of the channel on weathered, dusky yellow siltstones yielding current-strewn *Lingula lata*, with a two-inch band of bentonitic clay and calcite. The overlying layers are crowded with *Dayia navicula*, which show slight orientation of their longer dimension along a bearing of 60°.[1] Prodcasts (Dzulynski & others 1959, p. 1116), formed by current-transported *Dayia* skidding to rest in the soft sediment, show similar orientation. A slumped bed at this locality, its eroded top covered with many fossils, suggests movement from 40° to 220°.

At 'Martin's Shell', a disused quarry at 4109 7543 (Fig. 3), oriented *Sphaerirhynchia wilsoni* (a flat variety comparable with those at Church Hill) have their umbones pointing up-channel; grooves filled with shell-debris (including *C. knightii*) have a bearing of 73°. Here there are at least three horizons of rounded cobbles and small boulders between an inch and a foot in length, consisting of fairly pure limestone, (some with *C. knightii*, *S. wilsoni*, and fragments of other shells), derived from the Upper Bringewood Beds. These boulder-beds are separated by fine-grained laminated siltstones which are depressed below the boulders and curved over their tops. Before Alexander's paper of 1936 the boulder-beds were thought to be 'Aymestry Limestone' *in situ* and there was difficulty in interpreting the fauna (Lightbody 1863; Symonds *in* Woodward 1866–78; Hawkins & Hampton 1927). A reasonable explanation is that the Lower Leintwardine Beds present at 'Martin's Shell', as at Church Hill, occupy former submarine channels, and that the blocks are derived

---

[1] All bearings have been corrected for magnetic declination.

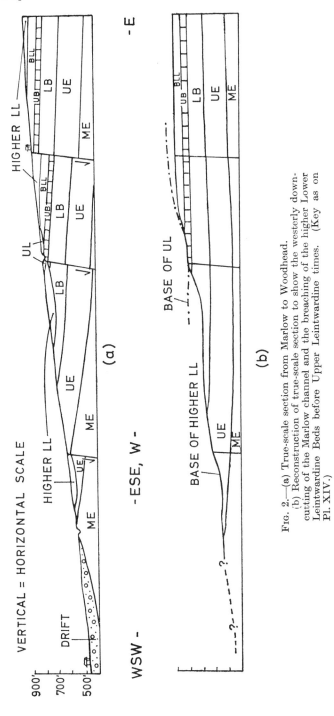

Fig. 2.—(a) True-scale section from Marlow to Woodhead. (b) Reconstruction of true-scale section to show the westerly down-cutting of the Marlow channel and the breaching of the higher Lower Leintwardine Beds before Upper Leintwardine times. (Key as on Pl. XIV.)

from nearby Upper Bringewood Beds of 'Aymestry Limestone' facies. Some of the flatter cobbles are set in 'imbricate' (pebble-stacked) fashion, best seen at 4107 7540. Only a few dips of the longer axes can be measured : these are towards the north-east.

At the small quarry north-east of Wassell Barn (4164 7561) (Fig. 3) a number of grooves filled with shell debris have a mean trend of 66° and a prod-cast has a similar trend.

A rich and remarkable fauna has been reported from the fine-grained siltstones between the boulder-beds at 'Martin's Shell' (Marston 1865; Hawkins & Hampton 1927). It includes many star-fish, crinoids, eurypterids, phyllocarids, and some echinoids and worms, as well as the usual *S. wilsoni*, *D. navicula*, *Camarotoechia nucula*, *M. leintwardinensis*, and trilobites. Phosphatic calculi of polyzoans (Oakley 1934) have also been found.

4. The Mocktree channel. Only two fragments of this channel remain, but they are of interest as they show conditions near the head of a channel. A shallow channel section in the most easterly fault-block of the large main-road quarries (4167 7537) (Fig. 3) was figured and described by Lightbody (1863) and photographed by Woodward & Dixon (1904, pl. 42, fig. 1 ; the base of the channel is seen two-thirds of the way up the photograph, a little below the line marked on the transparent overlay). The channel is 80 feet across and its axis trends at 51°. Below the channel basal Lower Leintwardine Beds crowded with *D. navicula* rest on massive Upper Bringewood Limestone. The lowest part of the channel cuts out all but two feet of the basal Lower Leintwardine Beds. The channel-fill consists of successively overlapping catenary-bedded higher Lower Leintwardine Beds ; they have the varied and characteristic fauna of the other channels, including a starfish bed. Similar but not very fossiliferous siltstones are faulted down to floor-level in the extreme east of the quarry.

A nearby small exposure on the north side of the main road at 4163 7531 displays a nearly vertical wall of Upper Bringewood lime-stone against which higher Lower Leintwardine Beds are banked with high original dip (23° at 140° after correcting for regional tilt), which falls off rapidly eastwards from the margin towards the centre of the channel. The channel has here cut down rather more deeply than at the previous locality and a steep 'canyon wall' is exposed (compare Shepard 1949, pl. 2, fig. 1). Unfortunately the lower parts of this channel are lost by erosion.

5. The Church Hill channel. Church Hill, half a mile east of Leintwardine, has long been famous for its fauna, but was not recognized as the site of a channel until the publication of Alexander's 1936 paper. Alexander inferred (p. 110) that *Dayia* beds (Lower Leintwardine Beds of this paper) filling a channel rested discordantly on the much lower *nilssoni-scanicus* mudstones (Middle Elton Beds of this paper), the actual contact being then obscured. An excellent contact has since come to light in Trippleton Lane (4115 7372) (Fig. 4) which fully confirms her deductions.

The northern margin is largely faulted out by an east–west fault, but outcrops of higher Lower Leintwardine Beds of the channel-fill

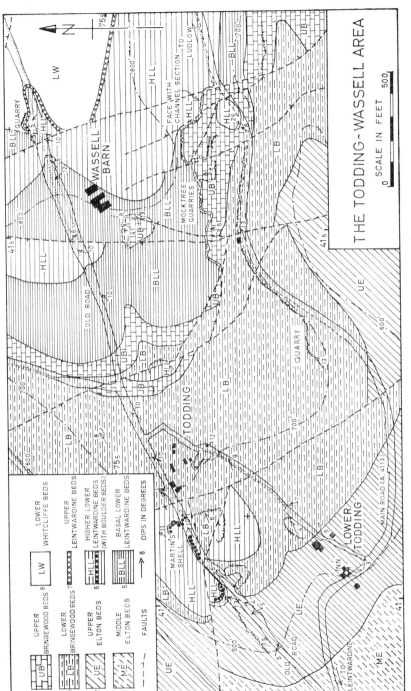

Fig. 3.—Detailed map of the Todding channel.

and the Middle Elton Beds are never far apart. The westerly prolongation of these beds from the main outcrop (Pl. XIV) is accentuated by the east–west faulting.

The marked downcutting of this channel was indicated by Alexander (1936, pl. viii). An estimated 580 feet of beds below the higher Lower Leintwardine Beds are cut out in a horizontal distance of only 3500 feet along the channel axis, giving a gradient of 10° or 17 per cent. Reconstructions across this channel are shown in Fig. 5.

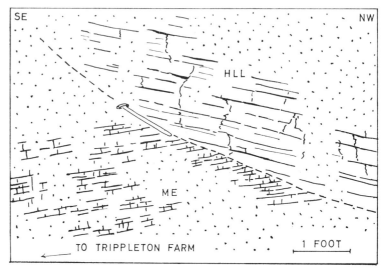

FIG. 4.—Southern margin of the Church Hill channel. Catenary-bedded higher Lower Leintwardine Beds (HLL) resting unconformably on Middle Elton Beds (ME). (4115 7372.)

The channel-fill consists of catenary-bedded, finely laminated calcareous siltstones with occasional slump-sheets. Well-developed cobble- and boulder-beds occur not far above the base of the channel at 4106 7374, with rounded boulders of Upper Bringewood limestone measuring up to 15 by 13 inches, and bearing *C. knightii* and *Ptilodictya lanceolata*. The cobbles and boulders lie mainly with their shortest axes vertical, but a few are imbricate as at 'Martin's Shell' (p. 330) and exhibit the same displacement of the siltstones above and below.

The bedding-planes of the laminated siltstones often show orientational features, such as ripple-marks, aligned *M. leintwardinensis* (Hawkins & Hampton 1927, p. 579), skip-casts (Dzulynski & others 1959, p. 1117), and specimens of a flat variety of *Sphaerirhynchia wilsoni* (compare 'Martin's Shell', p. 328) with their umbones pointing up-channel, bearing 76°.

6. The Tatteridge channel. This is the most extensive of the six channels in its present distribution. It is bounded to the north-north-west by a normal fault with a downthrow of 150 feet to the south and on the south-south-east by a steep erosional margin (Fig.

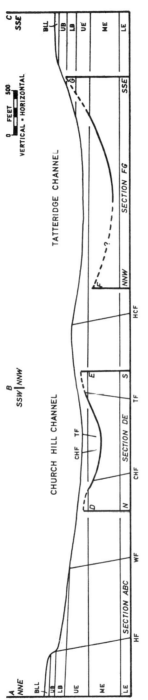

FIG. 5.—Approximately true-scale reconstructions across the Church Hill and Tatteridge channels at the end of basal Lower Leintwardine times (maximum submarine relief). For lines of section see Fig. 1. Key as on Pl. XIV. HF : Haregrove fault ; WF : Whitton fault ; CHF : Church Hill fault ; TF : Trippleton fault ; HCF : Hollybush cottage fault.

5). About 680 feet of lower beds are cut out in a horizontal distance of 3900 feet, giving an axial gradient of 10° or 17 per cent. The channel is filled by a thick series of monotonous calcareous siltstones, rather thickly bedded, sometimes crinkle-marked, and almost devoid of fossils except for an occasional *M. leintwardinensis*. These beds were exposed in 1958 and 1959 during the laying of the fourth pipe of the Elan–Birmingham aqueduct, when a deep trench was excavated across Brandon Hill, the alluvium of the River Teme, and Tatteridge Hill. Northerly dipping Lower Leintwardine Beds near the south margin of the Tatteridge channel were seen on the steep western side of Tatteridge Hill. The trench was a little too far to the north to cut through the margin and expose underlying formations, but detailed mapping at a time when deep ploughing had been carried out showed that the channel must cut out, in turn, Upper and Lower Bringewood Beds and Upper Elton Beds as these are traced westwards. The failure of several small features towards the west is most striking : the continuous exposures in the trench showed that this could not be due to faulting.

As in the other channels, slumps and boulder-beds are found. The slumps are directed into the channel from the northern fault-margin, as seen in the small roadside quarry between Trippleton and Nacklestone (4143 7297). The slumped horizons bear a fauna of slightly earlier date than the indigenous fauna of the undisturbed beds. The fossils in the slumps are often worn and limonitized and bear encrusting *Spirorbis*. The boulder-beds occur just above this disturbed horizon and again much farther down-channel, on the north-west side of Brandon Hill.

These Lower Leintwardine submarine channels are interpreted more fully on pp. 339–41.

[*Editor's Note:* Material has been omitted at this point.]

The Leintwardine channels.—Six parallel channels, trending ENE. –WSW., were initiated in mid-Lower Leintwardine times along a line from Woodhead to Downton-on-the-Rock. Their axial gradients were high (see Fig. 2 and pp. 328, 332, 334) and their side walls were sometimes steeply inclined (Fig. 5). The channels often have normal faults along one or both flanks parallel with the channel axes. These faults may have been initiated in Lower Leintwardine times and may have controlled the siting and trend of some of the channels. The size, shape, and spacing of the channels suggest close comparison with modern submarine canyon-heads. For comparable spacing see, for example, Kuenen (1950, fig. 211, and 1953, fig. 10) and for recon-structions of modern Californian submarine canyon-heads see Shepard 1949, fig. 2 and 1959, fig. 56). The gently sloping heads of modern canyons passing into steep rock walls, the high gradient of the floor, and talus fans with boulders among the fill all have their counterparts in the Leintwardine channels : but 'hanging valleys' and the joining of tributaries have not been found.

The continuation of the channels west of Leintwardine is not yet known in detail : erosion of the Clun and Teme valleys has removed some of the evidence, the Church Stretton disturbance introduces a complication, and beds of comparable age at Bucknell are buried under younger Ludlovian strata (Holland 1959B, p. 471, re-inter-preting Stamp 1919). But Brandon Hill lies along the continuation of the Tatteridge channel, and certain areas of disturbed beds east of Lingen and the boulders at Pedwardine (described as Aymestry Limestone by Cox [1912]) may prove to be sited along the continuations of these or similar channels. Whether the six canyon-heads described were tributaries of a full-scale submarine canyon is not known : but if this were so the canyon should be visualized as much shorter and less deep than many modern canyons, terminating in one of the axial zones of the Welsh geosyncline at no great depth.

The cutting of the canyon-heads, whether by turbidity currents aided by slumping or by other means, was completed in a small part of Lower Leintwardine times. An irregular shelf-edge area (see Fig 5 for a reconstruction of a section through the southern part) began to accumulate sediment of higher Lower Leintwardine age, conformably on the less deep areas between the channels and unconformably within the channels, usually with catenary bedding. Fig. 6 is an attempt to portray an idealized canyon-head with the infilling features found in the various channels. The finely laminated siltstones often formed overfolded slumps directed down-flank or down-channel. The poorly consolidated canyon walls, some of them possibly contemporaneous submarine fault-scarps, collapsed from time to time, resulting in spreads of boulders, sometimes imbricate down-flank or down-channel (compare Shrock 1948, p. 11, fig. 2 and pp. 254–7, and Pettijohn 1957, p. 78, but contrast Holland 1959A, p. 235). Marine currents moving down-channel with minor movements up-channel formed ripple-marks normal to the channel axis (compare Shepard 1949) and induced the formation of prod-casts, skip-casts, and grooves filled with shell debris parallel with the channel axis. Currents also oriented *Monograptus* stipes and brachiopods with their umbones pointing up-channel. The canyon-heads seem to have sheltered a greater variety of fauna than the more exposed shelf area.

Occasional rather drastic 'flushing-out' of the infilling sediments (compare Shepard 1932 and 1951 for modern canyons) is postulated from the evidence in the Marlow channel ; Fig. 2b shows Upper Leintwardine beds resting on Upper Bringewood Beds at the point where the latter made a step along the channel thalweg. Partial 'flushing-out' within the Todding channel is again deduced from Upper Leintwardine Beds resting on truncated higher Lower Leintwardine Beds with worn, derived fossils at the junction (p. 334) The largest channel (Tatteridge) does not seem to have become entirely filled until Upper Whitcliffe times (see p. 337).

Evidence for instability in nearby basin areas in Lower Leintwardine times is given by slumping at Kerry and south-west Clun (Earp 1938 and 1940), by some of the boulder-beds at Lingen and those at Pedwardine (see above), and by similar boulder-beds and slumps from the Presteigne area (Kirk 1951 and personal communication). At Knighton (Upper Bailey Hill Beds, Holland 1959A and 1959B) the only evidence for instability is the development of 'turbidites', while at Bucknell, where 'turbidites' might also be expected, the beds are buried under higher Ludlovian strata (Holland 1959B, p. 471).

After the channel-cutting and filling episode at Leintwardine, conditions became more uniform and the Upper Leintwardine siltstones, faunally linked with the basin areas, were laid down.

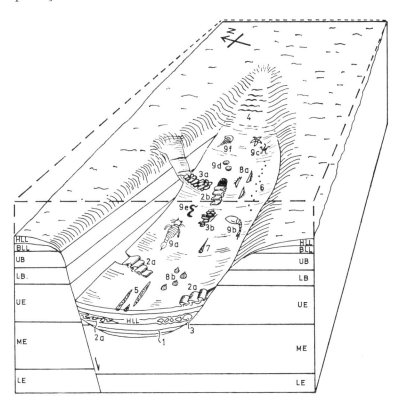

Fig. 6.—Diagrammatic reconstruction of a submarine canyon-head of higher Lower Leintwardine times. Data from various channels. Not to scale.

Key to beds as on Pl. XIV. Key to channel contents :

1 : Catenary-bedding of channel fill.
2 : Slump-structures : (a) Down-flank.
                    (b) Down-channel.
3 : Boulder-beds :   (a) Imbricate down-flank.
                    (b) Imbricate down-channel.
4 : Ripple-marks at right-angles to axis.
5 : Grooves filled with broken fossils, parallel to axis.
6 : Skip-casts parallel with axis.
7 : Prod-casts parallel with axis.
8 : Oriented fossils :   (a) *Monograptus leintwardinensis* parallel with
                            axis.
                   (b) *Sphaerirhynchia wilsoni* and *Camarotoechia*
                          nucula, umbones up-channel.
9 : Unusual fauna concentrated in channel :
                          (a) Eurypterida.
                          (b) Phyllocarida.
                          (c) Asterozoa.
                          (d) Echinoidea.
                          (e) Annelida.
                          (f) Xiphosura.

[*Editor's Note:* The "List of Ludlovian Fossils" has been omitted.]

VIII. LIST OF REFERENCES

ALEXANDER, F. E. S. 1936. The Aymestry Limestone of the main outcrop. *Q.J.G.S.* **92**, 103–14.

ALLENDER, R., C. H. HOLLAND, J. D. LAWSON, V. G. WALMSLEY & J. H. McD. WHITAKER. 1960. Summer field meeting at Ludlow, 1958. *Proc. Geol. Assoc.* **71**, 209–32.

BOSWORTH, T. O. [1912]. *The Keuper Marls around Charnwood*. Leicester.

COX, A. H. 1912. On an inlier of Longmyndian and Cambrian rocks at Pedwardine (Herefordshire). *Q.J.G.S.* **68**, 364–72.

CROSFIELD, M. C. & M. S. JOHNSTON. 1914. A study of ballstone and the associated beds in the Wenlock Limestone of Shropshire. *Proc. Geol. Assoc.* **25**, 193–224.

DENISON, R. H. 1956. A review of the habitat of the earliest vertebrates. *Fieldiana : Geology Memoirs*, **11** (8), 359–457.

DZULYNSKI, S., M. KSIAŻKIEWICZ & PH. H. KUENEN. 1959. Turbidites in flysch of the Polish Carpathian Mountains. *Bull. Geol. Soc. Am.* **70**, 1089–1118.

EARP, J. R. 1938. The higher Silurian rocks of the Kerry district, Montgomeryshire. *Q.J.G.S.* **94**, 125–57.

——. 1940. The geology of the south-western part of the Clun Forest. *Q.J.G.S.* **96**, 1–11.

ELLES, G. L. & I. L. SLATER. 1906. The highest Silurian rocks of the Ludlow district. *Q.J.G.S.* **62**, 195–221.

HAWKINS, H. L. & S. M. HAMPTON. 1927. The occurrence, structure, and affinities of *Echinocystis* and *Palaeodiscus*. *Q.J.G.S.* **83**, 574–601.

HILLS, E. S. 1953. *Outlines of structural geology*. (3rd ed.) London.

HOLLAND, C. H. 1959A. On convolute bedding in the Lower Ludlovian rocks of north-east Radnorshire. *Geol. Mag.* **96**, 230–6.

——. 1959B. The Ludlovian and Downtonian rocks of the Knighton district, Radnorshire. *Q.J.G.S.* **114** [for 1958], 449–78.

——. 1961. Origin of convoluted laminae. *Geol. Mag.* **98**, 168.

——, J. D. LAWSON & V. G. WALMSLEY. 1959. A revised classification of the Ludlovian succession at Ludlow. *Nature*, **184**, 1037–9.

KIRK, N. H. 1951. The Silurian and Downtonian rocks of the anticlinal disturbance of Breconshire and Radnorshire : Pont Faen to Presteigne. *Proc. Geol. Soc.* **1474**, 72–4.

KUENEN, PH. H. 1950. *Marine geology*. New York.

——. 1953. Origin and classification of submarine canyons. *Bull. Geol. Soc. Am.* **64**, 1295–1314.

LAWSON, J. D. 1955. The geology of the May Hill inlier. *Q.J.G.S.* **111**, 85–113.

LIGHTBODY, R. 1863. Notice of a section at Mocktree. *Q.J.G.S.* **19**, 368–71.

MARSTON, A. 1865. A short geological guide, 10 miles round Ludlow. *In* J. T. Irvine. *Handbook to Ludlow*, 143–68. (3rd ed.) Ludlow.

MORRIS, J., J. D. LA TOUCHE & R. LIGHTBODY. 1874. Excursion to Ludlow and the Longmynds, 1872. *Proc. Geol. Assoc.* **3** [1872–1873], 124–7.

MURCHISON, R. I. 1839. *The Silurian System*. London.

OAKLEY, K. P. 1934. Phosphatic calculi in Silurian polyzoa. *Proc. Roy. Soc.* B, **116**, 296–314.

PETTIJOHN, F. J. 1957. *Sedimentary rocks.* (2nd ed.) New York.

SANDERS, J. E. 1960. Origin of convoluted laminae. *Geol. Mag.* **97**, 409–21.

SHEPARD, F. P. 1932. Landslide-modifications of submarine valleys. *Trans. Am. Geophys. Union*, **13**, 226–30.

——. 1949. Terrestrial topography of submarine canyons revealed by diving. *Bull. Geol. Soc. Am.* **60**, 1597–1612.

——. 1951. Mass movements in submarine canyon heads. *Trans. Am. Geophys. Union*, **32**, 405–18.

——. 1959. *The earth beneath the sea.* London and Baltimore.

SHROCK, R. R. 1948. *Sequence in layered rocks.* New York.

SPENCER, W. K. 1914–1940. The British Palaeozoic Asterozoa, parts I–X. *Palaeont. Soc.* [*Monogr.*].

——. 1950. Asterozoa and the study of Palaeozoic faunas. *Geol. Mag.* **87**, 393–408.

SQUIRRELL, H. C. & E. V. TUCKER. 1960. The geology of the Woolhope inlier, Herefordshire. *Q.J.G.S.* **116**, 139–81.

STAMP, L. D. 1919. The highest Silurian rocks of the Clun-Forest district (Shropshire). *Q.J.G.S.* **74**, [for 1918], 221–44.

STRAW, S. H. 1937. The higher Ludlovian rocks of the Builth district. *Q.J.G.S.* **93**, 406–53.

TUCKER, E. V. 1960. Ludlovian biotite-bearing bands. *Geol. Mag.* **97**, 245–9.

WALMSLEY, V. G. 1959. The geology of the Usk inlier (Monmouthshire). *Q.J.G.S.* **114** [for 1958], 483–516.

WHITE, E. I. 1950. The vertebrate faunas of the Lower Old Red Sandstone of the Welsh Borders. *Bull. Brit. Mus.* (*Nat. Hist.*) **1** (3), 51–67.

WILLIAMS, B. J. & J. E. PRENTICE. 1958. Slump structures in the Ludlovian rocks of north Herefordshire. *Proc. Geol. Assoc.* **68** [for 1957], 286–93.

WILLIAMS, E. 1961. Origin of convoluted laminae. *Geol. Mag.* **98**, 168–70.

WILSON, V. 1958. In *Sum. Progr. Geol. Surv. G.B. for 1957*.

WOOD, E. M. R. 1900. The Lower Ludlow Formation and its graptolite-fauna. *Q.J.G.S.* **56**, 415–91.

WOODWARD, A. S. & E. E. L DIXON. 1904. Long excursion to the Ludlow district. *Proc. Geol. Assoc.* **18**, 487–91.

WOODWARD, H. 1866–78. British fossil crustacea belonging to the Order Merostomata. *Palaeont. Soc.* [*Monogr.*].

[*Editor's Note:* The "Discussion" has been omitted.]

19

SUBMARINE SLIDES IN THE JURASSIC OF BELLUNO AND FRIULI (*)

Alfonso Bosellini, presented (**) by P. Leonardi.

*This article was translated expressly for this Benchmark
volume by Dr. H. C. Jenkyns, Science Laboratories,
Durham, England, from "Frane Sottomarine nel Giurassico
del Bellunese e del Friuli,"* Accad. Nz. dei Lincei,
Ser. 8 **43**, *fasc. 6, 563–567 (Dec. 1967)*

Summary. - Widespread phenomena of submarine erosion and resedimentation, which are present in the Middle-Lower Jurassic of
Belluno and Friuli Prealps, are pointed out. The phenomena in
question are slump structures, slump conglomerates, slide breccias
with a lens geometry, sometimes filling submarine channels or
gullies; they occur near the hinge line which divides a neritic
carbonate platform from a pelagic bathyal basin. The writer refers
these synsedimentary perturbations to the tectonic activity which
affected during that time the structural elements of the Southern
Alps.

In this preliminary note extensive phenomena of submarine
erosion and resedimentation are documented from the Lower-Middle
Jurassic of a large region that embraces the eastern Belluno and
western Friuli.
   More precisely the zones concerned (1) are the low Zoldano,
the middle Piave valley and the whole depression of the Cellina
stream (Cimolais and Claut basins, Cimoliana Valley, and Settimana
Valley up to the environs of Monte Chiarescons at the locality
"la Pussa"). The front of the area thus delineated has a width
of 35-40 km and a length of 15-20 km (Fig. 1), but the general
palaeogeographical characters render probable a more widespread
distribution, particularly in longitudinal directions.
   In none of the various studies on the geology of this region
has there been any mention of what is described below.

(*)   Work carried out in the Geological Institute of Ferrara
      University directed by Prof. P. Leonardi, within the prog-
      ramme of the Research Group for the Geology of Sediments of
      the C.N.R.
(**) Meeting of December 9th, 1967.
(1)   I thank Prof. E. Semenza who courteously brought to my
      attention the presence of some of the phenomena described in
      this note.

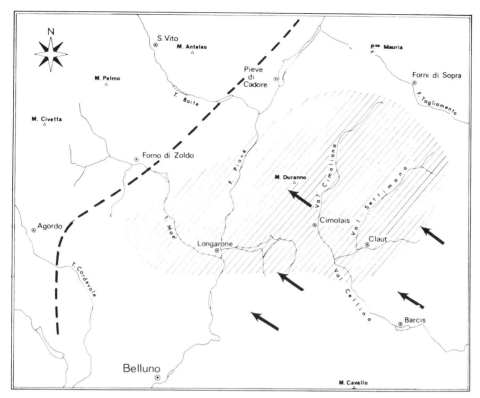

Fig. 1. - Delineation of the area (oblique cross-hatching) in
which submarine slides and their various associated
resedimentation phenomena are found. The thick discontinuous line indicates approximately the facies
threshold which, during the early Jurassic, divided
the shallow-water platform of the Dolomites from the
pelagic basin of the Belluno-Friuli Pre-Alps zone.
The arrows show the direction of migration of the
oolitic fan. (Modified from the original figure by
A. Bosellini).

From the palaeogeographical point of view we are dealing with
the northern border of the Belluno furrow (Aubouin, Bosellini and
Cousin, 1965) which, during the Lias, was characterized by miogeo-
synclinal sedimentation of pelagic type. This is now manifested
by dark grey micritic limestones, in very regular units of 20-30
cm, largely dolomitized and with much chert. The latter occurs
either as 10-20 cm beds intercalated between carbonate beds or in
amoeboid nodules within the limestones themselves. The presence
of dark grey or black partings or interbeds as millimetre laminae
is very common. In the upper part of the Lias (Toarcian) there is
a noticeable increase in the pelitic fraction which in certain

cases can constitute 50% of the lithological series: at this level
the limestones are always red and nodular and the chert can be the
same colour. From the micropalaeontological point of view the
Toarcian is also typical as it corresponds to a well-defined bio-
zone of pelagic bivalves.

At the top of the above-mentioned Toarcian facies comes a thick
(300-400 m) pile of oolitic limestones generally attributed to the
Dogger.

To the west and a little to the north of the studied zone the
Lower-Middle Jurassic contains many stratigraphic gaps; the upper
part of the Lias is missing, while the Dogger is represented by
very thin condensed pelagic nodular limestone. The part of the
Lias that remains is, moreover, developed in neritic facies and
lacks chert. This is the zone of the Atesina (Trento) Platform
(Ride tridentine or atesine of Aubouin, Bosellini and Cousin,
1965). The facies threshold that separates the two palaeogeographical provinces can only be determined approximately since
from Fiera di Premiero to Zoldano it coincides with the Valsugana
Line (hence it is probable there has been a change in the original
disposition of facies), and from Zoldano to Tagliamento there is a
vast zone in which no Jurassic rocks outcrop.

All the phenomena described in this note occur at a stratigraphical level that corresponds to the upper part of the Lias
(Toarcian), frequently at the contact with the overlying oolitic
formation (Longarone, Igne). It is thus evident that in this part
of the Jurassic significant disturbances of the sea bottom took
place, these reflecting a particular instability of the various
palaeostructural elements. There exists, in fact, a grade of
phenomena of progressively increasing intensity that range from
simple slumps to submarine slides of exceptional size.

We will here describe, albeit briefly and schematically, the
most common types of gravitational effects observed in the Jurassic
of Belluno and the Friuli Pre-Alps.

a) Intraformational folds (slumps)

By this term we mean all those gravitational phenomena which
are a few metres or, in exceptional cases, a few tens of metres
in scale, but which do not allow a fracturing of strata or allochthonous transport of the mass in question. They are generally
manifested as small-scale differential displacements between
strata, which results in a shortening or complicated refolding of
the beds: in some cases shear planes and faults with small throws
are present, which were caused by internal mechanical resistance
due to incipient consolidation of the sediments. The most important implication of this process is that it did not include resedimentation effects and that the sediments must still have been
plastic. The intraformational folds indicate the presence of an
unstable slope: they are generally localized near the palaeogeographical threshold (Zoldano).

b) Slide masses

By this term we mean conspicuous masses of material that have suffered transport en bloc. Such masses do not generally appear brecciated but comprise contorted and deformed strata which were heaped up during their sliding and at the act of stopping. This process implies allochthony, albeit of short distance (a few tens to a few hundreds of metres maximum), for the material involved in the slide.

These phenomena are particularly evident in the region around Igne where, associated with slumps, one may observe a mass some 50 m long discordantly cutting through the autochthonous beds.

c) Submarine slides of lenticular geometry

We here deal with lenticular accumulations of coarse detrital material: generally the components consist of the same material as the basinal sediments, that is, micritic limestones now largely dolomitized, and dark grey radiolarian cherts. The contacts with the stratified rocks are not very distinct (Plate I) demonstrating that the slides took place when the sediments were still not completely lithified. These submarine slides occur scattered through all the region studied: they can involve considerable transport of materials, but in many cases they have purely local significance. The principal characteristic of these accumulations is that of being laterally extensive but not very thick (10-20 m).

d) Detrital deposits filling submarine channels

Of all the phenomena observed, this is the most interesting: it is represented by coarse detrital material comprising the fill of more or less pronounced submarine channels. These masses, which generally have a width of 30-50 m, also extend vertically for some 70-80 m. Contacts with enclosing rocks are very sharp (Plate II, Fig. 1). These channels, albeit of relatively reduced dimensions, would seem to be analogous to the well-known submarine gullies and canyons off the coast of California.

These deposits are very common in the Maé Valley. Those that deserve mention for the most spectacular development occur at Soffranco (Plate III, Fig. 1) and Longarone (Plate III, Fig. 2): however, they are also present in the Cimolais depression, for example near Cellino, in front of the village on the left of the Cellina stream; in the mountains that surround the Cimoliana Valley up to the environs of Monte Chiarescons; and at the head of the Settimana Valley near "la Pussa". In this last locality all the matrix in which the submarine rock fall was emplaced became converted to a chaotic mixture of calcareous and silicified rocks of various shapes and sizes.

Certain of these channels are localized at the base of the Dogger oolitic formation: they cut down some tens of metres into the underlying Liassic sediments and are filled with blocks of

Plate I

Figure 1

Plate II

Figure 1

Figure 2

Plate III

Figure 1

Figure 2

oolitic limestone (Plate II, Fig. 2) which are set in a pelitic
fluidally structured and microlaminated matrix, often dolomitized,
that plastically surrounds the various fragments. In other localities (for example around Igne), the oolitic formation rests
with marked discordance on the Toarcian: I have noted erosion 30-
40 m deep.

Study of these phenomena (delineation, mode of formation,
causes, etc.) will be systematically extended and further studied
in all of the region. However, notwithstanding the preliminary
nature of this note, we may at this time draw some tentative conclusions of rather general nature.

First of all we can ascribe, with good approximation, these
synsedimentary disturbances to the Lias-Dogger boundary (Toarcian-
Aalenian). Given that submarine resedimentation phenomena have
already been described from the Lower-Middle Jurassic of the western edge of Lake Garda (Castellarin, 1964) and of the Giudicarie
(Bosellini, 1967) one can suppose that a common cause was responsible for the various phenomena described, this being tectonic
activity which affected the Atesine structural platform during the
Lias-Dogger transition period. Similar phenomena to those in the
Venetian Alps (of which the Belluno-Friuli Prealps are the southeastern part) have been recorded by Bernoulli (1964) in the Lower-
Middle Lias of the Lugano region suggesting the timing of the tectonic activity in this area and that of the Venetian Alps was
slightly out of phase.

REFERENCES

Aubouin, J., A. Bosellini and M. Cousin. 1965. Sur la paléogéo-
    graphie de la Vénétie au Jurassique. Mem. Geopal. Univ.
    Ferrara, 1, no. 2, 147-158.
Bernoulli, D. 1964. Zur Geologie des Monte Generoso. Mat. Carta
    Geol. Svizzera, N.F., 118 Lfg, 134 pp., Bern.
Bosellini, A. 1967. Torbiditi carbonatiche nel Giurassico delle
    Giudicarie e loro significato geologico. Ann. Univ. Ferrara,
    N.S., sez. IX, 4, No. 8, 101-109.
Castellarin, A. 1964. Geologia della zona di Tremosine e Tignale
    (Lago di Garda). Giorn. Geol., 32, fasc. II, 291-346.

EXPLANATIONS OF PLATES I-III

Plate I

Fig. 1 - Submarine slide in the Lias near Casoni (Maé Valley):
    the blocks and fragments that compose the slide are of the
    same material as the normal sediment seen on the left: the
    whole lot is dolomitized. Note that the passage between the
    detrital mass and the well-bedded cherty dolomite is gradual
    and that traces of stratification are still present in the
    breccia.

Plate II

Fig. 1 - Contact between the breccia filling the submarine channel
at Longarone (Plate III, Fig. 2) and the micritic cherty limestones of the Toarcian that form the walls of the small canyon.

Fig. 2 - The same breccia as in Fig. 1, a little nearer the centre
of the detrital mass: the large angular blocks, emphasized by
black surrounds, are composed of the oolitic limestones that
form the thick sequence occurring at the top of the Longarone
submarine channel.

Plate   III

Fig. 1 - Submarine channel at Soffranco (Mae Valley). Note the
massive aspect of the conglomeratic mass which contrasts with
the sharp and thin stratification of the Liassic sediments on
either side. In this case the channel fill is lithologically
the same as the matrix rocks.

Fig. 2 - Brecciated mass at Longarone: this occurs at the base of
the thick Dogger oolitic formation (it is in fact composed
mostly of large blocks of oolite) and cuts down several tens
of metres into the underlying Liassic sediments. Details of
this outcrop are shown in Plate II.

20

Copyright © 1968 by Elsevier Publishing Co.

Reprinted from *Marine Geol.*, **6**, 449–450, 457–461 (1968)

SUBMARINE CANYONS OF THE CONTINENTAL MARGIN, EAST BASS STRAIT (AUSTRALIA)

JOHN R. CONOLLY

Department of Geology and Geophysics, University of Sydney, Sydney, N.S.W. (Australia)

(Received November 21, 1967)
(Resubmitted March 14, 1968)

SUMMARY

Several 50–100 mile long canyons traverse a 100 mile by 100 mile amphitheatre-like depression that occurs from the shelf break to depths of over 2,000 fathoms on the continental slope of Australia east of Bass Strait. Bass Canyon heads in at least three major tributaries at depths of 60–90 fathoms, is about 100 miles long and has a 40 mile long and 2–5 mile wide flat floor flanked by 700 fathom high walls and occupies the central depths of the continental slope depression directly offshore from the main shelf graben. Flinders Canyon, lying south of Bass Canyon, is generally narrow and gorge-like, has two major tributaries, and extends to depths of 2,100 fathoms over a distance of 60 miles.

Following deposition of about 5,000–10,000 ft. of mostly deltaic and terrestial sediments in the Late Jurassic to Late Eocene in the Gippsland Basin there have been three main periods of canyon erosion and sedimentation. The first period occurred in the Late Eocene to Early Oligocene, the second in the Miocene and the third was the formation of the contempory canyon system which was initiated in the Pliocene.

Marine processes associated with these canyon systems have been responsible for prograding the continental slope 50–70 miles and removing about 20,000 cubic miles of sediment from the Australian mainland since the Late Eocene.

INTRODUCTION

During September 1967 the United States Coast and Geodetic Survey ship "Oceanographer" made three crossings (at depths of 1,800–2,100 fathoms) of a huge canyon east of Bass Strait off eastern Australia. Although there was insufficient time available to make a detailed survey, these tracks made with satellite precision navigation provided the necessary control for other available sounding data and permit the construction of a preliminary bathymetric chart of this little known region.

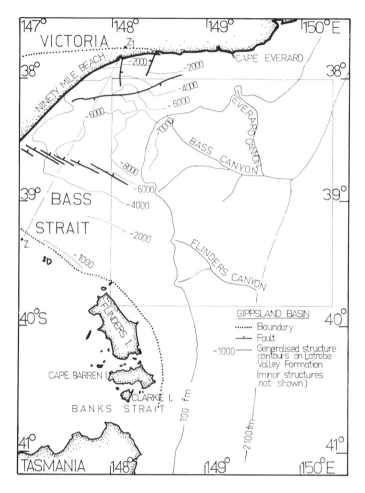

Fig.1. East Bass region showing the boundaries and main bounding faults of the Mesozoic–Tertiary Gippsland Basin, generalised structure and contours drawn on top of the Eocene Latrobe Valley Formation, and the major canyons on the continental slope. Detailed bathymetry of the inset area is shown on Fig.2. Structure contours after HAFENBRACK (1966). Section Z–Z_1 is shown on Fig.5.

Within the last 4 years there has been great activity in oil exploration in Bass Strait with consequent discovery of some very large oil and gas fields. The area discussed herein lies immediately seawards of the large oil and gas fields of the Gippsland Basin (Fig.1). Much of the geophysical and geological data gained from exploration of the Gippsland Basin has recently been made available and forms a framework for discussion of the origin of the continental margin in this region. Likewise a discussion of the modern processes of canyon erosion and sedimentation in deeper water directly offshore from the Gippsland Basin may have a direct bearing on the processes that occurred at some earlier time during the filling of the Basin.

[*Editor's Note:* Material has been omitted at this point.]

The distribution of the marine transgressive sediments of the Lakes Entrance Formation over a very uneven basement topography with local deep troughs (HOCKING and TAYLOR, 1964) and the variegated composition of the sediments (mixtures of gravel, sand and shale with plant fragments) suggests that these sediments were, in part, deposited in submarine valleys or canyons. The details of one of these possible submarine valley-fill accumulations are described by BOUTAKOFF (1964, fig.3) at Lakes Entrance (Fig.7), where thick accumulations of sand and gravel occur in a bayment-like trough within basement rocks. It is suggested that this embayment, called the Lakes Entrance Embayment by Boutakoff, was formed by headward erosion of a submarine canyon with deposition of quartz sands and gravels in the canyon.

TERTIARY CANYONS OFFSHORE

Detailed profiles and drilling offshore from the Gippsland Basin in Bass Strait indicate that two major canyon-fill sequences occur in the Tertiary. A northeast–southwest oriented seismic reflection profile (Fig.6) across one of the major anticlinal features of the offshore area, the "Marlin Dome", and a wildcat hole, Esso Gippsland—4, are used to illustrate these sequences (Figs.6, 7). A structural interpretation of the seismic profile shown on Fig.6 after completion of this well is shown on Fig.7 and is mainly based on data from a well completion report (ESSO EXPLORATION AUSTRALIA INC., 1966) now made available to the public through the Commonwealth of Australia Petroleum Search Subsidy Act.

Lower to Middle Miocene marine sediments form part of a canyon fill sequence which is characterised by an abundance of light grey, poorly sorted and porous calcarenites interbedded with finer grained sands and calcareous mudstones. The sands contain mixtures of both terrigeneous material and bioclastic debris along with plant remains and mudlumps.

The Miocene sequence overlies a thin Oligocene sequence which in turn overlies the uppermost Eocene to lowermost Oligocene marine sediments of the Lakes Entrance Formation. The Lakes Entrance Formation sediments also form part of a canyon fill sequence and consist mainly of light grey to light greenish grey calcareous and glauconite mudstones in the well but rapid lithofacies changes can be expected to occur eastwards into the main canyon sequence. The Lower Oligocene canyon has eroded all or part of the earlier Eocene and Cretaceous sequences (Fig.6, 7).

These sequences consist of a mixture of shallow water marine, estuarine, and fluvial to coastal plain sediments containing interbedded quartzose sands, silts and shales with occasional thin black to brown coal seams. The Upper Cretaceous sequence has similar lithologies to the Eocene sequence but is dominated by more quartzose sediments with minor dolomite and coal.

The well and seismic data indicate that during the uppermost Eocene to Lower Oligocene, a huge valley-like depression, 2–4 miles across and with at least 2,000 ft. of relief was cut and filled. This erosional depression extends landwards (Fig.7) and may be part of a canyon system similar in dimensions to the large canyons that occur on the continental slope off Bass Strait.

A canyon system of similar dimensions was formed during lower to Middle Miocene. This indicates that many of the structural highs such as the "Marlin Dome" found by seismic means (HAFENBRACK, 1966) are erosional remnants of flat-lying Eocene and Cretaceous sequences around which there have been periods of canyon erosion and filling during the Oligocene and Miocene. Detailed programs of seismic exploration and deep drilling in the potentially very rich oil and gas fields of the Bass Strait should eventually lead to the accurate delineation of both the Oligocene and Miocene canyons and their associated fill and fan sequences.

VON DER BORCH (1968) has suggested an Early Tertiary age for the initial subaerial cutting of the heads of large submarine canyons off southern Australia. In the Bass Strait, however, the present day canyons probably originated in the Pliocene and form the third major canyon cutting period since the Eocene.

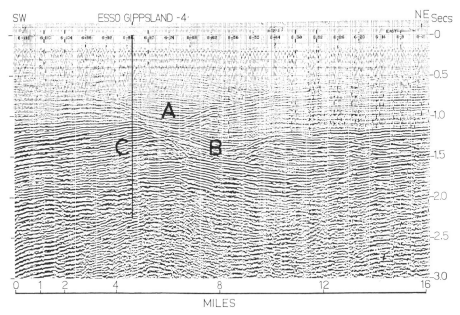

Fig.6. Seismic profile across the "Marlin Dome" in Bass Strait. Profile and position of wildcat well, Esso Gippsland–4 is shown on Fig.7A. *A* = Early to Middle Miocene canyon fill; *B* = Early Oligocene canyon fill; *C* = the "Marlin Dome" consisting mainly of Eocene deltaic sediments. A schematic geological interpretation along this profile is shown on Fig.7C.

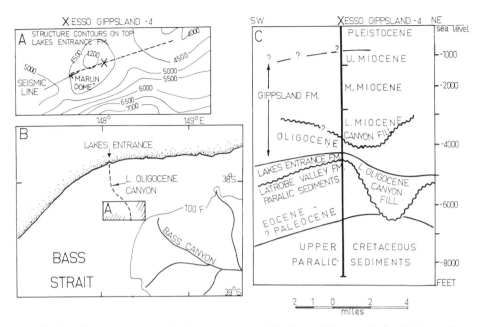

Fig.7A. Structure contours in feet on the top of the Lower Oligocene Lakes Entrance Formation indicating the position of the "Marlin Dome", seismic profile (Fig.6), and the wildcat well, Esso Gippsland—4. Location of area (A) is shown as an inset on Fig.7B.

B. Location of the modern Bass Canyon system on the continental slope and the probable location of a major Lower Oligocene canyon from its head at Lakes Entrance to a position east of the "Marlin Dome" (A).

C. Schematic geology along seismic line shown in Fig.7A, based mainly on the seismic profile (Fig.6) and a well report (Esso EXPLORATION AUSTRALIA INC., 1966).

DISCUSSION

The development of several exceptionally large canyons in the continental slope east of Bass Strait is not surprising for the continental slope of eastern Australia is generally steep and fairly narrow and in this way is somewhat similar to the continental slope off the Californian coast which is well known for its large canyons (SHEPARD, 1963; SHEPARD and DILL, 1966).

The Bass Strait shelf is generally 30–40 fathoms deep and provides a good access to an abundant sediment source, for it sits between fairly rugged hinterlands of Victoria and Tasmania each with river systems that would feed sediment across the Bass Strait and directly onto the present continental slope during times of low sea level. Now, and during former high stands of sea level, the amount of sediment supply to the edge of the shelf is probably much less. Nevertheless, the Bass Strait area is subjected to a continuing battering of the large swells originating from the Antarctic storms and is well known for its everchanging tidal currents of 1–3 knots. Large sand waves are known to occur within the Banks Strait region presumeably generated by strong tidal currents (R. Slater, personal communication, 1967). The effects of the currents on the divers working on the bottom of the Bass Strait has already been mentioned. Currents that move sandy sediment occur at depths of 40 fathoms at the outer edge of the continental shelf off southern Australia (CONOLLY and VON DER BORCH, 1967). Similar or even greater swell and tidal currents probably occur on the outer shelf edge of east Bass Strait indicating that sand could be fed to canyon heads during storms.

The present day configuration of the east Bass Strait canyons gives us a glimpse of part of a complex sedimentary history. There have been at least three major periods of canyon cutting and filling during the Tertiary and Quaternary. The first occurred from Late Eocene to Early Oligocene, the second in the Early to Middle Miocene and the third probably originated in the Pliocene and is still active.

Marine processes associated with canyon erosion and sedimentation have been responsible for prograding a pile of at least 10,000 ft. of sediment 50–70 miles seawards across the location of the present east Bass Strait since the Late Eocene. If it is assumed that at least twice this amount of sediment has been deposited further seaward on the adjacent continental slope and Tasman Abyssal Plain, then approximately 20,000 cubic miles of sediment have been removed from the highland of northern Tasmania and southern Victoria since the Late Eocene.

ACKNOWLEDGEMENTS

The writer is grateful to Dr. R. Dietz, personnel and crew of the "Oceanographer", to the staff of the Royal Australian Navy Hydrographic Office for their continuing help and consideration, to Dr. R. Dill for his helpful advice and criticism and to Mrs. Diana Conolly for drafting and typing. Financial support was received from a Queen Elizabeth II Fellowship and is gratefully acknowledged.

REFERENCES

BOUTAKOFF, N., 1964. Lakes Entrance oil and the continental shelf. *Australian Petrol. Exploration Assoc., J.*, 1964 : 99–110.
BUREAU OF MINERAL RESOURCES, 1966. Esso Gippsland Shelf No. 1 Well, Victoria. *Petrol. Search Subsidy Act, Publ.*, 76, 73 pp.
CONOLLY, J. R., 1967. The Quaternary Tasman Geosyncline. *Proc. Australian Inst. Mining Metallurgy*, in press.
CONOLLY, J. R. and VON DER BORCH, C. C., 1967. Sedimentation and physiography of the sea floor off southern Australia. *Sediment. Geol.*, 1 (2) : 181–220.
DILL, R. F., 1964. Sedimentation and erosion in Scripps Submarine Canyon head. In: R. L. MILLER (Editor), *Papers in Marine Geology*. Macmillan, New York, N.Y., pp.23–41.
DILL, R. F., 1967. Dynamic processes in submarine canyons. *U.S. Navy Electron. Lab. Center, San Diego, Calif., Film Rep. IA LSF 702, 16 mm Color Motion Picture*, 45 min.
ESSO EXPLORATION AUSTRALIA INC., 1966. Esso Gippsland shelf, 4.Victoria well completion report. *Australian Comm. Petrol. Subsidy Act Publ.*

FIRMAN, J., 1964. The Bakara Soil and other stratigraphic units of late Cainozoic age in the Murray Basin, South Australia. *Quart. Geol. Notes, Geol. Surv. S. Australia*, 10 : 2–5.

HAFENBRACK, J. H., 1966. Recent developments in the Bass and Gippsland Basins. *Australian Petrol. Exploration Assoc. J.*, 1966 : 47–49.

HOCKING, J. B. and TAYLOR, D. J., 1964. The initial marine transgression in the Gippsland Basin, Victoria. *Australian Petrol. Exploration Assoc. J.*, 1964 : 125–132.

SHEPARD, F. P., 1963. Submarine canyons. In: M. N. HILL (General Editor), *The Sea, Ideas and Observations on Progress in the Study of the Seas*. Interscience, New York, N.Y., 3 : 480–506.

SHEPARD, F. P. and DILL, R. F., 1966. *Submarine Canyons and other Sea Valleys*. Rand-McNally, Chicago, Ill., 381 pp.

VON DER BORCH, C. C., 1968. Submarine canyons of southern Australia: their distribution and ages. *Marine Geol.*, 6 (4) : 267–279.

WEEKS, L. G. and HOPKINS, B. M., 1967. Geology and exploration of three Bass Strait basins, Australia, *Bull. Am. Assoc. Petrol. Geologists*, 51 : 742–760.

21

Copyright © 1974 by the Society of Economic Paleontologists and Mineralogists

Reprinted from *Modern and Ancient Geosynclinal Sedimentation* (SEPM
Spec. Publ. 19). R. H. Dott, Jr., and R. H. Shaver, eds., 1974, pp. 106–109, 110–115, 121–125

ANCIENT SUBMARINE CANYONS AND FAN VALLEYS

J. H. McD. WHITAKER

University of Leicester, England

ABSTRACT

Many modern submarine canyons and deep-sea fans originated in pre-Pleistocene time. Similar submarine canyons, fans, and fan valleys are found in the geological record certainly as far back as the Precambrian. Criteria for recognizing ancient submarine channels include: (1) proved or inferred size comparable to modern canyons and fan valleys; (2) comparable geometry (e.g., high axial gradients diminishing seaward and steep wall slopes, some of which become vertical or overhanging); (3) similar locations (for submarine canyons) between shallow-marine (shelf) and deep-marine (basin) environments and (for fan valleys) incision into inferred deep-sea fans at the lower ends of canyons; (4) similarities in lithologies, grain sizes, and primary structures of the fills and their variations along the length and width of the canyons or valleys; (5) similarities in the observed or deduced processes of fast cutting and filling, together with clean-cut channel contacts, concave upward form, minor channels at the base of or within channel fill, compaction effects, and partial flushing out; and (6) similarities of multiple origin of faunas within the fill (indigenous, swept in from surrounding shelf areas, or derived from the canyon walls).

These criteria emerge from a study of all available data on ancient submarine canyons and fan valleys from 32 areas. The information is grouped, tabulated, and discussed under eight stratigraphic and geographic headings: (1) Lower Paleozoic of the Caledonian-Appalachian Geosyncline, (2) Carboniferous of the Pennine Basin, (3) Upper Paleozoic of the Variscan Geosyncline, (4) Permian of the Delaware Basin, (5) Mesozoic and Tertiary of the Tethyan Geosyncline, (6) Mesozoic and Tertiary of California, (7) Tertiary of the Gulf Coast Basin, and (8) other areas.

A study of these and future examples of ancient submarine canyons and fan valleys is important as an aid both in reconstructing the continental slopes and rises of geosynclinal and other basins and in the search for possible oil and gas traps. Further intensive studies of transitional areas between shelf and basin facies should reveal many more examples of ancient canyons and fan valleys.

INTRODUCTION

Two lines of evidence have led to acceptance of the antiquity of submarine canyons, deep-sea fans, and fan valleys: (1) evidence pointing to pre-Pleistocene and even early Tertiary origins for many present-day features and (2) evidence from the geological column of large filled channels that closely resemble modern examples in geometry and sedimentary evolution. Evidence from modern features is examined briefly here, and data from the geological column is presented in greater detail, including listing of well-documented ancient canyons and fan valleys, summarization of their characters, and discussion of their positions in ancient basins of deposition.

PRE-PLEISTOCENE ORIGINS FOR SOME MODERN SUBMARINE FEATURES

Recent work on present-day oceanic topographic features, such as continental shelves and slopes, submarine canyons, deep-sea fans, abyssal plains, and deep-sea channels, has led to the belief that many of these features may be older than at first suspected. For example, some deep-sea fans are so large that they may have originated long before the Pleistocene. Menard (1960) thought that the Delgada and Monterey Fans may date from pre-Pliocene time; Me-nard, Smith, and Pratt (1965) inferred that the Rhône Fan may have been forming since Oligocene time, and similar studies point to a long history of fan building and thus to a comparable antiquity for the submarine canyons that fed them. Daly (1936) realized that canyons may be preglacial, and there are many subsequent references to the antiquity of present-day canyons, such as by Fisk and McFarland (1955), Starke (1956), Bourcart (1958), Roberson (1964), Bourcart (1965), Martin and Emery (1967), Conolly (1968), and Von der Borch (1968).

FILLED CANYONS AND FAN VALLEYS IN THE GEOLOGICAL COLUMN

Hypothetical ancient canyons and fan valleys. —Increasing knowledge of present-day submarine canyons and deep fans together with their fan valleys, well summarized by Shepard and Dill (1966), led stratigraphers and sedimentologists to ask the question, "Where are the ancient submarine canyons and their fans?" Crowell (1952, p. 81), Shrock (1957, p. 1407), Kuenen (1958, p. 338), Knill (1959, p. 324), and others raised the matter in general terms, and more recently many workers on ancient basin environments have invoked lateral canyons through which sediment could be channelled from shallow to deeper water and spread out as submarine fans at the lower ends of the canyons.

Demonstrable ancient canyons and fan valleys.—Whilst for many authors the old canyons or fan valleys remained more or less hypothetical concepts, erosion or burial having precluded their discovery or revealed, at best, only fragmentary evidence, other workers have been able to locate a number of ancient submarine canyons and fan valleys and to demonstrate their geometry and the nature of their sedimentary fill. Examples from 32 areas (15 in Europe, 14 in North America, and 1 each in South America, Africa, and Australia) have been described in some detail, the first in 1948, the rest since 1954. Their degree of preservation varies considerably, from short stretches well exposed at outcrop to vast channel fills located entirely by subsurface methods. Outcrop studies usually give detailed information over limited parts of a channel, whereas subsurface data, less good for detailed study, tend to give overall dimensions more accurately. Both types of study may yield considerable data on channel fills. Present research is seeking to differentiate ancient canyons from fan valleys and to distinguish fan valleys on suprafan, inner, middle, and outer fans (see, for example, Nelson and Nilsen, this volume). The old canyons and fan valleys, although imperfectly known, may be matched in their dimensions, shape, and sedimentary histories with these various parts of modern canyon systems: canyon heads, the main, deeply cut canyons, and the fan valleys, which shallow, bifurcate, and migrate as they cross deep-sea fans.

The ages of ancient (pre-Pleistocene) submarine canyons and fan valleys range from Pliocene to Precambrian, and such features have known examples in every system except the Triassic and Devonian. With further search it is likely that canyons of these ages will be found in marine sequences.

CRITERIA FOR RECOGNIZING ANCIENT CANYONS
AND FAN VALLEYS

Before discussing the ancient examples in detail, the criteria used in deciding which channels in the geological column qualify as canyons and which as fan valleys must be stated on the basis of comparison with modern submarine canyons (Shepard and Dill, 1966) and fans and fan valleys (see recent studies by Normark and Piper, 1969; Shepard and others, 1969; Nelson and others, 1970; Normark, 1970; Curray and Moore, 1971; Haner, 1971; Normark and Nelson and Nilsen, this volume). The following six main criteria have proved useful for compari-

son, actual examples being indicated by table 1 entry number (in parentheses) or, if not in the table, by author reference:

(1) Proved or inferred comparable size to modern canyons and fan valleys in length, width, and depth; also, variation of width and depth along the length.

(2) Comparable geometry to modern canyons and fan valleys, such as high axial gradients diminishing seaward (6) (7) (10) and steep wall slopes, some of which become vertical or overhanging (1) (2) (5) (6) (7) (8) (11) (21) (27) (Bosellini, 1967, pl. 3). Tributaries joining canyons (20) (30) (32) and branching of fan-valley distributaries (10) (Mutti and Ghibaudo, 1972) (Mutti and Ricci Lucchi, 1972) and their migration (8) (14) have been noted, and occasionally a northern hemisphere left hook (14) (21) or a southern hemisphere right hook (after correcting for paleolatitude) (5) (6) (Schenk, 1970) have been found, which is consistent with the behavior of modern canyon and fan-valley systems.

(3) Similar positions of canyons on the slopes between shallow-marine (shelf) and deep-marine (basin) environments; fan valleys cut into inferred deep-sea fans at the lower ends of canyons. In favorable situations, inner (=upper, proximal), middle (=intermediate), and outer (=lower, distal) fan valleys and channels may be differentiated (8) (9) (10) (24) (28) (Mutti and Ghibaudo, 1972) (Mutti and Ricci Lucchi, 1972) (Mutti and Nelson and Nilsen, this volume).

(4) Similarities in lithologies, grain-size distributions, and primary structures of the fills and their variation along the length (proximal to distal) and width (axial to marginal) of the canyons or fan valleys. Characteristic features include the presence of olistoliths (11) (14) (21); boulder beds (4) (5) (6) (7) (10) (15) (16) (18) (27); cobble or pebble conglomerates and pebbly mudstones (3) (4) (5) (7) (8) (10) (13) (14) (16) (21) (22) (24) (25) (26) (27) (31), some having long-axis preferred orientation or imbrication (4) (6) (Unrug, 1963); breccias (1) (3) (10) (11), and, less commonly, sandstone spheroids, some of which have mud centers (13); and armored mud balls (5) (16) (Unrug, 1963), these clasts being set in a much finer grained matrix of sandstone or mudstone. The sandstones may be massive, amalgamated (8) (13) (17) (Stanley and Unrug, 1972), well bedded (some bedding slightly concave upward), or laminated (3) (4) (5) (6) (8) (9) (11) (13) (14) (26) (27) (28). Channel levees may be present (8) (10). Also characteristic are mass-flow phenomena, including slide conglomerates, slump folds (1) (2) (4) (6) (9) (13) (14) (26) (27) (28), and other channel-controlled primary structures, some being governed by remarkably consistent paleocurrents (6) (14) and others by downflank or downchannel paleoslopes (6) (13).

(5) Similarities in the observed or deduced pro-

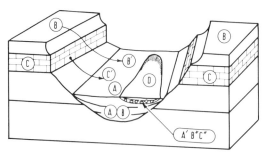

FIG. 1.—Channel fill bearing faunas of different origins. The letters A, B, C, D refer to fossil assemblages (commonly thanatocoenoses). A single suffix denotes derivation, a double suffix, rederived assemblages.

cesses of fast cutting and filling, some processes operating down active fault scarps (1) (4) (6) (15) (20) (28), on sedimentary cover tilted by buried faulting (4), or along active fault zones (6) (11) (29); clean-cut channel contacts that are concave upward (5) (6) (11) (13) (Bosellini, 1967), minor channels at the base of or within the main channel fill (3) (4) (5) (6) (14) (27), and compaction affecting the fill (6) (20) (23) (29), which itself may

undergo partial flushing out (6).

(6) Similarities in the multiple origin of the faunas within the fill, expressed diagrammatically in figure 1: A, an *indigenous* deeper water fauna, which may give useful depth indications (e.g., foraminifera), and some of which may be channel dwellers only; B, a contemporary shallow-water fauna washed in from surrounding shelf areas to give a derived fauna B′ termed *exotic;* and C′, an older *remanié* fauna (terms from Craig and Hallam, 1963) derived from erosion of the canyon walls, sometimes recognizable from the abraded, limonitized, or encrusted nature of the derived fossils (4) (6) (12) (13) (16) (19) (24) (28) and figure 2 or from their presence in fallen blocks of wall lithology (6) (11). Rarely, a derived (A′) or a rederived (B″ or C″) fauna may become concentrated at the base of a minor channel within the canyon fill, the result of flushing-out processes, as in (6) and figures 1 and 2. The minor channel sediments may have a fauna, D, slightly younger than A and B ((6) and fig. 2). Downchannel currents may align elongate fossils parallel with the channel axis (6) (10), but if such fossils become lodged in ripple-mark troughs, their orientation may be at right angles to the axis (10).

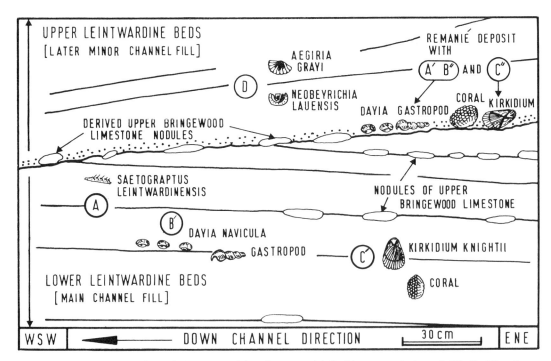

FIG. 2.—Derived and rederived faunas resulting from partial flushing out of channel fill, Todding Lane, Leintwardine, Welsh Borders (British National Grid Reference SO 417756). A, graptolites and indigenous fauna; B′, exotic current-oriented *Dayia navicula* and gastropods; C′, *remanié Kirkidium knightii* and corals derived from Upper Bringewood limestones of the canyon walls; A′B″, abraded and limonitized fossils (such as *Dayia navicula* and gastropods) in basal conglomerate of minor channel fill; C″, rederived *Kirkidium knightii* and corals, also abraded and limonitized; D, indigenous fauna of *Aegiria grayi* and *Neobeyrichia lauensis* (Upper Leintwardine Beds) slightly younger than the AB fauna (Lower Leintwardine Beds).

Differentiation between the lower ends of old submarine canyons and ancient deeply cut valleys on the inner fans is by no means easy; geometrical and spatial relationships are generally of most help. Canyons and inner fan valleys may be more readily distinguished from middle fan valleys, as the former are more deeply incised and contain more poorly bedded (Bouma *ae*), coarser, and more disturbed sediments than the middle fan valleys. These middle fan valleys, as well as the outer fan valleys, which may be wider but as little as 2 m deep, contain well-bedded complete *a-e* sequences. Fan valleys are flanked by dark silty overbank deposits characterized by *cde* sequences that are not found in association with canyons (Jacka, and others, 1968; Stanley and Unrug, 1972; Mutti, this volume; Nelson and Nilsen, this volume).

The criteria listed above exclude shelf channels (for these, see Passega, 1954, and Sedimentation Seminar, 1969) and the extensive and well-studied fluvial channels from many areas and horizons.

TABULATION OF DATA

The main dimensional and sedimentological data on all ancient canyons and fan valleys known to the author (excluding, for reasons of space, channels less than 15 m deep and those insufficiently described) are set out in table 1, so that comparisons and contrasts may be made readily. For a few entries, the writer had to deduce dimensions, trends, etc., from maps or sections. By comparing this table with the appendix in Shepard and Dill (1966), where data on modern canyons are tabulated, the reader may, if he wishes, seek the present-day equivalent to any ancient example.

The entries in table 1 are numbered consecutively and arranged in eight main groups. This arrangement conveniently deals with the older canyons and fan valleys first and the younger ones later. Within each group there is, where possible, a geographic arrangement, and if there are several papers on one channel or group of channels, these papers are grouped under a single entry number.

[*Editor's Note:* In the part omitted, points of interest from selected ancient canyons and fan valleys are discussed.]

TABLE 1.—SUMMARY OF INFORMATION ON ANCIE[NT]

Provincial Grouping	Entry number	Date of publication	Author	Location	Name of canyon (C) or fan valley (F)	Occurrence (surface or subsurface)	Straight (St) or sinuous (Si)	Length in km	Maximum width in km	Maximum axial gradient
I. Lower Paleozoic Channels of Caledonian-Appalachian Geosyncline	1	1969, pers. commun.	Dewey	Sops Arm, W White Bay, Newfoundland	C	Sur				0.03 seen; 0.06 inferred
	2	1971	Hamilton-Smith	NW Brunswick	C in N, F in S	Sur and sub		35+		
	3	1969, 1970, 1972	Piper	Co. Galway, Eire	C and F (1969 fig. 6.11),	Sur			1	
					F (1969 fig. 6.30)	Sur			0.35	
	4	1967, 1972 1969	James James & James	W-central Wales	Numerous channels, C and F	Sur	St	Up to 7	Up to 3	8° on Corri[s] Elerch slop[e]
	5	1969 and pers. commun.	Kelling & Woollands	Rhayader, central Wales	Caban Channel, C, and several smaller channels, C and F	Sur	Si	5	1.6	
	6	1962, 1963 1969	Whitaker Jones	Leintwardine, N Herefordshire, England	Marlow, Bagdon, Todding, Mocktree, Church Hill, and Tatteridge Channels, C	Sur	St	Up to 4	0.8+	175m/km =10°
II. Carboniferous of Pennine Basin, England	7	1964, 1970 1969, 1972, pers. commun.	Sadler Simpson & Broadhurst	Castleton, Derbyshire, England	Winnats Reef Channel, ?C	Sur and sub	Si	0.4	0.4	300m/km =16½°
	8	1966a, b	Walker	N Derbyshire, England	Grindslow Channels (17), C and F	Sur		3+	0.12 to 1	
III. Variscan Geosyncline	9	1966b	Walker	N Devon, England	Westward Ho! Channels (3), C	Sur			0.13	
		1969	Walker	do	Northam Channel, C	Sur			0.05+	
IV. Permian of Delaware Basin, Texas and New Mexico	10	1963	Pray & Stehli	Guadalupe Mountains, New Mexico and W Texas	Bone Spring Channel, C	Sur			0.1	
		1968	Jacka and others	do	Last Chance Canyon Channel, C	Sur	St	8+	3.2+	175m/[k]m. =10°, decreasing t[o] 35m/km, =2°
					W Dog Canyon-Cutoff Ridge Channel, C, continued as?:	Sur	St	12+ (2 and 3 together)	Fairly b[r]oad	
					Shumard and Bone Canyon Channel, C	Sur	St		do	
					Guadalupe Pass, Glover, and Lamar Canyons, W Chico Draw, Long Point, etc., Channels, F	Sur	St		0.4	

BMARINE CANYONS AND FAN VALLEYS

Geometry			Age		Regimen		
Downaxis direction or channel trend	Wall slopes	Maximum depth of axis below rims in meters	Beds cut by channel (inclusive)	Age of channel fill	Sedimentology of channel fill	Inferred site at time of formation	Inferred cutting and filling agents
	Up to 70°	30+	Arenigian (Lower Ordovician)		Limestone breccia, blocks up to 0.3 m, slumps in argillaceous calcilutites	Continental margin	
o S	To 90° and overhanging	50	Early Llandoverian (Carys Mills Fm.)	Early Llandoverian (Siegas Fm.)	Limestone-slate conglomerate in matrix of lithic wacke; clasts 0.6 to 1.5 m; slumps, pullapart, plastic deformation	Between Taconic folded region and Aroostook-Matapedia Trough	Sliding lenticules, turbulent high-density flow, wall collapse
?o S	15–40°	100	Connemara Schists	Late Llandoverian (Gowlaun member)	Conglomerates and coarse sandstones	Steep margin of turbidite basin	In part subaerial?
?o S	10–30°	100		Late Llandoverian	Conglomerates, fine breccias, and sandstones	Submarine fan fed by canyon	Turbidity currents
o W and ?W	11°	115	Ashgillian (Ordovician)		Arenites, conglomerates, and a variety of mass-flow deposits; flow rolls adjacent to steep channel margins	Stepped basin floor having slope-lip and slope-base channels	Turbidity currents, wall collapse, slumping
?o WNW ?n SW, to ?NW in N	30° to vertical and overhanging	110	Llandoverian (Silurian)		Pebbly grits to cobble conglomerates containing 2 m-long angular mudstone blocks, some armoured and contorted; periods of finer grained sedimentation	On submarine slope, cutting obliquely (Caban Channel); others are fan channels	Gravity flows of fluxoturbidities, undermining of walls
?o WSW	35°, locally to 90°	183	Mid-Eltonian to early Lower Leintwardine (Ludlovian, Silurian)	Later Lower and Upper Leintwardine (Ludlovian, Silurian)	Laminated calcareous siltstones, boulder beds, slumps, ripple marks, many paleocurrent indicators such as skip, prod and groove casts; derived and rederived fossils	Continental shelf edge and slope	?Turbidity currents aided by slumping; cut and fill, =1 Ma
To ESE, then ENE	Steep	?70+	Lower Carboniferous (Mid D$_1$=Mid B$_2$)		More or less *in situ* limestone boulders in calcareous matrix; fan of calcirudite	From shelf through back reef, down fore reef to basin	
To S and ?ESE inc	Gentle to 70°, Stepped	50	Namurian (Late Carboniferous), nine in Upper Shale Grit, one in Lower Shale Grit, and seven in Grindslow Shale		Sandstones (proximal turbidites), pebbly sandstones, mudstones	Inner fan at upcurrent edge of apron of basin turbidites On slope	Cut by fast underladen turbidity currents
	Less than 5°	14+	Westphalian (Late Carboniferous, Westward Ho! Fm.)		Two filled with turbidite sandstones, silty mudstones, and sandstone slump balls; one mudstone-filled channel complex	On slope (Haner, 1971, stated outer fan)	As above, or cut by permanent ocean currents or by slumping
		15	Westphalian (Northam Fm.)		Sandstones, siltstones	do	Cut in agitated water environment by turbidity currents
To E or SE	Sharp, concave upward	34	Leonardian (Permian)		Carbonate rudites, blocks up to 9×12×15 m; geopetal fabrics different in each block; matrix limestone, minor dolomite	Submarine slope between Diablo Platform and Delaware Basin	Submarine slides from shallow-water environment
To ESE	do	?460	Guadalupian (Permian) ?to Pennsylvanian	Guadalupian (Permian)	Very fine sandstone, giant foresets; siltstone and conglomeratic lime mudstone; many paleocurrent indicators; shallow-water fossils	Incised several km into northwestern shelf margin	High-density, high-velocity salinity currents, suction currents, sand flows, under-cutting and collapse of canyon walls
To SE	do	Deep	do	do	As above, downaxis thickening of formations	do	do
To SE	do	do	Leonardian and Guadalupian (Permian)		Conglomeratic mudstones, 2-m blocks derived from walls; contorted sandstones below	do	do
	do	15+ on proximal fan, less incised on distal fan	Guadalupian (Permian)		Proximal: minor flow units (Bouma a-d), conglomeratic mudflows, cross-bedded sandstones; intermediate: major flow units of cross-bedded sandstone followed by Bouma b-e intervals; distal: minor flow units	Cut into deep-sea fans in Delaware Basin	do

TABLE 1. (Contin

Provincial Grouping	Entry number	Date of publication	Author	Location	Name of canyon (C) or fan valley (F)	Occurrence (surface or subsurface)	Straight (St) or sinuous (Si)	Length in km	Maximum width in km	Maximum axial gradient
		This vol.	Harms & Pray	do	Several, C or F?	Sur		Many km	1	
V. Mesozoic and Tertiary Channels of Tethyan Geosyncline	11	1969, 1970	Van Hoorn	S-central Pyrenees, (Esera), Spain	C and F	Sur	Si	10	2+	Steep
	12	1965	Schoeffler	Aquitaine, SW France	Le gouf de Capbreton fossile, C	Sub	Si	1.160+	20	
								2.80+	10	
								3.72+	10	
	13	1964, 1967 1964 1968 1972	Stanley Stanley & Bouma Stanley & Mutti Stanley & Unrug	SE France	Annot,	Sur	Si	8	2	
					Contes, and	Sur	Si	6	3	
					Menton Channels, C	Sur	Si	2	2	
	14	1969	Ricci Lucchi	N Apennines, Italy	Several, F	Sur	St	To 105 km inferred	2	2°
	15	1968, pers. commun.	Sargent	Calabria, S Italy	C	Sur	St	6	2	
	16	1964 1967	Marschalko Koráb, Leško & Marschalko	Klenov-Suchá Dolina, Czecho-slovakia	C	Sur	St	10+	6 to 8	
	17	1969	Mutti	Island of Rhodes, Greece	Arnita, C Monte Schiati, C	Sur Sur	Si Si	Approx. 6 Approx. 5	Approx. 1.5 Approx. 1.2	
	18	1960	Neev	S Coastal Plain, N Negev, Israel	1. Kurnub-Beersheba-Saad Channel, C 2. Nahal-Shorek Channel, C	Sur and Sub	Si	100	4	14m/km, =1°
VI. Mesozoic and Tertiary Channels of California	19	1959	Frick, Harding & Marianos	N Sacramento Valley	C		Si	64	Narrow	
	20	1965 1967	Edmondson Dickas & Payne	Sacramento Valley	Meganos Channel C	Sur and sub	Si	80+	3.2 to 9.7	2° approx
	21	1972a	Lowe	Sacramento Valley	C and F	Sur	St	C approx. 8 F approx. 15	C 2.5 F to 3	
	22	1957	Almgren & Schlax	S Sacramento Valley	Markley Gorge, C	Sur and sub		96		
	23	1967	Martin & Emery	Monterey	Pajaro Gorge, C	Sub	Si	19	5.6	
	24	1963	Martin	Bakersfield	Rosedale Channel, C	Sub	Si	9.6 (perhaps 29)	1.1 in N to 2.1 in S	40m/km, =2½°

Geometry			Age		Regimen		
Downaxis direction or channel trend	Wall slopes	Maximum depth of axis below rims in meters	Beds cut by channel (inclusive)	Age of channel fill	Sedimentology of channel fill	Inferred site at time of formation	Inferred cutting and filling agents
	Steep sided	40	Guadalupian (Permian)		Sandstones, upper-flow regime features	On slopes between shelf and deep basin floor	Density currents in relatively deep submarine environment
To ESE	To nearly vertical	385	Late Cretaceous (Aguas Salenz Fm.)	Santonian (Late Cretaceous, Campo Breccia Fm.)	In west: coarse limestone breccias and immense olistoliths (Triassic to Albian and Santonian) derived from nearby diapiric uplift to west; in east: limestone microbreccias, turbidites, blue marls	Steep slope leading from shallow shelf to California-type basins separated by swells	Sudden catastrophic event; fault-controlled subsidence to form small graben; slumping from walls
To WNW		500 to 1,000		Lower Eocene	Epineritic facies, pelagic microfauna flanked by shallower facies		
				Oligocene	Clayey and sandy marls		
				Miocene	Sands and detrital limestones		
To N		200+	Priabonian (Eocene, Marnes Bleues Fm.)	Late Eocene to early Oligocene (Annot Sandstone Fm.)	Massive wedge-shaped coarse sandstones, thickening down-channel, sandstone spheroids, thin silts, and clays; channel tongues of gravel in axes and draping flanks; thick slumps; reworked fossils from canyon walls	On relatively steep slopes, deeper basin to north	Slide and sand flows, rapid but intermittent sedimentary transport leading to unstable fills
To NNW		350+					
To N							
To S and SW	5 to 20°		Tortonian (middle Miocene, Marnoso-arenacea Fm.)		Thick-bedded sandstones and polymictic conglomerates, blocks several hundred cubic meters in size; fining upward; as channels filled, they migrated to NE	Zone between slope and fans	Very dense and fast turbidity currents, mass flows and slumps, slipping of clays on channel walls
To NW		250	?Devonian granites	Miocene and Pliocene	Coarse boulder beds	Submarine fault scarp	
To N and NW		150 to 200+	Late Eocene		Conglomeratic flysch (coarse graded beds, 2-m boulders, slumps) and wildflysch (medium sandstones and coarse siltstones, thick slides); mud balls; fully marine (foraminifera) but reworked plant detritus	Cutting slopes of steep continental terraces	Fast-moving slide masses and turbidity currents
To E and SE			Middle-late Oligocene and Aquitanian (Messanagros Sandstone)		Sands, some amalgamated, and some slide conglomerates	Shelf-basin transitional area	
To SE							
To W	10° to 15°+	To 1,100 in W	Early Cretaceous to Oligocene	Neogene	1. Coarse clastics, gypsum, sandy shale, marls 2. Silts and clays, calcareous sandstones, large boulders	Subaerial river system subsequently drowned	Probably turbidity currents
To S		610+	Upper Cretaceous	Eocene	Deltaic, but foraminifera show these beds are all submarine		
To S and SW	5 to 15°	611	Late Cretaceous to late Paleocene	Late Paleocene	95% silty shale, some shaly siltstone and glauconitic units, 35–60% compaction	Neritic to upper bathyal on foraminifera; fault control	Extensive slumping, turbidity currents (rapid cutting)
To SW	Steepsided		Early Turonian to Albian sediments, Late Jurassic and older igneous and metamorphics	Turonian (Venaio Fm.)	Sandstones and pebbly mudstones, blocks up to 150 m long	Shelf-slope-rise area	Grain flow, mass movements, wall collapse
To SSW							
N–S		762	Late Cretaceous to late Eocene	Post-Eocene (Oligocene and ?Miocene)	Shales, sandstones, and conglomerates, lateral and vertical variation		
To W		1,525	Late Cretaceous or earlier	Early middle Miocene	Reddish sandstones and siltstones, few fresh-water gastropods washed in	Cut subaerially on Elkhorn erosion surface, then submerged	
To S	11° to 22°	366+	Middle Miocene	Middle Mohnian (late Miocene)	Mainly sandstones, some pebble conglomerates and siltstones, poorly sorted; fragments of shallow-water megafossils and mixed shallow and deep microfossils	Water depth probably greater than 400 m on foraminifera (Haner, 1971, stated inner fan)	Turbidity currents or gravity flows; cut and fill, =0.7 Ma

323

TABLE 1. (Contin...

Provincial Grouping	Entry number	Date of publication	Author	Location	Name of canyon (C) or fan valley (F)	Occurrence (surface or subsurface)	Straight (St) or sinuous (Si)	Length in km	Maximum width in km	Maximum axial gradient
	25	1960	Sullwold	Tarzana, Santa Monica Mts	?C	Sub		22.5 inferred		40m/km =2½°,
	26	1961	Sullwold	Los Angeles	East Sansinena Channel, C	Sub	Si	1.6	0.8	inferred 45m/km =2¼°
	27	1966 1969 1970 1971	Bartow Normark & Piper Komar Piper & Normark	Dana Point	Doheny Channel, F	Sur	St	0.2	0.2+	
VII. Tertiary Channel's of U.S. Gulf Coast	28	1948	Bornhauser	SW Louisiana, SE Texas	C	Sub				
		1960	Bornhauser		1. C	Sub	Si	1.2	0.3	122m/km =7°
					2. C	Sub	Si	0.8	0.8	90m/km, =5°
		1966	Bornhauser		?C	Sub		3.8+	3.2+	?74m/km =4°
		1966	Paine		Hackberry Channels, at least 3, C	Sub	Si	24	To 14	
	29	1959	Hoyt	Yoakum, Lavaca Co., Texas	Middle Wilcox or Yoakum Channel (Canyon, Gorge), C	Sub	Si	80	16, widening: down-channel	Gentle
VIII. Other Areas	30	1954	Tassega	Venezuela	C	Sub	St	4+	1.6	
	31	1968	Conolly	E Bass Strait, Australia	1. C	Sub	Si	64+	3.2 to 6.4	
					2. C	Sub			6.4	
	32	1967	Short & Stäuble	Niger delta, W Africa	Afam, C	Sub	Si	100	30	
		1972	Burke							

Geometry			Age		Regimen		
Downaxis direction or channel trend	Wall slopes	Maximum depth of axis below rims in meters	Beds cut by channel (inclusive)	Age of channel fill	Sedimentology of channel fill	Inferred site at time of formation	Inferred cutting and filling agents
To SW					Higher sand-shale ratio than in fan; cobble conglomerate between fan and source area		Probable turbidity currents, as these built fan
To W, then W	1½°	46+		Late Miocene (A-10 sand, upper part)	Sandy siltstones to conglomerates, thinly laminated to massive, arkosic, poorly sorted	Water depth 300–600 m	Turbidity currents
To SE	To 90°	40+		Late Miocene (Lower Capistrano Fm)	Lower part: gravel below to medium-grained massive sands above; siltstones and angular shale intraclasts up to 8×1.5 m. Upper part: bedded turbidite fine to coarse sands and silts	Channel cut through deep-sea submarine fan (at least 600 m deep)	Turbidity currents?, inertia flow?, mass flow?
		Nearly 1,100	Miocene (*Heterostegina* Shale)	Miocene	Thick sands and sandstones; deltaic sediments but entirely submarine	Slope of continental shelf	
To E	21½°	140+	Oligocene (lower Frio)	Oligocene-?Miocene (upper Frio)	Irregular thin-bedded sands and shales	Submarine fault scarps and steep fold limbs	Submarine slides and scouring
To SE	33½°	221+					
To S approx.	?5½°	183	do	do	No data—electric logs give chaotic pattern		
To S	?3½°	213+	Oligocene (Vicksburg to middle Frio)	Oligocene (upper Frio)	Lower part: thick sandstones spreading into fans at lower ends; high part: shales, sandstones, and shallow-water fauna (15–46 m deep)	On slope (Haner, 1971, stated inner and middle fan)	
To S and SE	Up to 20°	Approx. 914	Mid-Eocene (early Wilcox)	Mid-Eocene (middle Wilcox)	Silty shale; more sand up-channel; differential compaction	At edge of continental shelf	Largely by turbidity currents
To S		100+	Miocene (La Pica Fm.)	Miocene	Marine shales, rare strongly lenticular sands	Submarine slope	Powerful turbidity currents
	11°	600+	Late Cretaceous	Early Oligocene	Calcareous and glauconitic mudstones, quartz sands, gravels		
	6°	600+	Oligocene	Early to middle Miocene	Calcarenites, finer grained sands, calcareous mudstones		
To SSE		300 to ?800	?Upper Miocene (Benin Fm.)	(Afam Clay Member Benin Fm.)	Clays, sand or gravel at base	Heading in Niger delta	

DISCOVERY OF CANYONS AND FAN VALLEYS

A study of publication dates of papers on ancient canyons and fan valleys reveals that they are being discovered at an increasing rate. Most of the United States, Venezuelan, Australian, and African examples were found during the search for oil and gas, whereas the European, Canadian, and Japanese features were discovered mainly by academic research. The present almost total lack of knowledge (at least on the part of the writer) of comparable features from the rest of the world may result from a lack of detailed mapping and drilling. Very detailed work in shelf-basin transition areas is certainly required to find them, even in regions apparently well known such as the Texas Permian or the Welsh Borderland Silurian, where channels have been found only recently. This close scrutiny is needed because the requisite shelf-to-basin transitional zones are of limited area and the areal extent of channel fills is even more limited. In addition, these zones may be lost through later erosion or may be hidden by deep sedimentary burial or orogenic thrusting of basin sequences over shelf sediments. The search is worth while, however, as coarse channelized deposits may be used for distinguishing canyon and fan-valley environments and as an aid in recognizing slope and base-of-slope environments in ancient marine basins (Stanley and Unrug, 1972). This will help us to interpret more fully the general paleogeographies of ancient geosynclines, for we are understanding more clearly that canyons have been the major conduits for introducing coarse clastic sediments, via deep-sea fans, into deep-water basins within ancient geosynclinal belts.

CONCLUSIONS

There is now ample evidence that submarine canyons and their fans, deeply scored by fan valleys, are not unique features of Pleistocene and Holocene time. Evidence obtained from detailed studies of modern submarine canyons and fans (including estimates of the rates of cutting of canyons and building of fans in comparison with the rates of denudation of source areas) has shown that many of them must have begun during pre-Pleistocene time, some of them well back in the Tertiary. Different lines of evidence, from careful study of surface exposures and detailed subsurface drilling, have revealed not only filled canyons and fan valleys of Tertiary age but much older examples in nearly every system as far back as the Precambrian. Often the evidence is fragmentary as these transitory features of limited area are readily lost by erosion or buried by later sedimentation or overthrusting. But sufficient data have now been collected from both ancient and modern submarine canyons and fan valleys to provide support for the uniformitarian principle that "The Present is the Key to the Past." Additionally, they support the equally useful doctrine that "The Past is the Key to the Present," for some show evidence for their complete evolutionary histories such as one or more episodes of rapid downcutting followed by a period of channel filling that may be reversed from time to time by flushing out processes. The channel fills are composed of varied sediments (commonly very coarse grained) and show many different primary structures and complex faunal assemblages. For students of modern canyons and fan valleys this knowledge of the complete histories of some ancient examples may prove to be a useful stimulant to their researches on the origins and probable histories of the present-day canyons and fan valleys.

This brief review (for a fuller survey, see Whitaker, in press) shows also that the majority of ancient canyons and fan valleys occur along the margins of actively developing geosynclines. Most authors compare their channels with the well-studied modern canyons and fan valleys off California. The nongeosynclinal Pennine and Delaware Basins, however, have canyons and fan valleys of character similar to the others, both modern and ancient. Clearly, it is not yet possible to differentiate ancient continental margins into various types on the basis of canyons and fan valleys alone.

ACKNOWLEDGMENTS

For guidance in the field, I am indebted to F. P. Shepard, R. F. Dill, and G. Kelling. For permission to quote from theses and personal communications, I wish to thank F. M. Broadhurst, R. P. Coats, J. F. Dewey, A. D. Jacka, D. M. D. James, M. D. Jones, G. Kelling, D. J. W. Piper, W. A. Pryor, L. Redwine, G. E. G. Sargent, and I. M. Simpson. I have also been helped by D. S. Gorsline, A. Hallam, J. Helwig, J. D. Hudson, C. Lewis, R. Marschalko, D. F. Merriam, E. Mutti, H. E. Sadler, J. Schoeffler, P. C. Sylvester-Bradley, and D. Vass. A Fulbright Travel Award and grants from the William Waldorf Astor Foundation and the University of Leicester materially assisted the field work and are gratefully acknowledged. Finally, I wish to thank R. H. Dott, Jr., for inviting me to attend the "Conference on Modern and Ancient Geosynclinal Sedimentation" and to contribute to this volume in honor of Marshall Kay.

REFERENCES

ALMGREN, A. A., AND SCHLAX, W. N., 1957, Post-Eocene age of "Markley Gorge" fill, Sacramento Valley, California: Am. Assoc. Petroleum Geologists Bull., v. 41, p. 326–330.

BARTOW, J. A., 1966, Deep submarine channel in upper Miocene, Orange County, California: Jour. Sed. Petrology, v. 36, p. 700–705.

BIRD, J. M., AND DEWEY, J. F., 1970, Lithosphere plate-continental margin tectonics and the evolution of the Appalachian Orogen: Geol. Soc. America, Bull., v. 81, p. 1031–1060.

BORNHAUSER, M., 1948, Possible ancient submarine canyon in southwestern Louisiana: Am. Assoc. Petroleum Geologists Bull., v. 32, p. 2287–2290.

——, 1960, Depositional and structural history of Northwest Hartburg Field, Newton County, Texas: ibid., v. 44, p. 458–470.

——, 1966, Marine unconformities in the northwestern Gulf Coast: Gulf Coast Assoc. Geol. Societies Trans., v. 16, p. 45–51.

BOSELLINI, A., 1967, Frane sottomarine nel Giurassico del Bellunese e del Friuli: Accad. Naz. Lincei, v. 43, p. 563–567.

BOURCART, J., 1958, Problèmes de géologie sous-marine: Paris, Masson, 123 p.

——, 1965, Les canyons sous-marine de l'extremité orientale des Pyrénées, in SEARS, M. (ed.), Progress in oceanography, 3: London, Pergamon Press, p. 63–69.

BOUROULLEC, J., AND DELOFFRE, R., 1972, Esquisse paléogéographique de l'Albien supérieur a l'Yprésien en Aquitaine: Soc. Natl. Pétroles Aquitaine Centre Rech. Pau Bull., v. 6, p. 263–287.

BURKE, K., 1972, Longshore drift, submarine canyons, and submarine fans in development of Niger delta: Am. Assoc. Petroleum Geologists Bull., v. 56, p. 1975–1983.

——, AND WATERHOUSE, J. B., 1973, Saharan glaciation dated in North America: Nature, v. 241, p. 267–268.

CONOLLY, J. R., 1968, Submarine canyons of the continental margin, east Bass Strait (Australia): Marine Geology, v. 6, p. 449–461.

CRAIG, G. Y., AND HALLAM, A., 1963, Size-frequency and growth-ring analyses of *Mytilus edulis* and *Cardium edule,* and their palaeoecological significance: Palaeontology, v. 6, p. 731–750.

CROWELL, J. C., 1952, Submarine canyons bordering central and southern California: Jour. Geology, v. 60, p. 58–83.

——, 1955, Directional-current structures from the Prealpine Flysch, Switzerland: Geol. Soc. America Bull., v. 66, p. 1351–1384.

CURRAY, J. R., AND MOORE, D. G., 1971, Growth of the Bengal deep-sea fan and denudation in the Himalayas: ibid., v. 82, p. 563–572.

DALY, R. A., 1936, Origin of submarine "canyons": Am. Jour. Sci., ser. 5, v. 31, p. 401–420.

DAVIES, D. K., 1972, Mineralogy, petrography and derivation of sands and silts of the continental slope, rise and abyssal plain of the Gulf of Mexico: Jour. Sed. Petrology, v. 42, p. 59–65.

DEWEY, J. F., 1969, Evolution of the Appalachian/Caledonian Orogen: Nature, v. 222, p. 124–129.

——, AND BIRD, J. M., 1970, Mountain belts and the new global tectonics: Jour. Geophys. Research, v. 75, p. 2625–2647.

DICKAS, A. B., AND PAYNE, J. L., 1967, Upper Paleocene buried channel in Sacramento Valley, California: Am. Assoc. Petroleum Geologists Bull., v. 51, p. 873–882.

DILL, R. F., 1964, Sedimentation and erosion in Scripps submarine canyon head, in MILLER, R. L. (ed.), Papers in marine geology (Shepard Commemorative Vol.): New York, Macmillan Co., p. 23–41.

DZULYNSKI, S., KSIAZKIEWICZ, M., AND KUENEN, PH. H., 1959, Turbidites in flysch of the Polish Carpathian Mountains: Geol. Soc. America Bull., v. 70, p. 1089–1118.

EARP, J. R., AND OTHERS, in press, Map and explanation, Leintwardine-Ludlow sheet: Great Britain Inst. Geol. Sci.

EDMONDSON, W. F., 1965, The Meganos Gorge of the southern Sacramento Valley: San Joaquin Geol. Soc. Selected Papers, no. 3, p. 36–51.

FISK, H. N., AND MCFARLAND, E., 1955, Late Quaternary deposits of the Mississippi River, in POLDERVAART, A. (ed.), Crust of the earth, a symposium: Geol. Soc. America Special Paper 62, p. 279–302.

FRICK, J. D., HARDING, T. P., AND MARIANOS, A. W., 1959, Eocene gorge in northern Sacramento Valley (abs.): Am. Assoc. Petroleum Geologists Bull., v. 43, p. 255.

GOHEEN, H. C., 1959, Sedimentation and structure of the *Planulina*-Abbeville trend, South Louisiana: Gulf Coast Assoc. Geol. Societies Trans., v. 9, p. 91–103.

HAMILTON-SMITH, T., 1971, A proximal-distal turbidite sequence and a probable submarine canyon in the Siegas Formation (early Llandovery) of northwestern New Brunswick: Jour. Sed. Petrology, v. 41, p. 752–762.

HANER, B. E., 1971, Morphology and sediments of Redondo Submarine Fan, southern California: Geol. Soc. America Bull., v. 82, p. 2413–2432.

HELWIG, J., AND SARPI, E., 1969, Plutonic-pebble conglomerates, New World Island, Newfoundland, and history of eugeosynclines, in KAY, G. M. (ed.), North Atlantic—geology and continental drift: Am. Assoc. Petroleum Geologists Mem. 12, p. 443–466.

HOYT, W. V., 1959, Erosional channel in the middle Wilcox near Yoakum, Lavaca County, Texas: Gulf Coast Assoc. Geol. Societies Trans., v. 9, p. 41–50.

HSÜ, K. J., 1959, Flute- and groove-casts in the Prealpine Flysch, Switzerland: Am. Jour. Sci., v. 257, p. 529–536.

——, 1971, Origin of the Alps and western Mediterranean: Nature, v. 233, p. 44–48.

——, 1972a, The concept of the geosyncline, yesterday and today: Leicester Lit. and Philos. Soc. Trans., v. 66, p. 26–48.

——, 1972b, When the Mediterranean dried up: Sci. American, v. 227, p. 26–36.

HUBERT, C., LAJOIE, J., AND LÉONARD, M. A., 1970, Deep sea sediments in the lower Paleozoic Québec Supergroup, *in* LAJOIE, J. (ed.), Flysch sedimentology in North America: Geol. Soc. Canada Special Paper 7 p. 103–125.

JACKA, A. D., AND OTHERS, 1968, Permian deep-sea fans of the Delaware Mountain Group (Guadalupian), Delaware Basin: Soc. Econ. Paleontologists and Mineralogists, Permian Basin Sec. Pub. 68–11, p. 49–90.

JAMES, D. M. D., 1967, Sedimentary studies in the Bala of central Wales (Ph.D. thesis): Swansea, Univ Wales, 126 p.

——, 1972, Sedimentation across an intra-basinal slope: the Garnedd-wen Formation (Ashgillian), west central Wales: Sed. Geology, v. 7, p. 291–307.

——, AND JAMES, J., 1969, The influence of deep fractures on some areas of Ashgillian-Llandoverian sedimentation in Wales: Geol. Mag., v. 106, p. 562–582.

JONES, M. D., 1969, The palaeogeography and palaeoecology of the Leintwardine Beds of Leintwardine, Herefordshire (M.Sc. thesis): Leicester, England, Univ. Leicester, 67 p.

KAY, G. M. (ed.), 1969, North Atlantic—geology and continental drift: Am. Assoc. Petroleum Geologists Mem. 12, 1082 p.

KELLING, G., AND WOOLLANDS, M. A., 1969, The stratigraphy and sedimentation of the Llandoverian rocks of the Rhayader district, *in* WOOD, A. (ed.), The Pre-Cambrian and lower Palaeozoic rocks of Wales: Cardiff, Univ. Wales Press, p. 255–282.

KINDLE, C. H., AND WHITTINGTON, H. B., 1958, Stratigraphy of the Cow Head region, Newfoundland: Geol Soc. America Bull., v. 69, p. 315–342.

KNILL, J. L., 1959, Axial and marginal sedimentation in geosynclinal basins: Jour. Sed. Petrology, v. 29, p. 317–325.

KOMAR, P. D., 1970, The competence of turbidity current flow: Geol. Soc. America Bull., v. 81, p. 1555–1562.

KORÁB, T., LEŠKO, B., AND MARSCHALKO, R., 1967, Inner-Carpathian and Outer Flysch of the East Slovakia: 23rd Internat. Geol. Cong., Prague, 1968, Guide to Excursion 17AC, 38 p.

KRUIT, C., BROUWER, J., AND EALEY, P., 1972, A deep-water sand fan in the Eocene Bay of Biscay: Nature, v. 240, p. 59–61.

KUENEN, PH. H., 1953, Origin and classification of submarine canyons: Geol. Soc. America Bull., v. 64, p. 1295–1314.

——, 1958, Problems concerning source and transportation of flysch sediments: Geol. en Mijnb., v. 20, p. 329–339.

LAPWORTH, H., 1900, The Silurian sequence of Rhayader: Geol. Soc. London Quart. Jour., v. 56, p. 67–137.

LOCK, B. E., 1972, Lower Paleozoic history of a critical area; eastern margin of the St. Lawrence Platform in White Bay, Newfoundland, Canada: 24th Internat. Geol. Cong., Montreal, sec. 6, p. 310–324.

LOWE, D. R., 1972a, Implications of three submarine mass-movement deposits, Cretaceous, Sacramento Valley, California: Jour. Sed. Petrology, v. 42, p. 89–101.

——, 1972b, Submarine canyon and slope channel sedimentation model as inferred from Upper Cretaceous deposits, western California: 24th Internat. Geol. Cong., Montreal, sec. 6, p. 75–81.

McKERROW, W. S., AND ZIEGLER, A. M., 1971, The Lower Silurian paleogeography of New Brunswick and adjacent areas: Jour. Geology, v. 79, p. 635–646.

——, AND ——, 1972, Silurian paleogeographic development of the proto-Atlantic Ocean: 24th Internat. Geol. Cong., Montreal, sec. 6, p. 4–10.

MARSCHALKO, R., 1964, Sedimentary structures and paleocurrents in the marginal lithofacies of the central-Carpathian flysch, *in* BOUMA, A. H., AND BROUWER, A. (eds.), Turbidites: Amsterdam, Elsevier Pub. Co., Developments in sedimentology 3, p. 106–126.

MARTIN, B. D., 1963, Rosedale Channel—evidence for late Miocene submarine erosion in Great Valley of California: Am. Assoc. Petroleum Geologists Bull., v. 47, p. 441–456.

——, AND EMERY, K. O., 1967, Geology of Monterey Canyon, California: *ibid.,* v. 51, p. 2281–2304.

MENARD, H. W., 1960, Possible pre-Pleistocene deep-sea fans off central California: Geol. Soc. America Bull., v. 71, p. 1271–1278.

——, SMITH, S. M., AND PRATT, R. M., 1965, The Rhône deep-sea fan, *in* WHITTARD, W. F., AND BRADSHAW, R. (eds.), Submarine geology and geophysics: London, Butterworth and Co., p. 271–285.

MITCHELL, A. H., AND READING, H. G., 1971, Evolution of island arcs: Jour. Geology, v. 79, p. 253–284.

MUTTI, E., 1963, Confronto tra le direzioni d'apporto dei clastici entro il Macigno e il "Tongriano" dell'Appennino di Piacenza: Riv. Italiana Paleontologie e Stratigrafia, v. 69, no. 3, p. 235–258.

——, 1964, Schema paleogeografico del Paleogene dell'Appennino di Piacenza: *ibid.,* v. 70, p. 869–885.

——, 1969, Studi geologici sulle isole del Dodecaneso (Mare Egeo). X. Sedimentologia delle Arenarie di Messanagros (Oligocene-Aquitaniano) nell'isola di Rodi: Soc. Geol. Italiana Mem., v. 8, p. 1027–1070.

——, AND DE ROSA, E., 1968, Caratteri sedimentologici delle Arenarie di Ranzano e della formazione di Val Luretta nel basso Appennino di Piacenza: Riv. Italiana Paleontologia, v. 74, p. 71–120.

——, AND GHIBAUDO, G., 1972, Un esempio di torbiditi di conoide sottomarina esterna: le arenarie di San Salvatore (Formazione di Bobbio, Miocene) nell'Appennino di Piacenza: Accad. Sci. Torino Mem., Cl. Sci. Fis., Mat. e Nat., ser. 4a, no. 16, 40 p.

——, AND RICCI LUCCHI, F., 1972, Le torbiditi dell'Appennino Settentrionale: introduzione all'analisi di facies: Soc. Geol. Italiana Mem., v. 11, p. 161–199.

NASU, N., 1964, The provenance of the coarse sediments on the continental shelves and the trench slopes off the Japanese Pacific coast, *in* MILLER, R. L. (ed.), Papers in marine geology (Shepard Commemorative Vol.): New York, Macmillan Co., p. 65–101.

NEEV, D., 1960, A pre-Neogene erosion channel in the southern coastal plain of Israel: Israel Ministry Devel., Geol. Survey Bull. 25, Oil Div. Paper 7, 20 p.

NELSON, C. H., AND OTHERS, 1970, Development of the Astoria Canyon-Fan physiography and comparison with similar systems: Marine Geology, v. 8, p. 259–291.

NORMARK, W. R., 1970, Growth patterns of deep sea fans: Am. Assoc. Petroleum Geologists Bull., v. 54, p. 2170–2195.

——, AND PIPER, D. J. W., 1969, Deep-sea fan-valleys, past and present: Geol. Soc. America Bull., v. 80, p. 1859–1866.

PAINE, W. R., 1966, Stratigraphy and sedimentation of subsurface Hackberry wedge and associated beds of southwestern Louisiana. Gulf Coast Assoc. Geol. Societies Trans., v. 16, p. 261–274.

PASSEGA, R., 1954,. Turbidity currents and petroleum exploration: Am. Assoc. Petroleum Geologists Bull., v. 38, p. 1871–1887.

PIPER, D. J. W., 1969, Silurian sediments in western Ireland (Ph.D. thesis) : Cambridge, England, Univ. Cambridge, 213 p.

——, 1970, A Silurian deep sea fan deposit in western Ireland and its bearing on the nature of turbidity currents : Jour. Geology, v. 78, p. 509–522.

——, 1971, Sediments of the Middle Cambrian Burgess Shale, Canada: Lethaia, v. 5, p. 169–175.

——, 1972, Sedimentary environments and palaeogeography of the late Llandovery and earliest Wenlock of north Connemara, Ireland : Geol. Soc. London Quart. Jour., v. 128, p. 33–51.

——, AND NORMARK, W. R., 1971, Re-examination of a Miocene deep-sea fan and fan-valley, southern California : Geol. Soc. America Bull., v. 82, p. 1823–1830.

PRAY, L. C., AND STEHLI, F. G., 1963, Allochthonous origin, Bone Spring "patch reefs," west Texas (abs.) : Geol. Soc. America Special Paper 73, p. 218–219.

RADOMSKI, A., 1961, On some sedimentological problems of the Swiss flysch series: Eclogae Geol. Helvetiae, v. 54, p. 451–459.

RICCI LUCCHI, F., 1969, Channelized deposits in the middle Miocene flysch of Romagna (Italy) : Gior. Geologia., v. 36 (for 1968), p. 203–260.

ROBERSON, M. I., 1964, Continuous seismic profiler survey of Oceanographer, Gilbert and Lydonia submarine canyons, Georges Bank : Jour. Geophys. Research, v. 69, p. 4779–4789.

ROBINSON, F. M., 1964, Core tests, Simpson area, Alaska: U.S. Geol. Survey Prof. Paper 305-L, p. 645–730.

SADLER, H. E., 1964, The origin of the "Beach-Beds" in the Lower Carboniferous of Castleton, Derbyshire: Geol. Mag., v. 101, p. 360–372.

——, 1970, Boring algae in brachiopod shells from Lower Carboniferous (D_1) limestones in north Derbyshire, with special reference to the conditions of deposition: Mercian Geologist, v. 3, p. 283–290.

SCHENK, P. E., 1970, Meguma Group (lower Paleozoic), *in* LAJOIE, J. (ed.), Flysch sedimentology in North America : Geol. Soc. Canada Special Paper 7, p. 127–153.

SCHOEFFLER, J., 1965, Le "Gouf" de Capbreton, de l'Eocène inférieur a nos jours, *in* WHITTARD, W. F., AND BRADSHAW, R. (eds.), Submarine geology and geophysics : London, Butterworth and Co., p. 265–270.

SCOTT, K. M., 1966, Sedimentology and dispersal pattern of a Cretaceous flysch sequence, Patagonian Andes, southern Chile : Am. Assoc. Petroleum Geologists Bull., v. 50, p. 72–107.

SEDIMENTATION SEMINAR (Indiana Univ.), 1969, Bethel Sandstone of western Kentucky and south-central Indiana : Kentucky Geol. Survey, ser. 10, Rept. Inv. 11, p. 7–24.

SHEPARD, F. P., AND DILL, R. F., 1966, Submarine canyons and other sea valleys : Chicago, Rand McNally and Co., 381 p.

——, ——, AND VON RAD, U., 1969, Physiography and sedimentary processes of La Jolla submarine fan and fan-valley, California : Am. Assoc. Petroleum Geologists Bull., v. 53, p. 390–420.

SHIRLEY, J., AND HORSFIELD, E. L., 1940, The Carboniferous Limestone of the Castleton-Bradwell area, north Derbyshire : Geol. Soc. London Quart. Jour., v. 96, p. 271–299.

SHORT, K. C., AND STÄUBLE, A. J., 1967, Outline geology of Niger delta: Am. Assoc. Petroleum Geologists Bull., v. 51, p. 761–779.

SHROCK, R. R., 1957, New geological horizons : *ibid.,* v. 41, p. 1403–1408.

SIMPSON, I. M., AND BROADHURST, F. M., 1969, A boulder bed at Treak Cliff, north Derbyshire : Yorkshire Geol. Soc. Proc., v. 37, p. 141–151.

SMITH, A. G., 1971, Alpine deformation and the oceanic areas of the Tethys, Mediterranean, and Atlantic: Geol. Soc. America Bull., v. 82, p. 2039–2070.

STANLEY, D. J., 1964, Large mudstone-nucleus sandstone spheroids in submarine channel deposits: Jour. Sed. Petrology, v. 34, p. 672–676.

——, 1967, Comparing patterns of sedimentation in some modern and ancient submarine canyons: Earth and Planetary Sci. Letters, v. 3, p. 371–380.

——, AND BOUMA, A. H., 1964, Methodology and paleogeographic interpretation of flysch formations : a summary of studies in the Maritime Alps, *in* BOUMA, A. H., AND BROUWER, A. (eds.), Turbidites: Amsterdam, Elsevier Pub. Co., Developments in Sedimentology 3, p. 34–64.

——, AND MUTTI, E., 1968, Sedimentological evidence for an emerged land mass in the Ligurian Sea during the Palaeogene : Nature, v. 218, p. 32–36.

——, AND UNRUG, R., 1972, Submarine channel deposits, fluxoturbidites and other indicators of slope and base-of-slope environments in modern and ancient marine basins, *in* RIGBY, J. K., AND HAMBLIN, W. K. (eds.), Recognition of ancient sedimentary environments : Soc. Econ. Paleontologists and Mineralogists Special Pub. 16, p. 287–340.

STARKE, G. W., 1956, Genesis and geologic antiquity of the Monterey Submarine Canyon (abs.) : Geol. Soc. America Bull., v. 67, p. 1783.

SULLWOLD, H. H., 1960, Tarzana Fan, deep submarine fan of late Miocene age, Los Angeles County, California : *ibid.,* v. 44, p. 433–457.

——, 1961, Turbidites in oil exploration, *in* PETERSON, J. A., AND OSMOND, J. C. (eds.), Geometry of sandstone bodies : Tulsa, Oklahoma, Am. Assoc. Petroleum Geologists, p. 63–81.

TRETTIN, H. P., 1970, Ordovician-Silurian flysch sedimentation in the axial trough of the Franklinian Geosyncline, northeastern Ellesmere Island, Arctic Canada, *in* LAJOIE, J., (ed.), Flysch sedimentology in North America : Geol. Soc. Canada Special Paper 7, p. 13–35.

UNRUG, R., 1963, Istebna Beds—a fluxoturbidity formation in the Carpathian flysch: Soc. Géol. Pologne Annales, v. 33, p. 49–92.

Van Hoorn, B., 1969, Submarine canyon and fan deposits in the Upper Cretaceous of the south-central Pyrenees, Spain: Geol. en Mijnb., v. 48, p. 67–72.

———, 1970, Sedimentology and palaeogeography of an Upper Cretaceous turbidite basin in the south-central Pyrenees, Spain: Leidse Geol. Meded., v. 45, p. 73–154.

Von der Borch, C. C., 1968, Southern Australian submarine canyons: their distribution and ages: Marine Geology, v. 6, p. 267–279.

Walker, R. G., 1966a, Shale Grit and Grindslow Shales: transition from turbidite to shallow water sediments in the Upper Carboniferous of northern England. Jour. Sed. Petrology, v. 36, p. 90–114.

———, 1966b, Deep channels in turbidite-bearing formations: Am. Assoc. Petroleum Geologists Bull., v. 50, p. 1899–1917.

———, 1969, The juxtaposition of turbidite and shallow-water sediments: study of a regressive sequence in the Pennsylvanian of north Devon, England: Jour. Geology, v. 77, p. 125–143.

Warren, P. T., 1962, The petrography, sedimentation and provenance of the Wenlock rocks near Hawick, Roxburghshire: Edinburgh Geol. Soc. Trans., v. 19, p. 225–255.

Wezel, F. C., 1968, Osservazioni sui sedimenti dell'Oligocene-Miocene inferiore della Tunisia settentrionale: Soc. Geol. Italiana Mem., v. 7, p. 417–439.

Whitaker, J. H. McD., 1962, The geology of the area around Leintwardine, Herefordshire: Geol. Soc. London Quart. Jour., v. 118, p. 319–351.

———, 1963, The geology of the area around Leintwardine, Herefordshire: discussion: *ibid.,* v. 119, p. 513–514.

———, in press, Submarine canyons and deep-sea fans, modern and ancient: Stroudsburg, Pennsylvania, Dowden, Hutchinson and Ross.

Wilhelm, O., and Ewing, M., 1972, Geology and history of the Gulf of Mexico: Geol. Soc. America Bull., v. 83, p. 575–600.

Wilson, J. T., 1966, Did the Atlantic close and then re-open? Nature, v. 211, p. 676–681.

Ziegler, A. M., 1970, Geosynclinal development of the British Isles during the Silurian Period: Jour. Geology, v. 78, p. 445–479.

Part IV

ANCIENT DEEP-SEA FANS

Editor's Comments
on Papers 22 Through 26

In this part, examples of ancient deep-sea fans are described. We have selected two Tertiary examples and one each from the Upper and Lower Paleozoic. The final paper, published recently, usefully compares modern and ancient deep-sea fans and gives criteria by whcih ancient fan and fan-valley systems may be recognized.

Sullwold (22) made a pioneer study of a Miocene fan in California. In the first part of his paper, omitted here, he stressed the importance of studying turbidites and their directional structures. His own work on the Mohnian Stage (lower upper Miocene) of the Modelo Formation of California showed it to be a marine, mainly deep-water formation (1000 m), the depth estimate being based on foraminifera and fish remains.

Sullwold analyzed the Modelo sandstones and found them to be arkosic wackes (Gilbert) or labile graywackes (Packham). They have a very high feldspar content. Pebbles proved to be leucocratic acid igneous plutonic rocks, aplites, and perhaps some gneisses, again showing high feldspar content. Sullwold also studied the grain-size distribution, especially within the graded beds, and suggested that grain size may correlate with bed thickness. Sorting is poor, grain sizes spreading over seven or eight classes, and the grains are angular. This, together with the high feldspar content, indicates a first-cycle sediment.

The next section of the paper describes the many nonoriented primary structures in the Modelo Formation which are characteristic of turbidites. These structures are illustrated with photographs and a sketch. The oriented structures are of particular importance to the interpretation of the Modelo Formation as a deep-sea fan, so the map with rosettes of turbidity current directions is reproduced here (Figure 6). Sullwold points out that "the pattern clearly indicates a point source for the currents. This, plus the other evidence for rapid dumping in deep water, is convincing evidence, in the writer's opinion, that the lower Modelo Formation in this area was deposited as a fan at the mouth of a submarine canyon." This interpretation is developed in the second half of the paper, which is reprinted below.

Starting in the 1960s, Roger Walker produced a succession of often-cited papers on turbidites. The paper (23) reprinted here in shortened form shows how a careful study of facies changes in contemporaneous Namurian (Upper Carboniferous) strata of north Derbyshire, England, led to a clear reconstruction of the conditions of deposition in deltas, on a continental shelf and slope, and on a basin floor. What is especially relevant here is Walker's delimitation of slope feeder channels (see also Walker, 1966b) leading to laterally migrating submarine fans built up by proximal turbidites. Further from the slope, the turbidites became distal, and eventually only mudstones covered the basin floor. This study of slope channel (small submarine canyon)–inner fan–outer fan–basin floor transition and its migration southward with time was one of the landmarks along the path toward environmental syntheses which are a major goal of sedimentology at the present time.

The well-illustrated paper (24) by Hubert, Lajoie, and Léonard summarizes detailed sedimentological work on Cambro-Ordovician conglomerates and sandstones in southern Québec, Canada. The authors deduce that these sediments accumulated as channeled deep-sea fans at the mouths of submarine canyons. The fans described are two of several fans formed at the base of a slope that separated a carbonate shelf from a basin in which pelites accumulated. The paper is a pioneer effort in interpreting these ancient Lower Paleozoic rocks as fan deposits, and work on them is still continuing.

A short Addendum, kindly written for this volume by Jean Lajoie, deals with the transporting mechanism of the gravel fraction at L'Islet; he presents some new facts and concludes that the coarse fraction was transported by turbidity currents. Some new and improved photographs have been kindly supplied by Claude Hubert; the new ones are Pl. III, figs. 1, 2, and 3; Pl. IV, fig. 1; Pl. V. fig. 2; Pl. VI; Pl. VII, fig. 2; Pl. IX, figs. 1, 2, and 3. Figure 6 has been omitted as it was a large foldout; it gave a detailed stratigraphic section of the Cap Enragé Formation, with lithological description, at Cap-à-l'Orignal. Figure 7, however, shows a similar (but much thinner) section at Cap Enragé. Some corrections to this paper are: p. 359, Resumé, line 19, for semblages read semblable; p. 364, column 2, line 13, for one and one-half read one-half; p. 372, column 2, next-to-last line, a "less than" symbol has been omitted; p. 377, Plate IX, fig. 3, delete "ruler is 15 cm long"; p. 378, column 1, line 30, for Plate V read Plate IV; p. 378, column 1, line 53, for Plate IV read Plate V; p. 379, column 1, line 16, read "shelf has been eroded"; p. 381, delete reference to Léonard (1969).

In a brief paper (25), Kruit, Brouwer, and Ealey give criteria for considering a series of massive sandstones in northern Spain as an Eocene deep-water fan. The sandstones have a lens-shaped outline, thicken to about 600 m in the central area, and have a divergent paleocurrent pattern plotted from flutes, grooves, and parting lineation. Marl and shale intercalations have axial paleocurrent directions and yield deep-water benthonic foraminifera. Depositional depth is inferred to be at least 1000 m. The authors suggest that the fan could have been fed by a submarine canyon lying some kilometers north of the area. On p. 385, line 28, for 1,000 ml. read 1,000 m.

The final paper, by Nelson and Nilsen (26), is a good sequel to Normark's contribution (12) and gives a detailed comparison of an ancient fan system (the Eocene Butano Sandstone, described by Nilsen and Simoni, 1973) with a well-studied modern fan, part of the Astoria Canyon-Fan system off the coast of Washington (Carlson and Nelson, 1969; Nelson and others, 1970). This recently published paper gives a comparison of the depositional processes of modern and ancient fans, includes criteria for recognizing the different parts of ancient fan deposits, and discusses the tectonic settings of fans. It should prove invaluable to researchers seeking to differentiate the various parts of possible ancient fan systems. Additional references are given on pp. 9–10.

Biographical Notes

Harold H. Sullwold, Jr., was born in 1916 in St. Paul, Minnesota. He gained the B.A., M.A., and Ph.D. degrees in Geology at the University of California at Los Angeles, where for some years he was Instructor in Geology. He spent two years with the U.S. Geological Survey in the western states, but his major work has been as consulting petroleum geologist based in North Hollywood, and now in Carpinteria, California. He specializes in exploration for natural gas in the Sacramento Valley and in Tertiary and Cretaceous stratigraphy and structure. Sullwold is a Registered Geologist and Registered Petroleum Engineer, State of California.

Born in London, England, in 1939, Roger G. Walker studied at Oxford University, receiving the B.A. degree in 1961 and Ph.D. in 1964. After two years as NATO postdoctoral Fellow at Johns Hopkins University in Baltimore, he joined McMaster University, Hamilton, Ontario, where he is now Professor of Geology. Walker's many publications on turbidites are very well known and the paper that follows (23) won for the author a Journal of Sedimentary Petrology Honorable Mention. His invaluable historical review, "Mopping up the turbidite mess" (1973), has pertinent comments on submarine canyons and fans.

Claude Hubert received the B.Sc. degree at the Université de Montréal in 1958 and the Ph.D. at McGill University in 1965. He is currently Associate Professor at the Université de Montréal. His special interest is the structure, stratigraphy, and sedimentology of the northern Appalachians.

Jean Lajoie also received the B.Sc. degree at the Université de Montréal in 1958 and the Ph.D. at McGill University in 1963. He is Associate Professor at the Université de Montréal and his interests lie in the sedimentology of flysch sediments.

Marc Léonard was awarded the B.Sc. degree in 1967 at the Université de Montréal. Presently he is Geologist with Pickards Mahler Co., Vancouver, British Columbia.

Cornelis Kruit gained his Ph.D. in Geology in 1950 at Utrecht University, The Netherlands, and completed a thesis on recent sediments of the Rhône delta in 1955. He has been employed as a sedimentologist by the Royal Dutch Shell Group since 1951. During his time with Shell he has been in charge of research and services dealing with siliciclastic sediments, interrupted only by a posting to Shell-BP's operations at Port Harcourt, Nigeria, from 1964 to 1967. He commenced regional studies of the Southern Pyrenean Basin in 1969, which led to the discovery of the San Sebastian–Monte Jaizkibel deep-water fan in the spring of 1972.

Johannes Brouwer has been engaged as a micropaleontologist with Shell Companies in Colombia, Texas, and Indonesia, and during the last 15 years he has been dealing with paleontological research at Rijswijk, The Netherlands. He has published on Liassic foraminifera of northwestern Europe and foraminiferal faunas from graded sequences and from deep-sea sediments in the Gulf of Guinea.

Peter Ealey was born in England in 1941 and educated at Marlborough College and Birmingham University. He completed Master's and Doctoral degrees at the University of Illinois, where he was involved in deep-marine studies off South America. In 1969 he joined the Royal Dutch Shell Group and spent three years with the siliciclastic and deep-sea group at KSEPL in The Netherlands. He is currently involved in sedimentological studies with Shell UK in London.

C. Hans Nelson was born in 1937. His professional education was completed at Carleton College (B.A. 1959), University of Minnesota (M.S. 1962), and Oregon State University (Ph.D. 1968). He has taught at Carleton College (1959), Lehigh University (1961–1962), Portland State College (1962–1963), and at Chapman College World Campus Afloat (1966–1967). Nelson was visiting professor of geological oceanography at San Jose State University (1968–1969), Hayward State University (1970–1971), and Stanford University (1973). In 1967, he joined the U.S. Geological Survey as a marine geologist studying sediments, sedimentary processes, and history of marine placer deposition of the northern Bering Sea. Nelson's interests and experience are in Quaternary and Holocene sedimentation and ecology, especially the history of recent sedimentary systems and the characteristics of sedimentary processes and depositional trends for defining ancient environments. His specific research has covered: continental glacial and lake deposits of Minnesota; sedimentation as well as limnology of Crater Lake, Oregon; and deep-sea fan bathymetry, stratigraphy, sedimentary processes, and history of deposition. Present studies include history of glaciation, heavy and toxic trace metal dispersal, topographic development, sea-level changes, and bottom-current effects on the Bering Sea Shelf. Hans Nelson has authored or coauthored 45 publications on these topics.

Tor H. Nilsen was born in New York City in 1941. He gained the B.S. in 1962, the M.S. in 1964, and the Ph.D. in 1967 in Geology at the City College of New York. In 1967, he was Geologist with the Shell Development Company, and from 1967 to 1969, he was with the U.S. Army Corps of Engineers. In 1969, he joined the U.S. Geological Survey at Menlo Park, California, where he is currently employed as Geologist. He has worked on sedimentology, stratigraphy (especially on the Old Red Sandstone of Norway and the Tertiary of California), and environmental geology (especially on landslips). In 1974 he participated in DSDP Leg 38 on board D/S *Glomar Challenger*.

22

Copyright © 1960 by the American Association of Petroleum Geologists

Reprinted from *Amer. Assoc. Petrol. Geol. Bull.*, **44**(4), 433, 444–445,
447–457 (1960)

TARZANA FAN, DEEP SUBMARINE FAN OF LATE MIOCENE AGE LOS ANGELES COUNTY, CALIFORNIA[1]

HAROLD H. SULLWOLD, JR.[2]
North Hollywood, California

ABSTRACT

The sandstone beds in the Modelo formation (upper Miocene) exposed on the north flank of the Santa Monica Mountains were for the most part deposited from turbidity currents. Evidence of rapid deposition is provided by poor sorting, high clay-silt content, grain angularity, high feldspar content, and load deformation. Evidence for great depth of water (about 3,000 feet) is provided by abundant foraminifers and fish remains in the interbedded shales. That the mechanism for transporting this poorly sorted sand-silt-clay mixture into waters of this depth was that of turbidity currents is borne out by the multitude of syngenetic textures and structures which these rocks have in common with accepted turbidites and with structures which have been produced in laboratory studies.

Some of these structures are oriented with respect to the direction of travel of the current and thierefore provide a means of determining the direction of bottom slope along which the currents moved. When plotted on a map these directions reveal a pattern which is unmistakably that of a fan with its apex toward the north. The area studied constitutes a 16-mile-long section through this outcropping north-tlted fan. The point source was most logically the mouth of a submarine canyon. The canyon itself is thus far unrevealed, being concealed beneath the alluviated San Fernando Valley, but the eroding source area is tentatively identified on the basis of mineralogical studies in the San Gabriel Mountains about 22 miles northeast of the fan's apex.

Further studies of this sort can not fail to provide valuable data on earth history and thus aid petroleum exploration in similar rock types elsewhere in California.

[*Editor's Note:* Material has been omitted at this point.]

[1] Manuscript received, March 19, 1959.

[2] Consulting geologist.
This paper is a condensation of a dissertation submitted in partial satisfaction of the requirements for the Ph.D. degree at the University of California, Los Angeles. The writer is indebted to Professors J. C. Crowell, Cordell Durrell, and E. L. Winterer for stimulation, assistance, and criticism; to John de Grosse for preparation of petrographic thin sections; to W. T. Rothwell, R. L. Pierce, and R. L. Brooks of the Richfield Oil Corporation for micropaleontological control; and to the many graduate students of the University of California, Los Angeles, whose theses provided the geologic maps on which this study is partly based. These former students include F. W. Bergen, T. J. Brady, George Brown, W. W. Duarte, J. G. Elam, J. L. Elliott, T. P. Harding, G. C. Hazenbush, J. T. McGill, J. N. Terpenning, J. D. Traxler, and J. C. West. The critical reading of this manuscript by Professors Crowell and Winterer is greatly appreciated.

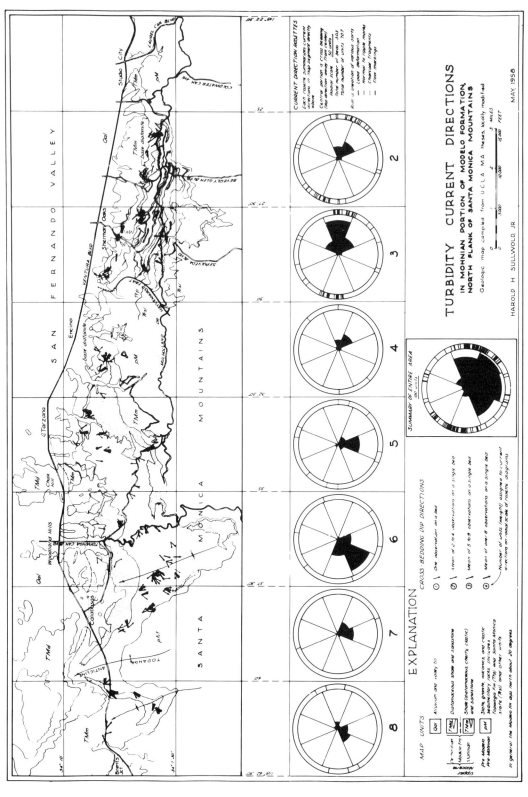

FIG. 6.—Geologic map and turbidity-current directions.

INTERPRETATION OF RESULTS AND DISCUSSION

The writer believes that he has offered ample evidence to suggest strongly that the Modelo sandstones in the Santa Monica Mountains are:

1. Wackes (dirty immature sandstones) containing an abundance of interstitial silt and clay.

2. Arkoses whose model size particles have an extremely high feldspar content.

3. Turbidites containing abundant textural and structural characteristics considered by Kuenen and others to be a result of deposition from turbidity currents.

4. Deep-water deposits (almost certainly bathyal; perhaps near 3,000 feet).

5. Deposits from currents traveling from a point source somewhere not far northwest of Tarzana.

These items are direct conclusions of this study and may be used as a basis for some speculation about environment and geography of the area in late Miocene time.

TARZANA FAN

The case for the radial pattern of current directions determined by cross-bedding dips is amply demonstrated on the map (Fig. 6). Since the sands are turbidity-current deposits which normally travel downslope, the conclusion is drawn that the sea bottom in the area investigated lay astride the south flank of a low cone. Projection backward of current directions of arrows or rosettes in map segments 4, 5, and 6 suggests that the point source may be $\frac{1}{2}$ mile north of Ventura Boulevard between Tarzana and Chalk Hill. Map segments 2, 3, 7, and 8 do not fit perfectly, but are not far out of line. Segments 2 and 3 lie east of the Hayvenhurst fault, which has had considerable strike slip, while segments 7 and 8 lie in a more complex structural area where the currents may have been diverted by warping of the sea floor as a prelude to the folding which is now so prominent. As the current directions here are not correlative precisely with the present fold axes, it is possible that the axes have migrated or that a second more westerly fan, whose main part is now eroded away, may have existed and interfingered with the Tarzana Fan.

Further evidence supporting the fan concept may be seen on the partly restored stratigraphic section (Fig. 7), where the marked thinning of the lower Modelo (Mohnian) part of the section both east and west of the Calabasas-Woodland Hills-

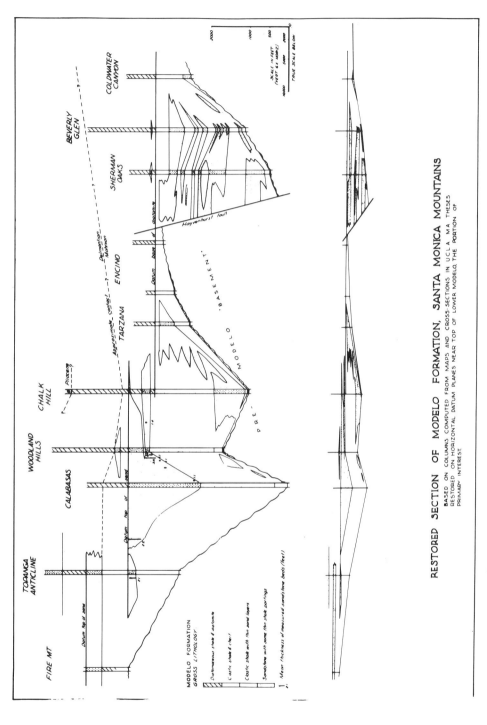

RESTORED SECTION OF MODELO FORMATION, SANTA MONICA MOUNTAINS

BASED ON COLUMNS COMPUTED FROM MAPS AND CROSS-SECTIONS IN UCLA MA THESES RESTORED ON HORIZONTAL DATUM PLANES NEAR TOP OF LOWER MODELO THE PORTION OF PRIMARY INTEREST

FIG. 7.—Restored section of Modelo formation, Santa Monica Mountains. Essentially a cross section of Tarzana Fan with its top surface horizontal and its eastern end offset by Hayvenhurst fault.

Chalk Hill columns is obvious. The columns were taken from cross sections or plotted from maps in Master of Arts theses at the University of California, Los Angeles, and from the published section by Hoots (1931) locally modified by the writer. The section has the usual shortcomings resulting from conversion of outcrop measurements in gently dipping strata to a pseudo-vertical stratigraphic section. Folding has been eliminated, but the Hayvenhurst fault is shown to emphasize the pronounced change in thickness on its two sides. As the only horizon that could be traced entirely through the area was the base of the Modelo, and, since this is a known erosional surface and probably had considerable relief, an attempt was made to establish a datum near the top of the rocks of interest to portray conditions near the end of Mohnian time. The datum planes are drawn horizontally rather than cone-shaped because the shape and steepness of the cone and location of its axis are not precisely known.

Gross shape of the Mohnian rocks in cross section is in agreement with the point-source concept in that the greatest thickness is near the point source indicated by current directions, and the thickness diminishes away therefrom. The coincidence of thickest section with current point-source is not perfect, but is close. The discrepancy can perhaps be explained by the 2-dimensional aspect of the outcrop area, the possible skewed position of the area on the fan, the "tilt" of the plane of the section, the possible asymmetric shape of the fan, and the irregularities of the basement surface upon which the fan was deposited. The section also reveals the greater abundance of sands in the areas of greater total thickness; that is, the excess thickness of the thicker columns is largely due to increased sand content.

An anomaly exists east of the Hayvenhurst fault where the thickness increases abruptly. This area could be a part of the same fan (or a different fan) brought to its present position by strike-slip movement along the Hayvenhurst fault as shown in Figures 8 and 9. Here the fan is shown elongate toward the east to account for the uniform easterly direction of currents in that area. The fault could have occurred either before or after tilting, but slip was nearly parallel with the bedding planes of the upper Modelo strata as shown by the relatively small separation at the base of the diatomite (Fig. 6).

Figure 8 also attempts to account for the lack of coincidence between current apex and maximum thickness by showing the unusually thick Calabasas sand lens in position west of the apex where it approximately doubles the total thickness. Also shown is the previously mentioned possibility of interference of current directions at the west end by a second fan postulated on the west, now largely eroded away.

The details of the relation of the large mapped sandstone lenses or bundles to the enclosing shales were not worked out in the field and would be a useful project for the future. Several partial sections were measured primarily to obtain individual bed thicknesses, and it is apparent that there is no consistent thinning of individual sand beds toward the edges of the bundles. It may be tentatively concluded that there are *more* sandstone beds in the thicker parts of the bundles. These data are shown in Figure 7.

The dirtiness and grain angularity of the Modelo sandstones are believed to be in accord with the fan concept. Rapid dumping and lack of winnowing would be expected in a fan or delta given the proper source material. Graded sands in modern deltas at the mouths of submarine canyons have been reported by Ericson et al. (1951) and Gorsline and Emery (1959). Cores averaging 20 feet in length in the delta of the Hudson submarine canyon show the sediment to be composed of 30 per cent sand, ranging in thickness from thin films to 25 feet, well sorted, graded, with sharp bases and sharp or gradational tops, containing both shallow-water and deep-water foraminifera. These sand bodies are not correlatable between cores and consist mostly of quartz with common feldspar and ferromagnesian grains, both angular and rounded, and vegetal matter. The parameters are not entirely in accord with the Modelo sandstones, especially as to sorting, but the Hudson delta is 400 miles from shore and at a depth of 15,000 feet! Fans offshore from Los Angeles have been recently identified at the mouths of four submarine canyons in the San Pedro and Santa Monica basins at depths of 3,000 feet. The fans and channels are clearly identifiable by their shape, topographic expression, and high sand-shale ratios. The thickest sand layer found at the top of a fan is 4 feet thick and grades from gravel at the base to very fine sand at the top. Sands farther down the flanks of the delta are thinner and finer-grained,

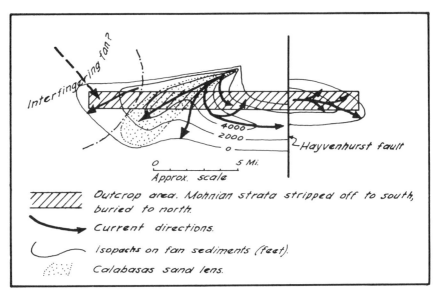

Fig. 8.—Diagrammatic map of Tarzana Fan showing proposed slip along Hayvenhurst fault to account for great change in thickness on its two sides. Presence of unusually thick Calabasas sand lens may account for non-coincidence of thickest part of fan with center as indicated by current directions. Also shown (in broken lines) is part of possible second fan on the west which may account for erratic current directions in that area. Only data are from shaded area of outcrop; hence, over-all configuration is highly conjectural.

and those in the basin floors are $\frac{1}{2}$–1 inch thick with median size of about 70 microns. All sands on the fans are graded, and the material is similar to that moving along the coast in shallow water. These two basins form a part of a group of thirteen offshore basins containing various amounts of sedimentary fill. Their onshore counterparts (Los Angeles Plain, Ventura Plain, and, presumably, the San Fernando Valley) have

been completely filled because of proximity to eroding mountains.

SOURCE AREA

It seems reasonably clear that the sandstone beds in the Modelo formation are first-cycle deposits, as borne out by the high feldspar content and grain angularity. The task is to locate an area of crystalline rocks that could (or preferably

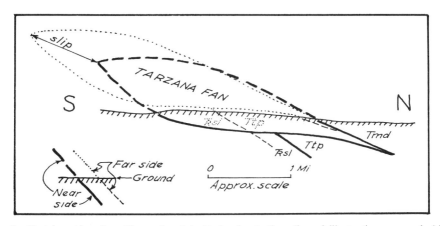

Fig. 9.—Sketch section along Hayvenhurst fault showing both walls and illustrating proposed oblique slip of Tarzana Fan to account for great change in thickness across fault. View westward. Actual cross-sectional shape of fan is unknown, for much of it is eroded away, and there is no well information on subsurface part.

must) have been the eroding source. This calls for a crystalline body which is not overlain by pre-Modelo strata. The regional map (Fig. 10) depicts possible Mohnian eroding areas.

A nearby source under the San Fernando Valley is possible but unlikely, as no evidence for this has been found in the exploratory holes drilled in the search for oil. Most wells have been in the northern and western parts of the valley where, although not uniform in thickness, the Mohnian rocks are everywhere present and overlie older sedimentary rocks.

Similarly, north and west of the San Fernando Valley (west of 118° 30') sedimentary rocks older than Mohnian are present above the basement. Furthermore, the Mohnian rocks which lie between the nearest possible northwestern crystalline source (in the vicinity of Ridge basin) and the Santa Monica Mountains are locally devoid of sand and contain other faunal and lithologic evidence suggesting the presence in this intervening region of several east-west-trending troughs which would constitute sand traps and prevent turbidity currents from reaching the Santa Monica Mountains (Winterer, 1954, and personal communication; Kew, 1924, cross sections). More specifically, the present Oak Ridge-Simi Valley trend seems to have been a submarine high in Mohnian time as evidenced by its relatively thin and sand-free strata.

At the north-central edge of the valley in the San Fernando-Tujunga area the Modelo formation has been mapped lying on crystalline rocks (Oakeshott, 1954, 1958), where it contains cobble and boulder conglomerate suggesting proximity to shore. A group of megafossil localities of late Miocene age contained 22 genera of mollusca, all of which could have lived in shallow water and 9 of which do not now live in depths greater than 180 feet. The fossils were reported by Beatie (1958), and the depth data were provided by Professor C. A. Hall of the University of California, Los Angeles. These localities are from a single horizon whose position with respect to the entire Mohnian section is uncertain because of faulting. Foraminifera in this area also suggest shallower water than in the Santa Monica Mountains, but the evidence is less conclusive. The writer gathered microfaunal data from several sources[3] and

prepared a depth chart which showed a common depth of 600–1,250 feet for all of the 8 species still living. This depth, though based on less complete faunas, corresponds with the 2,500–2,900-foot depth for the Tarzana Fan (Fig. 2). Thus the evidence suggests shallower Mohnian water on the north side of the San Fernando Valley and proximity to eroding land. This places a possible source area in the San Gabriel Mountains which are composed almost entirely of crystalline rocks. These rocks have not been everywhere mapped or studied in detail, but the distribution of certain major types is fairly well known, and only parts of the range appear to be possible source terrane for the Modelo sandstones.

The modal sand size of the Modelo sandstones has been shown to average about 15 per cent quartz, 22 per cent potash feldspar, and 63 per cent plagioclase, 96 per cent of which is albite or oligoclase. The average mineral content of the few small pebbles studied in thin section was 18 per cent quartz, 43 per cent potash feldspar, 37 per cent plagioclase, and 2 per cent accessories. Perthite was very common. Quartz diorite thus appears to be the ideal source rock, feldspar-rich gneiss might fit the picture, and granodiorite or quartz monzonite would not seriously alter it if present in subordinate amounts.

West of Longitude 118° 05' in the San Gabriel Mountains about half the exposed crystalline rocks consist of anorthosite, gabbro, and diorite, in which feldspar is too calcic to have contributed heavily to the Modelo sandstones. However, east of this meridian there are large quantities of granodiorite and diorite gneiss which could have been the original source area. These are the rocks labelled "gr" and "gn" on the map (Fig. 10) along the 118° meridian in the vicinity of Pacifico Mountain and Little Rock Creek north of Mt. Wilson. Detailed studies, especially of volumes of various rock types, are lacking in this area, but J. R. Walker, graduate student at University of California, Los Angeles, is currently studying them for the purpose of matching basement rock types across the San Andreas fault. He has looked at more thin sections in this area than anyone else to the writer's knowledge, and, after viewing the Modelo pebble thin sections, stated that they could have come from the San Gabriel Mountains east of the anorthosite body. Several characteristics common to the pebbles and the suspected source area are noteworthy: (1) microperthite is common and of similar aspect in both groups;

---

[3] Harold Lian, Union Oil Company, outcrop samples; W. T. Rothwell, Richfield Oil Corporation, outcrop samples; B. C. Jones, Union Oil Company, several wells; Howard Reynolds, consultant, several wells.

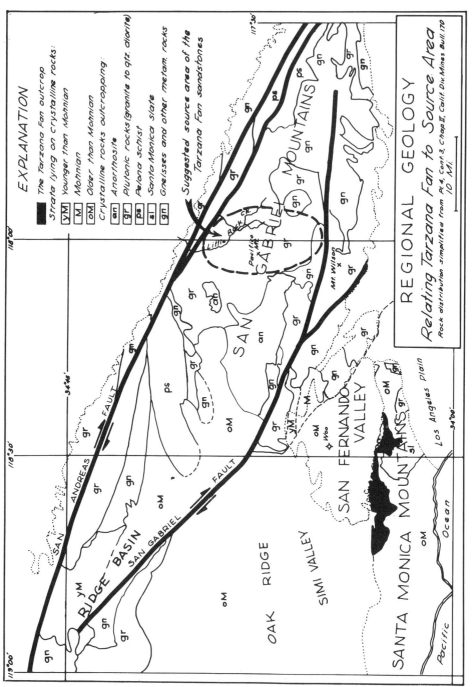

Fig. 10.—Regional geology. Physiographic features indicated by dotted lines, geology by solid and broken lines.

(2) strained quartz and strained feldspar twinning is common in both groups; (3) late quartz is common in both groups; (4) plagioclase in the crystalline rocks is usually about An 25–30 which fits reasonably well with the Modelo sand grains and very well with the pebbles. Farther east of the Pacifico Mountain-Little Rock Creek area the quartz percentage, An content of plagioclase, and degree of metamorphism increase in most of the crystalline rock greatly reducing the likelihood of this area as the source for the Tarzana Fan sediments (Hsu, 1954).

A possible source area is thus identified about 30 miles northeast of the apex of the fan. If the near-shore Mohnian facies here discussed were along the route of travel, the most logical geographic position for the source area would be the region now occupied by anorthosite and gabbro. Since these rocks are evidently not represented in the Tarzana Fan, we may consider that the stream supplying the proper debris by-passed the anorthosite area on the southeast or that during Mohnian time the present site of the anorthosite was occupied by proper lithologic types. The mechanism for this latter notion has already been provided by Crowell (1952) who showed that since Miocene time the San Gabriel fault has had about 20 miles of right strike-slip movement. "Unfaulting" would place the presently favored rock types into position north of the San Fernando-Sunland area and would not require special drainage conditions. It would also result in shorter subaerial transport, thus favoring the high feldspar content and grain angularity of the Tarzana Fan sandstones. Incomplete but highly significant is the cobble inspection of the Cenozoic conglomerates in the San Fernando-Sunland area by Beatie (1958) and Merifield (1958). They report anorthosite clasts rare or absent in the Modelo formation (upper Miocene) and common in the Saugus formation (Plio-Pleistocene).

It is also possible but far less likely that the source area may have been more distant, perhaps across the San Andreas fault. In this case the most abundant source rock would probably be biotite quartz monzonite which, in the Barstow Quadrangle (Bowen, 1954), has the mineral composition of quartz 25–35 per cent, microcline perthite 30–35 per cent, plagioclase (An 28) 25–30 per cent, biotite 2–4 per cent, and muscovite 0–2 per cent. This source area would necessitate a longer stream perhaps involving enrichment of

quartz through destruction of feldspars and possibly more grain-rounding than appears desirable.

In either case the low quartz content of the Modelo sandstones is difficult to explain. In view of the greater stability of quartz relative to feldspar, we should seek a source area with considerably less than 15 per cent quartz, but rocks with the combination of low quartz, high potash feldspar, and lack of calcic plagioclase are rare. Perhaps there is some mechanism by which some of the quartz did not reach the site of deposition (through being coarser as a result of being harder), or does not appear in its true proportions in the median size analyzed. The mineral analysis itself was admittedly inexact, and knowledge of the crystalline rocks is even more so. Hence the writer tentatively concludes that there were possible source areas present in the right direction to supply detritus for the Tarzana Fan, but that its precise identity awaits further study.

SUBMARINE CANYON

A submarine canyon seems required to provide a point source for the sands of the Tarzana Fan in the depth of water indicated. The precise location of this canyon must await further factual knowledge, especially the drilling of wells in the San Fernando Valley. Directional current data presented herein locate its mouth just northwest of the community of Tarzana, and a possible Miocene eroding source area has been tentatively located about 22 miles northeast at the present site of the San Gabriel anorthosite complex. If the Modelo on the north side of the Valley was deposited very near shore, as seems likely from rock distribution, coarseness of clasts, and faunal content, the length of the submarine canyon was about 14 miles with an average slope of about 4 per cent. This slope would increase if the canyon headed seaward from shore and would decrease if the water was relatively deep near shore. However, it appears to have a reasonable order of magnitude (B. C. Heezen, personal communication).

Additional supporting evidence for the canyon is the presence in a well (Standard Oil Company of California's Woo No. 1, Sec. 19, T. 2 N., R. 15 W., S.B.B.M.), roughly half-way between the fan and source area, of the highest Mohnian sand-shale ratio (about 1:1) of any of the central valley bore holes for which data were available to the writer. Much cobble conglomerate was present in this well. This was the easternmost of a series of exploratory holes drilled in the west-central valley, and its high sand and conglomerate content suggests the possibility that it may have penetrated the coarse fill of a depression, which might be a filled submarine canyon. Wells west of the Woo well have Mohnian sand-shale ratios of 1:4 to 1:12, but incompleteness of well data prevents accuracy in these determinations. At any rate the relative coarseness of the Mohnian in the Woo well corroborates the concept that the Tarzana Fan provenance was on the north.

<h3>SIGNIFICANCE OF DIRECTIONAL CURRENT STUDIES</h3>

Since the Modelo formation is not known to be unique in California, but rather representative of the type of clastic rocks that were deposited in the restricted troughs of the west coast throughout the Cenozoic era, it is a virtual certainty that similar studies in other west coast areas and in rocks of other ages will produce useful results. Oriented syngenetic structures have been observed by the writer in the Stevens sand (core from upper Miocene of San Joaquin Valley), Repetto and Pico formations (Pliocene of Ventura County), Puente formation (upper Miocene outcrops in the Puente Hills), Topanga formation (middle Miocene of Santa Monica Mountains), and Cretaceous outcrops in Sespe Creek and Simi Hills. They undoubtedly are very common, though unobtrusive, and must be sought to be found. Lack of mention of them in the literature does not prove their absence.

Use of oriented syngenetic structures in petroleum exploration in California (and probably elsewhere) is of great significance. The trend of linear shaped sand bodies could be predicted were current directions known. Channels could be predicted crossing anticlines where traps would result without structural closure. Masses of lenticular sand and shale bodies formerly thought to be randomly oriented may be found to have a predictable pattern. Magnitude of net slip on strike-slip faults can be determined by matching offset sedimentational patterns.

Establishment of the current patterns depends largely on well data, especially cores. In areas of known structure, cores whose top and bottom are known could be oriented with respect to the direction of dip, and their oriented features recorded. Where structure is not well known or dips are flat, cores oriented by other means would be required. After the current patterns are established, the reversal of these procedures could produce an oriented structural dip in a core containing a recognizable oriented syngenetic feature. In the case of cores, where a maximum amount of information must be squeezed from a small but expensive piece of rock, grain orientation might prove most useful. This technique has been successful in the Pliocene rocks of the Ventura Avenue oil field. Subsurface investigations would have the advantage of the third dimension of sand body shapes in contrast to the present study which was limited to the outcropping two-dimensional upturned cross section of the submarine fan. Because of the abundance of turbidites in the oil- and gas-bearing part of the geologic column in California (as yet largely undocumented), the writer feels that a whole new frontier of geologic knowledge is awaiting explorers, both academic and commercial, and that many oil pools and many answers to earth history problems will be found by detailed application of turbidity-current studies.

<h3>REFERENCES</h3>

AMERICAN GEOLOGICAL INSTITUTE, 1958, "Roundness of Sedimentary Particles," Data Sheet 7.

BANDY, O. L., 1953, "Ecology and Paleoecology of Some California Foraminifera; Part I: The Frequency Distribution of Recent Foraminifera off California," Jour. Paleon., Vol. 27, pp. 161–203.

BEATIE, R. L., 1958, "The Geology of the Sunland-Tujunga Area, Los Angeles County, California," unpublished Master's thesis, University of California, Los Angeles. 102 pp.

BERGEN, F. W., 1955, "A Restudy of the Upper Mohnian-Lower Delmontian Boundary near Calabasas, California," ibid.

*BOKMAN, JOHN, 1953, "Lithology and Petrology of the Stanley and Jackfork Formations," Jour. Geology, Vol. 61, pp. 152–70.

BOWEN, O. E., JR., 1954, "Geology and Mineral Deposits of Barstow Quadrangle, San Bernardino County, California," California Div. Mines Bull. 165.

BRADY, T. J., 1957, "Geology of Part of the Central Santa Monica Mountains East of Topanga Canyon, Los Angeles County, California," unpublished Master's thesis, University of California, Los Angeles.

BRAMLETTE, M. N., 1946, "Monterey Formation of California and Origin of Its Siliceous Rocks," U. S. Geol. Survey Prof. Paper 212. 57 pp.

BROWN, G. E., 1957, "Geology of Parts of the Calabasas and Thousand Oaks Quadrangles, Los Angeles and Ventura Counties, California," unpublished Master's thesis, University of California, Los Angeles. 75 pp.

* Papers dealing with recognition of turbidite structures.

BUTCHER, W. S., 1951, "Foraminifera, Coronado Bank and Vicinity, California," *Univ. California, Scripps Inst. Oceanography, Submarine Geol. Rept. 19*, pp. 1–9.

CROUCH, R. W., 1952, "Significance of Temperatures on Foraminifera from Deep Basins off Southern California," *Bull. Amer. Assoc. Petrol. Geol.*, Vol. 36, pp. 807–43.

CROWELL, J. C., 1952, "Probable Large Lateral Displacement on San Gabriel Fault, Southern California," *ibid.*, Vol. 36, pp. 2026–35.

*———, 1955, "Directional Current Structures from the Prealpine Flysch, Switzerland," *Bull. Geol. Soc. America*, Vol. 66, pp. 1351–84.

*DOREEN, J. M., 1951, "Rubble Bedding and Graded Bedding in Talara Formation of Northwestern Peru," *Bull. Amer. Assoc. Petrol. Geol.*, Vol. 35, pp. 1829–49.

DUARTE, W. W., 1954, "Sherman Oaks Area, Santa Monica Mountains," unpublished manuscript map for Master's thesis, University of California, Los Angeles.

DURHAM, J. W., 1954, "The Marine Cenozoic of Southern California," in "Geology of Southern California," *California Div. of Mines Bull. 170*, pp. 23–31.

DURRELL, C., 1954, "Geology of the Santa Monica Mountains, Los Angeles and Ventura Counties," *ibid.*, Map Sheet 8.

*DZULYNSKI, S., AND RADOMSKI, A., 1955, "Origin of Groove Casts in the Light of Turbidity Current Hypothesis," *Acta Geol. Polonica*, Vol. 5, pp. 47–66. (Ten-page English summary.)

Elam, J. G., 1948, "Geology of Seminole Quadrangle, Los Angeles County, California," unpublished Master's thesis, University of California, Los Angeles.

ELLIOTT, J. L., 1951, "Geology of Eastern Santa Monica Mountains between Laurel Canyon and Beverly Glen Boulevards, Los Angeles County, California," *ibid.*

EMILIANI, C., 1954, "Temperatures of Pacific Bottom Waters and Polar Superficial Waters during the Tertiary, *Science*, Vol. 119, pp. 853–55.

*ERICSON, D. B., EWING, MAURICE, AND HEEZEN, B. C., 1951, "Deep-Sea Sands and Submarine Canyons," *Bull. Geol. Soc. America*, Vol. 52, pp. 961–66.

*———, AND WOLLIN, G., 1955, "Sediment Deposition in Deep Atlantic," in "Crust of the Earth," *Geol. Soc. America Spec. Paper 62*, pp. 205–19.

*FAIRBRIDGE, R. W., 1946, "Submarine Slumping and Location of Oil Bodies," *Bull. Amer. Assoc. Petrol. Geol.*, Vol. 30, pp. 84–92.

*GILBERT, C. M., 1955, " 'Flow' Cast on Sandstone Beds" (abst.), Geol. Soc. America Cordilleran Section meeting, Berkeley, California.

GORSLINE, DONN S., AND EMERY, K. O., 1959, "Turbidity-Current Deposits in San Pedro and Santa Monica Basins off Southern California," *Bull. Geol. Soc. America*, Vol. 70, pp. 279–90.

HARDING, TOD P., 1952, "Geology of the Eastern Santa Monica Mountains between Dry Canyon and Franklin Canyon," unpublished Master's thesis, University of California, Los Angeles.

HAZENBUSH, G. C., 1950, "Geology of the Eastern Parts of the Dry Canyon and Las Flores Quadrangles, Los Angeles County, California," *ibid.*

*HILLS, E. S., AND THOMAS, D. E., 1954, "Turbidity Currents and the Graptolite Facies in Victoria," *Jour. Geol. Soc. Australia*, Vol. 1, pp. 119–33.

HOOTS, H. W., 1931, "Geology of the Eastern Part of the Santa Monica Mountains, Los Angeles County, California," *U. S. Geol. Survey Prof. Paper 165C*.

HSU, K. J., 1954, "Petrology of the Cucamonga Canyon-San Antonio Canyon Area, Southeastern San Gabriel Mountains, California," unpublished Ph. D. thesis, University of California, Los Angeles.

KEW, W. S. W., 1924, "Geology and Oil Resources of a Part of Los Angeles and Ventura Counties, California," *U. S. Geol. Survey Bull. 753*. 202 pp.

KLEINPELL, R. M., 1938, *Miocene Stratigraphy of California*, Amer. Assoc. Petrol. Geol. 450 pp.

*KOPSTEIN, F. P. H. W., 1954, "Graded Bedding of the Harlech Dome," *Rijksuniversiteit te Groningen, Geol. Inst. Pub. 81*. 97 pp. Groningen-Nederland.

*KRUMBEIN, W. C., AND SLOSS, L. L., 1951, *Stratigraphy and Sedimentation*, pp. 94–101 (Sedimentary structures). W. H. Freeman and Company, San Francisco.

*KSIAZKIEWICZ, M., 1954, "Graded and Laminated Bedding in the Carpathian Flysch," *Soc. Geol. Pologue Annales*, Vol. 22, Fasc. 4, pp. 442–49. English summary of paper in Polish (Polskiego Towarzystwa Geologicznego, SW., Anny 6, Krakow, Poland).

*KUENEN, P. H., 1948, "Slumping in the Carboniferous Rocks of Pembrokeshire," *Quar. Jour. Geol. Soc. London*, Vol. 104, pp. 365–85.

———, 1948a, "Turbidity Currents of High Density," *18th Internat. Geol. Cong.*, Pt. VIII, Proc. of Sec. G (1950), pp. 44–52.

*———, 1952, "Paleogeographic Significance of Graded Bedding and Associated Features," *Proc. Koninkl. Nederl. Akad. Van Wetenschappen-Amsterdam*, Ser. B, 55, No. 1, pp. 28–36.

*———, 1953, "Significant Features of Graded Bedding," *Bull. Amer. Assoc. Petrol. Geol.*, Vol. 37, pp. 1044–66.

*———, 1954, "Recent Deposits and the Interpretation of Sedimentary Rocks" (summary), *Adv. Sci.*, Vol. 11, pp. 65–66.

*———, 1957, "Sole Markings of Graded Graywacke Beds," *Jour. Geology*, Vol. 65, pp. 231–58.

*———, AND CAROZZI, A., 1953, "Turbidity Currents and Sliding in Geosynclinal Basins in the Alps," *ibid.*, Vol. 61, pp. 363–73.

*———, AND MIGLIORINI, C. I., 1950, "Turbidity Currents as a Cause of Graded Bedding," *ibid.*, Vol. 58, pp. 91–126.

*———, AND SANDERS, J. E., 1956, "Sedimentation Phenomena in Kulm and Flozleeres Graywackes, Sauerland and Oberharz, Germany," *Amer. Jour. Sci.*, Vol. 254, pp. 649–71.

*KUGLER, H. C., 1953, "Jurassic to Recent Sedimentary Environments in Trinidad, " *Bull. Ass. Suisse des Geol. et Ing. du Petrole*, Vol. 20, pp. 27–60.

MACNEIL, F. S., 1957, "Cenozoic Megafossils of Northern Alaska," *U. S. Geol. Survey Prof. Paper 294-C*, pp. 99–126.

McGILL, J. T., 1948, "Geology of a Portion of the Las Flores and Dry Canyon Quadrangles, Los Angeles County, California," unpublished Master's thesis, University of California, Los Angeles.

*MENARD, H. W., 1955, "Deep-Sea Channels, Topography, and Sedimentation," *Bull. Amer. Assoc. Petrol. Geol.*, Vol. 39, pp. 236–55.

MERIFIELD, PAUL, 1958, "Geology of a Portion of the Southwestern San Gabriel Mountains, San Fernando and Oat Mountain Quadrangles, Los Angeles County, California," unpublished Master's thesis, University of California, Los Angeles.

*MIGLIORINI, C. I., 1950, "Dati a conferma della risedimentazione delle arenarie del macigno," *Atti. Soc. Toscana, Sci. Nat. Mem.*, Vol. 57, Ser. A, pp. 3–15. Quoted by Natland and Kuenen (1951).

NATLAND, M. L., 1933, "Temperature and Depth Distribution of Some Recent and Fossil Foraminifera in the Southern California Region," *Bull. Scripps Inst. Oceanography*, Tech. Ser., Vol. 3, pp. 225–30.

———, 1957, "Paleoecology of West Coast Tertiary Sediments," Chap. 19, pp. 543–572, in "Treatise on Marine Ecology and Paleoecology, Vol. 2, Paleoecology," edited by H. S. LADD, *Geol. Soc. America Memoir 67*.

*———, AND KUENEN, P. H., 1951, "Sedimentary History of the Ventura Basin, California, and the Action of Turbidity Currents," *Soc. Econ. Paleon. and Min. Spec. Paper 2*, pp. 76–107.

———, AND ROTHWELL, W. T., JR., 1954, "Fossil Foraminifera of the Los Angeles and Ventura Regions, California," in "Geology of Southern California," *California Div. Mines Bull. 170*, pp. 33–42.

OAKESHOTT, G. B., 1954, "Geology of the Western San Gabriel Mountains, Los Angeles County," *ibid.*, Map Sheet No. 9.

———, 1959, "Geology and Mineral Deposits of San Fernando Quadrangle, Los Angeles County, California," *California Div. Mines Bull. 172*.

*PACKHAM, G. H., 1954, "Sedimentary Structures as an Important Factor in the Classification of Sandstones," *Amer. Jour. Sci.*, Vol. 252, pp. 466–76.

*PASSEGA, R., 1954, "Turbidity Currents and Petroleum Exploration," *Bull. Amer. Assoc. Petrol. Geol.*, Vol. 38, pp. 1871–87.

PAYNE, T. G., 1942, "Stratigraphical Analysis and Environmental Reconstruction," *ibid.*, Vol. 26, pp. 1697–1770.

PETTIJOHN, F. J., 1949, *Sedimentary Rocks.* Harper and Brothers, New York.

*———, 1950, "Turbidity Currents and Graywackes; a Discussion," *Jour. Geology*, Vol. 58, pp. 169–71.

PHLEGER, F. B, 1951, "Displaced Foraminifera Faunas," *Soc. Econ. Paleon. and Min. Spec. Pub. 2*, pp. 66–75.

PIERCE, R. L., 1956, "Upper Miocene Foraminifera and Fish from the Los Angeles Area, California," *Jour. Paleon.*, Vol. 30, pp. 1288–1314.

*PRENTICE, J. E., 1956, "The Interpretation of Flow Markings and Load Casts," *Geol. Mag.*, Vol. 93, pp. 393–400.

REICHE, PARRY, 1938, "An Analysis of Cross-Lamination: the Coconino Sandstone," *Jour. Geology*, Vol. 46, pp. 905–32.

RESIG, J. M., 1956, "Ecology of Foraminifera of Santa Cruz Basin, California," Master's thesis, University of Southern California. Later (1958) published under same title in *Micropaleontology*, Vol. 4, pp. 287–308.

*RICH, J. L., 1950, "Flow Markings, Groovings, and Intra-stratal Crumplings as Criteria for Recognition of Slope Deposits, with Illustrations from Silurian Rocks of Wales," *Bull. Amer. Assoc. Petrol. Geol.*, Vol. 34, pp. 717–41.

*———, 1951, "Three Critical Environments of Deposition and Criteria for Recognition of Rocks in Each of Them," *Bull. Geol. Soc. America*, Vol. 62, pp. 1–20.

*RIVEROLL, D. D., AND JONES, B. C., 1954, "Varves and Foraminifera of a Portion of the Upper Puente Formation (Upper Miocene), Puente, California," *Jour. Paleon.*, Vol. 28, pp. 121–31.

*SCHNEEBERGER, W. F., 1955, "Turbidity Currents, a New Concept in Sedimentation and Its Application to Oil Exploration," *Mines Magazine*, Vol. 45, No. 10, pp. 42–62.

*SHROCK, R. R., 1948, *Sequence in Layered Rocks.* McGraw-Hill Book Company, Inc., New York.

SOPER, E. K., 1938, "Geology of the Central Santa Monica Mountains, Los Angeles County, California," *California Jour. Mines and Geology*, Vol. 34, pp. 131–80.

STOKES, W. L., 1947, "Primary Lineation in Fluvial Sandstones, a Criterion of Current Direction," *Jour. Geology*, Vol. 55, pp. 52–53.

*SUJKOWSKI, ZB. L., 1957, "Flysch Sedimentation," *Bull. Geol. Soc. America*, Vol. 68, pp. 543–54.

*SULLWOLD, HAROLD H., JR., 1959, "Nomenclature of Load Deformation in Turbidites," *ibid.*, Vol. 70, pp. 1247–48.

*———, 1960, "Load Cast Terminology and Origin of Convolute Bedding: Further Comments," *ibid.*, Vol. 71, in press.

TERPENING, J. N., 1951, "Geology of Part of the Eastern Santa Monica Mountains," unpublished Master's thesis, University of California, Los Angeles.

TRASK, P. D., 1932, *Origin and Environment of Source Sediments of Petroleum.* Gulf Publishing Company, Houston, Texas.

TRAXLER, J. D., 1948, "Geology of East-Central Santa Monica Mountains," unpublished Master's thesis, University of California, Los Angeles.

TYRELL, G. W., 1950, *The Principles of Petrology*, 11th ed. Methuen and Co., London.

UCHIO, T., 1957, "Ecology of Living Benthonic Foraminifera from the San Diego, California, Area," unpublished Ph.D. thesis, University of California, Los Angeles (Scripps Institution of Oceanography).

*VAN HOUTEN, F. B., 1954, "Sedimentary Features of Martinsburg Slate, Northwestern New Jersey," *Bull. Geol. Soc. America*, Vol. 65, pp. 813–17.

WADELL, H., 1935, "Volume, Shape, and Roundness of Quartz Particles," *Jour. Geology*, Vol. 43, pp. 250–80.

WEST, J. C., 1947, "Geology of a Small Portion of the Central Santa Monica Mountains, California," Senior thesis in University of California, Los Angeles, library, unpublished.

WILLIAMS, H., TURNER, F. J., AND GILBERT, C. M., 1955, *Petrography.* W. H. Freeman and Company, San Francisco.

*WILSON, G., WATSON, J., AND SUTTON, J., 1953, "Current-Bedding in the Moine Series of Northwestern Scotland," *Geol. Mag.*, Vol. 90, pp. 377–87.

*WINTERER, E. L., 1954, "Geology of Southeastern Ventura Basin, Los Angeles County, California," unpublished Ph.D. dissertation, University of California, Los Angeles.

23

Copyright © 1966 by the Society of Economic Paleontologists and Mineralogists

Reprinted from *Jour. Sediment Petrol.*, **36**(1), 90, 106–114 (1966)

SHALE GRIT AND GRINDSLOW SHALES: TRANSITION FROM TURBIDITE TO SHALLOW WATER SEDIMENTS IN THE UPPER CARBONIFEROUS OF NORTHERN ENGLAND[1]

ROGER G. WALKER

Department of Geology and Mineralogy, Oxford, England[2]

ABSTRACT

The Shale Grit and Grindslow Shales (which together make up the Alport Group) lie between the Mam Tor Sandstones (turbidites) and the Kinderscout Grit (near-shore or coastal plain). All these formations are of Namurian (Upper Carboniferous) age and crop out in the North Derbyshire part of the Central Pennine Basin, England. The Shale Grit contains two main sandstone facies, 1, alternating parallel sided sandstones and mudstones interpreted as distal turbidites and 2, thick sandstones without mudstone partings interpreted as very near source, proximal turbidites. There are also three mudstone facies, silty mudstones, pebbly mudstones and thinly laminated black mudstones. The Grindslow Shales contain both of the Shale Grit sandstone facies, together with sandy mudstones, burrowed silty mudstones, parallel bedded silty sandstones and carbonaceous sandstones.

The Shale Grit facies sequence indicates that distal turbidites are more abundant below, and proximal turbidites are more abundant in the upper part of the formation. In the Grindslow Shales the facies become sandier upward with thin turbidites restricted to the lower part and horizontal burrows restricted to the upper part. The Shale Grit and Grindslow Shales contain at least seventeen channels from 10 to 50 feet deep, which appear to have been both cut and filled by turbidity currents. The presence of deep channels in association with proximal turbidites suggests that the environment of deposition of the Shale Grit was a submarine fan, similar in most respects to the fans at the foot of the Monterey and La Jolla canyons. The Grindslow Shales were probably deposited on the slope above the fan.

The sequence from the Mam Tor Sandstones (distal turbidites) via the lower Shale Grit (distal, with subordinate proximal turbidites) into the upper Shale Grit (proximal, with subordinate distal turbidites) suggests advance of a submarine fan southward into the North Derbyshire area. This southward advance of facies belts continued as the Grindslow Shales "slope" environment covered the fan, and was itself covered by the near-shore or coastal plain Kinderscout Grit.

[*Editor's Note:* The first part of the paper deals with the stratigraphy of North Derbyshire and facies analysis of the Shale Grit and Grindslow Shales. The following facies are described and their processes of deposition deduced: Facies A, turbidites; Facies B and C, sandstones; Facies D, mudstones; Facies E, pebbly mudstones; Facies F, paper laminated mudstones; Facies G, sandy and silty mudstones with horizontal burrows; Facies H, sandy and silty mudstones; Facies J, parallel bedded silty sandstones; and Facies K, parallel bedded carbonaceous sandstones. The facies relationships are then discussed and their vertical distribution tabulated.]

[1] Manuscript received April 8, 1965; revised October 25, 1965.

[2] Present address: Department of Geology, The Johns Hopkins University, Baltimore, Maryland.

MAJOR CHANNELS

The term "major channel" is used here to describe all channels with an eroded depth exceeding six feet. This is an arbitrary figure which nevertheless divides conveniently the smaller scours from the major channels. The deepest exposed channel is 50 feet (fig. 15), but mapping suggests that another channel may be at least 150 feet deep. Seventeen major channels have been observed, one in the lower Shale Grit, nine in the upper Shale Grit and seven in the Grindslow Shales. The channel fill facies are the same as those found elsewhere in the two formations, A, B, C, D and E. Facies F paper laminated mudstones have not been found within a channel.

Channels of this size have not previously been recognised in outcrop within turbidite formations, and a full description, together with interpretations of their processes of cutting and fill-

FIG. 15.—Channel at least 50 feet deep (figure top center for scale) cutting into mudstone marker band A-3 in Grinsbrook, SK 114872. The channel wall is dominantly steep, but flattens out half way down the exposure, perhaps suggesting two phases of cut and fill. The channel fill is entirely facies C sandstones, locally pebbly near the base.

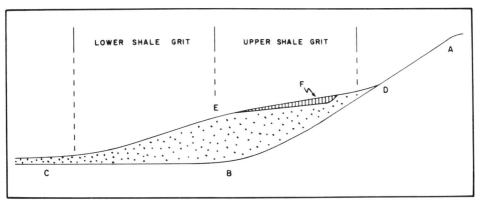

Fig. 16.—Partial reconstruction of Shale Griet nvironment of deposition. Letters fully explained in text.

ing, is given elsewhere (Walker, in press). The channel walls are usually fairly shallow, but can also be very steep (fig. 15). Cutting and filling took place in several stages in some channels, and mud flakes eroded from the walls are commonly found in the channel fill deposits. Internal scouring and amalgamation are observed in many places in the facies C sandstones which fill most of the channels. In one example (Section 1, horizon 455 feet, fig. 4; and Walker, in press, fig. 4), facies C sandstones in the deeper part of the channel pass laterally into thinner facies A turbidites on the levees, and in the same channel a broken mass of mudstone blocks and contorted sandstone beds is interpreted as a collapsed channel wall (Bersier, 1964). The channel fill is never graded, from coarse at the base to fine at the top, and the deposits are never crossbedded.

The origin of these channels is fully discussed elsewhere (Walker, in press). In that publication, it is concluded that permanent ocean currents and other ocean traction currents are normally too weak to erode such deep channels. There is no evidence of erosion by rivers or other subaerial processes, or by subaqueous mass movements of sediment (creep or slumping). The most likely possibility is erosion by turbidity currents, although at first it appears anomalous that turbidity currents can cut deep channels and deposit thick beds both in the same area. However, some turbidity currents are more likely to erode than others, the more erosive ones being fast-flowing without a traction carpet (Dzulynski and Sanders, 1962, p. 88), and being underladen. The importance of erosion by underladen currents was noted by Johnson (1962, p. 268), who states with reference to underladen currents that "rapid erosion and picking up of

sediment is likely, especially in view of the wide range in sizes that the current can carry in suspension. The rate of erosion per unit time would be many orders of magnitude faster than for rivers because of the very much higher velocities and high concentrations, including relatively large sizes." Thus once a small channel had been cut by a fast underladen current, it could be deepened by successive similar currents, or filled in by slower, fully laden, depositional currents. The paleogeographical significance of both powerful erosion and deposition of thick beds by turbidity currents in the same area is discussed below.

The channels are filled mainly with facies C sandstones, although facies A, B, D and E are also found. The muddy sediments fell from suspension, and the sandstones were introduced by turbidity currents. There is no evidence to suggest any filling by bottom traction currents, sand fall (Dill, 1964, p. 103), slumping (except minor movements forming facies E pebbly mudstones), "subaqueous mass flow" or creep.

If the Shale Grit channels were cut by turbidity currents flowing too fast to deposit their sediment, it implies first, that the currents encountered very little loose sediment between their source and the Shale Grit sediments, and second, that the currents had encountered no major flattening of slope until they reached the Shale Grit area. This area, therefore, probably lay near to the source of the turbidity currents. The environment is partially reconstructed in in figure 16. Before sedimentation, the slope AB curved gently into the basin floor BC. Gradually, facies A turbidites, facies C sandstones and mudstone facies were built up between DC. Fully laden turbidity currents flowing down AD were forced to deposit their load when slowed

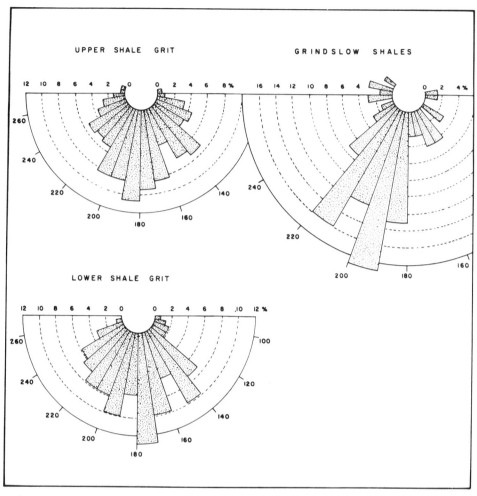

FIG. 17.—Paleocurrent directions in the lower and upper Shale Grit, and in the Grindslow Shales. Figures on east-west axis show percentage of readings. Total number of readings: lower Shale Grit, 149: upper Shale Grit, 221: Grindslow Shales, 69. Only readings from which actual current directions can be inferred are included in the diagram. Current directions roughly north to south in the Shale Grit, swinging toward southsouthwest in the Grindslow Shales.

down by the change of slope DE. Underladen turbidity currents, although also slowed down by the change of slope DE were able to erode loose sediment along DE, cutting channels, F. However, before reaching E the underladen currents had picked up sufficient sediment to make them fully laden with respect to their new, slower velocity along DE. Consequently, EC was mainly a depositional area, as opposed to DE, which was a more proximal depositional area but with accompanying channel erosion.

Only one major channel was observed in the

lower Shale Grit, and nine in the upper Shale Grit. The upper Shale Grit also has a higher proportion of facies C sandstones (relative to facies A turbidites). It appears, therefore, that the section EC corresponds to the lower Shale Grit, and DE to the upper Shale Grit. The Shale Grit as a whole represents the forward advance of DE onto EC.

PALEOCURRENTS

The three rose diagrams in figure 17 show paleocurrent directions measured in the lower and upper Shale Grit, and in the Grindslow Shales. Only readings in which both sense and direction are known are incorporated in the diagram. There is no significant difference between the lower and upper Shale Grit current patterns, but the directions in the Grindslow Shales show a much stronger grouping in the 180 to 215 degree range. All the readings show an unusually wide spread for turbidite formations (Potter and Pettijohn, 1963, p. 130), suggesting that the depositional surface was slightly irregular. It is also likely that there were several points of supply at the edge of the basin, and that successive currents flowing from different points of supply show different current directions (Potter and Pettijohn, 1963, p. 132; Hand and Emery, fig. 8, p. 540).

The trend of the major channels, where this can be measured, is also roughly north to south. Throughout the Alport Group, the dominant transport direction was north (or northeast) to south (or southwest), suggesting that the source of the currents lay to the north.

SEDIMENTARY ENVIRONMENT: SHALE GRIT

The sedimentary environment of the Shale Grit is characterised by the following features.

1. Distal turbidite deposition (facies A).
2. Proximal turbidite deposition (facies C).
3. Rapid succession of turbidity currents one after the other (inferred from amalgamated beds).
4. Long periods of quiet clay deposition (facies F paper-laminated mudstones).
5. Presence of deep channels cut by turbidity currents.
6. Irregular surface of deposition.
7. Rarity of slump deposits.
8. Upward change from facies C subordinate to A (lower Shale Grit) to facies A subordinate to C (upper Shale Grit).
9. Absence of sedimentary structures and facies associations characteristic of alluvial plain, or near-shore sedimentation.
10. No organic reworking of sediment.
11. No regular cyclic change of facies.

The complete absence of any shallow water features suggests a fairly deep water environment, while the positive features suggest very near-source deposition from turbidity currents. When the evidence of the deep channels is also considered, the most likely environment of deposition of the Shale Grit is a submarine fan. The term "fan" is used here to denote a thick wedge of sediment at the foot of a submarine slope (see Menard, 1955, p. 246 for synonyms of the term fan). The fan is part of the basin apron, "which can be further subdivided into subsea fans and more gently sloping features which may be bottomset beds of adjacent subsea fans" (Emery, 1960, p. 467).

Submarine fans were first recognized at the foot of submarine canyons by echo-sounding surveys. These surveys have shown that the fans usually die out into rather flat aprons of sediment on the deep sea floor (Emery, 1960, p. 470). By considering the work of several authors, it is possible to compile a list of seven characteristics of submarine fans (Bouma and Shepard, 1964; Bourcart, 1959; Bourcart, Gennessaux and Klimer, 1960; Dill, Dietz and Stewart, 1954; Emery, 1960; Heezen, Menzies and Ewing, 1959; Menard, 1960; Shepard, 1963a and 1963b; Shepard and Einsele, 1962; Hand and Emery, 1964). These characteristics are compared, item by item, with the features of the Shale Grit in table 4.

A very close similarity between the topographical and sedimentological features of modern fans and the Shale Grit can thus be demonstrated. One feature of the Shale Grit, the extensive sheets of facies F paper laminated mudstones, does not appear at first sight to fit into this picture. However, the various facies of the Shale Grit can be traced at least as far as the Todmorden area (Wright, Sherlock, Wray, Lloyd and Tonk, 1927; Reading, personal communication 1964; fig. 1), where they are known as the Todmorden Grit. If at one time the main supply of turbidity currents carried sediment into the Todmorden area, clays could accumulate quietly in the North Derbyshire area. A slight movement of the "Pennine Delta" in the northeast might then direct all the turbidity currents into North Derbyshire, and clays would then accumulate at Todmorden. The Shale Grit is therefore believed to have been deposited, not as one fan, but as a series of laterally migrating fans at the foot of the slope (cf. Hand and Emery, 1964, fig. 3, p. 529). Another possibility is that widespread transgressions of the sea would lead to deepening of the basin and trapping of all the coarse sediment in coastal waters further north, thus cutting off the supply of turbidity currents to the fan and allowing the formation of mudstone sheets. However, this is unlikely because there is no record of repeated transgressions in the three coarsening upward cycles of the R_{1c} zone in the Bradford area (Stephens, Mitchell and Edwards, 1953, p. 31; Reading, 1964).

SEDIMENTARY ENVIRONMENT: GRINDSLOW SHALES

Out of context, there are very few indications as to the depositional environment of the Grindslow Shales. The most significant features are:

TABLE 4.—*Comparison of modern submarine fans and the Shale Grit*

Sedimentary and topographic features of modern fans.	Sedimentary and topographic features of the Shale Grit.
1. Fans are cone-shaped wedges of sediment with their apices at the foot of a submarine canyon. They thus form a small part of the basin apron (Emery, 1960, p. 467).	The Shale Grit is a wedge-shaped mass of sediment. Recent mapping (in lit.) by the Geological Survey has shown that it thins out from the area studied by the author toward the southeast near Calver (fig. 1), and toward the southwest near Dove Holes (I. P. Stevenson, personal communication).
2. Fans are areas of near-source deposition from turbidity currents, at the first major flattening of slope.	Near-source turbidite complex, probably deposited at first major flattening of slope.
3. Fans often have channels, and the channels often have levees (Hand and Emery, 1964, p. 529, Fig. 3). They are straight or meandering, with steep or shallow walls.	Shale Grit has at least nine major channels, one of which has levees. Steep and shallow walled channels have been observed.
4. Fans often have an irregular, hummocky surface.	Surface believed to be irregular, from poor lateral extent of individual beds.
5. Cores show individual layers to be laterally impersistent.	Measured sections show individual beds to be laterally impersistent.
6. Fan sediments consist of alternating coarse and fine beds. Lower boundaries of sand layers are sharp, several superimposed graded sand layers sometimes present. Gravel has a patchy outcrop. Known sedimentary structures include parallel lamination, ripple cross-lamination, slumping and graded bedding.	Alternation of coarse and fine grades. Parallel lamination, ripple cross-lamination, slumping and graded bedding are all present.
7. Thicker sand layers found on upper parts of fan, and are less common on fan margin.	Thick composite sandstones (facies C) are more important in the upper than lower Shale Grit.

1. Thin turbidites with well developed sole marks in the lower part of the formation.
2. At least seven major channels, some filled with facies C sandstones and some with pebbly sandstones indistinguishable from the Kinderscout Grit.
3. The widespread occurrence of horizontal burrows in the upper part of the formation.
4. The occurrence of facies K carbonaceous sandstones in the highest part of the formation.
5. The facies sequence

G (burrowed sandy mudstones) with K
↑   (carbonaceous sandstones), rare C
| sandstones.
J (parallel bedded silty sandstones) with H
↑   (silty and sandy mudstones), some C
| sandstones, rare A turbidites.
H (silty and sandy mudstones), some C
    sandstones, many thin A turbidites.

These features, especially the increase in sand content upward and the restriction of thin turbidites to the lower part of the formation, suggest that the water gradually became shallower during the deposition of the Grindslow Shales. The horizontal burrows, therefore, appear to be confined to the shallower water sediments. In their context the Grindslow Shales appear to represent the sandy mudstones deposited above the Shale Grit fan, between it and the "near-shore or coastal plain" Kinderscout Grit (Reading, 1964).

Many of the features of the Grindslow Shales can be explained on this hypothesis. The thin turbidites are (with very few exceptions) restricted to the lower 100 feet of the formation. The thinness of the beds (1 to 3 inches), the well developed sole marks of high relief and the separation of individual beds by comparatively thick facies H sandy mudstones suggests that the turbidity currents flowing through the lower Grindslow Shales environment were strongly erosive but had not slowed down sufficiently to deposit thick beds. The environment may therefore have been on the slope immediately above the fan.

The structureless facies H sandy mudstones can be compared with the "massive silty sandy muds with no apparent bedding" which Gorsline and Emery (1959, p. 286) describe from the slopes above the Hueneme, Dume, Mugu and Redondo fans off Southern California. These sands and muds were apparently washed out to sea and settled from suspension onto the slope.

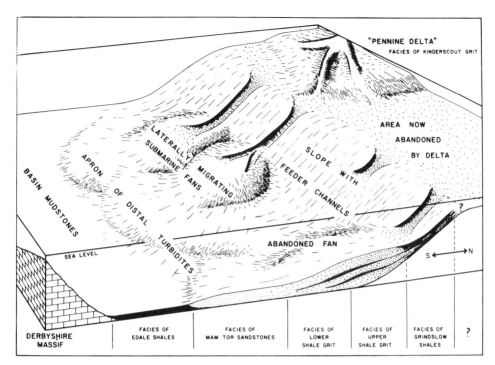

FIG. 18.—Hypothetical reconstruction of the Namurian R_{1c} zone facies of North Derbyshire, arranging the formations laterally rather than vertically and giving interpretations of their environments of deposition. The observed vertical sequence can thus be visualised in terms of a southward advance of facies belts.

The seven major channels in the Grindslow Shales were probably cut into the slope by turbidity currents and acted as feeder channels down which currents flowed to the fan. Channels are well known in slope environments (Hand and Emery, 1964, p. 533) and range in size from small delta-front valleys (Shepard, 1952, 1955) to very large submarine canyons (Shepard, 1963a, p. 311–348). The channels of the Grindslow Shales have a width of up to half a mile and are up to 100 to 150 feet deep. Most of the channels are filled with facies C sandstones, probably deposited by turbidity currents as the feeder channel gradually became disused.

Apart from some laminations and ripples in the turbidites at the base of the Grindslow Shales, there is no evidence for bed traction of sediment anywhere in the formation. It appears, therefore, that even the carbonaceous sandstones at the top of the formation were deposited in water too deep for traction current activity. The Kinderscout Grit in most places rests very sharply, often erosively, on the Grindslow Shales. Either these coarse pebbly grits were

transported into water too deep for traction current activity, or there was a shallowing of water accompanied by an erosive spread of Kinderscout Grit southward (fig. 18).

CONCLUSIONS

The Alport Group occupies a stratigraphic position between the turbidites of the Mam Tor Sandstones and the near shore or coastal plain pebbly sandstones of the Kinderscout Grit. In this paper the facies of the Alport Group have been examined with the object of determining the changes in paleogeography and depositional environment during the turbidite to shallow water transition. This paleogeographical reconstruction is built mainly on the turbidite interpretation of the two sandstone facies, A and C.

The facies A parallel-sided sharp based sandstones with interbedded mudstones possess most of the features now associated with turbidites, and possess no features indicative of any other process of deposition. They are identical to the turbidites of the Mam Tor Sandstones (Allen, 1960), though they lack the rare pelagic fauna.

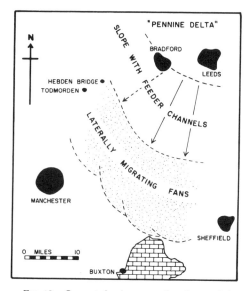

FIG. 19.—Suggested paleogeography of part of the Central Pennine Basin during the upper Shale Grit, when the submarine fans had advanced southward almost to the Derbyshire Massif.

graded division (Walker, 1965; Bouma, 1962 were seldom observed, suggesting that the turbidity currents carried all the finer material further into the basin. Since the proportion of facies C sandstones increases upward in the Shale Grit, it follows that proximal turbidite deposition gradually became more important, suggesting an advance of the source into the basin.

The presence of deep channels in a proximal turbidite environment invites comparison with a modern submarine fan. In general and in detail, this model accounts for the observed sedimentological features of the Shale Grit, except for the extensive sheets of quietly deposited mudstone. These sheets are believed to have formed at times when the locus of turbidity current deposition had moved to a neighboring fan at the foot of the slope. The Shale Grit was thus deposited from a series of laterally migrating fans (fig. 18). The Grindslow Shales are believed to have formed on the slope above the submarine fan. This slope was dissected by feeder channels down which turbidity currents flowed on their journey to the fan. According to the exact position of the "Pennine Delta" on the "delta platform," some of the feeder channels were used and others were abandoned, and became clogged either with facies C sandstones or pebbly sandstones very similar to those in the Kinderscout Grit.

All of these features are shown in the paleogeographical reconstruction of figure 18. However, this reconstruction is hypothetical. It shows the R_{1c} zone facies of North Derbyshire arranged laterally rather than vertically and gives the interpretation of the facies in terms of their environment. The observed vertical facies sequence can be visualized in terms of a southward advance of facies belts during R_{1c}, with successive facies lapping onto the Derbyshire Massif. During deposition of the upper part of the Shale Grit (fig. 19), the basin was entirely filled by the submarine fans and there was no space at the southern end for the formation of an apron of distal turbidites. The southward facies advance continued, until at the end of R_{1c} near-shore or coastal plain conditions extended across the whole of the North Derbyshire area.

The facies C sandstones are individually up to 10 feet thick, and groups of beds without well developed mudstone layers reach 100 feet. Crossbedding, complex lamination and cross-lamination and small scale irregular scouring are all absent, and it therefore appears that the thick sandstones were not deposited by traction currents. There is, however, a definite field relationship between the facies A turbidites and the facies C sandstones. In one case a facies A turbidite was observed to thicken laterally into a facies C sandstone. In many other places the thick facies C sandstones were observed to be composite in origin, formed by the amalgamation of several facies A turbidites. The amalgamation took place by the cutting out of the mudstone divisions and the annealing of the sandy portions (fig. 7). It is suggested that the very thick facies C sandstones are near-source, proximal turbidites, and that the facies A turbidites represent relatively distal deposition. Although many authors have described beds apparently similar to the facies C sandstones as "fluxoturbidites" (Dzulynski, Ksiazkiewicz and Kuenen, 1959, p. 1114; Unrug, 1964, p. 188–189, 191; Kuenen, 1964, p. 21), all the features of the facies C sandstones can be explained by proximal deposition from normal turbidity currents (Walker, 1965). Despite the fact that these beds are unusually thick, sedimentary divisions above the

ACKNOWLEDGMENTS

The author is indebted to Dr. H. G. Reading for his stimulating supervision of this study, which forms part of the author's Ph.D. thesis, University of Oxford. A Research Studentship from the Department of Scientific and Industrial Research is gratefully acknowledged. Professors P. Allen and F. J. Pettijohn, Doctors H. G. Reading and W. S. McKerrow and Mr. J. D. Collinson have kindly given helpful criticism of the manuscript. The author also wishes to thank the Geological Survey, Leeds Office, for their assistance. The present manuscript was prepared at The Johns Hopkins University during the tenure of a NATO Research Fellowship, which the author gratefully acknowledges. Assistance in the field was kindly given by Mr. J. P. B. Lovell.

REFERENCES

ALLEN, J. R. L., 1960, The Mam Tor Sandstones; a turbidite facies of the Namurian deltas of Derbyshire, England; Jour. Sedimentary Petrology, v. 30, p. 193–208.

———, 1963a, Asymmetrical ripple marks and the origin of water-laid cosets of cross-strata: Liverpool and Manchester Geol. Jour., v. 3, pt. 2, p. 187–236.

———, 1963b, Henry Clifton Sorby and the sedimentary structures of sands and sandstones in relation to flow conditions: Geol. en Mijnbouw., v. 25, p. 223–228.

———, 1964, Primary current lineation in the Lower Old Red Sandstone (Devonian), Anglo-Welsh Basin: Sedimentology, v. 3, p. 89–108.

ALLEN, P., 1959, The wealden environment; Anglo-Paris basin: Roy. Soc. [London] Phil. Trans., Ser. B, v. 242, p. 283–346.

BERSIER, A., 1964, Structure en créneaux et brèche cyclopéenne en milieu détritique paralique, *in* L.M.J.U. van Straaten (ed.), Deltaic and shallow marine deposits. Elsevier, Amsterdam, Netherlands, p. 35–38.

BIRKENMAJER, K., 1959, Classification of bedding in flysch and similar graded deposits: Studia Geol. Polon., v. 3, p. 1–133.

BISAT, W. S., AND HUDSON, R. G. S., 1941, The Lower "Reticuloceras" (R₁) goniatite succession in the Namurian of the North of England: Yorkshire Geol. Soc. Proc., v. 24, p. 282–440.

BLATT, H., AND CHRISTIE, J. M., 1963, Undulatory extinction in quartz of igneous and metamorphic rocks and its significance in provenance studies of sedimentary rocks: Jour. Sedimentary Petrology, v. 33, p. 559–579.

BOUMA, A. H., 1962, Sedimentology of some flysch deposits; a graphic approach to facies interpretation. Elsevier, Amsterdam, Netherlands, 168 p.

BOUMA, A. H., AND SHEPARD, F. P., 1964, Large rectangular cores from submarine canyons and fan valleys: Am. Assoc. Petroleum Geologists Bull., v. 48, p. 225–231.

BOURCART, J., 1959, Morphologie du précontinent des Pyrénées a la Sardaigne: Colloq. Intern. Centre Natl. Rech. Sci. (Paris), v. 83, p. 33–50.

BOURCART, J., GENNESSAUX, M., AND KLIMER, E., 1960, Ecoulements profonds de sables et de galets dans le grande vallée sous-marine de Nice: Comptes Rend. Acad. Sciences, v. 250, p. 3761–3765.

CROWELL, J. C., 1957, Origin of pebbly mudstones: Geol. Soc. America Bull., v. 68, p. 993–1010.

DE RAAF, J. F. M., READING, H. G., AND WALKER, R. G., 1965, Cyclic sedimentation in the Upper Carboniferous of North Devon, England: Sedimentology, v. 4, p. 1–52.

DILL, R. F., 1964, Features in the heads of submarine canyons; narrative of underwater film, *in* L.M.J.U. van Straaten (ed.), Deltaic and shallow marine deposits. Elsevier, Amsterdam, Netherlands, p. 102–104.

DILL, R. F., DIETZ, R. S., AND STEWART, H. B., 1954, Deep sea channels and delta of the Monterey Submarine Canyon: Geol. Soc. America Bull., v. 65, p. 191–193.

DOTT, R. H., JR., 1963, Dynamics of subaqueous gravity depositional processes: Am. Assoc. Petroleum Geologists Bull., v. 47, p. 104–128.

DZULYNSKI, S., 1963, Directional studies in flysch: Studia Geol. Polon., v. 12, p. 1–136.

DZULYNSKI, S., KSIAZKIEWICZ, M., AND KUENEN, PH. H., 1959, Turbidites in flysch of the Polish Carpathian Mountains: Geol. Soc. America Bull., v. 70, p. 1089–1118.

DZULYNSKI, S., AND SANDERS, J. E., 1962, Current marks on firm mud bottoms: Trans. Connecticut Acad. Arts Sci., v. 42, p. 57–96.

EMERY, K. O., 1960, Basin plains and aprons off Southern California: Jour. Geology, v. 68, p. 464–479.

FOLK, R. L., 1954, The distinction between grain size and mineral composition in sedimentary-rock nomenclature: Jour. Geology, v. 62, p. 344–359.

FOLK, R. L., AND WARD, W. C., 1957, Brazos River bar: Jour. Sedimentary Petrology, v. 27, p. 3–26.

GILLIGAN, A., 1920, The petrography of the Millstone Grit of Yorkshire: Geol. Soc. London Quart. Jour., v. 75, p. 251–294.

GORSLINE, D. S., AND EMERY, K. O., 1959, Turbidity current deposits in San Pedro and Santa Monica basins off Southern California: Geol. Soc. America Bull., v. 70, p. 279–290.

HAND, B. M., AND EMERY, K. O., 1964, Turbidites and topography of north end of San Diego trough, California: Jour. Geology, v. 72, p. 526–542.

HEEZEN, B. C., MENZIES, R. J., AND EWING, M., 1959, Influence of modern turbidity currents on abyssal productivity: Abstr. Int. Oceanographic Congr., Amer. Assoc. Adv. Sci., Washington, p. 375–377.

HSU, K. J., 1964, Cross-laminations in graded bed sequences: Jour. Sedimentary Petrology, v. 34, p. 379–388.

HUDSON, R. G. S., AND COTTON, G., 1943, The Namurian of Alport Dale, Derbyshire: Yorkshire Geol. Soc. Proc., v. 25, p. 142–173.

——— 1945, The carboniferous rocks of the Edale anticline, Derbyshire: Geol. Soc. London Quart. Jour., v. 101, p. 1–35.

JACKSON, J. W., 1927, The succession below the Kinderscout Grit in North Derbyshire: Jour. Manchester Geol. Soc., v. 6, p. 15–32.

JOHNSON, M. A., 1962, Physical oceanography: turbidity currents: Science Progress, v. 50, p. 257–273.

KUENEN, PH. H., 1964, Deep-sea sands and ancient turbidites, *in* A. H. Bouma and A. Brouwer (editors), Turbidites. Elsevier, Amsterdam, Netherlands, p. 3–33.

McKEE, E. D., 1957, Experiments on the production of stratification and cross-stratification: Jour. Sedimentary Petrology, v. 27, p. 129–134.

MENARD, H. W., 1955, Deep sea channels, topography and sedimentation: Am. Assoc. Petroleum Geologists Bull., v. 39, p. 236–255.

———— 1960, Possible pre-Pleistocene deep-sea fans off central California: Geol. Soc. America Bull., v. 71, p. 1271–1278.

PETTIJOHN, F. J., 1957, Sedimentary rocks. Harper, New York, 718 p.

PETTIJOHN, F. J., AND POTTER, P. E., 1964, Atlas and glossary of primary sedimentary structures. Springer-Verlag, New York, 370 p.

POTTER, P. E., 1963, Late Palaeozoic sandstones of the Illinois basin: Illinois State Geol. Surv., Report of Investigations No. 217, 92 p.

POTTER, P. E., AND GLASS, H. D., 1958, Petrology and sedimentation of the Pennsylvanian sediments in Southern Illinois: a vertical profile: Illinois State Geol. Surv., Report of Investigations No. 204, 60 p.

POTTER, P. E., AND PETTIJOHN, F. P., 1963, Palaeocurrents and basin analysis. Springer Verlag, Berlin, 296 p.

READING, H. G., 1964, A review of the factors affecting sedimentation of the Millstone Grit (Namurian) in the Central Pennines, *in* L.M.J.U. van Straaten (ed.), Deltaic and shallow marine deposits. Elsevier, Amsterdam, Netherlands, p. 340–346.

REINECK, H. E., 1960, Über die Entstehung von Linsen- und Flaser-schichten: Abhandl. Deutsche Akad. Wiss. Berlin, Kl. 3, Heft 1, p. 369–374.

SHEPARD, F. P., 1952, Composite origin of submarine canyons: Jour. Geology, v. 60, p. 84–96.

———— 1955, Delta-front valleys bordering the Mississippi distributaries: Geol. Soc. America Bull., v. 66, p. 1489–1498.

————, 1963a, Submarine geology, 2nd. edition. Harper, New York, 557 p.

————, 1963b, Submarine canyons, *in* M. N. Hill (ed.), The Sea, vol. 3. Wiley, New York, p. 480–506.

SHEPARD, F. P., AND EINSELE, G., 1962, Sedimentation in San Diego Trough and contributing submarine canyons: Sedimentology, v. 1, p. 81–133.

SIMONS, D. B., RICHARDSON, E. V., AND ALBERTSON, M. L., 1961, Flume studies using medium sand (0.45 mm): U. S. Geol. Survey, Water supply paper 1498-A, 76 p.

SORBY, H. C., 1859a, On the structure and origin of the Millstone Grit in South Yorkshire: Proc. Yorkshire Geol. Polytech. Soc., v. 3, p. 669–675.

———— 1859b, On the structures produced by the currents present during the deposition of stratified rocks: The Geologist, v. 2, p. 137–147.

———— 1908, On the application of quantitative methods to the study of the structure and history of rocks: Geol. Soc. London Quart. Jour., v. 64, p. 171–233.

STEPHENS, J. V., MITCHELL, G. H., AND EDWARDS, W., 1953, Geology of the country between Bradford and Skipton: Memoir Geol. Surv. Britain, H.M.S.O., London, 180 p.

STOKES, W. M. L., 1947, Primary current lineation: Jour. Geology, v. 55, p. 52–54.

TROTTER, F. M., 1951, Sedimentation facies in the Namurian of northwestern England and adjoining areas: Liverpool and Manchester Geol. Jour., v. 1, p. 77–112.

UNRUG, R., 1964, Turbidites and fluxoturbidites in the Moravia-Silesia Kulm zone: Bull. Acad. Polon. Sci., Sér. Sci. Géol. Géograph., v. 12, p. 187–194.

WALKER, R. G., 1963, Distinctive types of ripple-drift cross-lamination: Sedimentology, v. 2, p. 173–188.

———— 1964, Some aspects of the sedimentology of the Shale Grit and Grindslow Shales (Namurian R_{1e} of North Derbyshire) and the Westward Ho! and Northam Formations (Westphalian, North Devon): Ph.D. Thesis, Univ. of Oxford, 178 p.

———— 1965, The origin and significance of the internal sedimentary structures of turbidites: Yorkshire Geol. Soc. Proc., v. 35, p. 1–32.

———— Deep channels in turbidite-bearing formations: Am. Assoc. Petroleum Geologists Bull., in press.

WRIGHT, W. B., SHERLOCK, R. L., WRAY, D. A., LLOYD, W., AND TONKS, L. H., 1927, The geology of the Rossendale anticline: Mem. Geol. Surv. Gt. Britain, H.M.S.O., London, 182 p.

24

Copyright © 1970 by the Geological Association of Canada

Reprinted from *Flysch Sedimentology in North America* (Geol. Assoc. Canada Spec. Paper 7), J. Lajoie, ed., 1970, pp. 103–125

DEEP SEA SEDIMENTS IN THE LOWER PALEOZOIC QUÉBEC SUPERGROUP

C. Hubert, J. Lajoie and M. A. Léonard
Department of Geology, Université de Montréal, Montréal, Québec

ABSTRACT

Two conglomerate-sandstone assemblages that occur in the Cambro-Ordovician flysch of Québec, are interpreted as deep water, submarine fans that accumulated off a carbonate shelf at the mouths of canyons. The first, which occurs at L'Islet, is 293 feet thick and forms the fill of large channels cut in a normal limestone turbidite sequence. The second deposit, which occurs at St-Fabien in a thick pelite section, has the geometry of a flat half-cone with apical thickness near 1,000 feet.

Fragments of shallow-water limestones constitute the bulk of these conglomerates. At L'Islet as much as 40 per cent of some conglomerate beds are limestone-turbidite fragments. The sandstone are immature arkosic arenites and wackes. The conglomerate and sandstone beds show repetitive graded-bedding; the sandstones display divisions b, and, occasionally, c and d of the Bouma sequence as well as large-scale cross-bedding and flutes. Channels are common in the conglomerates and sandstones. The conglomerates are thought to be turbulent-flow deposits whereas the sandstones are turbidites. Paleocurrent data indicate that the gravels and sands were brought into the deep water from the north.

RESUME

Nous proposons que les deux assemblages de conglomérats et de grès du flysch Cambro-Ordovicien de la rive sud du Saint-Laurent que nous avons étudiés soient des dépôts alluvionnaires de milieu profond accumulés au pied d'une plate-forme carbonatée, à l'embouchure de "canyons" sous-marins. Le premier dépôt qui affleure à L'Islet atteint une puissance de 90 m (293 pieds). Les conglomérats et les grès qui le constituent représentent le remplissage d'un chenal creusé dans une séquence de turbidite calcaire. Le deuxième dépôt, situé à St-Fabien, est dans une séquence pélitique très épaisse; il possède la géométrie d'un cône alluvionnaire dont la hauteur apicale atteint 300 m (1000 pieds).

Les conglomérats sont composés surtout de fragments de grès et de calcaires néritiques mais certains conglomérats du dépôt de L'Islet renferment aussi jusqu'à 40 pour cent de fragments de calcaires intraclastiques semblables aux couches de turbidite calcaire encaissante. Les grès sont des arénites et des wackes arkosiques. Les bancs de conglomérats et de grès montrent un granoclassement et, de plus, les grès montrent les divisions b, et occasionnellement, c et d de la séquence de Bouma. Des flutes, des chenaux, et de grandes stratifications entrecroisées apparaissent dans plusieurs lits de conglomérats et de grès. Nous estimons que les conglomérats furent transportés par des coulées turbulentes et que les grès sont des turbidites. Les paléocourants indiquent que les graviers et les sables du flysch provenaient du Nord.

GENERAL STATEMENT

The purpose of this paper is to describe and interpret the geometry, material, sedimentary structures, and sedimentological history of two conglomerate-sandstone deposits in the Québec Supergroup (formerly the Québec Group, Logan, 1863). The two deposits occur in the flysch belt at L'Islet and St-Fabien on the south shore of the St. Lawrence River, 56 and 174 miles northeast of Québec City (Figure 1).

The flysch belt, forming the western front of the Québec Appalachian province, is contained in a series of imbricated slices and klippen of Cambrian, Cambrian to Lower Ordovician, Lower Ordovician, and Middle Ordovician rocks. Generally the sequence is successively older from Logan's Line southward; that is, the Middle Ordovician slice is structurally below the Lower Ordovician which, in turn, is covered by Cambrian rocks.

The Cambrian rocks comprise interbedded mudstones, feldspathic sandstones, and conglomerates locally intercalated with limestone beds. The Lower Ordovician rocks are characterized by black shale with oligomict and petromict limestone conglomerates. The wildflysch is found along the coast, chiefly in Middle Ordovician rocks.

Detailed descriptions of the geology of the Québec Appalachians and St. Lawrence Lowland are given in the literature (Ayrton, *et al.*, 1969; Cady, 1967; Houde and Clark, 1961; Hubert, 1967; Neale, *et al.*, 1961; Skidmore, 1967; St-Julien, 1967).

L'ISLET WHARF
GENERAL GEOLOGY

The section exposed at L'Islet is near the base of the St-Roch Formation (Hubert, 1965), and is Lower Cambrian as shown by *Bolboparia canadensis, Calodiscus theokritoffi,* and *Leptochilodiscus cuspunctulatus* (Rasetti, personal communication, 1967). The section is subdivided into three members which are, in ascending order, Members 1, 2, and 3 (Figure 2).

MEMBER 1

Member 1, 127 feet thick, consists of red mudstones intercalated with lenticular beds of coarse-grained siltstones and fine-grained sandstones. The mudstone beds are massive, and, in general, 2 feet thick. The siltstone and sandstone beds rarely exceed 6 inches in thickness; their

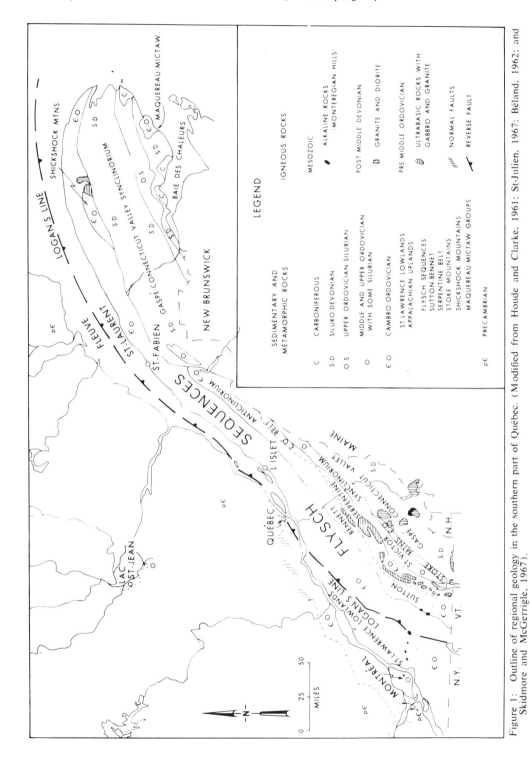

Figure 1: Outline of regional geology in the southern part of Québec. (Modified from Houde and Clarke, 1961; St-Julien, 1967; Béland, 1962; and Skidmore and McGerrigle, 1967).

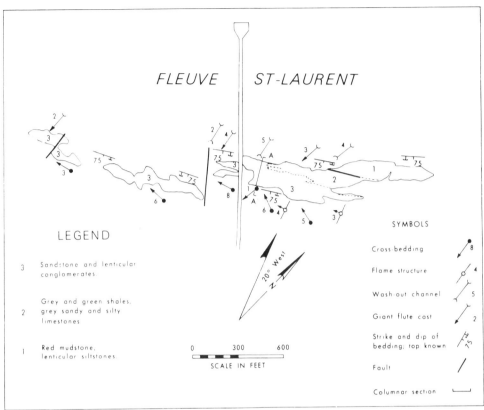

Figure 2: General geology at L'Ilet wharf and paleocurrent data in the conglomerates and sandstones of Member 3. The numbers shown beside each symbol represent the number of individual structures measured at each locality; the arrow on the flame structure points in the direction of overturning of the flame.

width ranges from 2 inches to more than 150 feet. Each bed is graded and shows the b, c, and, in some places, d divisions of the Bouma sequence. Tracks, burrows, flutes, and grooves are abundant on the bases of siltstone and sandstone beds. The upper contacts of these beds with the overlying red mudstone are always sharp.

Quartz is the chief constituent of the siltstones and sandstones; feldspar and glauconite are less than 15 per cent of the sand fraction. Detrital mica flakes and a suite of heavy minerals (garnet, rutile, tourmaline, apatite, zircon, hematite, and pyrite) make up to 5 per cent of the siltstones. The heavy minerals and mica are concentrated in laminae.

The siltstone and sandstone beds interbedded with the red mudstones show all the sedimentary features characteristic of proximal turbidites (Bouma, 1962; Walker, 1965) and hence are believed to be formed by turbidity currents which transported and deposited coarse silt and fine sand in the mudstone lithotope.

The red mudstones do not show evidence of resedimentation. In the St-Roch Formation, such red mudstone sequences are commonly interbedded with greenish grey mudstones that contain a marine fauna. The fauna consists of very small but complete valves of species of the inarticulate brachiopods *Botsfordia* and *Lingulella* (G. A. Cooper, personal communication, 1965). All other fossils that were collected from the other rocks of the St-Roch Formation are allochthonous and are always found in lithologies (limestone conglomerates) interpreted to be adventituous in the mudstone facies (Osborne, 1956; Hubert, 1965). They consist of algae, sponge spicules, trilobites, cephalopods, and articulate brachiopods, and are found in the limestone and collophane fragments that make up an important part of the petromict conglomerates. These fragments with their extensive fauna are believed to have been derived from a shallow marine platform or shelf located on the northern side of the flysch basin (Osborne, 1965; Hubert, 1965). The absence of a conventional neritic fauna in the indigenous mudstones of the St-Roch Formation strongly suggests that these

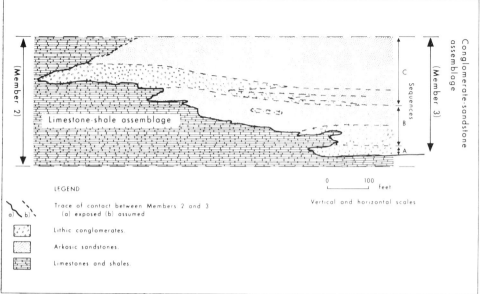

Figure 3: Diagrammatic cross-section showing the relationships of the conglomerate-sandstone and the limestone-shale assemblages in the area east of L'Islet wharf.

mudtones have accumlated off a neritic shelf in the deeper part of a marine basin.

MEMBER 2

Member 2 is composed of interbedded sandy limestones and green and grey shales. This member is generally 330 feet thick but immediately to the east and west of the wharf at L'Islet, its thickness ranges from 37 to 330 feet. At this locality, the limestone-shale assemblage is gradually cut off and replaced laterally by an assemblage of conglomerates and sandstones (Figure 3).

The limestones occur in regular, persistent beds, one-half inch to 8 inches thick. Graded bedding is faintly visible in most beds but parallel, current ripple, and convolute laminations are very well exposed. These structures occur in the order named, and commonly are followed by a layer of parallel laminations at the top of the bed. Tracks, burrows, flutes, grooves, and load casts, though locally very abundant, are generally absent. The contacts of the limestone beds with the overlying shales are sharp and planar. The interbedded shales range from one-eighth inch to 4 inches in thickness.

The limestone consists of silty and sandy calcisiltites and calcarenites. The terrigenous fraction (insoluble residue) is 50 per cent by weight of the rock in some beds, but generally constitutes less than 20 per cent. This fraction is composed of angular grains of quartz, glauconite, and a few heavy minerals. The shales interbedded with the limestones are non-calcareous.

The green and grey shales are marine as indicated by a few valves of species of *Botsfordia* and *Lingulella* (Hubert, 1965). The associated limestones have all the sedimentary characteristics of proximal turbidites (Bouma, 1962; Walker, 1965) and hence are believed to be formed by turbidity currents that transported and deposited carbonate silt and sand in the deeper water shale environment.

MEMBER 3

Member 3 consists of intercalated wedges of conglomerate and sandstone with rare interbedded shale. The member is lenticular and reaches a maximum thickness of 293 feet in the vicinity of the wharf. The conglomerate and sandstones intertongue with the limestones and shales of Member 2 and are, therefore, laterally equivalent facies (Figure 3). The contemporaneity of the two facies is also indicated, in that at several levels throughout this part of the section, thin shale beds can be traced laterally from the limestone-shale assemblage into the conglomerate-sandstone member.

THE CONGLOMERATE-SANDSTONE ASSEMBLAGE

GEOMETRY

The base of the conglomerate - sandstone assemblage is flat and conformable with the underlying shales and limestones. Upward, the conglomerate-sandstone assemblage intertongues with the limestone-shale facies and the contact transgresses conspicuously towards the east (Figure 3: Plate I). The transgression is not

PLATE I

Figure 1: Interdigitation of the conglomerate-sandstone assemblage with the limestone-shale facies. Sequences A and B of the conglomerate-sandstone assemblage are shown on the right and far center of the photo respectively. View is looking west toward L'Islet wharf. Beds are dipping 75 degrees to the left; top to the left.

Figure 2: View, looking west, of the lateral transgression of the intertonguing contact between the conglomerate-sandstone member and the limestone-shale facies. Distance to L'Islet wharf is 600 feet.

regular but occurs in marked jumps corresponding more or less with the three sequences A, B, and C identified in the conglomerate-sandstone unit. These sequences record three main successive stages in the deposition of the coarse-grained material. Each consists of a basal conglomerate filling a channel cut in the limestone-shale facies, overlain by a continuous sheet of conglomerate or a thick sandstone. The conglomerates tend to cluster in the eastern part of each sequence and it appears that the axis of deposition of the conglomerates throughout the section has followed closely the eastward migration of the interdigitated contact.

On a smaller scale, individual beds of conglomerate or sandstone are either gradational, interfingering, or discordant with those of the limestone-shale facies. The gradation and interfingering of lithologies takes place in a very narrow zone generally less than 5 feet wide. The graded beds are rare in the transition zone of the coarse- and fine-grained facies and they always occur just above a lenticular conglomerate that fills a channel cut in the limestone-shale assemblage. In these beds, small pebble limestone conglomerate grades very rapidly into sandy limestone with a diminutive train of small pebbles occurring between the two lithologies. Most of these beds thin rapidly in an eastward direction, and in these the conglomerate is generally 2 feet thick whereas the sandy limestone decreases from 10 to 2 inches in thickness. The small pebble conglomerate is massive; the sandy limestone displays, in order, all the superposed divisions of the classic turbidite: graded bedding, parallel laminations, current ripple and convolute laminations, and parallel laminations. These conglomerate-sandy limestone beds are interpreted as channelized gravel and levee deposits.

Discordant contacts occur most commonly where the conglomerates and sandstones fill channels in the limestone-shale facies. Other conglomerate and sandstone beds are also discordant with those of limestone. These always occur along the steep wall of cut-and-fill structures where part of a limestone bed has been

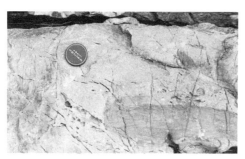

Plate II: Plan view of a bed of limestone eroded out and filled by conglomeratic sandstone along the east wall of a channel structure. Channel is off photo and to the right; L'Islet wharf.

undercut and filled by the coarse-grained material. Although these beds appear continuous with those of limestone, the contact is abrupt and vertical, and in many cases even inverted (Plate II).

The top of the conglomerate-sandstone assemblage is not exposed at L'Islet. Nearby, a sequence of thick, massive red mudstones conformably overlies Member 3, and Member 2 where the former is absent.

The conglomerates and sandstones of Member 3 constitute a deposit that, in cross section, is over one and one-half miles in width, and reaches a maximum observed thickness of 293 feet. In this deposit, the three coarse-grained sequences, A, B, and C, accumulated concurrently with the fine-grained, carbonate turbidite beds. Furthermore, observations indicate that, at the time of deposition, the difference in height between the concurrently deposited, coarse and fine-grained beds was not greater than 20 feet. These observations were made on some limestone and shale beds which have been traced laterally into the conglomerate-sandstone deposit. The widest spacing between such limestone and shale beds in the transition zone of the coarse- and fine-facies is 20 feet. Therefore, these conglomerates and sandstones have the characteristics of channelized deposits.

MINERALOGY OF THE CONGLOMERATES AND SANDSTONES

The sedimentological characteristics of the conglomerates and sandstones of Member 3 are summarized in Figure 4. The conglomerates and sandstones of the coarse-grained facies of the flysch are essentially similar throughout the section and are here grouped together for description. The mineralogical composition of the petromict conglomerates, arkosic arenites and wackes, and feldspathic arenites is summarized in Table I.

Petromict Conglomerates

There are four types of petromict conglomerates in sequences A, B, and C of Member 3. In the first (Plate III, figure 1), fragments of light grey micritic limestone constitute about two thirds of the clasts, the other third is made up of fragments of arkosic arenite and wacke. The second type of conglomerate is essentially similar in the proportion of micritic limestone and arkosic sandstones, but at least 20 per cent of the clasts are composed of slabs of grey, silty and sandy limestones (Plate III, figure 2). These slabs are very large rip-up clasts eroded from beds of limestone similar to those of Member 2. In the third conglomerate, equal sized fragments of light grey, micritic limestone make up almost all of the conglomerate fraction (Plate III, figure 3). In the fourth conglomerate, there are very few fragments of limestone. Instead, it is comprised of fragments of metaquartzite. quartz, and chips of black shale and collophane.

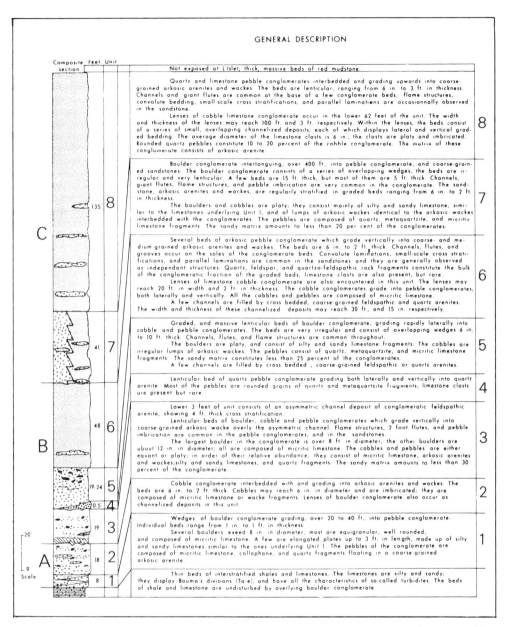

Figure 4: Columnar section and description of units of the L'Islet conglomerate-sandstone assemblage (Member 3).

TABLE I

Mineralogical composition of the conglomerates and sandstones of Member 3.

PETROMICT CONGLOMERATES	PER CENT OF THE ROCK	PER CENT OF FRACTION
Conglomerate fraction	60 to 90	
Limestone fragment (micrite)		40 to 60
Slabs of silty and sandy limestones		0 to 20
Fragments of arkosic sandstones		0 to 40
Fragments of collophane		trace to 1
Matrix (quartz and feldspars)	10 to 40	
ARKOSIC ARENITES AND WACKES		
Coarse-grained fraction	80 to 96	
Quartz		60 to 70
Feldspars		25 to 40
Rock fragments		trace to 8
Polycrystalline quartz		
Chert		
Granite and gneiss		
Shale		
Matrix (silt and clay)	6 to 20	
FELDSPATHIC ARENITES		
Coarse-grained fraction	100	
Quartz		75 to 90
Feldspars		6 to 15
Shale fragments		1 to 10
Matrix	0	

the order reflecting their relative abundance in the conglomerate fraction (Plate III, figure 4).

The sandy matrix of these petromict conglomerates is identical to the arenites and wackes that are interbedded with them.

Arkosic Arenites and Wackes

The arkosic arenites and wackes of Member 3 (Plate IV) have the same relative percentages of quartz, feldspar, and rock fragments and differ mostly in their relative amount of matrix and cement.

The feldspars constitute at least one third of the sandy fraction and comprise orthoclase, microcline, albite, and oligoclase. The grains of orthoclase are partly altered to kaolinite; those of microcline are very fresh, some grains are diagenetic, some clastic. The plagioclases are fresh.

The matrix of these sandstones consists of coarse and fine silt-size grains of quartz and feldspar with clay material composed of chlorite and illite. The arenites and a few of the wackes are cemented by carbonate and silica.

Feldspathic Arenites

The feldspathic arenites differ from the arkosic sandstones in that they are more mature both texturally and mineralogically, and always occur as channelized deposits (Plate V). These sandstones are composed of coarse, well rounded grains of quartz and feldspar cemented by abundant carbonate and silica. Chips of grey or green shale are present and tend to cluster along inclined laminae, or near the tops of the beds.

DIRECTION OF SEDIMENTARY TRANSPORT

The provenance of the grey micritic limestone clasts, and quartzo-feldspathic sand that constitute parts of the petromict conglomerates and sandstones of Member 3, will be discussed later. The other constituents, the slabs of silty and sandy limestones and fragments of arkosic sandstones, are interpreted as the product of erosion within the flysch sequence derived from beds of limestone and arkosic sandstone of Members 2 and 3 at or very near the site of deposition.

The general direction of sedimentary transport for the conglomerates and sandstones of Member 3 was established by measuring flutes, flame structures, wash-out channels, and cross-bedding

PLATE III

Figure 1: Petromict conglomerate interbedded with lenses of arkosic arenites and wackes. Sequence A, unit 2. View is looking east; top is to right; L'Islet wharf.

Figure 2: Petromict conglomerate with large slabs of silty and sandy limestones. Sequence B, unit 5; L'Islet wharf.

Figure 3: Graded bed of cobble conglomerate passing upwards into arkosic arenites. Sequence B, unit 6; L'Islet wharf.

Figure 4: Assymetric channel filled by a bed of lithic conglomerate grading to feldspathic arenite. Sequence B, unit 4. Coin is 1 inch in diameter; L'Islet wharf.

PLATE IV

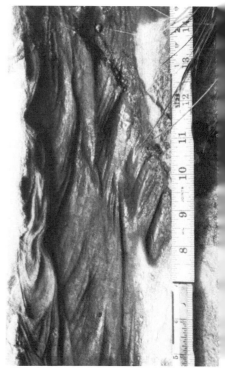

Figure 1: Bed of small-pebble conglomerate grading to arkosic wacke. Divisions a and b of the turbidite sequence are present; the laminations are faintly visible in the upper half of the bed. The bed is two and one-half feet thick. Sequence C, unit 8; L'Islet wharf.

Figure 2: Right side up bed of arkosic arenite displaying dune cross-stratification and b laminations. The cross beds are unusual in that they are discordant at the base and tangential at the top. Grading from small pebble to fine sand is conspicuous along some inclined laminae; the small pebbles and granules constitute very small wedges in the lower part of inclined laminae. Current was from left to right. Knife is 3 inches long; L'Islet wharf.

Figure 3: Arkosic arenite showing current-ripple laminae and convolute bedding. Current was from

PLATE V

Figure 1: Oblique view of large channel with feldspathic arenite fills displaying 4 foot-scale dune stratification; hammer is 11 inches long; L'Islet wharf.

Figure 2: Cross-stratification feldspathic arenite, filling cut-and-fill structure; L'Islet wharf.

orientations. The results are summarized in Figure 2.

SUMMARY

The conglomerate-sandstone assemblage that occurs at L'Islet in a limestone-shale facies, possesses the characteristics of a channelized deposit. The coarse-grained sediments consist of lenticular beds of petromict conglomerates and arkosic sandstones filling very large channels cut in the limestone-shale sequence. The conglomerate and sandstone beds display spectacular lateral and vertical graded bedding as well as parallel laminations and large cross-stratifica-tions. Small-, and large-scale channels also characterize the conglomerate-sandstone assemblage. Limestone fragments, quartzo-feldspathic sand, and sedimentary intraclasts are common. The direction of transport of most of the coarse-grained sediments was from the north.

THE ST-FABIEN AREA
GENERAL GEOLOGY

The St-Fabien area is located on the south shore of the St. Lawrence River, 130 miles northeast of L'Islet, and 10 miles southwest of Rimouski (Figure 5).

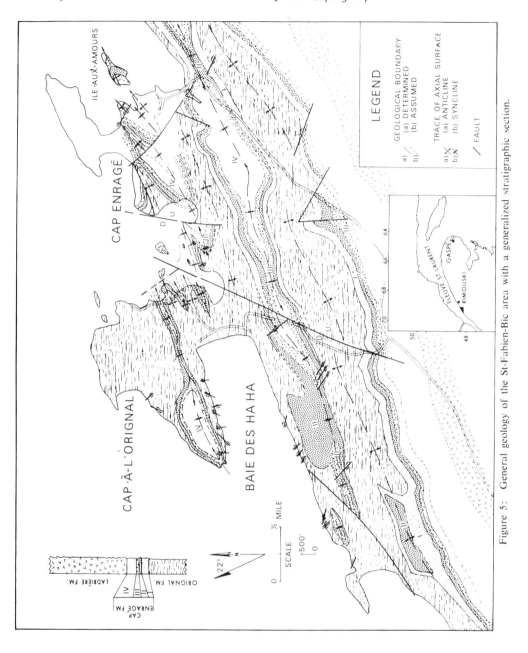

Figure 5: General geology of the St-Fabien-Bic area with a generalized stratigraphic section.

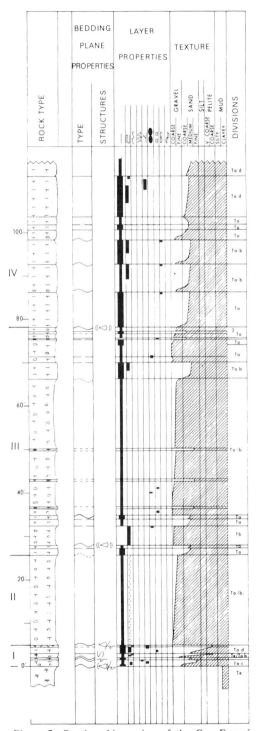

Figure 7: Stratigraphic section of the Cap Enragé Formation. The section is located at Cap Enragé.

The section described here consists of 1,000 feet of interbedded sandstones and conglomerates (the Cap Enragé Formation; Lajoie, in press), underlain by a minimum of 2,000 feet of claystones and siltstones (the Orignal Formation), and overlain by 5,000 feet of interbedded pelites and graded sandstones (the Ladrière Formation). The more than 8,000-foot section is folded into a series of northeast trending anticlines and synclines.

On the basis of grain size and composition, the Cap Enragé Formation can be subdivided into 4 members, I to IV, (Figures 5, 6, and 7). The fossils of the Cap Enragé Formation do not permit precise dating. Lower Cambrian and lower Middle Cambrian fossils were found in the conglomerate clasts (Rasetti, 1948 a, b: Lajoie, in press). The basal Ladrière beds have yielded Lower Ordovician fossils (Lespérance, personal communication). The Cap Enragé Formation could therefore range in age from Middle Cambrian to Lower Ordovician.

A brief description of the stratigraphic units underlying and overlying the Cap Enragé Formation is necessary for better understanding of the sedimentology.

The Orignal Formation which conformably underlies the Cap Enragé Formation consists mostly of thin-bedded green and black claystones and fine siltstones (Plate VI). The bedding is regular and does not show any erosional features. Thin films and small nodules of fluorapatite are commonly present in the bedding planes of the claystones. Rare coarse siltstones, and fine sandstones are also interbedded with the claystones. These coarse-grained beds generally show the a, b, and c divisions of the Bouma sequence.

The Ladrière Formation which conformably overlies the Cap Enragé Formation consists of interbedded pelites and sandstones. The pelites, which comprise 10 per cent of the section, are composed of green and black, laminated, thin-bedded claystones interbedded with siltstones in a 3 to 1 ratio. The sandstones (30 per cent of the section) are feldspathic arenites in 2-foot beds. The siltstone and sandstone beds show the a, b, and c divisions of the Bouma sequence (Lajoie, 1968).

THE CONGLOMERATE-SANDSTONE ASSEMBLAGE (THE CAP ENRAGE FORMATION)

General Description and Geometry of the Deposit

The Cap Enragé Formation is comprised of two conglomerate members (II and III) intercalated in sandstone members (I and IV). Although there is much pelitic material underlying and overlying the Cap Enragé Formation, there is very little within the formation itself.

The Cap Enragé Formation has a maximum thickness of 1,000 feet, but its thickness varies considerably both parallel and perpendicular to strike. From St-Fabien, the formation thins

Plate VI: Green, red, and black claystones of the Orignal Formation. South shore of Cap-à-l'Orignal; top to right; St.-Fabien.

gradually, northeastward for 12 miles as far as Rimouski where it pinches out, and also thins southwestward but its limit in this direction is not known. The formation pinches out perpendicular to strike, 6 miles south of the St. Lawrence River, and probably extends northward under the river, for an unknown distance. The general geometry of the deposit is that of a large wedge about 25 miles in length, a minimum of 6 miles in width, and a maximum of 1,000 feet in thickness. Detailed stratigraphic sections (Figures 6 and 7) show how much thinning can occur in the unit over a 2.5 mile distance. These two sections also serve to illustrate that very few beds, if any, can be traced from one section to the other.

Member I

The lower member (I) of the Cap Enragé Formation is composed of feldspathic sandstones in beds 4 inches to 30 feet thick, averaging 5 feet. Lenses of conglomerate are locally present; they have variable widths (20 inches to 60 feet), and thicknesses (up to 8 inches). The lower bedding planes generally undulate; sole marks and scours with graded fills are abundant.

Member I ranges in thickness up to 350 feet; its general geometry is that of a wedge which thins rapidly both northeast and southeast from Cap Enragé (Figures 6 and 7).

Member II and III

Members II and III are characterized by petromict conglomerates. Member II is composed entirely of conglomerate (Plate VII, figures 1, 2) whereas Member III consists of interbedded

sandstones and conglomerates (Plate VII, figure 3). Both members are lenticular with thicknesses ranging from 30 to 100 feet.

The conglomerate beds of Members II and III range in thickness from 8 inches to 30 feet. with a median at 4 feet. The basal contacts of beds are always sharp, but may be wavy. In Member III the upper contacts of the conglomerate beds are in places transitional with the overlying sandstones. Sole marks are rare in the conglomerates; graded (normal and reverse) and massive beds are equally abundant.

Member IV

Member IV consists of arkosic sandstones in beds ranging in thickness from 4 inches to 22 feet with a median at 4 feet. The lower bedding planes are sharp and planar. Sole marks (flutes and scour-and-fills) are abundant. The lower portion of the Bouma sequence (abc) is generally present; all beds in two measured sections begin with division a, 67 per cent of these also contain the b division, and 40 per cent, the a, b, and c divisions.

MINERALOGY AND TEXTURE OF THE SEDIMENTS

Sandstones

The sandstones of Member I are medium-grained ($1 \phi <$ Md $< 2.0 \phi$), moderately to poorly sorted ($.71 \phi <\sigma i < 2 \phi$), and contain less than 10 per cent matrix (fine fraction with grain size smaller than 4.5ϕ). The sandstones of Member IV are fine-grained ($2 \phi <$ Md $< 3 \phi$). poorly to very poorly sorted ($1 \phi < \sigma i < 4 \phi$). and on the average contain 10 per cent matrix.

PLATE VII

Figure 1: Petromict conglomerate (Member II) of Cap Enragé Formation; note quartz pebble foreset; St.-Fabien.

Figure 2: Petromict conglomerate of Cap Enragé Formation (Member II); St.-Fabien.

Figure 3: Quartz pebble conglomerate, interbedded with arkosic sandstones. Member III of Cap Enragé Formation; top is indicated by the pick of the hammer; St.-Fabien.

The composition of the sandstones of both members is summarized in Figure 8. The sandstones of Member I, are in a cluster which is astride the ranges of feldspathic and arkosic sandstones. All of the sandstones of Member IV are arkosic, the feldspar content being greater than 25 per cent.

The mineralogical composition of the sandstones is summarized in Table II. The quartz is commonly polycrystalline with bimodal size distribution of the crystals. The most common potash feldspar is microcline; the plagioclase is generally andesine, but albite is also relatively abundant (3 to 8 per cent in certain samples), and oligoclase is rare. Most rock fragments are composed of micritic limestone and fine-grained quartz arenite. Shale-mudstone fragments are common, particularly at the base of the beds.

	Member I (per cent)	Member IV (per cent)
Quartz	68 - 85	50 - 65
Feldspars		
andesine ⎫		
albite ⎬	6 - 15	12 - 23
oligoclase ⎭		
microcline ⎫	4 - 14	14 -25
orthoclase ⎬		
perthite	1	1
Rock fragments	1 - 4	1 - 6
limestone		
sandstone		
shale		
phosphate		
Glauconite	trace	trace
Pyrite	trace	trace
Rutile	trace	trace
Zircon	trace	trace
Apatite	trace	trace
Garnet (almandite)		trace
Leucoxene		trace

Table II—Mineralogical composition of the sandstones in Members I and IV.

The phosphate (fluoro-hydroxyloapatite) is in the form of oölites, which may be broken, or in small structureless fragments. The matrix of the sandstones consists of fine silt-size quartz and feldspar grains, and of clay material composed of chlorite and illite. The sandstones are cemented by silica and carbonates (calcite, dolomite, and ankerite; ankerite and calcite are the most abundant).

Conglomerates

The matrix, and the sandstones interbedded with the conglomerates of Member III, are similar to the arkosic sandstones of Member IV described previously.

The composition of the conglomerates in this part of the section is summarized in Figure 9. The gradual decrease in the percentage of limestone fragments upward is accompanied by a gradual increase of quartz. This relationship also holds for individual beds. In individual beds this is explained by a gradual decrease in grain size from base to top of the bed, the limestone clasts being generally much larger than the quartz grains. On the other hand, the up-section changes must be due to a variation of the composition at the source area. In order to provide an increase in quartz, new sources had to become available.

The conglomerate clasts range from granule to pebble with the median at - 2 ∅; a few boulders are present. They are well rounded and poorly sorted and composed mostly of carbonates and sandstones. The carbonate fragments are light grey, and contain little clay; they consist mostly of micritic, sparitic, oölitic, and pisolitic limestones. Large, but rare blocks of biosparite are also present; a few of these are composed 80 per cent of trilobite fragments. The sandstone clasts are quartz and glauconitic arenites, and rare quartz wackes. The quartz grains, in the fragments, are rounded and well sorted. Pebbles of granite, gneiss, and diabase are relatively rare.

PROVENANCE OF THE GRAVEL-SAND FRACTION

From the above description of the Cap Enragé Formation it can be concluded that the gravel and sand fractions of the formation were derived from sedimentary, metamorphic, and igneous terranes. This is also true of the conglomerates and sandstones at L'Islet.

The sedimentary source is easily demonstrated for the conglomerates by the composition of the clasts. The textures within the quartz sandstone clasts, and the structures present in the limestone fragments as described in the preceding chapter, coupled with the age of their trilobite fauna, suggest a shallow marine shelf-type environment, of Cambrian age as the locus of origin for these fragments.

The well-rounded quartz and glauconite grains in the sandstones, and in the matrix of the conglomerates, were probably derived from the same sandstones which provided the larger fragments. The other minerals found in these rocks could also have been derived partly or entirely from the same shelf sandstones; ultimately all of these minerals came from igneous and metamorphic terranes. That igneous and metamorphic parents contributed to the conglomerates is demonstrated by boulders and pebbles of granite, gneiss, and diabase (5 per cent of conglomerates). The gneiss fragments contain garnet and orthopyroxene crystals characteristic of the granulite facies (Martignole, personal communication). The mineral assemblage of the sandstones consisting of almandite ($N = 1.82$), leucoxene (sphene), rutile, andesine, perthite, and polycrystalline quartz was probably derived from the amphibolite- almandite, and granulite facies. These two facies are present over large areas of the Grenville province.

The relief of the source area must have been high, considering the mineralogical immaturity of the sandstones, and the relative coarseness of the feldspars. The climate at the source area was probably humid and warm because much hematite-bearing fine-grained sediments were also supplied to the trough. According to Hubert (1965), and Malihersik (1952) the red and maroon claystones found above and below the conglomerates were derived from red soils. In

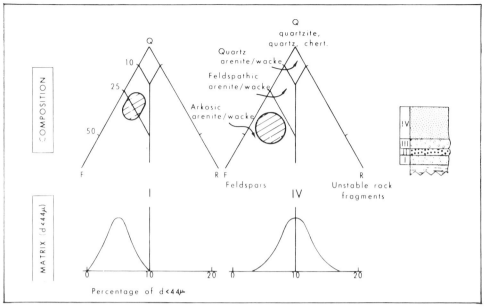

Figure 8: Classification of sandstone types of Members I and IV of the Cap Enragé Formation according to Gilbert (in Williams, *et al.*, 1955).

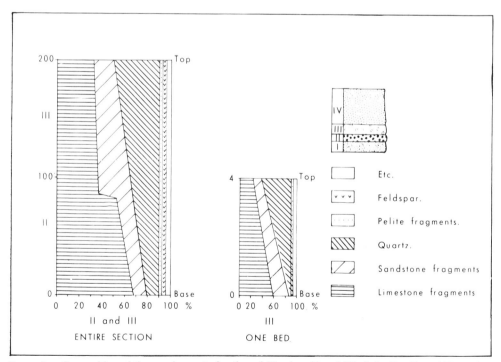

Figure 9: Composition of the conglomerate fraction of Members II and III of the Cap Enragé Formation.

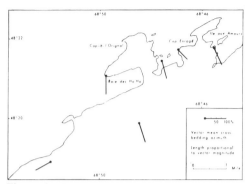

Figure 10: Vector means of 100 groups of measurements of flute casts and cross-bedding directions, in the Cap Enragé Formation; St-Fabien.

such a climate the feldspars could not have resisted chemical weathering unless the relief was high and the transport short and rapid.

DIRECTION OF TRANSPORT

The general direction of sediment transport was established by measuring sole marks (mostly flutes) and cross bedding orientations. The vector means of one hundred groups of measurements show that the general direction of sediment transport was from the north (Figure 10).

SUMMARY

The source area for the Cap Enragé sediments, was located north of the Appalachian trough, in a region of relatively high relief. The direction

of supply was mostly perpendicular to the elongation of the trough. The sediments were derived from sedimentary rocks of an early Paleozoic shelf-type environment, and, to a lesser extent, from igneous and metamorphic rocks of the Precambrian Shield.

SEDIMENTARY STRUCTURES OF THE CONGLOMERATE-SANDSTONE ASSEMBLAGES

The sedimentary structures in the coarse-clastic deposits of the L'Islet and St-Fabien areas consist of flutes and channels, large and small-scale cross-stratifications, syndepositional deformation ("flames"), and all the structures of the Bouma model.

SCOURS, CHANNELS, FLUTES, AND THEIR FILL

Small channels (Plate III, figure 4; Plate VIII) were observed at the base of the sandstone and conglomerate beds; they range in width from 4 inches to 10 feet. The channels are cut in beds of fine sand and gravel. The grain sizes of the fills vary considerably from channel to channel but are relatively constant in individual channels. The fills are generally graded and their median grain size is either coarser or equivalent to that of the bed in which they are cut.

Flute marks are not very common in the conglomerate-sandstone assemblages; they range from 3 inches to 4 feet in length (Plate IX. figure 1), but are generally less than 8 inches. The flutes are filled by material similar to the one already described for the channels.

Plate VIII: Channel in upper Member of Cap Enragé Formation. Note cross-stratification underlying the channel, and the content-graded fill; St.-Fabien.

PLATE IX

Figure 1: Large flutes in Cap Enragé Formation (Member I); St-Fabien.
Figure 2: Trough cross-bedding in Member IV of Cap Enragé Formation; St.-Fabien.
Figure 3: Dune cross-stratification in Member IV of Cap Enragé Formation; ruler is 15 cm long; St.-Fabien.
Figure 4: Flame structures at the base of a quartz pebble conglomerate; Member IV, Cap Enragé Formation; St.-Fabien.

CROSS-STRATIFICATIONS

Tabular - and trough-type cross-stratifications with scale ranging from 1 inch to 5 feet, are commonly observed in the conglomerate - sandstone assemblages. In a given bed, the cross-strata may occur either, 1) above a division of parallel laminations or, 2) at the base of the bed, below a division of parallel laminations.

In the first, more common association, the cross-laminae occur in a vertical sequence above a division of parallel laminations and correspond to the c division of Bouma, 1962. In these beds the parallel laminations are underlain by a division of graded bedding. Such cross-laminations were observed in 35 per cent of the sandstone beds of the Cap Enragé Formation but are relatively rare in the sandstone beds at L'Islet (Plate IV, figure 3). The division of cross-laminations does not exceed 6 inches in thickness. This association of cross-laminations above a division of parallel laminations and grading has been interpreted by Walker (1965), and Harms and Fahnestock (1965) in terms of the decrease in the flow velocity.

The second type of association is far less common (only 1.5 per cent of the beds show this feature in the Cap Enragé Formation but it is relatively common in conglomerate and sandstone beds at L'Islet). The cross-strata occur at the base of the bed, below a division of parallel laminations (Plate V, figure 2), in a few cases, it overlies a graded division (Plate IX, figure 3). The scale of the cross-stratified division ranges from 3 inches to 5 feet. The cross-laminations are generally observed in coarse to very coarse sandstones, and in fine-grained conglomerates. Although present (Plate VII, figure 1), they are rare in the coarse petromict conglomerates. The foreset laminae in these cross-strata are well defined, with average angles of slope between 17 and 20 degrees; it has never been observed to be less than 10 degrees. Similar cross-strata have been described in bioclastic deposits in Scotland by J. Hubert (1966) and in the Marathon region by Thomson and Thomasson (1969), and were interpreted by these authors to represent the dune phase of the lower flow regime. The difficulty with this interpretation is that the division of parallel laminations b should occur below the cross-stratifications and not above it, as it is in most of the dunes described by Thomson and Thomasson, and in this paper. At L'Islet, dunes were also observed in channel fills (Plate IV, figures 2, 3). The channels are asymmetric, and the observed orientation of the foreset beds shows a strong and consistent discrepency with the long axis of the channels. In these cases, the cross-strata are thought to represent the lateral construction of point bars filling the channels.

DISH STRUCTURES

Dish-type structures, as defined by Wentworth (1967) and described by Stauffer (1967) are found in a few sandstone beds of the Cap Enragé Formation. The structure consists of thin, dark laminations, concave side up. The margins of the structures are generally truncated by adjacent dishes. The dishes are small with lengths not exceeding 4 inches, and thicknesses less than 1 inch. In the Cap Enragé sandstones the dish structures always occur above a division of parallel laminations underlain by a division of graded bedding.

Wentworth (1967) suggested that the dishes were formed by antidunes, and Stauffer (1967) used the structure in his model for grain flow. It is difficult to use grain flow to explain the dish structures in the sandstones of the Cap Enragé Formation; excellent grading such as displayed in the sandstone beds, passing into a division of well defined parallel laminations is not what one would expect to be charactertistic of grain flow deposits. Grain flow as defined by Bagnold (1956) should not form traction structures. However, a transition may exist from turbidity currents into grain flow (see Middleton, this volume). The position of the dish structure division in the Cap Enragé sandstones, above a division of parallel laminations and, in rare cases, below a division of very faint parallel laminations, suggests formation in the lower flow regime. These structures look like, and could perhaps be, small trough sets.

SYNDEPOSITIONAL DEFORMATION

A few flame structures (Plate IX, figure 4) are present at the base of coarse-grained sandstone beds and some pebble conglomerate beds. The flames typically consist of a pointed curved wisp of fine argillaceous sand drawn up into fine gravels. The geometry of the flames illustrated in Plate IX, figure 4 can only be explained by flowage in one direction and is interpreted to be primary, and not the result of subsequent differential loading. The structure could be formed by current drag from a turbulent flow at the gravel-mud interface of the cohesive but hydroplastic bottom.

THE BOUMA SEQUENCE

The sandstone beds of the Cap Enragé Formation and those at L'Islet exhibit the sedimentary structure sequence characteristic of turbidites (Bouma, 1962; Walker, 1965; and others).

Of the sandstone beds of the Cap Enragé Formation 10 per cent have the a, b, c, and d divisions of the Bouma sequence (Figure 11): 30 per cent have the a, b, and c divisions, and 25 per cent the a and b divisions. The pelitic division e has not been considered because pelites are absent at the top of the beds.

Over 90 per cent of the coarse-grained beds (sandstones and conglomerate) start with the lower a division (T_1 of Bouma, 1962), and 60 per cent of the beds show either ab, abc, or abcd.

378

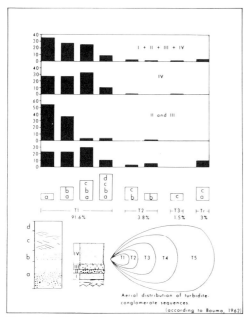

Figure 11: The turbidite sequence in the St-Fabien region. The vertical scale of the histograms is in per cent; I, II, III, and IV refer to members of the Cap Enragé Formation.

SEDIMONTOLOGICAL MODEL

The coarse-grained sediments described at L'Islet and St-Fabien are exotic in the stratigraphic column; they are underlain and overlain by thick sequences of pelites. Most of the gravels and sands that constitute the exotic rocks must have been derived from a marine, shallow carbonate-sandstone shelf situated to the north of the pelite lithosome. Although this carbonate shelf is not presently observed in Québec, it must have existed during all of the Cambrian Period because fossils of Early, Middle, and Late Cambrian and Early Ordovician ages are found in the limestone fragments contained in the conglomerates at L'Islet, St-Fabien, Kamouraska (Hubert, 1965), and Lévis (Osborne, 1956). This carbonate shelf has eroded or is underlying part of the Appalachian cover which was thrust in late Ordovician time toward the Canadian Shield.

The conglomerate-sandstone assemblages at St-Fabien and L'Islet have the geometry of fans. Similar gravel and sand assemblages also occur at Lévis (Osborne, 1956), Kamouraska (Hubert, 1965), and Lac Matapédia (Ollerenshaw, 1967) and are relatively common in other parts of the Cambro-Ordovician flysch sequence of Quebec. All these coarse-grained deposits are concentrated along a narrow belt that lies near and parallel to the St. Lawrence River. The recurrent geographically widespread nature of these coarse-grained deposits throughout the

flysch sequence and their concentration along a narrow belt in the pelite lithosome strongly suggest that the occurrence and development of such deposits were probably related to a major geomorphic feature, a slope that existed on the northern side of the trough between the shallow carbonate shelf and the deeper pelite lithotope. The slope and the carbonate shelf must also have been dissected by canyon-like channels in order to explain the localised distribution of the gravels and sands, and to expose to erosion carbonate rocks ranging in age from Early Cambrian to Early Ordovician. At times the Precambrian basement probably also underwent erosion. Recurrent uplift of the Precambrian Shield and the shelf produced coarse-grained clastic debris which was transported southward, off-shelf, into a mud lithotope.

The sandstone beds of the St-Roch and Cap Enragé Formations generally display typical turbidite characteristics; they are graded, commonly show the b and c divisions, and contain sole marks such as flutes. Repetitive graded bedding in a vertical sequence of nearly 1,000 feet is difficult to explain by mechanisms other than turbidity currents. The sandstone beds beginning with dune cross-stratification are difficult to explain in this context; these sands were probably transported by strong turbulent bottom currents. They represent, however, less than 2 per cent of the section.

The transport of the conglomerate fraction is much more difficult to explain in view of our present knowledge of the transport of coarse gravel into deep water. It is proposed that the gravels first accumulated in a neritic environment, where the clasts were rounded by surf and strong currents, followed by resedimentation into the deeper mud lithotope where they are found interbedded with the turbidites. The gravels are either graded (Plates IV, figure 1; VII, figure 3), massive (Plates VII, figure 2; IX, figure 4), or are incorporated into thick beds showing large cross-bedding (Plate VII, figure 1). Some authors may refer to massive deposits as fluxoturbidites, but the mechanism behind this term is not too well understood. Aalto and Dott (this volume) favor grain-flow to explain their ungraded, well sorted conglomerates. The imbrication observed in some conglomerates coupled with large-scale cross-stratification suggest that strong turbulent currents are responsible for the transportation of some of the coarse fraction. Whatever the mechanism, it is possible to move much coarse gravel into deep water as in deep marine troughs to-day.

The model proposed here is summarized in Figure 12. An unstable source area composed of shelf type limestones and sandstones, and the Precambrian Shield provided coarse debris which were channelized and deposited on submarine fans by turbidity currents and other mechanisms. The tectonic movements at the source areas, although frequent, must have been short-lived because within each deposit the grain size of

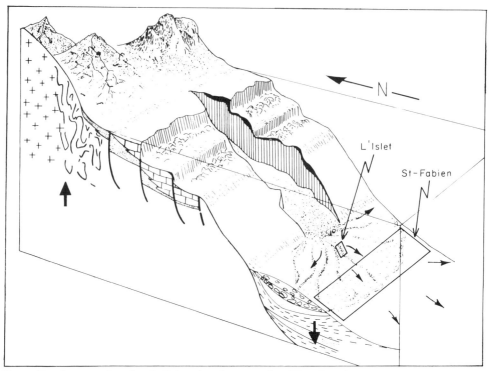

Figure 12: Interpretation of early Paleozoic paleogeography. The submarine fan sediments are located at the base of a slope, derived from a tectonically unstable shelf composed of Paleozoic limestones (blocks) and sandstones (stippling), and Precambrian metamorphic (folded) and igneous (cross) terranes. The two areas discussed in this paper are not located on the same fan.

the gravel rapidly decreases towards the top of the section; the mineralogical maturity of the sands also increases steadily from the base to the top of the coarse lithosome.

ACKNOWLEDGEMENTS

We express our gratitude to the Québec Department of Natural Resources for financial support of the field work at L'Islet and St-Fabien. The cost of the laboratory work, the preparation, and publication of this manuscript was defrayed by grants from the National Research Council of Canada (NRCC Grant No. A-5242 and A-2635). Miss Denise Leboeuf drafted the diagrams; her technical aid is greatly appreciated. Finally, we thank Hans Hoffmann and R. G. Walker for critically reviewing the manuscript; their constructive criticisms helped greatly to improve it.

REFERENCES

Aalto, K. R. and Dott, R. H. (1970)—Late Meszoic conglomeratic flysch in southwestern Oregon, and the problem of transport of coarse gravel in deep water; Geol. Assoc. Canada, Spec. Paper 7.

Ayrton, W., *et al.*, (1969)—Lower Llandovery of the Northern Appalachians and adjacent regions; Geol. Soc. Amer. Bull., vol. 80, pp. 459-484.

Bagnold, R. A., (1956)—The flow of cohesionless grains in fluids; Roy Soc. London, Phil. Trans; vol. 249, Ser. A, pp. 235-297.

Barley, L. M. and McInnes W., (1890)—Rapport sur certaines parties de la province de Québec et les régions adjacentes du Nouveau Brunswick et du Maine et traitant plus particulièrement des comtés de Témiscouata et de Rimouski; Com. Géol., Can. Rap. Ann. vol. 5, pt M, pp. 1-30.

Bouma, A. H. (1962)—Sedimentology of some flysch deposits; a graphic approach to facies interpretation; Elsevier, Amsterdam, Netherlands, 168 p.

Cady, W. M. (1967)—Geosynclinal setting of the Appalachian mountains in southeastern Québec and northwestern New England; Roy. Soc. Can., Spec. Publ. No. 10, pp. 57-69.

Harms, J. C. and Fahnestock, R. K. (1965)— Stratification, bed forms, and flow phenomena (with example from the Rio Grande); in, Primary Sedimentary Structures and their Hydrodynamic Interpretation (G. M. Middleton, Editor); Soc. Econ. Paleont. and Mineral., Spec. Publ. No. 12, pp. 84-116.

Houde, M. and Clark, T. H. (1961)—Geological map of the St. Lawrence Lowlands; Québec Department of Mines, Map No. 1407.

Hubert, C. (1965)—Stratigraphy of the Quebec Complex in the L'Islet-Kamouraska area, Quebec; Ph.D. dissertation, McGill University, Montreal.

........(1967)—Tectonics of part of the Sillery Formation in the Chaudière-Matapédia segment of the Québec Appalchians; Roy. Soc. Can., Spec. Publ. No. 10, pp. 33-41.

Hubert, J. F. (1966)—Sedimentary history of Upper Ordovician geosynclinal rocks, Girvan, Scotland; J. Sed. Pet., vol. 36, pp. 677-699.

Lajoie, J. (1968)—Turbidites sans matrice: produits de diagénèse; Nat. Can., vol. 95, pp. 1243-1255.

........(in press)—Régions de Rimouski et de Lac des Baies (Moitié Ouest); Qué. Dept. Nat Res., Rap. Géol.

Léonard, M. A. (1969)—Le flysch de Saint-Fabien; Mémoire de M. Sc., Université de Montréal, 80 p.

Logan, W. E. (1863)—The Québec Group; in, The Geology of Canada; Geol. Surv. Can., Rept. Prog. to 1863.

Malihersik, S. J. (1952)—Petrology of the Charny Formation; Unpubl. D. Sc. thesis, Université Laval, Québec, 130 p.

Middleton, G. V. (1970)—Experimental studies related to problems of flysch sedimentation; Geol. Assoc. Canada, Spec. Paper 7.

Neale, E. R., *et al.* (1961)—A preliminary tectonic map of the Canadian Appalachian region based on age of folding; Bull. Inst. Mining Metal., vol. 54, pp. 687-694.

Ollerenshaw, N. C. (1967) — Région de Cuoq-Langis; Min. des Richesses Nat., Québec, R. G. 121, 230 p.

Osborne, F. F. (1956)—Geology near Québec City; Nat. Can., vol. 83, No. 83, pp. 157-223.

Rasetti, F. (1948a) — Lower Cambrian trilobites from the conglomerates of Quebec (exclusive of the Ptychopariidea); J. Pal., vol. 22, pp. 1-24.

........(1948b)—Middle Cambrian trilobites from the conglomerates of Quebec (exclusive of the Ptychopariidea); J. Pal., vol. 22, pp. 315-339.

Skidmore, B. (1967)—The Taconic unconformity in the Gaspé Peninsula and neighbouring regions; Roy. Soc. Can., Spec. Publ., No. 10, pp. 25-33.

Skidmore, B. and McGerrigle, H. W. (1967) — Geological map, Gaspé Peninsula; Quebec Dept. Nat. Res., Map No. 1642.

St. Julien, P. (1967)—Tectonics of part of the Appalachian region of southeastern Québec; Roy. Soc. Can., Spec. Publ. No. 10, pp. 41-48.

Stauffer, P. H., (1967)—Grain-flow deposits and their implications, Santa Ynez Mountains, California; J. Sed. Pet., vol. 37, pp. 487-508.

Thomson, A. F. and Thomasson, M. R. (1969)—Shallow to deep water facies development in the Dimple limestone (Lower Pennsylvanian), Marathon Region, Texas; in, Depositional Environments in Carbonate Rocks (G.M. Friedman, editor), Soc. Econ. Paleont. and Mineral., Spec. Publ. No. 14, pp. 51-77.

Walker, R. G. (1965)—The origin and significance of the internal sedimentary structures of turbidites; Yorkshire Geol. Soc. Proc., vol. 35, pp. 1-32.

........(1967)—Turbidite sedimentary structures and their relationship to proximal and distal depositional environments; J. Sed. Pet., vol. 37, No. 1, pp. 25-43.

Williams, H., Turner, F. J., and Gilbert, C. M., (1955)—Petrography; W. H. Freeman and Co., San Francisco, 406 p.

Wentworth, C. M. (1967)—Dish structure, a primary sedimentary structure in coarse turbidites (abstract); Am. Assoc. Petrol. Geol. Bull., vol. 51, p. 485.

TRANSPORTING MECHANISM OF THE CONGLOMERATE

ADDENDUM BY JEAN LAJOIE

In the preceding paper, the transport of the conglomerate is said to be difficult to explain. Further work at L'Islet (Rocheleau and Lajoie, 1974) shows that the sedimentary structures found in the conglomerate beds (grading, oblique and parallel stratifications) occur vertically in transitional divisions, forming sequences in the beds. It suggests that within each bed the structures were formed by single depositional events.

Figure 1 summarizes the various sedimentary structure sequences observed at L'Islet. The figure shows that 84 percent of the studied beds have at their base a graded division; in 50 percent of the beds grading is the only structure present. All sizes, up to 256 mm, are graded, and both the coarsest and mean diameters decrease regularly from base to top of the beds. In the conglomerate the graded division is overlain either by a division of parallel stratification (in 16 percent of the beds) or by a division of oblique stratification (in 14 percent of the beds). Both types of stratifications are ill-defined because they are composed of alternating coarse diameters, averaging 8 and 2 mm. In sandstone

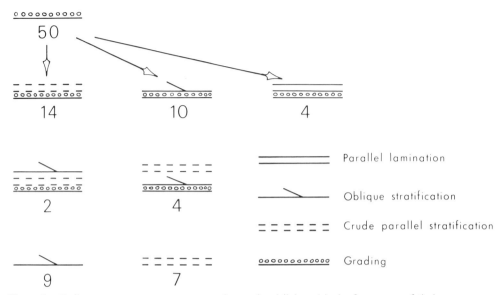

Figure 1. Sedimentary structure sequences observed at L'Islet with the frequency of their occurrence expressed as percentages.

beds, the graded division is overlain by parallel lamination (4 percent of the beds).

Two sedimentary structure sequences are found in the L'Islet conglomerate beds: (1) a graded division overlain by a division of parallel stratification, in turn overlain by a division of oblique stratification, or (2) a graded division overlain by a division of oblique stratification, followed by a division of parallel stratification. Both sequences are found in about the same number of beds. The beds in which the oblique stratification overlies the graded division, also have the coarsest diameter of the fill, the mean sizes at the base of the beds being 32 mm.

Turbidity current is the mechanism that best explains all the observed features in the L'Islet conglomerate. Grading at the base of most beds evinces that at least part of the bed material was transported in suspension. The presence of traction structures such as parallel and oblique stratifications suggests that bed load existed during the deposition of the upper part of many beds. Grain flow, laminar by defi-

nition, cannot explain the traction structures found in many of the L'Islet conglomerates. Other mass flows, such as slump and subaqueous debris flow, have characteristics that are much different from those observed in the L'Islet beds.

Komar (1970) showed that turbidity currents could have the competency to transport large gravel clasts. If the equations used by Komar (1970, p. 1557) are applied to the L'Islet clasts, the minimum velocity needed to roll the coarsest diameters would be in the order of 800 cm/sec. This figure is comparable to the velocity obtained by Komar for the clasts in the Doheny Channel.

REFERENCES

Komar, P. D. (1970). The competence of turbidity current flow. *Geol. Soc. America Bull.*, **81**, 1555–1562.

Rocheleau, Michel, and Lajoie, Jean (1974). Sedimentary structures in resedimented conglomerate of the Cambrian flysch, L'Islet, Québec Appalachians. *Jour. Sediment. Petrol.*, **44**, 826–836.

25

Copyright © 1972 by Macmillan Journals Ltd.

Reprinted from *Nature Phys. Sci.,* **240**(99), 59-61 (1972)

A Deep-water Sand Fan in the Eocene Bay of Biscay

C. Kruit, J. Brouwer, and P. Ealey

THE thick, massive sand deposits, which characterize the Monte Jaizkibel (east of San Sebastian, Guipúzcoa, Spain), seems to represent the topographical expression of a major "fossil" deep-water fan of Early Eocene age. This fan is approximately 15 km wide, and reaches a maximum thickness of about 600 m in its central area (Fig. 1).

The Monte Jaizkibel occupies the eastern portion of a coastal belt of Lower Eocene deep-water deposits, which belongs to the Mesozoic/Tertiary cover of the northwest flank of the westernmost Hercynian massif of the Pyrenees.

Deep-water conditions, within an "early Bay of Biscay", have characterized the area from the Albian onward up into the Early Eocene[1-3]. Field observations in the Maastrichtian/ Lower Tertiary interval demonstrate the prevalence of west-ward-flowing turbidity currents, apparently in line with the axial trend of the deep-water basin.

The Jaizkibel fan, however, was produced by a lateral influx of sand from a focal point to the north. Leeward of the fan, the sand-transporting currents diverted into the axial-basinal direction. Typical metres thick bedded coarse grained sand deposits can be observed along the entire outcrop area of the Lower Eocene, which extends about 20 km westwards.

The structural development of the region originates from the (Oligocene) Pyrenean folding phase and from subsequent block-faulting[4]. Some insight into its present structural relation to the adjacent area of the Bay of Biscay can be obtained from seismic "flexotir" profiles[5,6], which show the folding and rise of the Mesozoic/Eocene on the southern Bay of Biscay continental slope to be in sharp contrast to the apparently little affected Upper Cretaceous/Lower Tertiary sediments of the Landes marginal plateau in the north.

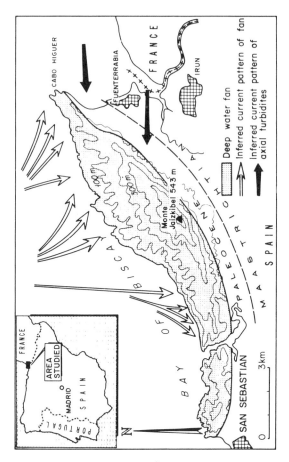

Fig. 1 The Monte Jaizkibel deep-water fan.

The principal criteria applied in considering the entire Monte Jaizkibel as a "fossilized" deep-water fan are the following: the massive sandstone development, reaching a thickness of approximately 600 m in its central area; its divergent palaeo-current pattern; its lens-shaped outlines; and the deep-marine benthonic foraminiferal fauna characterizing the marl and shale intercalations.

The deep-water fan deposits rest on an interval of thin-bedded turbidite sandstones (bed thickness 5–15 cm; sand/"shale" 25%) alternating with marls and lime-mudstones of Palaeocene age, which are well exposed in the coastal-cliff section off Fuenterrabia beach. The scarce flutes here follow the axial basin strike.

The sand influx of the deep-water fan began rather suddenly with the deposition of massive sand beds, a few metres thick and separated by thin clay interbeddings. Faunal analysis of the interbedded clay indicates that the build-up had already started in the Palaeocene (*Globorotalia pseudomenardii* zone), although full development was not attained before late Early Eocene (*Globorotalia formosa-aragonensis* zone).

The benthonic Palaeocene and Lower Eocene faunas are predominantly agglutinated, and of the "*Rhabdammina*" fauna group. This fauna, in combination with such calcareous components as *Nuttallides carinotrümpyi* Finlay, *Oridorsalis umbonatus* (Reuss) and *Osangularia mexicana* (Cole)/*navarroana* (Cushman) and a rich planktonic fauna, points towards a depth of deposition ranging between 200 and 3,000 m, with an average of 1,000 ml. (ref. 7).

The sporadic occurrence of *Palaeodictyon* Meneghini, a typical burrow structure in deep-water "flysch" deposits, corroborates these conclusions.

The occurrence of horizontal branching burrows, in this deep-water fauna association is remarkable. They are observed in the coastal area, east of Cabo Higuer, in the top of some sandstone beds. These burrows are identical to the ones shown by Gomez de Llarena[8] (Lám. 54, Fig. 2) and look very similar to *Thalassinoides* Ehrenberg, which is common in shallow water deposits and is produced by crustaceans.

Oscillation ripple marks have not been observed, which sets a minimum depth for these deposits below "wave-ripple base", which may have been of the order of at least 50 m.

We conclude that the faunal evidence in particular points to "deep-water" conditions. Although precise definition of "deep" would be difficult here, it seems likely that a depositional depth of at least 1,000 m has prevailed.

A survey of the coastal cliff exposures indicated that the average thickness of the individual sandstone packs is of the order of 4 m, with an estimated maximum thickness of about

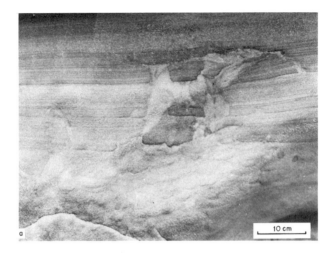

Fig. 2 Secondary liquefaction structures in submarine fan sandstones, Monte Jaizkibel. *a*, Intrusion of liquefied sand into parallel laminated sand, and development of dish structures in the liquefied interval. *b*, Dish structures in liquefied sand.

15 m. The majority of these sandstone units are fundamentally parallel laminated, with occasional evidence of flowage and/or scour-and-fill in their uppermost part. Finely parallel laminated intervals are often disturbed by a secondary liquefaction process, which may lead to intrusions of liquefied sand into intervals of well laminated sand (Fig. 2a). The liquefied intervals often show "dish structures"[9,10] (Fig. 2a and b), which suggest that secondary liquefaction was followed by upward flow within the sandbody, this flow proceeding no more than a few centimetres before being arrested.

Other thick sandstone beds have a largely homogeneous appearance due to lack of lamination. This may be a primary feature (for example, the a-interval of turbidite deposits[11]), but it is also conceivable that the lack of structure merely represents an advanced phase of the observed liquefaction phenomena.

The origin of the pressure variations initiating the secondary flow processes remains a matter of speculation. The effect of bypassing turbidity streams is just one of the possibilities that could merit consideration.

Flutes, grooves and parting lineations are directional current marks that contributed substantially to a reconstruction of the regional palaeo-flow pattern. The general pattern of flute directions shows beyond doubt that the deep-water fan deposits were derived from the north, and the additional information from grooves and parting lineations has helped to provide a coherent regional picture. The divergent palaeo-current pattern suggests that a main point source of sand supply, for example, the distal termination of a submarine canyon, would have been situated 6 or 7 km north of Jaizkibel.

[1] Rat, P., *Publ. Univ. Dijon*, 18 (1959).
[2] Wiedmann, J., *Livre mém. Prof. Paul Fallot*, **1**, 351 (1962).
[3] Saavedra, J. L., *I Congreso Hispano-Luso-Americano de Geología Económica*, Sección 1, Tomo I, 403 (1971).
[4] Boillot, G., Dupeuble, P. A., and Hennequin, I., *CR Acad. Sci. Paris*, D, **274**, 1147 (1972).
[5] Montadert, L., Damotte, B., Debyser, J., Fail, J. P., Delteil, J. R., and Valéry, P., *The Geology of the East Atlantic Continental Margin* (Rep. Inst. Geol. Sci., 70/15, 1971).
[6] Damotte, B., Debyser, J., Montadert, D., and Delteil, J. R., *Rev. Inst. Franc. Pétrol.*, **24**, 1061 (1969).
[7] Brouwer, J., *Proc. Kon. Ned. Acad. Wet*, B, **68**, 5, 309 (1965).
[8] Gomez de Llarena, J., *Monogr. Inst. "Lucas Mallada" de Invest. Geol.*, 13 (Madrid, 1954).
[9] Stauffer, P. H., *J. Sed. Petrol.*, **37**, 487 (1967).
[10] Wentworth, C. M., *Bull. Amer. Assoc. Petrol. Geol.*, **51**, 485 (1967).
[11] Bouma, A. H., *Sedimentology of Some Flysch Deposits* (Elsevier, 1962).

26

Copyright © 1974 by the Society of Economic Paleontologists and Mineralogists

Reprinted from *Modern and Ancient Geosynclinal Sedimentation* (SEPM Spec. Publ. 19), R. H. Dott, Jr., and R. H. Shaver, eds., 1974, pp. 69–91

DEPOSITIONAL TRENDS OF MODERN AND ANCIENT DEEP-SEA FANS

C. HANS NELSON AND TOR H. NILSEN

U.S. Geological Survey, Menlo Park, California

ABSTRACT

Many flysch, turbidite, fluxoturbidite, and grain-flow sequences from ancient geosynclines probably have been deposited in deep-sea fans adjacent to continental margins. We have obtained stratigraphic and sedimentologic criteria for recognizing ancient fan deposits by comparing the Astoria Fan, a large open-ocean fan off the coast of northern Oregon, with the Eocene Butano Sandstone, an ancient continental borderland fan deposit of similar size in the Santa Cruz Mountains, California.

Deep-sea fan deposits consist of channel and interchannel facies. Both facies change significantly downfan and laterally across the fan as a result of decreasing current velocities during each turbidity current and as a result of the lateral migrations of channels through time. Geologic mapping reveals thick-bedded, coarser grained, and lens-shaped channel deposits intermixed with thin-bedded and finer grained interchannel deposits. Sand-shale ratios are high within channels and low within upper fan interchannel and distal fan areas. The coarsest grained and thickest bedded gravels and sands are deposited by channelized sediment gravity flows in the submarine canyons and upper fan valleys. These ungraded, poorly sorted, and massive channel sediments change by midfan to thinner bedded, finer grained, vertically graded, and better sorted turbidite sands that contain sedimentary structures in Bouma sequences. Turbidites in interchannel areas are formed by overbank spilling and consist of thin-bedded fine-grained sands and silts characterized by Bouma *cde* and *de* sequences.

The delineation of fan margins and paleogeography is aided by the lateral and downfan changes in thickness, texture, composition, and paleocurrent directions of fan sediments. High contents of terrigenous debris are present in the sand fractions of hemipelagic muds deposited near the continental margin, and this may help delimit the shoreward boundary of the fan; in contrast, gradation to high contents of pelagic material indicates the direction of the seaward edge of the fan. Radially oriented paleocurrent patterns define the fan apex but are typically complex because of lateral overflow out of and away from channels and because of meandering and lateral shifting of fan channels. The outlining of fan geometry and major channels also is assisted by the decrease of the maximum clast size and of thickness of turbidite beds both downfan and laterally from channels.

Variations in morphology, stratigraphy, sedimentary facies patterns, grain-size distribution, sediment composition, and sediment dispersal patterns help identify fans from different geosynclinal settings such as restricted borderland or marginal sea basins, open ocean continental rises, and deep-sea trenches.

INTRODUCTION

General

Descriptions of both modern deep-sea fans (Shepard and others, 1969; Nelson and others, 1970; Normark, 1970a; Piper, 1970a; Haner, 1971; Normark and Piper, 1969, 1972; Nelson, 1974; Nelson and Kulm, 1973) and inferred ancient deep-sea fan deposits (Sullwold, 1960; Walker, 1966, 1970; Jacka and others, 1968; Piper, 1970b; Stanley, 1969b; Hubert and others, 1970; Mutti and Ricci Lucchi, 1972) have become more common in the geologic literature. Modern fan systems have been shown to vary greatly in size and shape, but the character of the sediments and nature of the depositional processes on them seem comparable. Similarly, although many different tectonic and paleogeographic settings have been ascribed to ancient deep-sea fan deposits, the stratigraphic sequences, sedimentary structures, and character of the sedimentary rocks seem comparable. The common occurrence of modern deep-sea fans in presumed geosynclinal settings such as the intersection of continental margins with spreading oceanic plates (Dietz, 1963; Dewey and Horsfield, 1970), marginal-sea basins, and continental borderland basins makes a comparison of the modern and ancient depositional systems a worthwhile exercise for understanding geosynclinal sedimentation.

Some characteristics of deep-sea fan sedimentation are best examined in modern deposits, others in ancient deposits. In this paper we will compare the morphologic and sedimentary features of the modern Astoria Fan, located off the coast of Oregon and studied by Nelson (1968), with the Butano Sandstone, an inferred Eocene deep-sea fan deposit located in the central Coast Ranges of California and studied by Nilsen (1971). Six important characteristics of the deep-sea fans and their deposits will be compared: (1) morphology and physiography, (2) stratigraphy and sedimentary facies, (3) sedimentary structures, (4) sediment grain-size distributions, (5) sediment compositions, and (6) sediment-dispersal systems. Our objectives are to outline the depositional processes acting in deep-sea fan systems and to establish some criteria for the recognition of ancient deep-sea fan deposits and their geosynclinal settings.

Butano Sandstone

The Butano Sandstone is found in the Santa Cruz Mountains south of San Francisco, California (fig. 1). It was named by Branner and others (1909) for sandstones and conglomerates that crop out on Butano Ridge. Foraminiferal studies indicate that the Butano ranges in age from the Penutian (early Eocene) to the Narizian (late Eocene?) provincial foraminiferal stages of Mallory (1959), although most of it apparently accumulated during Narizian time (Brabb, 1960; Sullivan, 1962; Clark, 1966; Fairchild and others, 1969). The formation was deposited at lower bathyal to abyssal depths in a basin that had unrestricted access to the ocean (Cummings and others, 1962). It unconformably overlies the marine Locatelli Formation and is conformably overlain by the marine Two-bar Shale Member of the San Lorenzo Formation (Brabb, 1964). No complete section of the Butano is exposed, so its total thickness is not known; its minimum thickness is approximately 1,500 m, the maximum perhaps 3,000 m.

Sedimentological studies by Nilsen (1970, 1971) and Nilsen and Simoni (1973) indicate that the Butano Sandstone forms the southwestern part of an Eocene deep-sea fan. Evidence that it originated as a deep-sea fan includes deposition in deep marine waters, abundant grain-flow and turbidite deposits, radially oriented and fan-shaped paleocurrent patterns, strongly defined proximal to distal stratigraphic and sedimentologic relations, and prominent development of channel and interchannel facies.

The Butano was deposited in an elongate basin within the early Tertiary continental borderland formed by the Salinian block (fig. 1). The northeastern part of the ancient fan has been displaced by the San Andreas Fault and is represented by the Eocene Point of Rocks Sandstone of the Temblor Range, southern California Coast Ranges. The two segments of the fan are separated by about 305 km of right-lateral offset along the San Andreas Fault (Clarke and Nilsen, 1972). The reconstructed fan in the Butano-Point of Rocks Sandstone is approximately 120 to 160 km long and about 80 km wide.

Astoria Fan

The modern Astoria Fan is comparable in size, shape, and thickness to the reconstructed ancient Butano fan but has a different basin setting because of its location on the continental rise off the coast of Oregon (fig. 1). The wedge of fan sediments radiates asymmetrically southward from the mouth of Astoria Canyon, which heads off the Columbia River. From the canyon mouth, the fan extends about 100 km to its western boundary, Cascadia Channel. The morphology of the fan ends 160 km south of the canyon mouth, although the depositional basin extends southward for another 150 km to Blanco Trough (Nelson and others, 1970). The Astoria Fan is 284 m thick at its lower end, 1,000 m thick at the base of the continental slope, and more than 1,000 m thick at its upper end (von Huene, Kulm, and others, 1973). The Pleistocene sediments of the Astoria Fan rest unconformably on a thin sequence of Pliocene abyssal plain turbidites (Kulm, von Huene, and others, 1973; Kulm and Fowler, this volume).

CHARACTERISTICS OF THE BUTANO AND ASTORIA FAN SYSTEMS

Physiographic Setting and Morphology

The main sediment source for the Butano Sandstone probably was an island or peninsula to the south underlain by granitic rocks; a less important source area may have been located to the northwest (fig. 1). No record remains of possible connections such as submarine canyons between the fan deposits and the source areas. However, the distribution of thick sequences of coarse-grained conglomerates, which are interpreted to represent channel deposits, indicates that at least two major channels were present in the proximal part of the ancient Butano and Point of Rocks fan (fig. 2).

Astoria fan also contains two major channel systems, the Astoria and slope-base fan valleys (fig. 1). Both fan valleys connect to Astoria Canyon at the present time; however, in the past, Willipa Canyon may have alternated with Astoria Canyon in funneling sediment from the edge of the continental shelf to the continental rise. A variety of sediments have been transported to the canyon by the Columbia River, the third largest river of North America. Source areas included the Coast Range, Cascade Range, Columbia Plateau, and the Rocky Mountains.

Tectonic movements, paleogeographic features, and depositional processes have influenced the shapes of the Astoria Fan and of the Butano and Point of Rocks fan. The Astoria Fan fills and covers an apparent trench at the eastern edge of the subducting Gorda-Juan de Fuca plate (Silver, 1972; Kulm and Fowler, this volume). The fan thus constitutes a thick north-south oriented wedge of sediment along the margin of the continental slope, rather than a simple cone-shaped body. The shape of the reconstructed Butano and Point of Rocks fan is similarly elongate; paleogeographic reconstructions indicate that upland areas of the con-

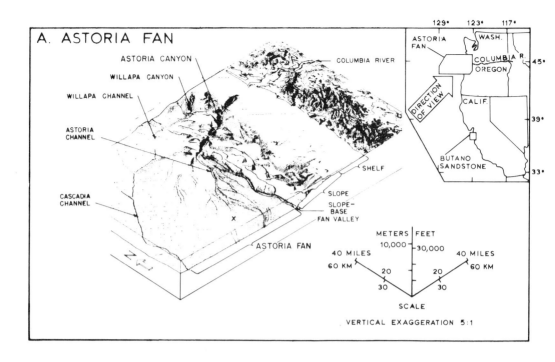

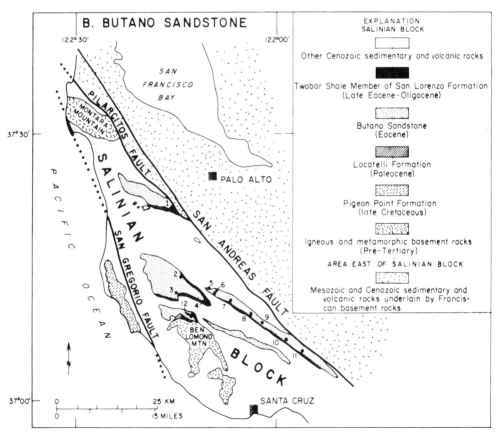

Fig. 1.—Location and setting of Astoria Fan and Butano Sandstone. X on Astoria Fan denotes location of Deep Sea Drilling Site 174 described by Kulm, von Huene, and others (1973).

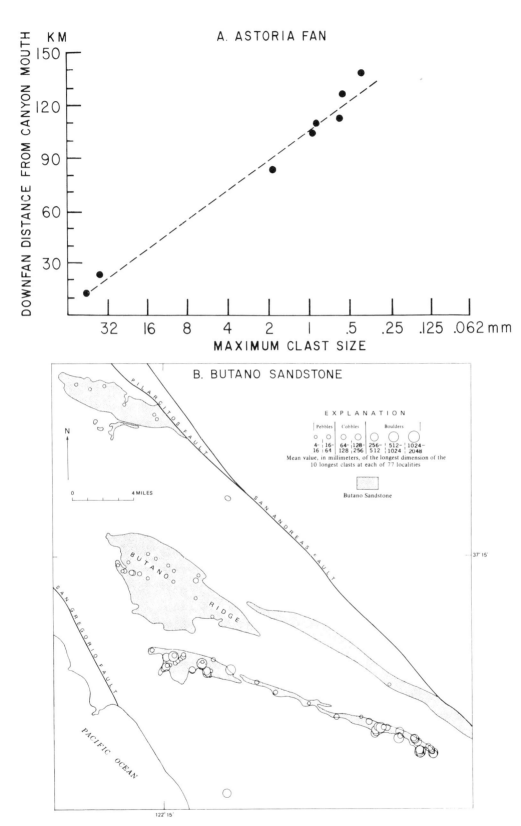

Fig. 2.—Distribution of largest clasts in Astoria Fan and Butano Sandstone turbidites.

tinental borderland to the west and a west-facing submarine slope to the east restricted the growth of the fan to the west and east (Nilsen, 1973; Nilsen and Clarke, 1973).

Cascadia Channel, located along the northern and western margins of the Astoria Fan, appears to have restricted the seaward and radial growth of the fan by limiting and capturing the flow of sediment westward from Astoria Canyon. In addition, the elongate growth of the Astoria Fan to the south was encouraged by the progressive leftward shift of the Astoria fan-valley systems through time (fig. 1; Nelson and others, 1970). When the fan-valley systems reached a position parallel to the base of the continental slope, all the sediments on the fan were transported southward. The youngest sedimentary rocks of the Butano and Point of Rocks Sandstones are located to the southwest, or on the left side of the ancient fan, suggesting a similar leftward growth pattern for this system. Menard (1955) has explained this type of growth pattern for deep-sea fans by postulating that the Coriolis effect causes sediment-laden currents to build higher channel levees on the right, thus encouraging deep-sea fan channels in the northern hemisphere to shift leftward with time.

Limited observations of channels within the Butano Sandstone suggest that the morphology of its fan valleys was similar to those of the Astoria Fan. A few large and deep channels characterize the proximal part of the fan; these are at least 1 to 3 km wide in the Butano and are 2 to 5 km wide on the upper part of the Astoria Fan. Prominent levees and levee deposits border these proximal channels in the Astoria Fan; similar deposits are found adjacent to channels in the proximal portions of the Butano Sandstone. The 200-m relief of the channels on the upper part of the Astoria Fan rapidly flattens to about 80 m in the middle part of the Astoria Fan. The main fan valleys generally break into numerous distributary channels in the middle fan, although some main channels continue throughout the length of the Astoria Fan, maintaining channel depths of 25 to 50 m and widths of 5 to 20 km.

Lateral gradation to smaller and shallower distributary channels continues to the distal fringes of both fan systems. It is difficult to establish the presence of small channels less than 200 m wide and 10 m deep on the lower part of the Astoria Fan because of poor resolution of sounding instruments. However, numerous small channels less than 10 m wide and 2 m deep are present in distal deposits of the ancient Butano and Point of Rocks fan.

In addition to the changes in the number and size of channels, the shapes of channels in both fans also change in the downfan direction. In the upper part of the Astoria Fan and locally in the proximal parts of the Butano Sandstone channel walls are steep; in the Astoria Fan channels have U-shaped profiles with flat floors. The walls gradually become less steep, and main channels assume a shallow saucer-shaped profile toward the lower fan.

Stratigraphy and Sedimentary Facies

Sources of data.—Forty piston cores provide a means of establishing Holocene and late Pleistocene stratigraphy for the Astoria Fan. The cores contain unique, correlative coarse-grained beds containing ash from the Mt. Mazama eruption 6,600 years ago (fig. 3; Nelson and others, 1968). Only limited information is available from beneath the upper several meters of the Astoria Fan, including data from site 174 of the Deep Sea Drilling Project (von Huene, Kulm, and others, 1971) and a few previously unpublished reflection profiles (Kulm and Fowler, this volume); this information suggests that the facies relationships seen in the cores persist at depth.

The stratigraphy of the Butano Sandstone is defined by 12 columnar sections measured from the upper contact with the Twobar Shale Member of the San Lorenzo Formation (fig. 3). The thickest section totals 855 m. Ten of the sections were measured in the middle fan, one each in the lower and upper fan. Unfortunately, no marker beds similar to the Mazama ash horizons of the Astoria Fan are available for correlation between the Butano sections (fig. 3).

Channel and interchannel facies.—Similar channel and interchannel facies that change from upper to lower parts of the fans are observed in both the Astoria Fan and the ancient Butano fan, even though there are correlation difficulties and stratigraphic data limit our comparison to the upper 6 m of Astoria Fan with the upper 855 m of the Butano Sandstone. It is generally impossible to correlate adjacent measured sections or cores bed by bed except for the tuffaceous layers on the Astoria Fan (fig. 3). However, the coarse-grained layers from the upper fan channels, middle to lower fan channels, and interchannel areas of the Astoria Fan have distinctive lithologic characteristics that are also present in apparent channel and interchannel facies of the Butano Sandstone.

In the Astoria Fan, the channels contain thicker layered gravels and sands whereas the interchannel areas include thinner layered fine-grained sands and silts (fig. 4). The lateral gra-

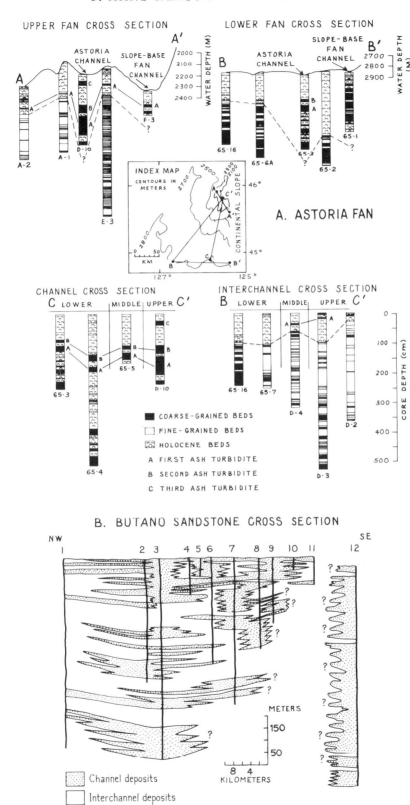

FIG. 3.—Stratigraphic cross sections of Astoria Fan (A-D) and Butano Sandstone (1-12). Location of 12 measured sections of Butano Sandstone shown in figure 1; location of core samples from Astoria Fan shown on index map for this figure.

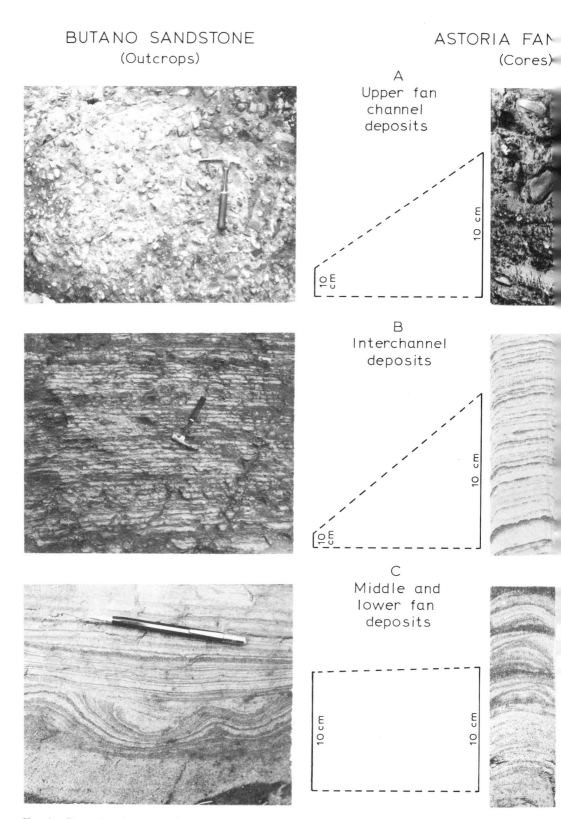

BUTANO SANDSTONE
(Outcrops)

ASTORIA FAN
(Cores)

A
Upper fan
channel
deposits

10 cm

10 cm

B
Interchannel
deposits

10 cm

10 cm

C
Middle and
lower fan
deposits

10 cm

10 cm

Fig. 4.—Comparison between sediment types and sedimentary structures of Astoria Fan and Butano Sandstone.

394

dation from channel to interchannel facies is most abrupt in the upper fan where thick irregular layers on the channel floor change into thin regularly layered levee deposits; these in turn grade laterally into more thinly layered and less regular interchannel deposits (figs. 3, A-A'; 4A,B).

The channel-to-interchannel facies changes are not as evident on the lower part of the Astoria Fan; there, prominent thinly layered levee deposits are not found, and thick sand and silt layers are widespread throughout channel and interchannel areas (fig. 3, B-B'). These thick layers found throughout the lower fan may have resulted from deposition in numerous shifting distributary channels or possibly from major overbank flows during the Pleistocene (Nelson and Kulm, 1973).

The channel deposits of the Butano Sandstone have been defined arbitrarily in the measured sections as those containing very thick-bedded conglomerates and medium- and coarse-grained sandstones. Interchannel deposits have been delineated by thick sequences of thinly interbedded fine-grained sandstones, siltstones, and shales. These two types of deposits are not mutually exclusive because some coarse-grained sequences are found in apparent interchannel facies and the reverse. The presumed channel and interchannel deposits alternate irregularly within the measured sections, resulting in an irregular distribution of channel deposits generally enclosed by interchannel deposits (fig. 3). This is expected because channels on modern fans are known to meander and migrate laterally, and new channels replace or capture old ones (Shepard, 1966; Normark, 1970b).

Proximal and distal facies.—The channel deposits of both the Astoria Fan and the Butano Sandstone change character from proximal to distal regions. The thickness and mean grain size of the tuffaceous sand and gravel layers in the upper part of Astoria Channel are twice that of the correlative sand layers in the channel in the midfan area (fig. 3—see layers A,B—and fig. 4). Downchannel from this point, however, the correlative beds are variably thicker or thinner from one location to the next but, in general, gradually thin from middle to distal parts of Astorial Channel. Mean grain size continues to decrease nearly linearly down the entire channel (Nelson and Kulm, 1973).

Thick conglomerate sequences are present in the two apparent channels located in the proximal part of the ancient Butano and Point of Rocks fan (figs. 2, 4) and beds as much as 20 m thick have been observed in these apparent channel deposits. In the distal channel deposits of the reconstructed ancient fan, only medium- to coarse-grained sandstone as much as 5 m thick is locally present. These data, plus the occurrence of a sand bed 8 m thick on the lower part of the Astoria Fan (Kulm, von Huene, and others, in press, 1973), suggest that relatively thick coarse-grained sandstones may be deposited in proximal, middle, and distal channels but that gravels appear to be restricted to proximal channels.

Sand-shale ratios of channels decline similarly downfan in both the modern and ancient systems even though thick coarse-grained beds are present in channel deposits of all fan areas. Proximal channel deposits of the Astoria Fan have sand-shale ratios[1] of about 4:1, and those of the Butano Sandstone have ratios greater than 5:1 (fig. 3, sec. 12). Downfan, ratios of 2:1 are maintained in the lower 75 km of Astoria Channel, and similar ratios are present in the comparable paleogeographic region of the ancient Butano and Point of Rocks fan.

The proximal to distal variations in sand-shale ratios from interchannel deposits of the Astoria and Butano and Point of Rocks fans are different, but the difference may result from insufficient data. Sand and silt layers are thicker and sand-shale ratios change from less than 1:9 to more than 3:1 from the proximal to distal parts of the Astoria Fan. In contrast, the ratios grade from 1:1 in the midfan region to 1:10 in the distal fringe of the Butano Sandstone, where deposits consist primarily of thinly interbedded sandstones and shales. The lack of distal gradation to finer and thinner sands and silts in the Astoria Fan may be explained by the prevention of distal growth on the western fan margin because of the presence of Cascadia Channel (fig. 5A). The distal facies may be present south of the fan margin, but no sampling has been done there.

Sedimentary Structures

The cores from the Astoria Fan provide sufficient data for a useful comparison of sedimentary structures in the ancient and modern fans even though they are only 6 cm wide and 600 cm long (table 1). However, the variety, lateral variation, and distribution of structures are more easily seen in the Butano Sandstone because of its extensive exposures. Except for the proximal channel deposits, both modern and ancient coarse-grained layers are typically graded and have sharp commonly erosional basal con-

[1] Astoria Fan calculations are based on the assumption that compaction of mud to shale is approximately 35 percent of the former thickness (Emery and Bray, 1962).

ASTORIA FAN

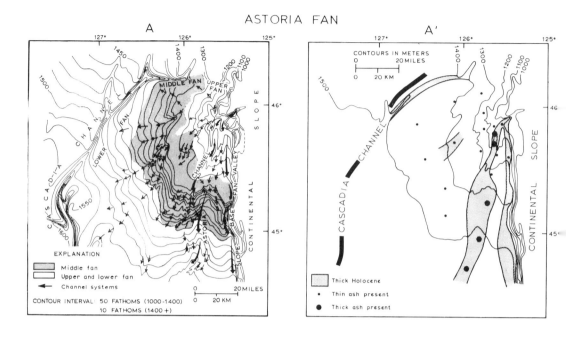

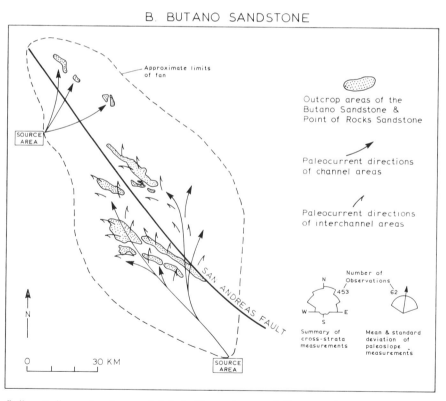

FIG. 5.—Sediment dispersal patterns of Astoria Fan and restored Butano and Point of Rocks fan. Data on Point of Rocks Sandstone from Clarke (1973).

TABLE 1.—COMPARISON OF TYPES AND DISTRIBUTION OF SEDIMENTARY STRUCTURES[1]
IN ASTORIA FAN SEDIMENTS (A) AND THE BUTANO SANDSTONE (B)

Sedimentary structure	Channel deposits			Interchannel deposits
	Upper fan	Middle fan	Lower fan	
Bouma sequences	A, B	A, B	A, B	A, B
a and ae	A, B	B		
abcde		A, B		
cde-de			A, B	A, B
Gravels, conglomerates	A, B	B		
Ungraded bedding	A, B	B		
Amalgamated sandstones	B	B		
Dish structures	B	B		
Mudstone ripups	A, B	A, B	B	
Flat stratification	A, B	A, B	A, B	A, B
Flute casts, groove casts, load casts	B	B	B	
Graded bedding		A, B	A, B	A, B
Cross stratification, current-ripple markings, convolute laminations		A, B	A, B	A, B
Flame structures		B	B	A, B
Bioturbation	A, B	A, B	A, B	A, B
Contorted stratification		B	B	B
Rotational slumps	A, B			B
Structures indicating sand-flow movements	B	B		
Sandstone dikes		B	B	

[1] Only the most common occurrence of the sedimentary structures and sequences is indicated in the chart; most of the structures may be found locally in other depositional environments. Evidence for Astoria Fan features mainly from photographs in Carlson and Nelson (1969).

tacts. Structures typical of Bouma sequences (Bouma, 1962) are the most abundant in these layers and also have distinctive vertical and lateral distribution patterns (fig. 4; table 1; Nelson and Kulm, 1973). Complete Bouma sequences may be found in channel deposits, whereas incomplete sequences are characteristic of interchannel deposits.

The proximal channel deposits of the two fans consist of Bouma a and ae[2] sequences. The Bouma a unit in the channels of upper Astoria Fan is generally massive but may contain some laminations and evidence of scour. It commonly contains mudstone ripup clasts. In the Butano and Point of Rocks fan, many types of conglomerate are present in the proximal channel deposits, but all have matrices that are composed primarily of sandstone. In contrast, the matrices of gravels in channel floors of the upper part of the Astoria Fan are commonly muddy, and gravels are irregularly layered (figs. 4A, 6A— see massive sand and gravel; Carlson and Nelson, 1969).

The conglomerates in the Butano range from well-bedded finely conglomeratic sandstones

[2] We refer to thick, ungraded, and internally structureless sandstone beds separated from one another by thin mudstone or shale layers as Bouma ae sequences in this paper, even though the sandstones are not graded and probably were not deposited by turbidity currents.

having well-defined clast orientations to massively or indistinctly bedded boulder conglomerates having chaotic clast orientations. The conglomerates grade upward into massively bedded sandstones of the Bouma a unit and flat-stratified sandstones of the Bouma b unit, or they may be overlain by thin Bouma e mudstones comparable to those separating a units in the proximal deposits of Astoria Channel (fig. 4A).

The Bouma a units from the Butano, that are composed wholly of sandstone, range up to 60 feet or more in thickness. They commonly are amalgamated or separated from overlying similar Bouma a sandstones only by thin discontinuous mudstones of the Bouma e unit. These sandstones are generally ungraded and massive but may contain diffuse parallel laminations, dish structures, irregular erosional surfaces, and scattered pebbles. Many are characterized by delayed grading, in which only the uppermost part of the bed grades abruptly upward into finer grained sediment. Mudstone and siltstone ripup clasts that are contorted and irregular in size and shape are commonly found in the a units of the Butano. Locally many feet long, they may be concentrated in the upper parts of the a units, randomly distributed, or segregated into irregular layers. The basal contacts of the Bouma a sequences in the Butano commonly are erosional and are either gently disconformable with or channeled into underlying sediments;

large irregular flute- and groove-cast sole markings and load casts may be present at the contacts. Many basal contacts, however, are remarkably smooth, flat, and conformable.

A greater variety of sedimentary structures is present in the middle and distal channel deposits than in the proximal channel deposits of both fans. Complete Bouma *abcde* sequences are most often found in the middle channel deposits, although the individual sequences are generally not as thick or as coarse grained as the Bouma *ae* sequences of the proximal channel deposits. The *a* unit of the midfan deposits is generally thinner and graded. Smaller scale flute and groove casts, bounce marks, tool marks, prod marks, and other sole markings characteristic of turbidites are more common at the base of the *a* units in the Butano. The *b* unit also becomes thinner downfan and may be missing in the lower fan of both systems. Consequently, the Bouma *c* unit is dominant in the distal channel deposits. It exhibits current-ripple markings, small-scale cross strata, convolute laminations, flame structures and contorted bedding (Carlson and Nelson, 1969); in addition to these structures, ball-and-pillow structure has been noted in the Butano *c* units. The *d* unit in both fans is generally thin and consists of flat-laminated fine-grained sandstones, siltstones, and mudstones.

The interchannel deposits of the two fans consist primarily of Bouma *cde* sequences of fine sand and silt that rhythmically alternate with interbedded muds (fig. 4B). The fine sands of the *c* unit generally contain small-scale cross strata and are overlain by flat-laminated to unlaminated silts and muds of the *d* and *e* units. The small-scale cross strata in the Butano Sandstone were produced by migrating current ripples and are locally associated with or replaced by convolute laminae. Locally, entire sequences of beds in the Butano are contorted, probably as a result of downslope movement under the influence of gravity. Toward the distal part of the Butano fan, in interchannel deposits, Bouma *de* deposits become more common, and the *e* unit increases in thickness and is more widely distributed. Hemipelagic mudstones commonly comprise 50 percent or more of the interchannel deposits at the distal end (fig. 3B).

The Holocene hemipelagic deposits of the Astoria Fan as well as most interchannel deposits of the Butano Sandstone were typically extensively bioturbated, probably reflecting slow rates of sedimentation. In the Astoria Fan this resulted in a mottled structureless sediment containing scattered fecal pellets (Carlson and Nelson, 1969). Bioturbation of the interchannel deposits of the Butano Sandstone consists of burrows and borings of variable shape and size, most of which are oriented roughly parallel to the bedding planes. Reworking commonly extends downward from hemipelagic beds to disrupt the upper Bouma *de* layers in both fans.

Grain-Size Distributions

Similar grain-size distributions were determined for the following three sediment types of both fans: (1) thick to massive-bedded, ungraded sands and gravels from the *a* unit of Bouma *ae* sequences that are characteristic of proximal channel deposits; (2) sands from basal *ab* units of more complete Bouma sequences that are characteristic of middle and lower channel deposits; and (3) fine sands and silts from upper *cd* units of more complete Bouma sequences that are present everywhere except in proximal channel deposits (fig. 6; table 2). The deposits of incomplete Bouma *ae* sequences from the Astoria Fan are more variable in grain size than those from the Butano and consequently have been divided into two subgroups, those that are very poorly sorted and contain abundant silt- to clay-sized matrix, and those that are better sorted and contain only a small amount of silt- to clay-sized matrix. Only the better sorted variety was noted in outcrops of the Butano; however, because the grain-size distribution of the gravel and clay-sized fractions was not determined, the more poorly sorted variety may not have been detected if present.

Sands and gravels from Bouma *ae* sequences of both fans are characterized by positive skewness (skewed toward finer grain sizes) and poor sorting (fig. 6; table 2). The extremely poor sorting and high positive skewness of some Bouma *ae* sands of the Astoria Fan can be attributed to the very high content of silt- and clay-sized matrix (table 2).

Sediments from *ab* units of the more complete Bouma sequences on both fans are finer grained, better sorted, and less positively skewed than those from the Bouma *ae* sequences.

The grain-size distributions of the Bouma *cd* units from both fans are very similar, although the content of silt- and clay-sized matrix again is higher in the Astoria Fan deposits than in those from the Butano (table 2). Sediments from the Bouma *cd* units generally are finer grained and contain more silt- and clay-sized material than the coarser grained *ab* sediments.

The interbedded hemipelagic muds from the Astoria Fan have been analyzed but not those from the Butano Sandstone. They are charac-

A. ASTORIA FAN

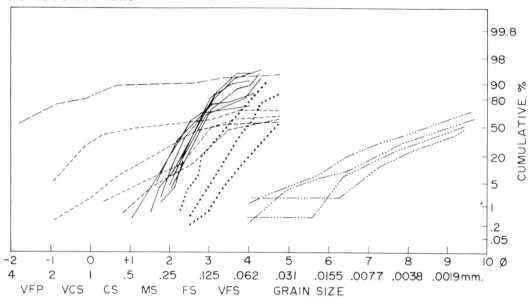

B. BUTANO SANDSTONE

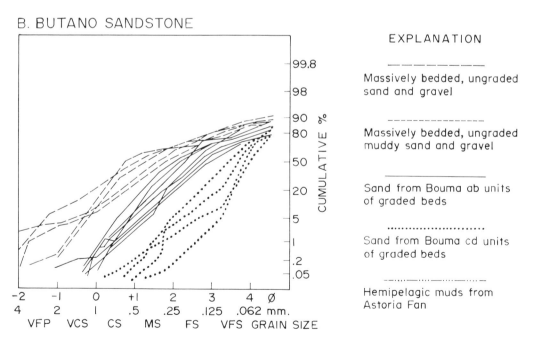

EXPLANATION

— — — — — — —

Massively bedded, ungraded
sand and gravel

– – – – – – – – – –

Massively bedded, ungraded
muddy sand and gravel

———————

Sand from Bouma ab units
of graded beds

· · · · · · · · · · · · · · · · · ·

Sand from Bouma cd units
of graded beds

—··—··—··—··—

Hemipelagic muds from
Astoria Fan

Fig. 6.—Log-probability plots of grain-size distributions from Astoria Fan and Butano Sandstone, plotted in straight line segments after method of Visher (1969).

terized by extremely fine-grain size and poor sorting, which also may be characteristic of hemipelagic mudstones of the Butano.

Because the Bouma *ae, ab,* and *cd* units typically occur in deposits from particular fan environments, the sorting, skewness, and other parameters (table 2) grade laterally over the whole fan. The hemipelagic muds also show lateral gradations over the Astoria Fan from silty clays at the base of the continental slope to fine clays at the western margin (Nelson and Kulm, 1973). As the channel deposits grade downfan from Bouma *ac* sequences to *abcde* sequences and laterally to *cde* sequences in interchannel deposits, the mean grain size decreases, the sorting increases, and the skewness becomes less positive.

Composition

The sediments of both fans have been derived from two separate sources: (1) the erosion of coarser grained clastic detritus from adjacent land areas and (2) the slow settling out in ocean waters of pelagic biogenic shell debris and suspended clay-sized terrigenous material. The different land-source areas of the two fans are indicated by the composition of the silt, sand, and gravel components of arkosic arenites in the Butano and lithic wackes or arenites in the Astoria Fan (fig. 7).

The finer grained hemipelagic sediments contain primarily *in situ* benthic fauna, whereas the coarse-grained sediments commonly contain reworked and displaced shallow-water biota such as shell fragments of megafauna and tests of benthic foraminifera. Consequently, biota of the hemipelagic muds best define the time-stratigraphic relations for fans. For example, the pelagic composition of the hemipelagic deposits of the Astoria Fan changed very abruptly from planktonic foraminifera to radiolarians during late Pleistocene and Holocene time, although no change in biota of the coeval coarse-grained sediments was detected.

Lateral changes in composition are present in both the hemipelagic and coarse-grained sediments of the Astoria Fan; similar changes probably are present in the Butano Sandstone although they have not been documented. The wackes of the upper part of the Astoria Fan typically change downfan to arenites as the amounts of matrix and lithic material become less (fig. 7). In addition, the content of platy and lower density constituents (i.e., mica, plant fragments, volcanic glass) decreases downchannel but progressively increases in interchannel fan deposits more distal from the channels (Nelson, 1974). The amount of terrigenous de-

TABLE 2.—SUMMARY OF GRAIN-SIZE DISTRIBUTION PARAMETERS[1] FOR ASTORIA AND BUTANO FAN DEPOSITS

Parameter	Massive-bedded ungraded sand and gravel			Bouma ab intervals		Bouma cd intervals		Holocene hemipelagic muds
	Astoria		Butano	Astoria	Butano	Astoria	Butano	Astoria
	High matrix	Low matrix	Low matrix					
	3 samples	1 sample	3 samples	8 samples	11 samples	3 samples	6 samples	4 samples
Median (mm)	0.07 to 0.27	4.03	0.29 to 0.59	0.15 to 0.19	0.13 to 0.23	0.042 to 0.085	0.07 to 0.12	.0033 to .0012
Mean (ϕ)	3.29 to 5.24	−1.84	1.53 to 1.89	2.48 to 3.01	2.34 to 3.15	3.55 to 5.12	3.25 to 4.01	8.0 to 9.56
Graphic standard deviation (sediment sorting)	3.03 to 4.08	2.47	1.30 to 1.70	0.52 to 1.26	1.06 to 1.31	0.72 to 1.94	0.48 to 0.87	1.48 to 2.91
Inclusive graphic skewness	+0.51 to +0.70	+0.35	+0.32 to +0.79	+0.62 to +0.15	+0.43 to −0.08	+0.56 to −0.04	+0.47 to −0.81	.11 to −.13
Graphic kurtosis	0.68 to 0.083	1.58	1.92 to 2.99	0.90 to 2.82	1.96 to 2.40	0.83 to 3.30	2.27 to 5.74	.90 to 1.23
Size (mm) of coarsest 1%	0.5 to 2.5	25.0	1.60 to 1.65	0.30 to 0.46	2.65 to 1.15	0.12 to 0.19	0.18 to 0.40	.14 to .062
Percent finer than 4ϕ	32.5 to 52.0	7	8 to 13	5 to 28	13 to 25	2.5 to 80	21 to 44	98.3 to 99.2
Velocity (cm/sec)[2] to transport coarsest 1%	125		25	30	50	25	30	

[1] The grain-size distribution characteristics of 350 samples from piston cores of the Astoria Fan were determined by (1) wet sieving the samples to separate the >62μ fractions from the silt- and clay-size fractions, (2) sieving the coarse fraction in 1φ screen sizes or analysis in an Emery (1938) sedimentation tube as modified by Poole (1957), and (3) analyzing the fine fraction using the American Society for Testing Materials (ASTM) hydrometer method (1964). Twenty-one samples of the Butano Sandstone were sieved to ½φ intervals by T. R. Simoni and R. H. Wright of the U.S. Geological Survey. The samples were first disaggregated by means of water, hydrochloric acid, and mechanical devices. Because of both extensive calcite cementing, which formed concretions in which the detrital grains are infiltrated and broken, and the extensive weathering, which altered the feldspars to clay minerals, relatively few samples could be used for the sieving analyses. Statistical parameters for both data sets were derived utilizing the methods of Folk and Ward (1957).

[2] Velocities estimated by utilizing Sundborg's (1956) erosion-transport-deposition velocity fields.

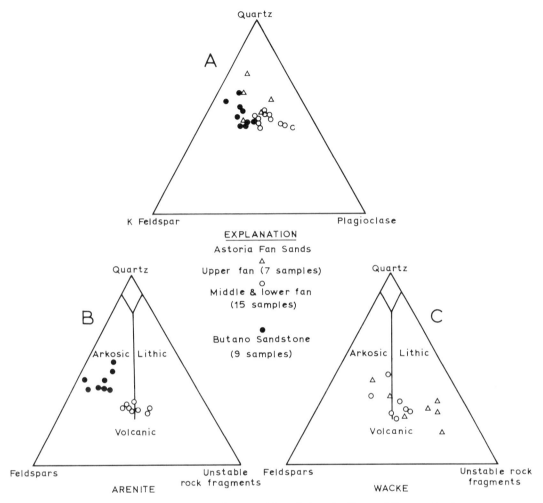

Fig. 7.—Comparison between sand composition of Astoria Fan and Butano Sandstone.

bris lessens and pelagic constitutents increase in hemipelagic deposits away from the continental slope (Nelson and Kulm, 1973).

Individual coarse-grained layers of both the Butano and Astoria fan systems display systematic vertical changes in composition. Detailed investigations on the Astoria Fan show that the most current-sensitive material, the platy and lower density debris, collects at the top of a coarse-grained layer. Consequently, when present, mica, plant fragments, volcanic glass shards, smaller foraminiferal species, and light minerals increase in abundance toward the top of a layer (Nelson, 1974). Plant debris appears to be more common in the upper parts of sandstone beds in the Butano, but information is lacking on the other platy constituents.

Because of the vertical and lateral gradations in composition, the lower parts of coarse-grained beds in the middle and lower fan contain less matrix and constituents of low density than the upper parts of the beds. Consequently, they are more likely to be arenites. In contrast, upper parts of coarse-grained beds in the middle and lower fan beds, in addition to all parts of upper fan interchannel beds, commonly are wackes because of their relatively high content of matrix and platy constituents. Lack of recognition of these compositional and textural trends may account for some of the past arguments about comparability of modern deep-sea sands versus ancient turbidites (Kuenen, 1964; Emery, 1965).

Sediment Dispersal Patterns

Directions of sediment transport on the Astoria Fan are inferred from the location, orientation, and distribution of fan channels (fig. 5).

Although sediment has not been directly observed in transport, the distribution of sedimentary facies (fig. 3) indicates that sediments are transported from the mouth of the Columbia River directly down Astoria Canyon to the head of the Astoria Fan and from there out onto the fan via the system of fan valleys. Many channels now exist on the fan surface, but development of present morphology (fig. 1) and distribution of the most recent tuffaceous sands (fig. 5A) suggest that only a few channels actively transport or fill with sediments at any particular time. The orientation of channel morphology and distribution of channel sediments, therefore, outline the pattern of sediment dispersal.

In ancient deep-sea fan deposits such as the Butano Sandstone, however, it is generally difficult to ascertain the location, orientation, and distribution of the fan channels because of incomplete exposures. Therefore, measurements of paleocurrent directions from both channel and interchannel deposits must be relied upon to determine the directions of sediment transport responsible for growth of the fan. An extensive paleocurrent study of the Butano, based on measurements of 565 paleocurrent indicators and 105 paleoslope indicators, provides a mappable pattern of sediment transport comparable to that inferred from the Astoria Fan (fig. 5; Nil-

sen, 1970; Nilsen and Simoni, 1973).

The dominant feature of both fan systems is the radial orientation of channels and apparent dispersal of sediment away from the fan apex This pattern is clearly indicated by channel orientations on the Astoria Fan and by paleocurrent directions in the Butano Sandstone (figs. 1 5). The dominant transport of sediments down the slope of the fans is indicated by the orientation of the surficial channels parallel to the slope of the Astoria Fan and by similar orientation of paleocurrents and paleoslopes in the Butano Sandstone (fig. 5). These data, give no indication that contour or bottom currents,[3] flowing at an angle to the fan slope, have had any effect on sedimentation in these two fans.

Directions of sediment transport from the interchannel deposits of the ancient Butano fan vary in azimuth orientation from about 270° to about 90° (fig. 5). Paleocurrent directions perpendicular to channel orientations suggest that interchannel deposition resulted from transport of sediments out of the channels toward the interchannel areas. Distribution of correlative thin beds of ash points to a similar type of dispersal pattern in interchannel areas of the Astoria Fan (fig. 5A).

DISCUSSION
Depositional Processes

The sedimentary processes responsible for deposition of the Astoria Fan and the Butano Sandstone are inferred to be similar. The fan shapes and morphology result primarily from channeled flow that has migrated laterally through time. Sediment dispersal patterns indicate that the coarsest debris is transported downfan along channels and that some of the associated finer grained sediment is transported out and away from the channels. Repeated alternations of thick coarse-grained sediment grading upward into thin interbeds of hemipelagic mud indicate that intermittent depositional events are mainly responsible for sedimentation on the fans. These events cause the episodic movement of coarse-grained sediments from shallower areas to the deep-sea fan.

The following five-point model for deep-sea fan depositional processes is based on our comparison of the modern Astoria Fan and the ancient Butano fan:

(1) The chaotic, massive-bedded, and very poorly sorted sands and gravels found at the canyon mouth and in upper channels of the As-

TABLE 3.—GENERAL MORPHOLOGIC CHARACTERISTICS OF MODERN AND ANCIENT DEEP-SEA FANS

Feature	Characteristics
Shape	Shape is more fanlike adjacent to straight open margins or if fan is too small to be disrupted by geographic elements; most fans are elongate or irregular because of growth constrictions by other geographic elements, multiple sources, dominant channel migration right or left, and(or) bottom-current patterns
Size	In modern fans ranges from 1,000 km ×2,500 km to 7×11 km; commonly about 150 km×300 km in open ocean basins and less than 100 km ×100 km in restricted basins
Thickness	In modern fans ranges from 0.3 km to 12 km; variable, probably within same limits for ancient fans
Channel and interchannel dispersal and paleocurrent patterns	Channel patterns generally oriented radially outward from fan apex; interchannel and levee patterns generally lateral away from channels; channel systems may shift to right or to left in southern or northern hemispheres respectively; all patterns may be modified by bottom and contour currents

[3] The term "bottom current" is used to indicate a traction current flowing because of the potential energy in a water mass, and not because of sediment load (Piper, 1970a).

toria Fan appear to have been deposited by debris flows[4] from the continental slope and canyon walls (fig. 3B).

(2) The massive-bedded, better sorted, and low-matrix Bouma *ae* sands and gravels found primarily in upper fan channel deposits probably result from fluidized sediment flows or grain flows;[5] these may be associated with sediment funneled directly into submarine canyons by longshore drift.

(3) The basal *ab* units of well-developed Bouma sequences appear to be deposited by turbidity currents[6] flowing downfan in channels. They represent bedload deposited under upper flow regime conditions (Harms and Fahnestock, 1965; Walker, 1965; Walton, 1967) and generally are restricted to middle and lower fan channels.

(4) The *cd* units of well-developed Bouma sequences appear to be deposited by turbidity currents flowing both in channels and outside of channels. They represent suspended load deposited under lower flow regime conditions (Harms and Fahnestock, 1965; Walker, 1965; Walton, 1967).

(5) Clay-sized materials containing a sand fraction of pelagic constituents are deposited by the continuous vertical settling of fine suspended debris from within the water column; hemipelagic sediment also may be contributed near the base of the continental slope by turbid layer transport[7] from the continental terrace.

From the data presented, the following history of an individual turbidity current can be postulated: The coarsest sediments of incipient turbidity currents are deposited in the most proximal channels as the massive *a* unit of incomplete Bouma *ae* sequences. The bedload of the turbidity current continues on in the channel and is deposited as Bouma *ab* units in the channels of the middle and lower fan. The finer grained debris of the turbidity current is enriched in low-density and platy material (mica, plant fragments, and volcanic glass shards); most of it is carried upward by the turbulence and lags behind the main channeled flow. Part of this cloud of fine debris spills out of the channel and flows laterally to deposit Bouma *cde* sequences in levee and interchannel areas. Some of this material remains behind in the channel to form Bouma *cde* units. Eventually, only the finest debris of the tail of the turbidity current is deposited in Bouma *cde* and *de* sequences on the distal fringes of the fan.

Resedimentation processes such as turbidity currents and grain flows interrupt the continuous rain of suspended debris accumulating as hemipelagic deposits. The suspended debris close to the source of sediments is enriched in terrigenous material, whereas toward the seaward edge of the fan it is enriched in pelagic materials. Movement of sediment by turbid-layer flow off the continental terrace onto the shoreward part of fans and down main channels is suggested by changes in clay-mineral composition along the axis of Astoria Channel and by greater thickness of the Holocene clays in the channel areas (fig. 5; Duncan and others, 1970; Nelson, 1974).

Comparison of Deposits and Depositional Processes

Both fans have similar facies changes and grain-size distribution variations in the downfan direction as well as from channel to interchannel areas (tables 3 and 4, fig. 3). Very definitive changes from thick-bedded conglomerates and sandstones to thin-bedded siltstones and mudstones are found in the downfan stratigraphy of the Butano Sandstone. A similar but less emphatic change may be present on the Astoria Fan, but Cascadia Channel on the west and the Gorda Rise and Blanco Gap on the south seem to have limited the development of the distal fringe facies. The youngest channel facies, in both fans, however, is generally parallel the basin slopes. This appears to be related to the tendency for a leftward shift of main channels through fan history (Menard, 1955; Nelson and Kulm, 1973).

The average grain size of deposits of the Astoria Fan and the ancient Butano fan is different, and it has varied from time to time within

---

[4] Debris flows are mixtures of granular solids and water that, in general, are of greater density and more sluggishly moving than turbidity currents (Hampton, 1970). For the most thorough discussion of this and other resedimentation processes described in footnotes 5 and 6, see Middleton and Hampton (1973).

[5] A grain flow is a mass flow of sand (Stauffer, 1967) in which the sediment is supported by direct grain-to-grain interactions (collisions or close approaches). Fluidized sediment flow is a similar flow in which the sediment is supported by the upward flow of fluid escaping from between the grains as the grains are settled out by gravity (Middleton and Hampton, 1973).

[6] We use the term turbidity currents in the context of Kuenen (*in* Sanders, 1965, p. 217) for "a current flowing in consequence of the load of sediment it is carrying and which gives it excess density" and in the context of Middleton (1969, p. GS-A-6) for "a density current in which the difference in density between the current and the ambient fluid (commonly sea water) is due to the presence of dispersed sediment" (see footnote 4 also).

[7] We use turbid layer transport in the context of Moore (1969, p. 83) as low velocity (<10 cm/sec) and low density turbidity currents of fine-grained debris flowing down the continental terrace and sometimes through canyons and fan valleys.

both fan systems. Deposits of the ancient Butano and Point of Rocks fan are generally coarser grained, thicker bedded, and contain larger amounts of Bouma *a* units than do deposits of the Astoria Fan. This is probably the result of deposition in a borderland basin, where only very narrow marine shelves are present between the source area and the deep-sea basin. Without an intervening shelf, sands and gravels from littoral drift are carried more directly from the source area down subsea slopes to deep water; the resulting fan developed at the base of the slope contains large amounts of thick-bedded coarse-grained deposits that are probably deposited primarily by grain flow or fluidized sediment-flow mechanisms rather than by turbidity currents. The narrow shelf probably also contributes to the formation of hemipelagic sediments that are rich in terrigenous debris and that are deposited mainly by turbid-layer transport.

The Astoria Fan, on the other hand, has been deposited at the base of the continental slope along the western margin of the North American continent. The continental shelf or terrace is wider than those typically surrounding borderland basins, so that less of the coarsest sediment reaches the fan. Because it has been fed primarily by a single major river, the Columbia, detritus available for deposition on the Astoria Fan is finer grained than littoral debris available for deposition in borderland fans. The Astoria Fan turbidites are thus typically finer grained, more thinly layered, and contain thinner and more complete Bouma sequences than do the Butano turbidites. This is particularly true for the turbidites of the Astoria Fan that were formed during Holocene time, when the continental shelf or terrace has been widest and glaciation in the source area has been at a minimum. Because of the narrower shelf during the Pleistocene, when sea level was lower, less coarse-grained sediment was trapped on the shelf as well as in estuaries, and more bypassed to the Astorian Fan.

The borderland setting of the Butano Sandstone and the lower sea levels of the Pleistocene for the Astoria Fan thus resulted in funneling of coarser grained, better sorted, and texturally more mature sand more directly from the shoreline to the deep sea floor. The higher sea level of the Holocene caused longer transport across a wider shelf and more fine-grained debris; consequently, more fine-grained matrix was incorporated during transport to the deep sea, and wackes became dominant in turbidites deposited on the Astoria Fan during the Holocene. At the same time, more debris flows resulted on the Astoria Fan apparently because of the fine grained sediments present on the continental slope and on the Astoria Canyon walls.

Both the Astoria and ancient Butano fan systems changed from growth regimes characterized by rapid sedimentation of coarse-grained sediment to nongrowth regimes characterized primarily by hemipelagic deposition. This occurred from the late Pleistocene to the Holocene in the Astoria Fan (Nelson, 1974) and during the late Eocene in the ancient Butano fan. The alternation from glacial to interglacial climates in the Pacific Northwest resulted in a reduced production of sediment and a rise of sea level that disrupted river transport of debris directly to canyon heads (Nelson and Kulm, 1973). Deposition of sand on the Butano and Point of Rocks fan ended abruptly with deposition of the Twobar Shale Member of the San Lorenzo Formation and of the Kreyenhagen Shale. (The Twobar and the Kreyenhagen overlie the Butano and Point of Rocks Sandstones respectively.) This rapid change may have resulted from tectonic activity, sea level fluctuation, climatic variation, or combinations of these effects.

Criteria for Recognizing Ancient Deep-Sea Fan Deposits

Most fans may be composed primarily of turbidites, but not all turbidites form deep-sea fans. The floor of some small local basins like Crater Lake, Oregon, may be covered by turbidites, but fan morphology is not developed (Nelson, 1967). Turbidite fills in some deep-sea trenches and other basins with extremely complex shapes also lack fan or cone shapes. Turbidite deposits of deep-sea channels are very linear and extend from hundreds to several thousand kilometers in length (Nelson and Kulm, 1973), whereas their abyssal plain terminations have irregular shape (Horn and others, 1971). On the other hand, fan shapes may not be associated with distal parts of extremely large fans (Curray and Moore, 1971) or coalesced fans. Nevertheless, deep-sea fans can be distinguished from other types of turbidite deposits on the combined basis of characteristic patterns of morphology, stratigraphy, sediment dispersal, sedimentary structures, texture of coarse-grained layers, and composition in hemipelagic muds as well as turbidites (tables 3, 4).

Distinguishing a radiating fan-valley system and(or) shape throughout an entire deposit is a fundamental key to recognizing deep-sea fans. If, as is common, the outcrop pattern does not permit this, the internal stratigraphy, facies, and lithology of deep-sea fan deposits can be identified in accord with the following six ob-

TABLE 4.—CHARACTERISTICS OF MODERN AND A..ENT DEEP-SEA FANS AND CRITERIA FOR RECOGNITION OF ANCIENT DEEP-SEA FAN DEPOSITS[1]

Characteristic	Submarine canyon	Physiographic regions of fans				Depositional environments of fan regions		
		Upper fan	Middle fan	Lower fan	Fan fringe	Channel	Levee	Interchannel
A Physiography	Generally straight, V-shaped, and deep (hundreds of meters); commonly steep-walled with cliffs and overhangs	Few main channels (generally one or two) that are very deep (100+ m in large fans), steep-walled and wide (several kilometers); levees are well developed; most irregular channel topography of any fan region	Few main channels branching into tens of distributaries; (up to tens of meters deep); width variable (tens of meters to tens of kilometers); steep to gently sloping walls; channels may meander	Numerous shallow, narrow, braided, randomly shifting, distributary channels having gently sloping walls; limited or lack of levee development; size intermediate between middle and fan-fringe channels	Flat surface having a limited number of gentle channels of few meters width and depth in open basin fans or may lack channels in fans of restricted basins	U-shaped, leveed, flat-floored; typically depositional-erosional features on the fan surface	Sediment wedges up to tens of meters thick flanking channels in upper to middle fan areas; range from tens to hundreds of kilometers wide depending on fan size	Generally smooth, flat, gently sloping, and concave upward surfaces except in the middle fan where convex suprafan bulges may occur
B Stratigraphy and sedimentary facies	Thick (up to tens of meters) coarse-grained beds having highly variable and poorly developed bedding; fining upward to fewer and thinner coarse-grained beds in muds	Abrupt vertical and lateral distribution of channel and interchannel facies; channel bodies enclosed within interchannel facies, and isolated blocks of levee facies enclosed within channel facies	Like upper fan region evolving to distinctly bedded and structured channel facies having thin well-developed mud interbeds associated with interchannel facies	Dominant channel facies throughout section evolving to dominant interchannel facies containing fewer and thinner coarse-grained beds; channel beds as much as 5 to 10 m thick	Thin (5 to 20 cm) coarse-grained beds of even thickness over great lateral extent and a greater amount of mud beds	Like canyon but thick coarse-grained, lenticular beds dominate; mud interbeds cut out or incompletely developed except in fining upward sequence	Thin (<15 cm) well-bedded lenticular coarse-grained layers alternating with equal amounts of fine-grained beds in rhythmic sequence	Very thin and evenly bedded coarse-grained layers alternating with greater amounts of interbedded mudstone
C Sediment types	Dominantly sandstone conglomerate, or pebbly mud; silty mud important on canyon walls or in fining upward fill of canyon fill	Conglomerates and sandstones in channel beds flanked by fine sandstone and siltstone in levee and interchannel areas; silty mudstones limited in channel environments, but dominant in interchannel	Sandstone in channels; interchannel having sandstone and siltstone; limited silt and clay-sized mudstone	Medium- to fine-grained channel sandstone and interchannel sandstones and siltstones; clay-sized mudstone most common in interchannel areas	Clay-size mudstones dominant but numerous fine-grained sandstones and siltstones throughout area	Dominantly sandstones, conglomerates, or muddy conglomerates and limited mudstones	Fine-grained sandstone, siltstone, and mudstone in equal amounts	Sandstones, siltstones, and mudstones in variable amounts in different regions of the fan
D Sedimentary structures (see also table I)	Bouma ae; typically lacking or poorly developed internal structure; may be highly bioturbated in muds; commonly very highly contorted; slump structures common; beds typically amalgamated in lower part of valley fill sequence	Channel is like canyon; Bouma ae predominant, some ab possible; coarsely laminated beds having dish structure, mudstone ripups and clasts, very highly contorted beds and slump structures common; channel sands and gravels injected into interchannel beds; interchannel like levee environment	Full Bouma abcde sequences common in channels; typically less complete Bouma sequences in interchannels; muds most bioturbated in interchannel	Bouma bcde sequences common in channels and Bouma cde in interchannels; base of beds flat and typically lacking sole marks; mud most highly bioturbated in interchannel	Bouma cde grading downfan to de; lamination dominant; cross lamination limited; muds most highly bioturbated of any fan environment	Bouma cde, typically poorly developed if present; beds display-ing well-developed sole marks and channeled basal contacts typical	Some Bouma cde or de units convoluted and typically cross laminated, showing rippling and starved ripples characteristically at top of sandstone and siltstone beds	Similar to levee but generally thinner bedded and laminated in upper fan; in midlower fan, like channel deposits but thinner bedded and more base cutouts

405

TABLE 4.—Continued

Characteristic	Submarine canyon	Physiographic regions of fans				Depositional environments of fan regions		
		Upper fan	Middle fan	Lower fan	Fan fringe	Channel	Levee	Interchannel
E Grain-size distribution in coarse-grained beds	Very poorly sorted, very positively skewed	Very poorly sorted, positively skewed; size grading vertically is generally slight or lacking	Poorly to moderately sorted, positive to neutral skewness, size graded vertically	Moderately sorted, slightly positive to negative skewness, size graded vertically	Positively skewed, poorly sorted, size graded vertically	Varies downchannel; see description from upper to middle fan areas	Moderately sorted, not highly skewed	Moderately to poorly sorted, not highly skewed
F Sandstone/shale ratio	Generally very high except in upper part of total channel fill	Very high in channel facies, very low in interchannel facies, and intermediate in levee facies	High in channel facies, low to intermediate in interchannel facies	High in channel, intermediate in interchannel facies	Low to very low throughout facies	High throughout but may become lower in upper part of channel fill	Intermediate and higher than interchannel and distal fringe areas	Low except in suprafan areas
G Conglomerate/sandstone ratio	Very high to 0	May be very high to 0 in channels	Generally low to 0 in channels	Generally 0 in channels		High to 0	Generally 0	Generally 0
H Maximum possible size	Boulders	Boulders in channels; sand in levee	Pebbles in channels; coarse sand in interchannel	Fine pebbles to coarse sand in channels; medium sand in interchannel	Fine sand in channel or interchannel	Boulders	Coarse sand	Medium sand
I Composition and textural maturity	Displaced shelf fauna characterizes coarse-grained beds; unique in situ species characterize hemipelagic mud; wackes dominate	See channel and levee; high % heavy minerals; high terrigenous content in sand fraction of hemipelagic mud; wackes typical	Composition of minerals and fauna graded vertically in each sand layer; wackes and arenites present in channel sands	Sand composition and maturity like middle fan; high pelagic content in sand fraction of hemipelagic beds	High content of mica and plant fragments and some displaced biota in coarse-grained beds but in situ fauna dominates; hemipelagic like lower fan, but pelagic oozes possible; sands typically wackes	Dominance of displaced biota and minerals from shelf occurs in coarse-grained beds; in situ pelagic and terrigenous debris forms hemipelagic mud	In coarse-grained beds high mica and plant fragments and fewer, smaller sized, and deeper environment displaced biota; hemipelagic like channel; wackes dominate	Coarse-grained beds like levee; hemipelagic muds like channel
J Depositional processes	Slumps, slides, grain flow, fluidized sediment flow, debris flow, and bottom currents dominant; turbidity currents limited; turbid layer flow dominant for hemipelagic deposition	Dominant downfan grain flow, fluidized sediment flow, and debris flow; some turbidity-current bed load in channels; suspension load typical in levees and interchannel areas; turbid layer flow most important in interchannel hemipelagic deposition	Turbidity-current bed load dominant in channels and suspension load throughout channel and interchannel areas; continuous particle by particle fall most important in interchannel hemipelagic deposition	Turbidity-current bed load and suspension load throughout channel and interchannel areas; hemipelagic like middle fan	Turbidity-current suspension load throughout entire area; continuous particle by particle hemipelagic deposition most important throughout area	Slumps, slides, grain flow, fluidized sediment flow, debris flow; turbidity currents and bottom currents; see canyon for hemipelagic deposition	Mainly turbidity-current suspension load; hemipelagic deposition mostly from continuous particle by particle fall except near continental margin where turbid layer flow may dominate	Mainly turbidity-current suspension load; hemipelagic deposition like levee

[1] The characteristics have been synthesized from our research and the following studies: Haner (1971), Mutti and Ricci Lucci (1972), Normark (1970), Normark and Piper (1972), Piper (1970a, b), and Stanley and Unrug (1972).

servations: (1) lack of stratigraphic correlation between adjacent areas because of channel deposition and migration; (2) major abrupt lateral facies changes between channel, channel wall, levee, and interchannel deposits; (3) broad lenticularity of thick channel deposits and sequences of thick-bedded sands, channeling and channel walls being locally recognizable; (4) thickest bedding in the upper fan-channel deposits, the thinner bedding being found in channels downfan and laterally in interchannel areas; (5) turbidites containing Bouma sequences in which the completeness of the sequence, coarseness, thickness, sedimentary structures, and other parameters vary sharply from channel to interchannel deposits and from proximal to distal fan areas (table 1); and (6) decrease in grain size of hemipelagic deposits and turbidites in the downfan direction, the coarsest materials being deposited in the upper fan valleys and in submarine canyons.

The typical radial growth pattern of fans (Normark, 1970a) in addition to common tectonism in the region of fan development (Haner, 1971) results in the concept of stratigraphic superposition generally not being applicable to deep-sea fan deposits, just as it is not applicable to alluvial fan deposits (Bull, 1964). The youngest sediment of the fan may be deposited in the upper, middle, or lower fan region; this depends upon the energy of depositing currents, the tectonic framework of the fan, and whether or not the sediments bypass the upper and middle fan region to be deposited on the lower fan. Because channels tend to shift leftward in the northern hemisphere (Menard, 1955), younger fan deposits may be to the left of the older deposits of the fan, (and the reverse in the southern hemisphere) and yet may be at approximately the same topographic level.

Deep-Sea Fans and Geosynclines

Requirements for fan development include availability of a substantial source of sediment, a submarine canyon to conduct the sediment to deeper water, and a decrease in slope at the lower end of the canyon. The sediments are deposited in the area where the slope decreases, and a fan-shaped wedge of sediment forms that generally radiates outward from the lower end of the submarine canyon. Deep-sea fans already have been noted in a variety of modern continental margin settings, including continental borderlands, marginal seas adjacent to island arcs, mediterranean seas, continental rises, and deep-sea trenches. Inasmuch as these settings are the same as those inferred to have existed in ancient geosynclines, it should be possible to determine, in part, the nature of the continental margin from detailed studies of ancient deep-sea fan deposits.

Attempts should be made to define the local and regional paleogeographic setting of the fan and its relationship to the tectonic and sedimentologic history of a particular mobile belt or geosyncline. Although few such reconstructions have been accomplished, we can suggest some guidelines based on our comparative study. For example, in a tectonically stable basin, without major changes in climate or circulation of marine waters, relatively simple fan systems should develop. In geosynclines, where tectonic instability is generally the rule, a great variety of different types and styles of fan growth and development might be expected. Fans eventually may be wholly or partly destroyed by subduction in continental-rise and deep-sea trench settings.

There appear to be two categories of deep-sea fans with slightly differing characteristics and separate geosynclinal settings. The first type develops in more or less restricted basins such as continental borderlands, small marginal seas, and mediterranean seas. These fans are generally small in areal extent and relatively thin. The smallest may develop nearly perfect fan shapes, but larger ones in restricted basins may be sinuous, elongate, and irregularly shaped because of disruption by surrounding topographic features. The fans of restricted basins are generally characterized by (1) influx of sediments from multiple-source areas located at variable distances and directions from the basin floor, (2) well-developed proximal facies ("base of slope" facies of Stanley and Unrug, 1972) or "proximal exotic" facies of Walker and Mutti, 1973), (3) prominent depositional bulges in the middle fan region (suprafan of Normark, 1970a), and (4) formation of distinct distal facies because of ponding of distal flows in the restricted basin. The multiple sources and distal ponding result in abrupt lateral facies changes, distinct proximal-to-distal gradations in turbidite beds, and relatively coarse-grained sands, gravels, and interbedded hemipelaic muds.

The second category of fan develops in open ocean basins, typically on the continental rise. Because such basins are not restricted, larger and thicker fans develop that generally are regularly shaped but typically somewhat elongate. Generally, a large submarine canyon and a single source area control deposition, so that base-of-slope deposition along the continental margin is minor. The midfan or suprafan region commonly is not well developed. The distal parts of the fan may extend for many miles and grade imperceptibly into abyssal plains without a clearly defined fan edge. As a result, changes in proximal and distal facies in these fans are more gradual and take place over long distances. Also, hemipelagic muds and sands are generally finer grained because of the greater distances from source areas. Bottom currents generated by circulation in large bodies of water may be more effective in redistributing sediments.

Turbidite fills in elongate deep-sea trenches may be a third recognizable fanlike sequence derived from apparent geosynclinal settings (Piper and others, 1973). These deposits are characterized by a long, linear channel facies oriented parallel to the trench axis and by sediments of smaller fans built out perpendicularly to the continental-slope side of the trench. Because the sea floor may be spreading, so to speak, into the continent, the main channel facies tends to be pushed against and parallel to the base of the continental slope. Consequently, overbank-spill deposits of levee and interchannel facies are minimal on the slope side of the channel facies but are well developed on the seaward side of the channel facies. Such a linear yet asymmetric system of channel and interchannel turbidites may distinguish trench fill from the deep-sea fan deposits of restricted basins and open continental-rise settings.

ACKNOWLEDGMENTS

Work by Nilsen on the Butano Sandstone was funded by a U.S. Geological Survey Postdoctoral Research Associateship. Data collection by Nelson on the Astoria Fan was supported by the Office of Naval Research (contract Nonr 1286(10)) and by the Oceanography Department of Oregon State University. We thank Samuel H. Clarke, Jr., Department of Geology and Geophysics, University of California at Berkeley, for providing data on the Point of Rocks Sandstone, and Bradley Larsen and Tully Simoni of the U.S. Geological Survey for assistance in compiling data and preparing figures. Beneficial review comments were provided by Peter Barnes, H. Edward Clifton, R. H. Dott, Jr., and R. G. Walker. Extensive discussions on field trips with Franco Ricci Lucchi added to, strengthened, and corroborated our data for the summary table 4.

Publication has been authorized by the Director, U.S. Geological Survey.

REFERENCES

BOUMA, A. H., 1962, Sedimentology of some flysch deposits, a graphic approach to facies interpretation: Amsterdam, Elsevier Publishing Co., 168 p.

BRABB, E. E., 1960, Geology of the Big Basin area, Santa Cruz Mountains, California (Ph.D. thesis): Stanford, California, Stanford Univ., 191 p.

———, 1964, Subdivision of the San Lorenzo Formation (Eocene-Oligocene), west-central California: Am. Assoc. Petroleum Geologists Bull., v. 48, p. 670–679.

BRANNER, J. C., NEWSOME, J. F., AND ARNOLD, R., 1909, Description of the Santa Cruz Quadrangle, California: U.S. Geol. Survey Geol. Atlas, Folio 163, 12 p.

BULL, W. B., 1964, Geomorphology of segmented alluvial fans in western Fresno County, California: erosion and sedimentation in a semiarid environment: *ibid.*, Prof. Paper 352-E, p. 89–128.

CARLSON, P. R., AND NELSON, C. H., 1969, Sediments and sedimentary structures of the Astoria submarine canyon-fan system, northeast Pacific: Jour. Sed. Petrology, v. 39, p. 1269–1282.

CLARK, J. C., 1966, Tertiary stratigraphy of the Felton-Santa Cruz area, Santa Cruz Mountains, California (Ph.D. thesis): Stanford, California, Stanford Univ., 179 p.

CLARKE, S. H., JR., 1973, The Eocene Point of Rocks Sandstone—provenance, mode of deposition and implications for the history of offset along the San Andreas Fault in central California (Ph.D. thesis): Berkeley, Univ. California, 302 p.

———, AND NILSEN, T. H., 1972, Postulated offsets of Eocene strata along the San Andreas Fault zone, central California: Geol. Soc. America Abs. with Programs, v. 4, p. 137–138.

CUMMINGS, J. C., TOURING, R. M., AND BRABB, E. E., 1962, Geology of the northern Santa Cruz Mountains, California, *in* BOWEN, O. E. (ed.), Geologic guide to the gas and oil fields of northern California: California Div. Mines and Geology Bull. 181, p. 179–220.

CURRAY, J. R., AND MOORE, D. G., 1971, Growth of the Bengal deep-sea fan and denudation in the Himalayas: Geol. Soc. America Bull., v. 82, p. 563–572.

DEWEY, J. F., AND HORSFIELD, B., 1970, Plate tectonics, orogeny and continental growth: Nature, v. 225, p. 521–525.

DIETZ, R. S., 1963, Collapsing continental rises: an actualistic concept of geosynclines and mountain building: Jour. Geology, v. 71, p. 314–333.

DUNCAN, J. R., KULM, L. D., AND GRIGGS, G. B., 1970, Clay mineral composition of late Pleistocene and Holocene sediments of Cascadia Basin, northeastern Pacific Ocean: *ibid.*, v. 78, p. 213–221.

EMERY, K. O., 1938, Rapid method of mechanical analysis of sands: Jour. Sed. Petrology, v. 8, p. 105–111.

———, 1965, Turbidites—Precambrian to present, *in* YOSHIDA, Kozo (ed.), Studies on oceanography, dedicated to Professor Hidaka: Seattle, Univ. Washington Press, p. 486–493.

———, AND BRAY, E. E., 1962, Radiocarbon dating of California basins sediments: Am. Assoc. Petroleum Geologists Bull., v. 46, p. 1839–1856.

FAIRCHILD, W. W., WESENDUNK, P. R., AND WEAVER, D. W., 1969, Eocene and Oligocene foraminifera from the Santa Cruz Mountains, California: California Univ. Pub. Geol. Sci., v. 81, 93 p.

FOLK, R. L., AND WARD, W. C., 1957, Brazos River bar—a study in the significance of grain-size parameters: Jour. Sed. Petrology, v. 27, p. 3–26.

HAMPTON, M. A., 1970, Subaqueous debris flow and generation of turbidity currents (Ph.D. thesis): Stanford, California, Stanford Univ., 180 p.

HANER, B. E., 1971, Morphology and sediments of Redondo submarine fan, southern California: Geol. Soc. America Bull., v. 82, p. 2413–2432.

HARMS, J. C., AND FAHNESTOCK, R. K., 1965, Stratification, bed forms, and flow phenomena (with an example from the Rio Grande), in MIDDLETON, G. V. (ed.), Primary sedimentary structures and their hydrodynamic interpretation, p. 84–115: Soc. Econ. Paleontologists and Mineralogists Special Pub. 12, p. 84–115.

HORN, D. R., AND OTHERS, 1971, Turbidites of the Hatteras and Sohm Abyssal Plains, western North Atlantic: Marine Geology, v. 11, p. 287–323.

HUBERT, C., LAJOIE, J., AND LEONARD, M. A., 1970, Deep-sea sediments in the lower Paleozoic Quebec Supergroup, in LAJOIE, J. (ed.), Flysch sedimentology in North America: Geol. Assoc. Canada Special Paper 7, p. 103–125.

JACKA, A. D., AND OTHERS, 1968, Permian deep-sea fans of the Delaware Mountain Group, Delaware Basin, in field trip guidebook for 1968 symposium, Guadalupian facies, Apache Mountain area, West Texas: Soc. Econ. Paleontologists and Mineralogists, Permian Basin Sec., Special Pub. 68, 11 p.

KUENEN, PH. H., 1964, Deep-sea sands and ancient turbidites, in BOUMA, A. H., AND BROUWER, A. (ed.), Turbidites: Amsterdam, Elsevier Publishing Co., p. 3–33.

KULM, L. D., PRINCE, R. A., AND SNAVELY, P. D., 1973, Site survey of the northern Oregon continental margin and Astoria Fan, in KULM, L. D., VON HUENE, R. E., AND OTHERS, Initial reports of the Deep-Sea Drilling Project, vol. 18: Washington, D.C., U.S. Govt. Printing Office, p. 979–986.

———, VON HUENE, R. E., AND OTHERS, 1973, Initial reports of the Deep Sea Drilling Project, v. 18: ibid., 1077.

MALLORY, V. S., 1959, Lower Tertiary biostratigraphy of the California Coast Ranges: Tulsa, Oklahoma, Am. Assoc. Petroleum Geologists, 416 p.

MENARD, H. W., 1955, Deep-sea channels, topography, and sedimentation: ibid., Bull., v. 39, p. 236–255.

MIDDLETON, G. V., 1969, Turbidity currents, in STANLEY, D. J. (ed.), The new concepts of continental margin sedimentation: Washington, D.C., Am. Geol. Inst., Short Course Lecture Notes, p. GM-A-1 to GM-A-20.

———, AND HAMPTON, M. A., 1973, Part I: Sediment gravity flows—mechanics of flow and deposition, in Turbidites and deep-water sedimentation: Anaheim, California, Soc. Econ. Paleontologists and Mineralogists, Pacific Sec., Short Course Lecture Notes, p. 1–38.

MOORE, D. G., 1969, Reflection profiling studies of the California continental borderland: Structure and Quaternary turbidite basins: Geol. Soc. America Special Paper 107, 142 p.

MUTTI, E., AND RICCI LUCCHI, F. R., 1972, Le torbiditi dell'Appennino settentrionale: introduzione all'analisi di facies: Soc. Geol. Italiana Mem. 11, p. 161–199.

NELSON, C. H., 1967, Sediments of Crater Lake, Oregon: Geol. Soc. America Bull., v. 79, p. 833–848.

———, 1968, Marine geology of Astoria deep-sea fan (Ph.D. thesis): Corvallis, Oregon State Univ., 287 p.

———, 1974, Late Pleistocene and Holocene depositional trends, processes, and history of Astoria deep-sea fan: Marine Geology.

———, AND KULM, L. D., 1973, Part II: Submarine fans and deep-sea channels, in Turbidites and deep-water sedimentation: Anaheim California, Soc. Econ. Paleontologists and Mineralogists, Pacific Sec., Short Course Notes, p. 39–70.

———, AND OTHERS, 1968, Mazama Ash in the northeastern Pacific: Science, v. 161, p. 47–49.

———, AND ———, 1970, Development of the Astoria Canyon-Fan physiography and comparison with similar systems: Marine Geology, v. 8, p. 259–291.

NILSEN, T. H., 1970, Paleocurrent analysis of the flysch-like Butano Sandstone (Eocene), Santa Cruz Mountains, California: Geol. Soc. America Abs. with Programs, v. 2, p. 636.

———, 1971, Sedimentology of the Eocene Butano Sandstone, a continental borderland submarine fan deposit, Santa Cruz Mountains, California: ibid., v. 3, p. 660.

———, 1973, Facies relations in the Eocene Tejon Formation of the San Emigdio and western Tehachapi Mountains, California, in Sedimentary facies changes in Tertiary rocks—California Transverse and southern Coast Ranges: Anaheim, California, Soc. Econ. Paleontologists and Mineralogists, Pacific Sec. Field Trip Guidebook, Field Trip 2, p. 7–23.

———, AND CLARKE, S. H., JR., 1973, Sedimentation and tectonics in the early Tertiary continental borderland of central California: Am. Assoc. Petroleum Geologists Bull., v. 57, p. 797.

———, AND SIMONI, T. R., JR., 1973, Deep-sea fan paleocurrent patterns of the Eocene Butano Sandstone, Santa Cruz Mountains, California: U.S. Geol. Survey Jour. Research, v. 1, no. 4, p. 439–452.

NORMARK, W. R., 1970a, Growth patterns of deep-sea fans: Am. Assoc. Petroleum Geologists Bull., v. 54, p. 2170–2195.

———, 1970b, Channel piracy on Monterey deep-sea fan: Deep-Sea Research, v. 17, p. 837–846.

———, AND PIPER, D. J. W., 1969, Deep-sea fan-valleys, past and present: Geol. Soc. America Bull., v. 80, p. 1859–1866.

———, AND ———, 1972, Sediments and growth pattern of Navy deep-sea fan, San Clemente Basin, California borderland: Jour. Geology, v. 80, p. 198–223.

PIPER, D. J. W., 1970a, Transport and deposition of Holocene sediment on La Jolla deep-sea fan, California: Marine Geology, v. 8, p. 211–227.

———, 1970b, A Silurian deep-sea fan deposit in western Ireland and its bearing on the nature of turbidity currents: Jour. Geology, v. 78, p. 509–522.

————, VON HUENE, R. E., AND DUNCAN, J. R., 1973, Sedimentation in a modern trench: Geology, v. 1, no. 1, p. 19–22.

POOLE, D. M., 1957, Size analysis of sand by a sedimentation technique: Jour. Sed. Petrology, v. 27, p. 460–468.

SANDERS, J. E., 1965, Primary sedimentary structures formed by turbidity currents and related resedimentation mechanisms, *in* MIDDLETON, G. V. (ed.), Primary sedimentary structures and their hydrodynamic interpretation: Soc. Econ. Paleontologists and Mineralogists Special Pub. 12, p. 192–219.

SHEPARD, F. P., 1966, Meander in valley crossing a deep ocean fan: Science, v. 154, p. 385–386.

————, DILL, R. F., AND VON RAD, U., 1969, Physiography and sedimentary processes of La Jolla submarine fan and fan-valley, California: Am. Assoc. Petroleum Geologists Bull., v. 53, p. 390–420.

SILVER, E. A., 1972, Pleistocene tectonic accretion of the continental slope off Washington: Marine Geology; v. 13, p. 239–249.

STANLEY, D. J., 1969b, Submarine channel deposits and their fossil analogs ('fluxoturbidites'), *in* STANLEY, D. J. (ed.), The new concepts of continental margin sedimentation: Washington, D.C., Am. Geol. Inst. Short Course Lecture Notes, p. DJS-9-1 to DJS-9-17.

————, AND UNRUG, R., 1972, Submarine channel deposits, fluxoturbidites, and other indicators of slope and base-of-slope environments in modern and ancient marine basins: Soc. Econ. Paleontologists and Mineralogists Special Pub. 16, p. 287–340.

STAUFFER, P. H., 1967, Grain flow deposits and their implications, Santa Ynez Mountains, California: Jour. Sed. Petrology, v. 37, p. 481–508.

SULLIVAN, F. R., 1962, Foraminifera from the type section of the San Lorenzo Formation, Santa Cruz County, California: California Univ. Pub. Geol. Sci., v. 37, p. 233–352.

SULLWOLD, H. H., JR., 1960, Tarzana Fan, deep submarine fan of late Miocene age, Los Angeles County, California: Am. Assoc. Petroleum Geologists Bull., v. 44, p. 433–457.

SUNDBORG, ÅKE, 1956, The river Klarälven— a study of fluvial processes: Geog. Annalav, Stockholm, v. 38, p. 125–316.

VISHER, G. S., 1969, Grain size distributions and depositional processes: Jour. Sed. Petrology, v. 39, p. 1074–1106.

VON HUENE, R., KULM, L. D., AND OTHERS, 1971, Deep Sea Drilling Project leg 18: Geotimes, v. 16, p. 12–15.

WALKER, R. G., 1965, The origin and significance of the internal sedimentary structures of turbidities: Yorkshire Geol. Soc. Proc., v. 35, p. 1–32.

————, 1966, Shale grit and Grindslow shales; transition from turbidite to shallow-water sediments in the Upper Carboniferous of northern England: Jour. Sed. Petrology, v. 36, p. 90–114.

————, 1970, Review of the geometry and facies organization of turbidites and turbidite-bearing basins, *in* LAJOIE, J. (ed.), Flysch sedimentology in North America: Geol. Assoc. Canada Special Paper, 7, p. 219–251.

WALKER, R. G., AND MUTTI, E., 1973, Part IV—Turbidite facies and facies associations, *in* Turbidites and deep-water sedimentation: Anaheim, California, Soc. Econ. Paleontologists Mineralogists, Pacific Sec. Short Course Lecture Notes, p. 119–157.

WALTON, E. K., 1967, The sequence of internal structures in turbidites: Scottish Jour. Geology, v. 3, p. 305–317.

REFERENCES

(See also the bibliographies at the end of Papers 8, 21, and 26.)

Agarate, C., Got, H., Monaco, A., and Pautot, G. (1967). Éléments structuraux des canyons sous-marins et du plateau continental catalans, obtenus par "sismique continue". *C. R. Acad. Sci. Paris*, **265**, sér. D, 1278–1281.

Allen, G. P., Castaing, P., and Klingebiel, A. (1971). Preliminary investigation of the surficial sediments in the Cap-Breton Canyon (southwest France) and the surrounding continental shelf. *Marine Geol.*, **10**, M27–M32.

Almgren, A. A., and Schlax, W. N. (1957). Post-Eocene age of "Markley Gorge" fill, Sacramento Valley, California. *Amer. Assoc. Petrol. Geol. Bull.*, **41**, 326–330.

Andrews, J. E., Shepard, F. P., and Hurley, R. J. (1970). Great Bahama Canyon. *Geol. Soc. America Bull.*, **81**, 1061–1078.

Ansfield, V. J. (1972). Eocene submarine cone-fan deposits and their significance, northern Olympic Peninsula, Washington (abs.). *Geol. Soc. America, Abstracts with programs*, 1972 *Ann. Meetings*, **4**, no. 7, p. 435.

Arleth, K. H. (1968). Maine Prairie Gas Field, Solano County, California. In *Natural Gases of North America* (B. W. Beebe and B. F. Curtis, eds.), Vol. 1, pp. 79–84. Amer. Assoc. Petrol. Geol. Mem. 9.

Baker, E. T., Sternberg, R. W., and McManus, D. A. (1973). Characteristics of the bottom nepheloid layer over Nitinat Deep-Sea Fan (abs.). *Eos (Amer. Geophys. Union Trans.)*, **54**, no. 4, p. 334.

Baldwin, B. (1971). Ways of deciphering compacted sediments. *Jour. Sed. Petrol.*, **41**, 293–301.

Ballard, J. A., Patton, D., Cacchione, D. A., Carlmark, J. W., and Sanders, J. E. (1973). The structure of Oceanographer Canyon from reflection profiles (abs.). *Eos (Amer. Geophys. Union Trans.)*, **54**, no. 4, p. 335.

Beer, R. M. (1969). Suspended sediment over Redondo Submarine Canyon and vicinity, southern California. Unpubl. Thesis, Univ. Southern Calif., Los Angeles, 94 pp.

——, and Gorsline, D. S. (1971). Distribution, composition and transport of suspended sediment in Redondo Submarine Canyon and vicinity (California). *Marine Geol.*, **10**, 153–175.

411

Belderson, R. H., and Stride, A. H. (1969). The shape of submarine canyon heads revealed by Asdic. *Deep-Sea Res.,* **16,** 103–104.

——, Kenyon, N. H., Stride, A. H., and Stubbs, A. R. (1972). *Sonographs of the Sea Floor,* 185 pp. Elsevier/Excerpta Medica/North Holland, Amsterdam.

Benson, P. H. (1971). Geology of the Oligocene Hackberry trend, Gillis English Bayou, Manchester area, Calcasieu Parish, Louisiana. *Gulf Coast Assoc. Geol. Socs. Trans.,* **21,** 1–14.

Bergantino, R. N. (1971). Submarine regional geomorphology of the Gulf of Mexico. *Geol. Soc. America Bull.,* **82,** 741–752.

Bornhauser, M. (1948). Possible ancient submarine canyon in southwestern Louisiana. *Amer. Assoc. Petrol. Geol. Bull.,* **32,** 2287–2290.

Bosellini, A. (1967). Frane sottomarine nel Giurassico del Bellunese e del Friuli. *Accad. Nz. dei Lincei,* **43,** fasc. 6, 563–567.

——, and Masetti, D. (1972). Ambiente e dinamica deposizionale del Calcare del Vajont (Giurassico medio, Prealpi Bellunese e Friulane). *Ann. Univ. Ferrara,* sez. 9, Sc., Geol., e Paleont., **5,** n. 4, 87–100.

Bouma, A. H. (1965). Sedimentary characteristics of samples collected from some submarine canyons. *Marine Geol.,* **3,** 291–320.

—— (1974). Detrital deposits of fan and basin in Gulf of Mexico (abs.). *Ann. Meeting Abstracts, Amer. Assoc. Petrol. Geol. and Soc. Econ. Paleontol. Mineral.,* **1,** p. 11.

——, and Shepard, F. P. (1964). Large rectangular cores from submarine canyons and fan valleys. *Amer. Assoc. Petrol. Geol. Bull.,* **48,** 225–231.

——, Bryant, W. R., and Antoine, J. W. (1968). Origin and configuration of Alaminos Canyon, northwestern Gulf of Mexico. *Gulf Coast Assoc. Geol. Socs. Trans.,* **18,** 290–296.

——, Chancey, O., and Merkel, G. (1971). Alaminos Canyon area. In *Texas A & M Univ. Oceanographic Studies* (R. Rezak and V. J. Henry, eds.), Vol 3, pp. 153–179. Gulf Publishing Co., Houston, Tex.

——, Rezak, R., Antoine, J. W., Bryant, W. R., and Fahlquist, D. A. (1972). Deep sea sedimentation and correlation of strata off Magdalena River and in Beata Strait. In *VI Conferencia Geologica del Caribe-Margarita, Venezuela, Memorias,* pp. 430–438.

Bourcart, J. (1965). Les canyons sous-marins de l'extremité orientale des Pyrénées. In *Progress in Oceanography* (M. Sears, ed.), Vol 3, pp. 63–69. Pergamon Press, London.

Bouysse, Ph., Horn, R., and Leclaire, L. (1968). Données nouvelles sur le Gouf de Cap Breton (Golfe de Gascogne, France). *C. R. Acad. Sci. Paris,* **267,** sér. D, 827–830.

Buchanan, J. Y. (1887). On the land slopes separating continents and ocean-basins, especially those on the west coast of Africa. *Scottish Geog. Mag.,* **3,** 217–238.

Burke, K. (1972). Longshore drift, submarine canyons, and submarine fans in development of Niger Delta. *Amer. Assoc. Petrol. Geol. Bull.,* **56,** 1975–1983.

Busch, D. A. (1974). *Stratigraphic Traps in Sandstones—Exploration Techniques.* Amer. Assoc. Petrol. Geol. Mem. 21, 174 pp.

Bush, S. A., and Bush, P. A. (1969). Trincomalee and associated canyons, Ceylon. *Deep-Sea Res.,* **16,** 655–660.

Cannon, G. A. (1972). Wind effects on currents observed in Juan de Fuca Submarine Canyon. *J. Phys. Oceanog.,* **2,** 281–285.

Carlson, P. R., and Nelson, C. H. (1969). Sediments and sedimentary structures of the Astoria Submarine Canyon-Fan system, northeast Pacific. *Jour. Sediment. Petrol.,* **39,** 1269–1282.

Carter, R. M., and Lindqvist, J. K. (1975). Sealers Bay submarine fan complex, Oligocene, southern New Zealand. *Sedimentology*, **22**, 465–483.

——, and Lindqvist, J. K. (in press). Balleny Group, Chalky Island, southern New Zealand: an inferred Oligocene submarine canyon and fan. *Pacific Geology*.

Chipping, D. H. (1972). Sedimentary structure and environment of some thick sand-stone beds of turbidite type. *Jour. Sediment. Petrol.*, **42**, 587–595.

Clarke, S. H. (1973). The Eocene Point of Rocks Sandstone: provenance, mode of deposition and implications for the history of offset along the San Andreas Fault in central California. Ph.D. Thesis, Univ. Calif., Berkeley, Calif. 302 pp.

Clifton, H. E. (1974). Estuarine and deep-sea fan deposits—similarities and con-trasts (abs.). *Ann. Meeting Abstracts, Amer. Assoc. Petrol. Geol. and Soc. Econ. Paleontol. Mineral.*, **1**, p. 19.

Colburn, I. P. (1968). Grain fabrics in turbidite sandstone beds and their relation-ship to sole mark trends on the same beds. *Jour. Sediment. Petrol.*, **38**, 146–158.

Conolly, J. R. (1968). Submarine canyons of the continental margin, east Bass Strait (Australia). *Marine Geol.*, **6**, 449–461.

——, and Cleary, W. J. (1971). Braided deep sea deltas and the origin of turbidite sands (abs.). *Eos (Amer. Geophys. Union Trans.)*, **52**, no. 4, p. 244.

——, and Von der Borch, C. C. (1967). Sedimentation and physiography of the sea floor south of Australia. *Sediment. Geol.*, **1**, 181–220.

——, Flavelle, A., and Dietz, R. S. (1970). Continental margin of the Great Aus-tralian Bight. *Marine Geol.*, **8**, 31–58.

Conrey, B. L. (1967). Early Pliocene sedimentary history of the Los Angeles Basin, California. *Calif. Div. Mines Geol., Spec. Rept.* 93, San Francisco, pp. 1–63.

Coulbourn, W. T., Campbell, J. F., and Moberly, R. (1974). Hawaiian submarine terraces, canyons, and Quaternary history evaluated by seismic-reflection pro-filing. *Marine Geol.*, **17**, 215–234.

Creager, J. S. (1958). A canyon-like feature in the Bay of Campeche. *Deep-Sea Res.*, **5**, 169–172.

Cruz, P., Barcenas, R., and Verdugo, R. (1974). Influence of Laramide orogeny on Tertiary conglomerate distribution, Veracruz Basin, Mexico (abs.). *Ann. Meeting Abstracts, Amer. Assoc. Petrol. Geol. and Soc. Econ. Paleontol. Mineral.*, **1**, 21–22.

Curray, J. R., and Moore, D. G. (1971). Growth of the Bengal Deep-Sea Fan and denudation in the Himalayas. *Geol. Soc. America Bull.*, **82**, 563–572.

——, and Moore, D. G. (1974). Sedimentary and tectonic processes in the Bengal Deep-Sea Fan and Geosyncline. In *The Geology of Continental Margins* (C. A. Burk and C. L. Drake, eds.), pp. 617–627, Springer-Verlag, New York.

Daly, R. A. (1936). Origin of submarine "canyons." *Amer. Jour. Sci.*, ser. 5, **31**, 401–420.

—— (1942). *The Floor of the Ocean—New Light on Old Mysteries*, 177 pp., Univ. North Carolina Press, Chapel Hill, N. C.

Dana, J. D. (1863). *A Manual of Geology*, 798 pp., Trubner, London.

Davies, D. K. (1972). Deep sea sediments and their sedimentation, Gulf of Mexico. *Amer. Assoc. Petrol. Geol. Bull.*, **56**, 2212–2239.

Davis, D. M. (1953). Markley Gorge, Sacramento County, California (abs.). *Amer. Assoc. Petrol. Geol. Bull.*, **37**, p. 186.

Davis, J. R. (1971). Sedimentation of Pliocene sandstones in Santa Barbara Chan-nel, California (abs.). *Amer. Assoc. Petrol. Geol. Bull.*, **55**, p. 335.

Dickas, A. B., and Payne, J. L. (1967). Upper Paleocene buried channel in Sacra-mento Valley, California. *Amer. Assoc. Petrol. Geol. Bull.*, **51**, 873–882.

Dietz, R. S., and Knebel, H. J. (1971). Trou sans Fond Submarine Canyon: Ivory Coast, Africa. *Deep-Sea Res.*, **18**, 441–447.

——, Knebel, H. J., and Somers, L. H. (1968). Cayar Submarine Canyon. *Geol. Soc. America Bull.*, **79**, 1821–1828.

Dill, R. F. (1964a). Contemporary submarine erosion in Scripps Submarine Canyon. Ph.D. Thesis, Univ. Calif., Scripps Inst. Oceanog., privately printed, 269 pp.

—— (1964b). Sedimentation and erosion in Scripps Submarine Canyon head. In *Papers in Marine Geology* (Shepard Commemorative Volume), (R. L. Miller, ed.), pp. 23–41, Macmillan, New York.

—— (1969). Earthquake effects on fill of Scripps Submarine Canyon. *Geol. Soc. America Bull.*, **80**, 321–328.

Dillon, W. P., and Zimmerman, H. B. (1970). Erosion by biological activity in two New England submarine canyons. *Jour. Sediment. Petrol.*, **40**, 542–547.

Drake, D. E., and Gorsline, D. S. (1973). Distribution and transport of suspended particulate matter in Hueneme, Redondo, Newport, and La Jolla Submarine Canyons, California. *Geol. Soc. America Bull.*, **84**, 3949–3968.

Dulemba, J. L. (1970). Quelques remarques sur l'origine des canyons sous-marins situés au large des côtes ouest de la Corse. *Geol. Rundschau*, **59**, 601–604.

Duncan, J. R., and Kulm, L. D. (1970). Mineralogy, provenance, and dispersal history of late Quaternary deep-sea sands in Cascadia Basin and Blanco Fracture Zone off Oregon. *Jour. Sediment. Petrol.*, **40**, 874–887.

Duplaix, S. (1972). Les minéraux lourds de sables de plages et de canyons sous-marins de la Méditerranée Française. In *The Mediterranean Sea: A Natural Sedimentation Laboratory* (D. J. Stanley, ed.), pp. 293–303, Dowden, Hutchinson & Ross, Stroudsburg, Pa.

——, and Olivet, J. L. (1969). Etude sédimentologique et morphologique de la tête du rech Bourcart (Golfe du Lion). *Cahiers Océanographiques*, **22**, 127–146.

Edmondson, W. F. (1965). The Meganos Gorge of the southern Sacramento Valley. *San Joaquin Geol. Soc. Selected Papers*, **3**, 36–51.

—— (1972). Geologic effects produced by compaction of the Meganos Gorge fill. *San Joaquin Geol. Soc. Selected Papers*, **4**, 13–20.

Emery, K. O. (1969). Continental rises and oil potential. *Oil Gas Jour.*, **67**, 231–243.

——, and Uchupi, E. (1972). *Western North Atlantic Ocean: Topography, Rocks, Structure, Water, Life and Sediments.* Amer. Assoc. Petrol. Geol. Mem. 17, 532 pp.

Enos, P. (1973). Channelized submarine carbonate debris flow, Cretaceous, Mexico (abs.). *Amer. Assoc. Petrol. Geol. Bull.*, **57**, p. 777.

Ewing, M., and Lonardi, A. G. (1971). Sediment transport and distribution in the Argentine Basin. 5. Sedimentary structure of the Argentine Margin, Basin, and related provinces. In *Physics and Chemistry of the Earth* (L. H. Ahrens, F. Press, S. K. Runcorn, and H. C. Urey, eds.), Vol. 8, pp. 123–251, Pergamon Press, Oxford.

Fairbridge, R. W. (1966). Submarine cones or fans. In *Encyclopedia of Earth Sciences*, Vol. 1, *Encyclopedia of Oceanography* (R. W. Fairbridge, ed.), pp. 870–872, Van Nostrand Reinhold, New York.

Felix, D. W., and Gorsline, D. S. (1971). Newport Submarine Canyon, California: an example of the effects of shifting loci of sand supply upon canyon position. *Marine Geol.*, **10**, 177–198.

Fenner, P., Kelling, G., and Stanley, D. J. (1971). Bottom currents in Wilmington Submarine Canyon. *Nature Phys. Sci.*, **229**, 52–54.

Field, M. E., and Pilkey, O. H. (1971). Deposition of deep-sea sands: comparison of two areas of the Carolina continental rise. *Jour. Sediment. Petrol.*, **41**, 526–553.

Fischer, P. J. (1971). An ancient (upper Paleocene) submarine canyon and fan: the Meganos Channel, Sacramento Valley, California (abs.). *67th Ann. Geol. Soc. America Cordilleran Section Meeting Program*, **3**, no. 2, p. 120.

—— (1973). Evolution of Santa Barbara Basin, western Transverse Ranges, California (abs.). *Amer. Assoc. Petrol. Geol. Bull.*, **57**, 778–779.

Forel, F. -A. (1885). Les ravins sous-lacustres des fleuves glaciaires. *C. R. Acad. Sci. Paris*, **101**, 725–728.

—— (1887). Le ravin sous-lacustre du Rhône dans le lac Léman. *Bull. Soc. Vaudoise des Sciences Naturelles*, **23**, 85–107.

Friedmann, G. M. (1972). *Sedimentary Facies:* products of sedimentary environments in Catskill Mountains, Mohawk Valley, and Taconic sequence, eastern New York State. Guidebook, Soc. Econ. Paleontol. Mineral. Eastern Section, 48 pp.

Galloway, W. E., and Brown, L. F. Jr. (1973). Depositional systems and shelf-slope relations on cratonic basin margin, uppermost Pennsylvanian of north-central Texas. *Amer. Assoc. Petrol. Geol. Bull.*, **57**, 1185–1218.

Gennesseaux, M. (1966). Prospection photographique des canyons sous-marins du Var et du Paillon (Alpes-Maritimes) au moyen de la Troïka. *Rev. de Géog. Phys. et de Géol. Dynamique*, sér. 2, **8**, fasc. 1, 3–38.

——, Guibout, P., and Lacombe, H. (1971). Enregistrement de courants de turbidité dans la vallée sous-marine du Var (Alpes Maritimes). *C. R. Acad. Sci. Paris*, **273**, sér. D, 2456–2459.

Gershanovich, D. E. (1968). New data on geomorphology and recent sediments of the Bering Sea and the Gulf of Alaska. *Marine Geol.*, **6**, 281–296.

Ghibaudo, G., and Mutti, E. (1973). Facies ed interpretazione paleoambientale delle Arenarie di Ranzano nei dintorni di Specchio (Val Pessola, Appennino Parmense). *Soc. Geol. Italiana Mem.*, **12**, 251–265.

Glangeaud, L., Bellaiche, G., Gennesseaux, M., and Pautot, G. (1968). Phénomènes pelliculaires et épidermiques du rech Bourcart (golfe du Lion) et de la mer hespérienne. *C. R. Acad. Sci. Paris*, **267**, sér. D, 1079–1083.

Goedicke, T. R. (1972). Submarine canyons on the central continental shelf of Lebanon. In *The Mediterranean Sea: A Natural Sedimentation Laboratory* (D. J. Stanley, ed.), pp. 655–670, Dowden, Hutchinson & Ross, Stroudsburg, Pa.

Gonthier, E., and Klingebiel, A. (1973). Facies et processus sédimentaires dans le canyon sous-marin Gascogne I. *Bull. Inst. Géol. Bassin d'Aquitaine*, **13**, 163–262.

Got, H., Monaco, A., and Reyss, D. (1969). Les canyons sous-marins de la mer Catalane. Le Rech du Cap et le Rech Lacaze-Duthiers. II—Topographie de détail et carte sédimentologique. *Vie et Milieu*, sér. B, **29**, 257–277.

Gvirtzman, G. (1969). The Saqiye Group (late Eocene to early Pleistocene) in the coastal plain and Hashephela regions, Israel. *Geol. Survey Israel Bull.*, **51** (2 vols.).

Halbouty, M. T. (1969). Hidden trends and subtle traps in Gulf Coast. *Amer. Assoc. Petrol. Geol. Bull.*, **53**, 3–29.

Hall, B. A., and Stanley, D. J. (1973). Levee-bounded submarine base-of-slope channels in the Lower Devonian Seboomook Formation, Northern Maine. *Geol. Soc. America Bull.*, **84**, 2101–2110.

Hand, B. M., and Emery, K. O. (1964). Turbidites and topography of north end of San Diego Trough, California. *Jour. Geol.*, **72**, 526–542.

Haner, B. E. (1971). Morphology and sediments of Redondo Submarine Fan, south-ern California. *Geol. Soc. America Bull.,* **82,** 2413-2432.

Harbison, R. N. (1968). Geology of De Soto Canyon. *Jour. Geophys. Res.,* **73,** 5175-5185.

Harland, W. B., and Gayer, R. A. (1972). The Arctic Caledonides and earlier oceans. *Geol. Mag.,* **109,** 289-314.

Heezen, B. C., and Hollister, C. D. (1971). *The Face of the Deep,* 659 pp. Oxford Univ. Press, London.

——, Tharp, M., and Ewing, M. (1959). *The Floors of the Ocean: I. North Atlantic.* Geol. Soc. America Spec. Paper 65, 122 pp.

——, Menzies, R. J., Schneider, E. D., Ewing, W. M., and Granelli, N. C. L.(1964). Congo Submarine Canyon. *Amer. Assoc. Petrol. Geol. Bull.,* **48,** 1126-1149.

Herman, Y. (ed.). (1974). *Marine Geology and Oceanography of the Arctic Seas.* Springer-Verlag, Berlin.

Hoover, R. A., and Bebout, D. G. (1974). Structural and topographic control of sediment deposition, Magdalena Slope, offshore northern Colombia (abs.). *Ann. Meeting Abstracts, Amer. Assoc. Petrol. Geol. and Soc. Econ. Paleontol. Mineral.,* **1,** p. 46.

Hopkins, B. M. (1966). Submarine canyons. *Broken Hill Proprietary Tech. Bull., no. 26,* 39-43.

Horn, D. R., Ewing, M., Delach, M. N., and Horn, B. M. (1971). Turbidites of the northeast Pacific. *Sedimentology,* **16,** 55-69.

Houbolt, J. J. H. C. (1973). The deep-sea canyons in the Gulf of Guinea near Fer-nando Póo. *Trans. Roy. Geol. Mining Soc. Netherlands,* **30,** 7-18.

Houtz, R., Ewing, J., Ewing, M., and Lonardi, A. G. (1967). Seismic reflection pro-files of the New Zealand Plateau. *Jour. Geophys. Res.,* **72,** 4713-4729.

Hoyt, W. V. (1959). Erosional channel in the Middle Wilcox near Yoakum, Lavaca County, Texas. *Gulf Coast Assoc. Geol. Socs. Trans.,* **9,** 41-50.

Hrabar, S. V. Deep-water sedimentation in the Ravalli Group (Late Precambrian Belt Megagroup), northwestern Montana. Belt Symposium Trans., Idaho Bureau of Mines and Geology (in press).

Hsü, K. J., Ryan, W. B. F., and Cita, M. B. (1973). Late Miocene desiccation of the Mediterranean. *Nature,* **242,** 240-244.

Huang, T. -C., and Goodell, H. G. (1970). Sediments and sedimentary processes of eastern Mississippi Cone, Gulf of Mexico. *Amer. Assoc. Petrol. Geol. Bull.,* **54,** 2070-2100.

Hubert, C., Lajoie, J., and Léonard, M. A. (1970). Deep sea sediments in the Lower Paleozoic Québec Supergroup. In *Flysch Sedimentology in North America* (J. Lajoie, ed.), pp. 103-125. Geol Assoc. Canada Spec. Paper 7.

Hulsemann, J. (1968). Morphology and origins of sedimentary structures on sub-marine slopes. *Science,* **161,** 45-47.

Il'yin, A. V., and Lisitsyn, A. P. (1969). The origin of submarine canyons as related to their actual extent in the Atlantic Ocean. *Doklady Acad. Sci. U.S.S.R., Earth Sci. Sec. Amer. Geol. Inst.,* **183,** 221-224.

Inman, D. L. (1970). Strong currents in submarine canyons (abs.). *Eos (Amer. Geo-phys. Union Trans.),* **51,** no. 4, p. 319.

——, and Murray, E. (1964). Transportation of sand and water in submarine can-yons. In *Progress Report on Research Sponsored by O.N.R., Earth Sciences Div., Scripps Institution of Oceanography* (W. A. Nierenberg, ed.), pp. 85-89.

Jacka, A. D., Beck, R. H., St. Germain, L. C., and Harrison, S. C. (1968). Permian deep-sea fans of the Delaware Mountain Group (Guadalupian), Delaware Basin. In *Guadalupian Facies, Apache Mountain Area West Texas.* Permian Basin Section, Soc. Econ. Paleontol. Mineral. Publ. 68-11, pp. 49-90.

James, D. M. D. (1967). Sedimentary studies in the Bala of central Wales. Unpubl. Ph.D. Thesis, Univ. Wales (Swansea), 126 pp.

—— (1971). The Nant-y-moch Formation, Plynlimon inlier, west central Wales. *Jour. Geol. Soc. London,* **127,** 177-181.

—— (1972). Sedimentation across an intra-basinal slope: the Garnedd-wen Formation (Ashgillian), west central Wales. *Sediment. Geol.,* **7,** 291-307.

, and James, J. (1969). The influence of deep fractures on some areas of Ashgillian-Llandoverian sedimentation in Wales. *Geol. Mag.,* **106,** 562-582.

Jones, M. D. (1969). The palaeogeography and palaeoecology of the Leintwardine Beds of Leintwardine, Herefordshire. Unpubl. M.Sc. Thesis, Univ. Leicester, England, 67 pp.

Jordan, G. F. (1951). Continental slope off Apalachicola River, Florida. *Amer. Assoc. Petrol. Geol. Bull.,* **35,** 1978-1993.

——, and Stewart, H. B. Jr. (1961). Submarine topography of the western Straits of Florida. *Geol. Soc. America Bull.,* **72,** 1051-1058.

Kanie, Y. (1969). Sedimentary structures observed in the Tertiary systems in Akiya, in the northern part of the Sajima area, Miura Peninsula. *Sci. Rept. Yokosuka City Mus., no. 15,* 37-43. (In Japanese, English summary.)

Keller, G. H., Lambert, D., Rowe, G., and Staresinic, N. (1973). Bottom currents in the Hudson Canyon. *Science,* **180,** 181-183.

Kelling, G., and Stanley, D. J. (1970). Morphology and structure of Wilmington and Baltimore Submarine Canyons, eastern United States. *Jour. Geol.,* **78,** 637-660.

Kenyon, N. H., and Belderson, R. H. (1973). Bed forms of the Mediterranean undercurrent observed with side-scan sonar. *Sediment. Geol.,* **9,** 77-99.

Kepferle, R. C. (1968). Geologic map of the Shepherdsville Quadrangle, Bullitt County, Kentucky. *U. S. Geol. Survey GQ-740.*

—— (1969). Geologic map of the Samuels Quadrangle, northcentral Kentucky. *U. S. Geol. Survey GQ-824.*

Kieken, M. (1973). Evolution de l'Aquitaine au cours du Tertiaire. *Bull. Géol. Soc. France,* sér. 7, **15,** no. 1, 40-50.

Komar, P. D. (1969). The channelized flow of turbidity currents with application to Monterey deep-sea fan channel. *Jour. Geophys. Res.,* **74,** 4544-4558.

—— (1970). The competence of turbidity current flow. *Geol. Soc. America Bull.,* **81,** 1555-1562.

Kruit, C., Brouwer, J., and Ealey, P. (1972). A deep-water sand fan in the Eocene Bay of Biscay. *Nature Phys. Sci.,* **240,** 59-61.

Kuenen, Ph. H. (1937). Experiments in connection with Daly's hypothesis on the formation of submarine canyons. *Leidsche Geol. Meded.,* **8,** 327-351.

—— (1938). Density currents in connection with the problem of submarine canyons. *Geol. Mag.,* **75,** 241-249.

—— (1953). Origin and classification of submarine canyons. *Geol. Soc. America Bull.,* **64,** 1295-1314.

Lajoie, J., and Chagnon, A. (1973). Origin of red beds in a Cambrian flysch sequence, Canadian Appalachians, Quebec. *Sedimentology,* **20,** 91-103.

La Fond, E. C. (1964). Andhra, Mahadevan, and Krishna submarine canyons and other features of the continental slope off the east coast of India. *Jour. Indian Geophys. Union,* **1,** no. 1, 25-32.

LeBlanc, R. J. (1972). Geometry of sandstone reservoir bodies. In *Underground Waste Management and Environmental Implications* (T. D. Cook, ed.), pp. 133–190. Amer. Assoc. Petrol. Geol. Mem. 18.

Litvin, V. M. (1965). Origin of the bottom configuration of the Norwegian Sea. *Oceanology, Acad. Sci. U.S.S.R.* (translated for Amer. Geophys. Union), **5**, 90–96.

Lonardi, A. G., and Ewing, M. (1971). Sediment transport and distribution in the Argentine Basin. 4. Bathymetry of the continental margin, Argentine Basin and other related provinces. Canyons and sources of sediments. In *Physics and Chemistry of the Earth* (L. H. Ahrens, F. Press, S. K. Runcorn, and H. C. Urey, eds.), Vol. 8, pp. 79–121, Pergamon Press, Oxford.

Lowe, D. R. (1972). Implications of three submarine mass-movement deposits, Cretaceous, Sacramento Valley, California. *Jour. Sediment. Petrol.*, **42**, 89–101.

McKerrow, W. S., and Ziegler, A. M. (1972). Silurian paleogeographic development of the proto-Atlantic Ocean. 24 Int. Geol. Congr., Montreal, Sec. 6, 4–10.

Mansfield, C. F. (1972). Petrofacies units and sedimentary facies of the late Mesozoic strata west of Coalinga, California. Guidebook, Cretaceous of the Coalinga area, Oct. 21, 1972. Pacific Section, Soc. Econ. Paleontol. Mineral., 8 pp.

—— (1974). Late Mesozoic deep-marine fan deposits in the southern Diablo Range, California (abs.). *Ann. Meeting Abstracts, Amer. Assoc. Petrol. Geol. and Soc. Econ. Paleontol. Mineral.*, **1**, p. 58.

Marlowe, J. L. (1965). Probable Tertiary sediments from a submarine canyon off Nova Scotia. *Marine Geol.*, **3**, 263–268.

Martin, B. D., and Emery, K. O. (1967). Geology of Monterey Canyon, California. *Amer. Assoc. Petrol. Geol. Bull.*, **51**, 2281–2304.

——, and Rex, R. W. (1967). Submarine weathering as an aid to submarine erosion, California (abs.). Abstracts for 1966, Geol. Soc. America Ann. Meeting, San Francisco. *Geol. Soc. America Spec. Paper 101.*

Mathewson, C. C. (1970). Submarine canyons and the shelf along the north coast of Molokai Island, Hawaiian Ridge. *Pacific Sci.* **24**, 235–244.

Matsumoto, T., and Okada, H. (1973). Saku Formation of the Yezo geosyncline. *Sci. Rep. Dept. Geol., Kyushu Univ.*, **11**, 275–309. (In Japanese, English summary.)

Mauffret, A., and Sancho, J. (1970). Etude de la marge continentale au nord de Majorque (Baléares, Espagne). *Rev. Inst. Franc. Pétrole*, **25**, no. 6, 714–730.

——, Fail, J. P., Montadert, L., Sancho, J., and Winnock, E. (1973). Northwestern Mediterranean sedimentary basin from seismic reflection profile. *Amer. Assoc. Petrol. Geol. Bull.*, **57**, 2245–2262.

Menard, H. W. (1955). Deep-sea channels, topography and sedimentation. *Amer. Assoc. Petrol. Geol. Bull.*, **39**, 236–255.

—— (1960). Possible pre-Pleistocene deep-sea fans off central California. *Geol. Soc. America Bull.*, **71**, 1271–1278.

——, Smith, S. M., and Pratt, R. M. (1965). The Rhône deep-sea fan. In *Submarine Geology and Geophysics* (W. F. Whittard and R. Bradshaw, eds.), pp. 271–285, Butterworth, London.

Miller, R. L. (ed.). (1964). *Papers in Marine Geology*, 531 pp., Macmillan, New York.

Mitchell, A. H. G., and Reading, H. G. (1971). Evolution of island arcs. *Jour. Geol.*, **79**, 253–284.

Mizutani, S. (1964). Superficial folding of the Palaeozoic system of central Japan. *Jour. Earth Sci. Nagoya Univ.*, **12**, 17–83.

Montadert, L., Damotte, B., Debyser, J., Fail, J. P., Delteil, J. R., and Valéry, P. (1971). The continental margin in the Bay of Biscay. In *The Geology of the East Atlantic Continental Margin* (F. M. Delaney, ed.), pp. 49–74, ICSU-SCOR Working Party 31 Symp., Cambridge 1970. Rept. Inst. Geol. Sci., London, 70/15.

Moore, D. G. (1969). Reflection profiling studies of the California continental borderland: structure and Quaternary turbidite basins. *Geol. Soc. America Spec. Paper 107.*

——, Curray, J. R., Raitt, R. W., and Emmel, F. J. (1974). Stratigraphic-seismic section correlations and implications to Bengal Fan history. In *Initial Reports of the Deep Sea Drilling Project* (C. C. Von der Borch and J. G. Sclater, eds.), Vol. 22, pp. 403–412, Government Printing Office, Washington, D. C.

Morelock, J., Maloney, N. J., and Bryant, W. R. (1972). Manzanares Submarine Canyon. *Acta Cientifica Venezolano*, **23**, 143–147.

Morris, R. C. (1973). Sedimentary and tectonic history of Ouachita Mountains (abs.). *Amer. Assoc. Petrol. Geol. Bull.*, **57**, p. 796.

Morrison, R. R., Brown, W. R., Edmondson, W. F., Thomson, J. N., and Young, R. J. (1971). Potential of Sacramento Valley Gas Province, California. In *Future Petroleum Provinces of the United States—Their Geology and Potential* (I. H. Cram, ed.), pp. 329–338, Amer. Assoc. Petrol. Geol. Mem. 15.

Moyes, J., Caralp, M., Gonthier, E., Klingebiel, A., Latouche, C., Prud'homme, R., and Vigneaux, M. (1972). Etude géologique du canyon Gascogne I. *Bull. Soc. Géol. France*, sér. 7, **14**, 261–280.

Mutti, E. (1969). Studi geologici sulle isole del Dodecaneso (Mare Egeo). X. Sedimentologia delle Arenarie di Messanagros (Oligocene-Aquitaniano) nell'isola di Rodi. *Soc. Geol. Italiana Mem.*, **8**, 1027–1070.

—— (1974). Examples of ancient deep-sea fan deposits from circum-Mediterranean geosynclines. In *Modern and Ancient Geosynclinal Sedimentation* (R. H. Dott, Jr., and R. H. Shaver, eds.), pp. 92–105, Soc. Econ. Paleontol. Mineral. Spec. Publ. 19.

——, and Ghibaudo, G. (1972). Un esempio di torbiditi di conoide sottomarina esterna: le arenarie di San Salvatore (Formazione di Bobbio, Miocene) nell'Appennino di Piacenza. *Accad. delle Sci. Torino Mem., Cl. Sci., Fis., Mat. e Nat.*, ser. 4a, 40pp.

——, and Ricci Lucchi, F. (1972). Le torbiditi dell'Appennino Settentrionale: introduzione all'analisi di facies. *Soc. Geol. Italiana Mem.*, **11**, 161–199.

Naini, B. R., and Leyden, R. (1973). Ganges Cone: a wide angle seismic and refraction study. *Jour. Geophys. Res.*, **78**, 8711–8720.

Nasu, N. (1964). The provenance of the coarse sediments on the continental shelves and the trench slopes off the Japanese Pacific coast. In *Papers in Marine Geology* (R. L. Miller, ed.), pp. 65–101, Macmillan, New York.

Nelson, C. H., and Kulm. L. D. (1973). Submarine fans and deep-sea channels. In *Turbidites and Deep-Water Sedimentation* (G. V. Middleton and A. H. Bouma, eds.), pp. 39–78, Short Course Notes, Soc. Econ. Paleontol. Mineral. Pacific Section, Anaheim, Calif.

——, and Nilsen, T. H. (1974). Depositional trends of modern and ancient deep-sea fans. In *Modern and Ancient Geosynclinal Sedimentation* (R. H. Dott, Jr., and R. H. Shaver, eds.), pp. 69–91, Soc. Econ. Paleontol. Mineral. Spec. Publ. 19.

——, Carlson, P. R., Byrne, J. V., and Alpha, T. R. (1970). Development of the Astoria Canyon-Fan physiography and comparison with similar systems. *Marine Geol.*, **8**, 259–291.

Nesteroff, W. D., Duplaix, S., Sauvage, J., Lancelot, Y., Melieres, F., and Vincent, E. (1968). Les dépôts récents du Canyon de Cap-Breton. *Bull. Soc. Géol. France,* sér. 7, **10**, 218–252.

Newton, J. G., and Pilkey, O. H. (1969). Topography of the continental margin off the Carolinas. *Southeastern Geol.,* **10**, 87–92.

Nilsen, T. H. (1973). Continental margin sedimentation of Eocene Tejon Formation, western Tehachapi and San Emigdio Mountains, California (abs.). *Amer. Assoc. Petrol. Geol. Bull.,* **57**, p. 797.

——, and Clarke, S. H. (1973). Sedimentation and tectonics in early Tertiary continental borderland of central California (abs.). *Amer. Assoc. Petrol. Geol. Bull.,* **57**, p. 797.

——, and Simoni, T. R., Jr. (1973). Deep-sea fan paleocurrent patterns of the Eocene Butano Sandstone, Santa Cruz Mountains, California. *U. S. Geol. Survey Jour. Res.,* **1**, 439–452.

——, Dibblee, T. W., Jr., and Simoni, T. R., Jr. (1974). Stratigraphy and sedimentology of the Cantua Sandstone Member of the Lodo Formation, Vallecitos area, California. In *Paleogene of the Panoche Creek–Cantua Creek Area, Central California* (M. B. Payne, Chmn.), pp. 38–68. Pacific Section, Soc. Econ. Paleontol. Mineral. Field Trip Guidebook, Fall 1974.

Normark, W. R. (1970a). Growth patterns of deep-sea fans. *Amer. Assoc. Petrol. Geol. Bull.,* **54**, 2170–2195.

—— (1970b). Channel piracy on Monterey Deep-Sea Fan. *Deep-Sea Res.,* **17**, 837–846.

—— (1974). Submarine canyons and fan-valleys: factors affecting growth patterns of deep-sea fans. In *Modern and Ancient Geosynclinal Sedimentation* (R. H. Dott, Jr., and R. H. Shaver, eds.), pp. 56–68, Soc. Econ. Paleontol. Mineral. Spec. Publ. 19.

——, and Curray, J. R. (1968). Geology and structure of the tip of Baja California, Mexico. *Geol. Soc. America Bull.,* **79**, 1589–1600.

——, and Piper, D. J. W. (1969). Deep-sea fan-valleys, past and present. *Geol. Soc. America Bull.,* **80**, 1859–1866.

——, and Piper, D. J. W. (1972). Sediments and growth pattern of Navy deep-sea fan, San Clemente basin, California borderland. *Jour. Geol.,* **80**, 198–223.

Northrop, J. (1953). A bathymetric profile across the Hudson Submarine Canyon and its tributaries. *Jour. Marine Res.,* **12**, 223–232.

——, Frosch, R. A., and Frassetto, R. (1962). Bermuda-New England Seamount Arc. *Geol. Soc. America Bull.,* **73**, 587–594.

Palmer, H. D. (1971). Observations on the erosion of submarine outcrops, La Jolla Submarine Canyon, California (abs.). *Geol. Soc. America, Abstracts with programs, 1971 Ann. Meetings,* **3**, no. 7, p. 666.

Parker, J. R. (1975). Lower Tertiary sand development in the central North Sea. In *Petrol. Continental Shelf Northwest Europe* (A. W. Woodland, ed.), Vol. 1, pp. 447–453. Applied Science Publishers, London.

Payne, R. R., and Conolly, J. R. (1972). Turbidite sedimentation off the Antarctic Continent. *Amer. Geophys. Union, Antarctic Res. Ser.,* **19**, 349–364.

Pequegnat, W. E., James, B. M., Bouma, A. H., Bryant, W. R., and Fredericks, A. D. (1971). Photographic study of deep-sea environments of the Gulf of Mexico. In *Texas A & M University Oceanographic Studies* (R. Rezak and V. J. Henry, eds.), Vol. 3, pp. 67–128, Gulf Publishing Co., Houston, Tex.

Pícha, F. (1974). Ancient submarine canyons of the Carpathian Miogeosyncline. In

Modern and Ancient Geosynclinal Sedimentation (R. H. Dott, Jr., and R. H. Shaver, eds.), pp. 126-127, Soc. Econ. Paleontol. Mineral. Spec. Publ. 19.

——, and Niem, A. R. (1974). Distribution and extent of beds in flysch deposits, Ouachita Mountains, Arkansas and Oklahoma. *Jour. Sediment. Petrol.*, **44**, 328-335.

Piper, D. J. W. (1970). Transport and deposition of Holocene sediment on La Jolla deep-sea fan, California. *Marine Geol.*, **8**, 211-227.

—— (1971). Sediments of the Middle Cambrian Burgess Shale, Canada. *Lethaia*, **5**, 169-175.

——, and Marshall, N. F. (1969). Bioturbation of Holocene sediments on La Jolla deep-sea fan, California. *J. Sediment. Petrol.*, **39**, 601-606.

——, and Normark, W. R. (1971). Re-examination of a Miocene deep-sea fan and fan-valley, southern California. *Geol. Soc. America Bull.*, **82**, 1823-1830.

Pratt, R. M. (1967). The seaward extension of submarine canyons off the northeast coast of the United States. *Deep-Sea Res.*, **14**, 409-420.

Pryor, R. (1970). Formation of a deep-water submarine canyon head in the Tongue of the Ocean. *Bull. Marine Sci.*, **20**, 813-829.

Redwine, L. E. (1972). The Tertiary Princeton submarine valley system beneath the Sacramento Valley, California. Unpubl. Ph. D. Thesis, Univ. California, Los Angeles, approx. 900 pp.

Rees, A. I., von Rad, U., and Shepard, F. P. (1968). Magnetic fabric of sediments from the La Jolla submarine canyon and fan, California. *Marine Geol.*, **6**, 145-178.

Reimnitz, E. (1971). Surf-beat origin for pulsating bottom currents in the Rio Balsas Submarine Canyon, Mexico. *Geol. Soc. America Bull.*, **82**, 81-89.

——, and Gutiérrez-Estrada, M. (1970). Rapid changes in the head of the Rio Balsas Submarine Canyon system, Mexico. *Marine Geol.*, **8**, 245-258.

Remane, J. (1970). Die Entstehung der resedimentären Breccien im Obertithon der subalpinen Ketten Frankreichs. *Eclogae Geol. Helv.*, **63**, 685-740.

Reyss, D. (1964). Observations faites en soucoupe plongeante dans deux vallées sous-marines de la mer Catalane: le rech du Cap et le rech Lacaze-Duthiers. *Bull. Inst. Océanographique, Monaco*, **63** (1308), 8 pp.

Ricci Lucchi, F. (1969). Channelized deposits in the Middle Miocene flysch of Romagna (Italy). *Giorn. Geol.*, ser. 2, **34**, 203-282.

—— (1973). Sequential analyses of turbidite basins in north-central Apennines (abs.). *Amer. Assoc. Petrol. Geol. Bull.*, **57**, p. 801.

——, and Pialli, G. (1973). Apporti secondari nella Marnoso-arenacea: 1: Torbiditi di conoide e di pianura sottomarina a est-nord-est di Perugia. *Soc. Geol. Italiana Boll.*, **92**, 669-712.

Robb, J. M., Schlee, J., and Behrendt, J. C. (1973). Bathymetry of continental margin off Liberia, West Africa. *U. S. Geol. Survey Jour. Res.*, **1**, 563-567.

Robinson, F. M. (1964). Core tests, Simpson area, Alaska. Exploration of Naval Petroleum Reserve No. 4 and adjacent areas, northern Alaska, 1944-1953. Pt. 5, Subsurface geology and engineering data. *U. S. Geol. Survey Prof. Paper 305-L*, 645-730.

Rona, P. A. (1970). Submarine canyon origin on upper continental slope off Cape Hatteras. *Jour. Geol.*, **78**, 141-152.

——, Schneider, E. D., and Heezen, B. C. (1967). Bathymetry of the continental rise off Cape Hatteras. *Deep-Sea Res.*, **14**, 625-633.

Ross, D. A. (1968a). Current action in a submarine canyon. *Nature,* **218,** 1242–1245.

—— (1968b). Geological observations from *Alvin. Woods Hole Oceanog. Inst.,* Ref. No. 69-13, 71–72.

Rowe, G. T. (1971). Observations on bottom currents and epibenthic populations in Hatteras Submarine Canyon. *Deep-Sea Res.,* **18,** 569–581.

——, Keller, G., Edgerton, H., Staresinic, N., and MacIlraine, J. (1974). Time lapse photography of the biological reworking of sediments in Hudson Submarine Canyon. *Jour. Sediment. Petrol.,* **44,** 549–552.

Ruddiman, W. F., Bowles, F. A., and Molnia, B. (1972). Maury Channel and Fan. 24th Int. Geol. Congr., Montreal, sec. 8, 100–108.

Ryan, W. B. F., and Heezen, B. C. (1965). Ionian Sea Submarine Canyons and the 1908 Messina turbidity current. *Geol. Soc. America Bull.,* **76,** 915–932.

Sacramento Petroleum Association. (1962). Maine Prairie Gas Field, California. *Calif. Div. Mines Geol. Bull. 181,* 132–139.

Safonov, A. (1962). The challenge of the Sacramento Valley. *Calif. Div. Mines Geol. Bull. 181,* 77–97.

—— (1968). Stratigraphy and tectonics of Sacramento Valley. In *Natural Gases of North America* (B. W. Beebe and B. F. Curtis, eds.), Vol. 1, pp. 611–635. Amer. Assoc. Petrol. Geol. Mem. 9.

Sato, T., and Koike, K. (1957). A fossil submarine canyon near the southern foot of Mt. Kano, Tiba Prefecture. *Jour. Geol. Soc. Japan,* **63,** 100–116. (In Japanese, English summary.)

Schlager, W., and Schlager, M. (1973). Clastic sediments associated with radiolarites (Tauglboden-Schichten, Upper Jurassic, eastern Alps). *Sedimentology,* **20,** 65–89.

Scholl, D. W., Buffington, E. C., and Hopkins, D. M. (1968). Geologic history of the continental margin of North America in the Bering Sea. *Marine Geol.,* **6,** 297–330.

——, Buffington, E. C., Hopkins, D. M., and Alpha, T. R. (1970). The structure and origin of the large submarine canyons of the Bering Sea. *Marine Geol.,* **8,** 187–210.

Schüpbach, M. A. (1973). Comparison of slope and basinal sediments of marginal cratonic basin and geosyncline (abs.). *Amer. Assoc. Petrol. Geol. Bull.,* **57,** p. 804.

——, and Morel, R. (1974). Fans and channels in classical flysch, central Alps (abs.). *Ann. Meeting Abstracts, Amer. Assoc. Petrol. Geol. and Soc. Econ. Paleontol. Mineral.,* **1,** 80–81.

Seibold, E., and Hinz, K. (1974). Continental slope construction and destruction, West Africa. In *The Geology of Continental Margins* (C. A. Burk and C. L. Drake, eds.), pp. 179–196, Springer-Verlag, New York.

Shepard, F. P. (1934). Canyons off the New England coast. *Amer. Jour. Sci.,* **27,** 24–36.

—— (1937). Daly's submarine-canyon hypothesis. *Amer. Jour. Sci.,* **33,** 369–379.

—— (1952). Composite origin of submarine canyons. *Jour. Geol.,* **60,** 84–96.

—— (1965). Importance of submarine valleys in funneling sediments to the deep sea. In *Progress in Oceanography* (M. Sears, ed.), Vol. 3, pp. 321–332, Pergamon Press, Oxford.

—— (1966). Meander in valley crossing a deep-ocean fan. *Science,* **154,** 385–386.

—— (1969). Relation of submarine canyons to continental slope (abs.). *Amer. Assoc. Petrol. Geol. Bull.,* **53,** 741–742.

—— (1973a). *Submarine Geology* (3rd ed.), 517 pp., Harper & Row, New York.

—— (1973b). Sea floor off Magdalena delta and Santa Marta area, Colombia. *Geol. Soc. America Bull.*, **84**, 1955-1972.

——, and Buffington, E. C. (1968). La Jolla Submarine Fan-valley. *Marine. Geol.*, **6**, 107-143.

——, and Dill, R. F. (1966). *Submarine Canyons and Other Sea Valleys*, 381 pp., Rand McNally, Chicago.

——, and Emery, K. O. (1973). Congo Submarine Canyon and Fan Valley. *Amer. Assoc. Petrol. Geol. Bull.*, **57**, 1679-1691.

——, and Marshall, N. F. (1969). Currents in La Jolla and Scripps Submarine Canyons. *Science*, **165**, 177-178.

——, and Marshall, N. F. (1973a). Currents along floors of submarine canyons. *Amer. Assoc. Petrol. Geol. Bull.*, **57**, 244-264.

——, and Marshall, N. F. (1973b). Storm-generated current in La Jolla Submarine Canyon, California. *Marine Geol.*, **15**, M19-M24.

——, Dill, R. F., and von Rad, U. (1969). Physiography and sedimentary processes of La Jolla Submarine Fan and Fan-valley, California. *Amer. Assoc. Petrol. Geol. Bull.*, **53**, 390-420.

——, Marshall, N. F., and McLoughlin, P. A. (1974). "Internal waves" advancing along submarine canyons. *Science*, **183**, 195-197.

Silcox, J. H. (1962). West Thornton and Walnut Grove Gas Fields, California. *Calif. Div. Mines Geol. Bull.*, *181*, 140-148.

—— (1968). Thornton and Walnut Grove Gas Fields, Sacramento and San Joaquin Counties, California. In *Natural Gases of North America* (B. W. Beebe and B. F. Curtis, eds.), Vol. 1, pp. 85-92, Amer. Assoc. Petrol. Geol. Mem. 9.

Simpson, E. S. W., and Forder, E. (1968). The Cape submarine canyon. *Fish. Bull. S. Africa*, **5**, 35-37.

Sonnenfeld, P. (1974). The Upper Miocene evaporite basins in the Mediterranean Region—a study in paleo-oceanography. *Geol. Rundschau*, **63**, 1133-1172.

—— (1975). The significance of Upper Miocene (Messinian) evaporites in the Mediterranean Sea. *Jour. Geol.*, **83**, 287-311.

Sprigg, R. C. (1964). The South Australian continental shelf as a habitat for petroleum. *Australian Petrol. Exploration Assoc. Jour.*, 53-63.

Stanley, D. J. (1967). Comparing patterns of sedimentation in some modern and ancient submarine canyons. *Earth Planet. Sci. Letters*, **3**, 371-380.

—— (1969). Submarine channel deposits and their fossil analogs ("fluxoturbidites"). In *The New Concepts of Continental Margin Sedimentation* (D. J. Stanley, ed.), pp. DJS-9-1–DJS-9-17, Short Course Notes, Amer. Geol. Inst., Washington, D. C.

—— (1971). Bioturbation and sediment failure in some submarine canyons. *Vie et Milieu* (Troisième Symposium Européen de Biologie Marine). Supplément 22, 541-555.

—— (1974). Pebbly mud transport in the head of Wilmington Canyon. *Marine Geol.*, **16**, M1-M8.

——, and Fenner, P. (1973). Underwater television survey of the Atlantic outer continental margin near Wilmington Canyon. *Smithsonian Contribs. Earth Sci.*, no. 11, 54 pp.

——, and Kelling, G. (1968a). Photographic investigation of sediment texture, bottom current activity, and benthonic organisms in the Wilmington Submarine Canyon. U.S. Coast Guard Oceanographic Rept. No. 22 (CG-373-22), 95 pp.

——, and Kelling, G. (1968b). Sedimentation patterns in the Wilmington Submarine Canyon area. In *Ocean Sciences and Engineering of the Atlantic Shelf*, pp. 127–142, Trans. Natl. Symp., Marine Technol. Soc., Philadelphia, Mar. 19-20, 1967.

——, and Kelling, G. (1970). Interpretation of a levee-like ridge and associated features, Wilmington Submarine Canyon, eastern United States. *Geol. Soc. America Bull.*, **81**, 3747–3752.

——, and Unrug, R. (1972). Submarine channel deposits, fluxoturbidites and other indicators of slope and base-of-slope environments in modern and ancient marine basins. In *Recognition of Ancient Sedimentary Environments* (J. K. Rigby and W. K. Hamblin, eds.), pp. 287–340, Soc. Econ. Paleontol. Mineral. Spec. Publ. 16.

——, Sheng, H., and Pedraza, C. P. (1971). Lower continental rise east of the Middle Atlantic States: predominant sediment dispersal perpendicular to isobaths. *Geol. Soc. America Bull.*, **82**, 1831–1840.

——, Fenner, P., and Kelling, G. (1972). Currents and sediment transport at the Wilmington Canyon shelfbreak, as observed by underwater television. In *Shelf Sediment Transport: Process and Pattern* (D. J. P. Swift, D. B. Duane, and O. H. Pilkey, eds.), pp. 621–644, Dowden, Hutchinson & Ross, Stroudsburg, Pa.

——, Got, H., Leenhardt, O., and Weiler, Y. (1974). Subsidence of the Western Mediterranean Basin in Pliocene-Quaternary time: further evidence. *Geology*, **2**, no. 7, 345–350.

Starke, G. W., and Howard, A. D. (1968). Polygenetic origin of Monterey Submarine Canyon. *Geol. Soc. America Bull.*, **79**, 813–826.

Stassano, E. (1886). La foce del Congo. *Randico delle R. Accad. dei Lincei*, June 6, 1886, 510–513.

Stride, A. H., Curray, J. R., Moore, D. G., and Belderson, R. H. (1969). Marine geology of the Atlantic continental margin of Europe. *Phil. Trans. Roy. Soc. London*, ser. A, **264**, 31–75.

Sullwold, H. H., Jr. (1960). Tarzana Fan, deep submarine fan of late Miocene age, Los Angeles County, California. *Amer. Assoc. Petrol. Geol. Bull.*, **44**, 433–457.

Tanaka, K., and Teraoka, Y. (1973). Stratigraphy and sedimentation of the Upper Cretaceous Himenoura Group in Koshiki-jima, southwest Kyushu, Japan. *Geol. Survey Japan Bull.* **24**, 157–184. (In Japanese, English summary.)

Travers, W. B. (1972). A trench off central California in late Eocene-early Oligocene time. In *Studies in Earth and Space Sciences* (R. Shagam, R. B. Hargraves, W. J. Morgan, F. B. Van Houten, C. A. Burk, H. D. Holland, and L. C. Hollister, eds.), pp. 173–182, Geol. Soc. America Mem. 132.

Trimonis, E. S., and Shimkus, K. M. (1970). Sedimentation at the head of a submarine canyon. *Oceanology, Acad. Sci. U.S.S.R.* (English translation, Amer. Geophys. Union), **10**, 74–85.

Trumbull, J. V. A., and Garrison, L. E. (1973). Geology of a system of submarine canyons south of Puerto Rico. *U. S. Geol. Survey Jour. Res.*, **1**, 293–300.

——, and McCamis, M. J. (1967). Geological exploration in an east coast submarine canyon from a research submersible. *Science*, **158**, 370–372.

Uchupi, E. (1968a). Atlantic continental shelf and slope of the United States—physiography. *U. S. Geol. Survey Prof. Paper 529C*, 30pp.

—— (1968b). Seismic profiling survey of the east coast submarine canyons. Part 1. Wilmington, Baltimore, Washington and Norfolk Canyons. *Deep-Sea Res.*, **15**, 613–616.

Van Hoorn, B. (1969). Submarine canyon and fan deposits in the Upper Cretaceous of the south-central Pyrenees, Spain. *Geol. Mijnbouw,* **48,** 67–72.

Varadachari, V. V. R., Nair, R. R., and Murty, P. S. N. (1967). Submarine canyons off the Coromandel coast: Proc. Symp. Indian Ocean. *Bull. Natl. Inst. Sci. India,* no. 38, 457–462.

Vasiliev, B. I., and Markov, Yu. D. (1973). Submarine canyons on the continental slope of the Peter the Great Bay (the Sea of Japan). *Okeanologiya,* **13,** 658–661. (In Russian, English abstract.)

Veatch, A. C., and Smith, P. A. (1939). Atlantic submarine valleys of the United States and the Congo Submarine Valley. Geol. Soc. America Spec. Paper 7, 101 pp.

Von der Borch, C. C. (1967). South Australian submarine canyons. *Aust. Jour. Sci.,* **30,** no. 2, 66–67.

—— (1968). Southern Australian submarine canyons: their distribution and ages. *Marine Geol.,* **6,** 267–279.

—— (1969). Submarine canyons of southeastern New Guinea: seismic and bathymetric evidence for their modes of origin. *Deep-Sea Res.,* **16,** 323–328.

——, Conolly, J. R., and Dietz, R. S. (1970). Sedimentation and structure of the continental margin in the vicinity of the Otway Basin, southern Australia. *Marine Geol.,* **8,** 59–83.

Walker, J. R., and Massingill, J. V. (1970). Slump features on the Mississippi Fan, northeastern Gulf of Mexico. *Geol. Soc. America Bull.,* **81,** 3101–3108.

Walker, R. G. (1966a). Shale Grit and Grindslow Shales: transition from turbidite to shallow water sediments in the Upper Carboniferous of northern England. *Jour. Sediment. Petrol.,* **36,** 90–114.

—— (1966b). Deep channels in turbidite-bearing formations. *Amer. Assoc. Petrol. Geol. Bull.,* **50,** 1899–1917.

—— (1973). Mopping up the turbidite mess. In *Evolving Concepts in Sedimentology* (R. N. Ginsburg, ed.), pp. 1–37, Johns Hopkins Univ. Press, Baltimore, Md.

——, and Mutti, E. (1973). Turbidite facies and facies associations. In *Turbidites and Deep Water Sedimentation* (G. V. Middleton and A. H. Bouma, eds.), pp. 119–157, Soc. Econ. Paleontol Mineral., Short Course, Anaheim, Calif.

Webb, G. W. (1965). The stratigraphy and sedimentary petrology of Miocene turbidites in the San Joaquin Valley (abs.). *Amer. Assoc. Petrol. Geol. Bull.,* **49,** p. 362.

Weiler, Y., and Stanley, D. J. (1973). Sedimentation on Balearic Rise, a foundered block in western Mediterranean (abs.). *Amer. Assoc. Petrol. Geol. Bull.,* **57,** 811–812.

Wezel, F. C. (1968). Osservazioni sui sedimenti dell'Oligocene-Miocene inferior della Tunisia settentrionale. *Soc. Geol. Italiana Mem.,* **7,** 417–439.

Whitaker, J. H. McD. (1962). The geology of the area around Leintwardine, Herefordshire. *Quart. Jour. Geol. Soc. London,* **118,** 319–351.

—— (1963). The geology of the area around Leintwardine, Herefordshire: Discussion. *Quart. Jour. Geol. Soc. London,* **119,** 513–514.

—— (1974). Ancient submarine canyons and fan valleys. In *Modern and Ancient Geosynclinal Sedimentation* (R. H. Dott, Jr., and R. H. Shaver, eds.), pp. 106–125, Soc. Econ. Paleontol. Mineral. Spec. Publ. 19.

Wilde, P. (1965). Recent sediments of the Monterey Deep-sea Fan. Univ. California Hydraulics Engineering Lab. Tech. Rept., 155 pp.

Wilhelm, O., and Ewing, M. (1972). Geology and history of the Gulf of Mexico. *Geol. Soc. America Bull.*, **83**, 575–600.

Winterer, E. L. (1970). Submarine valley systems around the Coral Sea Basin (Australia). *Marine Geol.*, **8**, 229–244.

——, Curray, J. R., and Peterson, M. N. A. (1968). Geological history of the intersection of the Pioneer Fracture Zone with the Delgada Deep-Sea Fan, northeast Pacific. *Deep-Sea Res.*, **15**, 509–520.

Wong, H. K., Zarudski, E. F. K., Phillips, J. D., and Giermann, G. K. F. (1971). Some geophysical profiles in the eastern Mediterranean. *Geol. Soc. America Bull.*, **82**, 91–100.

Yeats, R. S., Cole, M. R., Merschat, W. R., and Parsley, R. M. (1974). Poway Fan and Submarine Cone and rifting of the inner southern California borderland. *Geol. Soc. America Bull.*, **85**, 293–302.

Yegorov, Ye. N., and Galanov, L. G. (1966). Loss of near-shore deposits into submarine canyons. *Okeanologiya*, **6**. (English translation, 95–98).

Yerkes, R. F., Gorsline, D. S., and Rusnak, G. A. (1967). Origin of Redondo Submarine Canyon, southern California. *U. S. Geol. Survey Prof. Paper 575-C*, C97–C105.

Zenkovich, V. P. (1958). *Shores of Black and Azov Seas*, 371 pp., State Geographic Editions, Moscow.

AUTHOR CITATION INDEX

427

SUBJECT INDEX

The terms "canyon" and "fan" are used to denote submarine canyons and deep-sea fans, respectively. Land canyons and fans are specifically identified.

437

459

About the Editor

JOHN H. McD. WHITAKER was born in 1921 in Cambridge, England. He received the B.A. and M.A. degrees at Cambridge University, the B.Sc. from London University, and the Ph.D. from Leicester University. After three years as Assistant Lecturer at the University of Manchester, he moved to the University College (later the University) of Leicester in 1951 to start a new Department of Geology; he is currently Senior Lecturer there. His main research interests are in the Silurian and lower Devonian stratigraphy and sedimentology of the Welsh Borderlands and southern Norway, where he worked for 14 months as a Postdoctoral Fellow with the Royal Norwegian Council for Scientific and Industrial Research. He is also working on the lower Tertiary flysch of Japan, where he recently spent three months on a Royal Society Study Visit. Finding ancient submarine canyon heads in the Silurian of the Welsh Borderlands, as outlined in Paper 18 of this volume, led him to a general study of ancient canyons and fan valleys, and he has had the opportunity of visiting examples of these features in California and Japan. His review of pre-Pleistocene submarine canyons and fan valleys appears in this book as Paper 21.